High-Resolution Microwave Imaging

Ruliang Yang · Haiying Li
Shiqiang Li · Ping Zhang
Lulu Tan · Xiangwu Gao
Xueyan Kang

High-Resolution Microwave Imaging

国防工業出版社
National Defense Industry Press

Springer

Ruliang Yang
Chinese Academy of Sciences
Beijing
China

Haiying Li
Chinese Academy of Sciences
Beijing
China

Shiqiang Li
Chinese Academy of Sciences
Beijing
China

Ping Zhang
Chinese Academy of Sciences
Beijing
China

Lulu Tan
East China Research Institute of Electronic Engineering (ECRIEE)
Chinese Academy of Sciences
Beijing
China

Xiangwu Gao
China Academy of Launch Vehicle Technology
Beijing
China

Xueyan Kang
Chinese Academy of Sciences
Beijing
China

ISBN 978-981-10-7136-2   ISBN 978-981-10-7138-6 (eBook)
https://doi.org/10.1007/978-981-10-7138-6

Jointly published with National Defense Industry Press, Beijing

The print edition is not for sale in China Mainland. Customers from China Mainland please order the print book from: National Defense Industry Press.

Library of Congress Control Number: 2017958618

Printed on acid-free paper

This Springer imprint is published by Springer Nature
The registered company is Springer Nature Singapore Pte Ltd.
The registered company address is: 152 Beach Road, #21-01/04 Gateway East, Singapore 189721, Singapore

Preface

High-resolution microwave imaging radar technology is one of the key issues in the high-resolution earth observation system technology.

Synthetic aperture radar (SAR) is a kind of high-resolution microwave imaging radar. It can image the ground static and moving targets in all-weather, day-and-night, penetrates the natural vegetation and artificial camouflage, and finds important military targets. Representing by spaceborne SAR, the microwave imaging earth observation satellite technology has become an important means of earth observation, and drawn wide attention of all countries. It shows wide applications and great potential in disaster monitoring, resource survey, military reconnaissance, mapping, attacking effect reconnaissance, sea research, and other fields. Strong military and civilian demands have driven the rapid development of SAR, many countries are competing to develop high-resolution microwave imaging radar technology. Various airborne, spaceborne, shuttle and space station microwave imaging radars are being used for different military and civilian purposes.

With the wide application of SAR imaging technology, the high-resolution SAR is still one of the research hotspots in the field of microwave remote sensing. High-resolution microwave imaging contains high range resolution imaging and high azimuth resolution imaging.

To improve the range resolution, the transmission signal with large bandwidth is needed. The signal bandwidth is wider and the resolution is higher. The signal form of high range resolution mainly includes impulse waveform, linear frequency modulation waveform, stepped-frequency imaging waveform, stepped-chirp SAR imaging burst waveform and linear frequency modulation continuous wave signal, etc.

Improvement on the azimuth resolution can be achieved by increasing the synthetic aperture length, viz., the Doppler bandwidth. Spotlight SAR can achieve longer synthetic aperture time and higher azimuth resolution than strip map SAR by continuous irradiation at a small piece of fixed area with a long time.

A fundamental limitation in the design of spaceborne SAR system is the contradiction between the swath width and resolution. Single Phase Center Multiple Beams (SPCMB) technology, Multiple Phase Center Multiple Beams (MPCMB) technology, and multiple-input multiple-output (MIMO) technology can be used to obtain high-resolution and wide-swath microwave imaging.

Over the past decade, super-resolution information processing technology has developed rapidly in the microwave imaging radar technology. SAR super-resolution information processing technology adopts super-resolution imaging algorithm, using the modern spectrum estimation instead of Fourier transform, effective spectrum extrapolation, adaptive weighting and regularization, and other super-resolution imaging algorithm to deal with SAR data. Without the increase in the bandwidth and the synthetic aperture length, Fourier imaging algorithm performance limitations are broken through, and the imaging resolution and radar image quality are improved.

This book is divided into two parts, which are high-resolution microwave imaging and super-resolution information processing. The book totally has 17 chapters. The structure arrangement of the book can be seen in the configuration of "High Resolution Microwave Imaging".

High-resolution microwave imaging technology includes microwave imaging radar basics, radar resolution theory, linear frequency modulated signal, stepped-frequency imaging waveform, stepped-chirp SAR imaging burst waveform spotlight SAR, spotlight SAR imaging algorithm, sliding spotlight SAR, SPCMB SAR, MPCMB SAR, MIMO SAR, SAR autofocusing algorithm, and so on.

On the super-resolution information processing technology part, the super-resolution processing technology basics and microwave imaging super-resolution processing technology are introduced. The classical Fourier spectrum analysis and super-resolution processing signal model are discussed. Microwave imaging super-resolution processing technology is studied systematically, including the deconvolution method using the system point spread function, super-resolution spectrum estimation methods including minimum variance spectrum estimation method, amplitude and phase spectrum estimation algorithm, subspace orthogonal decomposition method, estimation of signal parameters via rotational invariance techniques, relaxation-based spectrum estimation algorithm, maximum likelihood estimation method, and autoregressive spectrum estimation method. Super-resolution spectrum bandwidth extrapolation imaging algorithm includes SAR super-resolution autoregressive linear predictive extrapolation imaging algorithm, weighting adaptive minimum norm extrapolation method, adaptive weighting side lobe suppression method including spatially variant apodization method and adaptive side lobe suppression algorithm, super-resolution regularization imaging algorithm including Tikhonov regularization, non-secondary regularization, total variant difference method, other SAR super-resolution regularization method, etc.

In the recent years, we have organized a number of doctors and masters to systematically and deeply study the high-resolution microwave imaging and super-resolution information processing technology. This book is a summary of the research on this group of doctoral and master's degree theses. This book attempts to systematically discuss high-resolution microwave imaging radar systems and super-resolution information processing techniques. The book tries to link theory

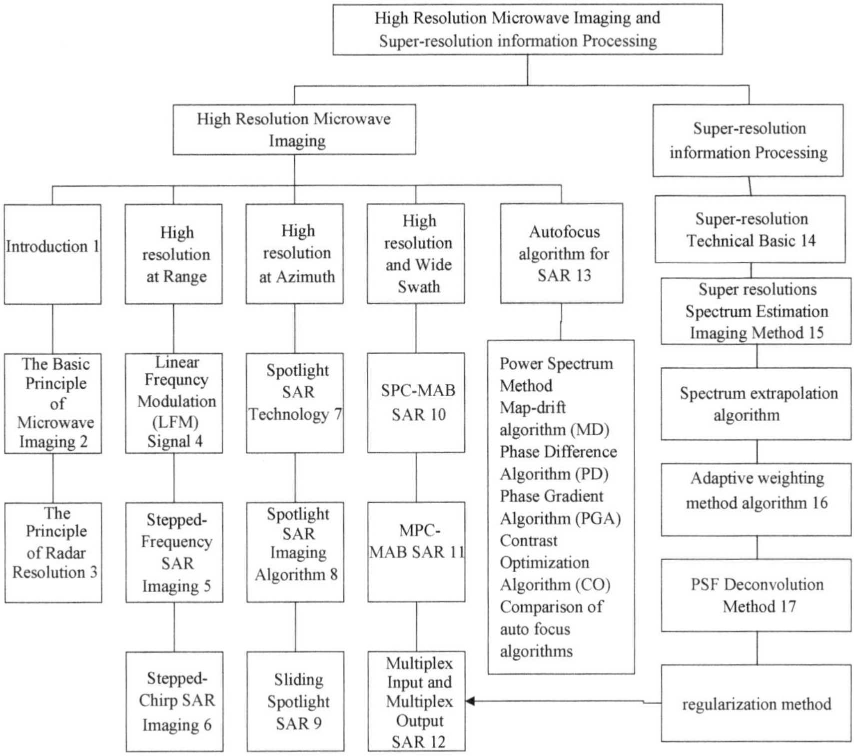

The configuration of "High Resolution Microwave Imaging"

with practice, as far as possible to reflect the most updated theories, methods, and achievements recently of this field in recent years.

This book is planned and designed by Prof. Ruliang Yang. The following doctors participated in writing, including Prof. Ruliang Yang (Chaps. 1, 4), Dr. Haiying Li (Chaps. 2–6), Dr. Xiangwu Gao (Chaps. 7, 9), Dr. Tan Lulu (Chaps. 7–9), Dr. Shiqiang Li (Chaps. 10–12), Dr. Xueyan Kang (Chap. 13), Dr. Ping Zhang (Chaps. 14–17), Dr. Feng Fan (Chap. 12), Dr. Song Yuepeng (Chap. 12), Wang Pei (Chap. 4) and Tian Xuwen (Chaps. 13, 15), Prof. Ruliang Yang completes the final draft of this book.

In this book, high-resolution microwave imaging and super-resolution information processing technology is studied deeply and systematically. It can be used for the researchers, engineers, and technical personnel in the electronic system, especially for microwave remote sensing and microwave imaging radar. And it can also be used as a related research or course teaching reference book for teachers and students in the course of communication and information system, electronic technology, computer application, etc.

Prof. Peng Yingning and Prof. Yang Jian, of Tsinghua University, Prof. Mao Shiyi, of Beihang University, etc., provided many valuable advices. Sincere thanks in this table.

The author would like to thank Dr. Xiaolei Han who translated Chap. 1 and Chap. 6 from Chinese to English in this book, Dr. Liying Xu who translated Chap. 11, and Dr. Yuepeng Song who translated Chap. 12 in this book.

The author would like to thank the people who are concerned with and helped this book's publication in Institute of Electronics Chinese Academy of Sciences.

High-resolution microwave imaging and super-resolution information processing is a rapidly developing technology. Since the ability and knowledge is limited, the authors apologize in advance for any errors that may be present in the book, and would be very grateful to have them brought to their attention.

Beijing, China
August 2013

Prof. Ruliang Yang

Contents

About the Authors

Prof. Ruliang Yang is Ph.D. supervisor, graduated from Radar System Engineering Department of the University of Electronics Science and Technology of China at 1965. He was supported by K. C. Wang Foundation, Hong Kong and is a senior visiting scholar in the Engineering Department of Aberdeen University UK.

Prof. Ruliang Yang has been engaging in research of airborne and spaceborne electronics system engineering for 50 years.

From 1965 to 1980, he was engaged in research for satellite-borne and missile-borne electronics equipment of the ballistic trajectory measurement and security system in satellites and missiles. He has taken part in satellite assembly and satellite launched for the Satellites "Orient Red" No1 and "Practice" No1 in China.

After joining the Institute of Electronics Chinese Academy of Sciences at 1980, he was engaged in research of system collectivity engineering for satellite-borne and airborne synthetic aperture radar (SAR) systems. As a chief general designer, he developed the satellite-borne SAR and airborne SAR systems. Successively, he was engaged in researches of new technology and new systems in microwave imaging radars, such as the high-resolution microwave imaging radar system, the polarization synthetic aperture radar, the interferometric microwave imaging, the super-resolution information processing, the radar target identify technology, GMTI, the chaos signal SAR and SAR radar control and guide technology, etc.

He was awarded the six items invent patents. He has published 240 papers, wrote one book, and translated one book. He has supervised more than 50 students, who have received their Ph.D. and master degrees.

He was the Deputy Director of Institute of Electronics Chinese Academy of Sciences (IECAS), and the Director of department of Imaging Radar System IECAS. He is the director member of Unite Remote Sensing Centre of Chinese Academy of Sciences, the commissioner of Radio communication committee of Chinese Communication Academy, the commissioner of Remote Sensing committee of Chinese Space Sciences Academy, the committeeman of Radar and Navigation Committee of Chinese Aviation Academy, the vice-chief editor of "Remote Sensing Technology and Application" journal, the committeeman of "Journal of Radars" editor committee. He was awarded with the government special allowance from the State Council.

Dr. Haiying Li Chapters 2–6—High range resolution technology of SAR system.

Dr. Haiying Li , female, Ph.D., Associate Professor. She received her Ph.D. in communication and information system from the Institute of Electronics, Chinese Academy of Sciences in July 2002. She has been engaging in research of the SAR satellite missions in IECAS from 2002 to 2013; her research interest was spaceborne SAR system design and simulation. She has published 10 papers and came out with one item invent patents. Now Dr. Haiying Li works in National Astronomical Observatory, Chinese Academy of Sciences (NAOC), her research interests are radar technology and applications on the moon and deep space exploration.

Dr. Shiqiang Li Chapters 10–12—Multiple-phase center multiple azimuth beam SAR.

Dr. Shiqiang Li, Ph.D., Professor. He was born in 1967 and received B.S. from the Beijing Institute of Technology, Beijing, China, in 1989 and M.S. from The 54th Research Institute of China Electronics Technology Group Corporation(CETGC), Shijiazhuang, Hebei province, in 1992 and Ph.D. from the Institute of Electronics, Chinese Academy of Sciences(IECAS), Beijing, in 2004. He was an engineer at (CETGC) during 1992–2001, where his research interest was spread spectrum communication technique. Since 2004, he has been with IECAS, where his research interests are system design and simulation of spaceborne SAR, distributed spaceborne SAR technique. Currently, he is a Professor and Graduate Supervisor in IECAS.

Dr. Ping Zhang Chapters 14–17—Super-resolution information processing of SAR.

Dr. Ping Zhang Ph.D., associate professor, born in Cangzhou, China, in 1979. She received Ph.D. in communication and information system from Institute of Electronics, Chinese Academy of sciences (IECAS) on 2009. At the same time, she joined the Institute of Remote Sensing and Digital Earth, Chinese Academy of sciences (IRSDECAS). Her current research activities concentrate in signal processing and calibration of SAR.

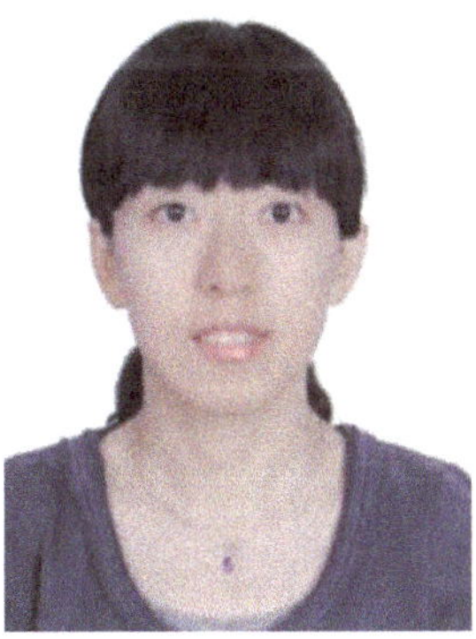

Dr. Lulu Tan Chapters 7–9—Spotlight SAR imaging algorithm.

Dr. Lulu Tan Ph.D., senior engineer, born in Anqing, Anhui Province. She received B.S. degree in electrical engineering from the Dalian University of Technology in 2004, and Ph.D. in communication and information system from Institute of Electronics, Chinese Academy of Sciences (IECAS) in 2010. From 2006 to 2010, she focused on the study of polarimetric SAR Interferometry. From 2010 till now, she is with East China Research Institute of Electronic Engineering (ECRIEE), where her research activity has been focusing on the design of spaceborne SAR system.

Dr. Xiangwu Gao Chapters 7, 9—Spotlight SAR imaging technology.

Dr. Xiangwu Gao Ph.D., Professor, was born in Gaizhou City of Liaoning Province in June 1976, and received the Ph.D. degree in communication and information system from the Institute of Electronics Chinese Academy of Sciences (IECAS) in 2004. His research interest was spaceborne spotlight synthetic aperture radar technology. Since 2005, he is working at the R&D center of China Academy of Launch vehicle Technology (CALT), and engaged in development of the missiles measuring system of the long March series for six models. Now he is a Deputy Chief Designer of a project, the director member of Institute of command and control of China, a member of the flight vehicle TT&C Professional Committee of Chinese Astronautics Academy, and the senior specialist of CALT. He has published more than 20 papers.

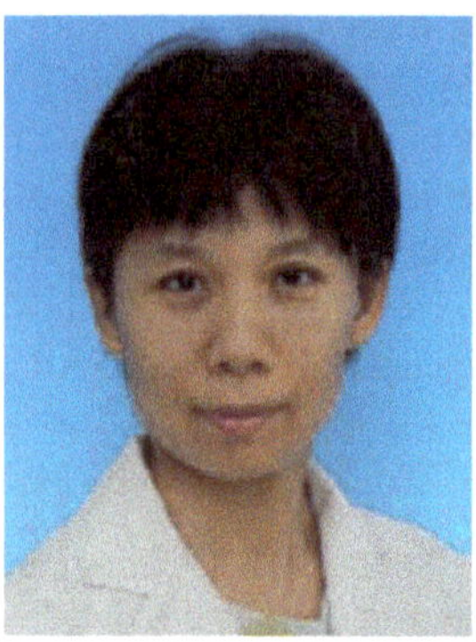

Dr. Xueyan Kang Chapter 13 Autofocus algorithm for SAR.

Dr. Xueyan Kang Ph.D., associate professor. She received the Ph.D. degree in communication and information system from the Institute of Electronics Chinese Academy of Sciences (IECAS) in 2004. She is now with the Center for Space Science and Applied Research, Chinese Academy of Sciences (CSSARCAS). Her current research interests include SAR imaging processing, GMTI and array signal processing, etc.

Acronyms

AASR	Azimuth Ambiguity-to-Signal Ratio
ACF	Autocorrelation Function
ADC	Analog-to-Digital Converter
AGC	Automatic Gain Control
APES	Amplitude and Phase Estimation of a Sinusoid
AR	Autoregressive
ARLP	Autoregressive Linear Prediction
ARMA	Autoregressive and Moving Average
ASR	Adaptive Side lobe Reduction
AWNE	Adaptive Weighted Norm Extrapolation
BAS	Baseband Azimuth Scaling
BT	Blackman–Tukey
CAT	Computer-Aided Tomography
CAWNE	Adaptive Weighted Norm Extrapolation based Capon's Method
CDA	Complex Double Apodization
CO	Contrast Optimization
CSA	Chirp Scaling Algorithm
CTM	Conversion Transpose Memory
D/A	Digital to Analog
DA	Double Apodization
dB	Decibel
DBF	Digital BeamForming
DDS	Direct Digital Synthesis
DOA	Direction Of Arrival
DTFT	Discrete-Time Fourier Transform
ECSA	Extended Chirp Scaling Algorithm
EPROM	Erasable and Programmable Read Only Memory
ESPRIT	Estimation of Signal Parameters Via Rotational Invariance Techniques

FFT	Fast Fourier Transform
FIFO	First-In First-Out
FIR	Finite Impulse Response
FMCW	Frequency-Modulated Continuous Wave
FPGA	Field-Programmable Gate Array
FSA	Frequency Scaling Algorithm
GMTI	Ground Moving Target Indication
HRWS	High-Resolution Wide Swath
I/Q	In-phase/ Quadratic
ICTM	Inverse Conversion Transpose Memory
IDFT	Inverse Discrete Fourier Transform
IF	Intermediate Frequency
IFFT	Inverse Fast Fourier Transform
IFOV	Instantaneous Field Of View
IPR	Impulse Response
ISLR	Integrated Side Lobe Ratio
LOS	Line Of Sight
LOSPI	Line-Of-Sight Polar Interpolation
MA	Moving Average
MAM	Multiple Aperture Mapdrift
MD	Mapdrift
MIMO	Multiple-Input Multiple-Output
ML	Main Lobe
MLE	Maximum Likelihood Estimation
MNM	Mini-Norm Method
MPCMAB	Multiple-Phase Center Multiple Azimuth Beam
MPCMB	Multiple-Phase Center Multi-Beam
MTRC	Motion Through Resolution Cells
MUSIC	Multiple Signal Classification
MVM	Minimum Variance Method
$\mathrm{NE}\sigma^0$	Noise equivalent backscatter coefficient
NTNR	Narrow beam Transmit-Narrow beam Receive
PD	Phase Difference
PFA	Polar Format Algorithm
PGA	Phase Gradient Autofocus
PR	Perfect Reconstruction
PRF	Pulse Repetition Frequency
PRI	Pulse Repetition Interval
PSD	Power Spectral Density
PSLR	Peak Side Lobe Ratio
PSM	Power Spectrum Method
RAM	Random Access Memory
RASR	Range Ambiguity-to-Signal Ratio
RCM	Range Cell Migration
RCMC	Range Cell Migration Correction

RD	Range-Doppler
RMA	Range Migration Algorithm
ROM	Read Only Memory
RRMVM	Reduced-Rank MVM
RVP	Residual Video Phase
SAC	Shift and Correlation
SAR	Synthetic Aperture Radar
SLC	Single Look Complex Image
SPC-MAB	Single-Phase Center Multiple Azimuth Beam
SPECAN	SPECtran Analysis
SQNR	Signal-to-Quantization Noise Ratio
SRAM	Static Random Access Memory
SRC	Second Range Compression
SSLR	Second Side Lobe Ratio
SSPI	Stabilized Scene Polar Interpolation
STAP	Space-Time Adaptive Processing
STC	Sensitivity Time Control
SVA	Spatially Variant Apodization
TSP	Two-Step Processing
UWB	Ultra-Wide Band
WTNR	Wide beam Transmit-Narrow beam Receive

Brief Introduction

The book expounds comprehensively the high-resolution microwave imaging and super-resolution information processing technology. The two parts of the book have been divided into 17 chapters: The high-resolution microwave imaging technology includes basic principle of microwave imaging, principle of radar resolution, linear frequency modulation (LFM) signal, steep-frequency SAR imaging, steep-chirp SAR imaging, spotlight SAR technology, spotlight SAR imaging algorithm, sliding spotlight SAR, single-phase center multiple azimuth beams SAR, multiple-phase center multiple azimuth beams SAR, multiple-input multiple-Output SAR, etc. New theories and new methods of the super-resolution information processing technology include autofocus algorithm for SAR, super-resolution technology basic of microwave imaging, super-resolution spectrum estimation imaging method, super-resolution imaging algorithm based on spectrum extrapolation and adaptive weighting method, SAR super-resolution based on regularization method and PSF deconvolution method, etc.

The novel offers a close and systematic framework. Theories are integrated with practices, which possess strong application background. It reflects new theories, methods, and achievements in the high-resolution microwave imaging and super-resolution information processing fields in recent years. This book can be supplied to researchers and engineers in electronics systems, especially in microwave remote sensing and microwave radar fields. It can also be used as a reference book for teachers and postgraduate students of related departments such as the communication and information system, electronics technology, computer applications, during their subjects studies or educations in colleges and universities.

Chapter 1
Introduction

High-resolution microwave imaging radar technology is one of the key issues in the high-resolution earth observation technology. High-resolution microwave imaging radar provides a wealth of target information, not only for terrain mapping and target imaging but also for target detection in complicated surveillance environment [1]. In the civilian field, it is widely applied to the land resource survey, disaster monitoring, vegetation growth observation, ocean observation, terrain mapping, oceanographic research, glacier research, etc. High-resolution technology can be used to navigate for aircraft and ground transportation vehicles in low visibility conditions. Military applications require higher resolution. In recent years, United States Department of Defense (USDOD) always regards the ultra-wideband radar and high-resolution radar as important research contents in the plan of the defense key technology [2].

Synthetic aperture radar (SAR) is a kind of high-resolution microwave imaging radar. It combines synthetic aperture and pulse compression technique [3]. High azimuth resolution is obtained by using the synthetic aperture technique, and high range resolution is obtained by using wide bandwidth or narrow pulse signal. Thus, a large area 2D high-resolution radar image is obtained.

Strong military and civilian demands have driven the development of SAR rapidly. Many countries are competing to develop high-resolution microwave imaging radar technology. The countries that have mastered this technology include the United States, Russia, Canada, Japan, Germany, Israel, Britain, France, and Italy. These countries already have a variety of airborne, spaceborne, space shuttle, and space station SAR, which is used for different purposes. At present, the resolution of foreign military high-resolution SAR has reached 0.3 m and that of spotlight mode can reach 0.1 m. And 5 cm $\times$ 2 cm resolution experimental verification has been completed. The use of high-resolution SAR remote sensing image in battlefield reconnaissance greatly improves the ability of prewar battlefield perception and postwar attack effect assessment. It provides a safe technical support for the in-depth enemy reconnaissance [4].

R. Yang et al., *High-Resolution Microwave Imaging*,
https://doi.org/10.1007/978-981-10-7138-6_1

1.1 The History and Future of Synthetic Aperture Radar

SAR has been developing for 60 years. In 1951, Carl Wiley was employed by the American Goodyear Aerospace Corporation, first proposed to improve the angular resolution of radar by the frequency analysis method. In 1952, a simple experimental device was developed to demonstrate the feasibility of the proposed method by Wiley Carl. In 1953, through the non-coherent radar experiments, the control system laboratory of Illinois University proved that the frequency analysis method can improve the angular resolution of radar and manufactured the first airborne SAR system. Another workgroup at Illinois University had been developing SAR at the same time and obtained the first SAR image by the unfocused aperture synthesis method in July 1953. Cooperating with the United States Military, Michigan University researched SAR experiment system and obtained the first fully focused SAR image by the optical film recording and optical signal processing method in August 1957. They completed the development of the trip map airborne SAR in 1958. In 1962, Jet Propulsion Laboratory (JPL) developed an L band SAR system and applied to the missile. In 1966, JPL developed an L band airborne SAR and installed it into CV-900 airplane of National Aeronautics and Space Administration (NASA).

American military airborne SAR developed rapidly. AN/APQ-102A radar, with the azimuth and range resolution of 15 m × 15 m, represented the first-generation products. AN/APD-10 radar, with the resolution of 3 m × 3 m, represented the second-generation products. AN/APD-11 radar, with the resolution of 1.5 m × 1.5 m, represented the third-generation products. AN/UPD-6 radar, with the resolution of 0.7 m × 0.7 m, represented the fourth-generation products. AN/APY-3 SAR equipped in E-8A early warning airplane had two modes. The first mode is used for the detection of fixed target, with the resolution of 1 m × 1 m. The second mode is used for the detection of moving target, with the resolution 0.3 m × 0.3 m for the moving target. It can embed the moving target signal into the image of the fixed object, and obtain a complete 160 km × 180 km image of battlefield.

With the development of electronic technology, computer technology, and aerospace technology, SAR technology has gradually matured. Many countries take great efforts to develop their own spaceborne SAR system.

In June 27, 1978, NASA launched the SEASAT—An ocean satellite equipped with a SAR—which created the history of spaceborne SAR. It worked in L band, HH polarization, the fixed incidence angle, and 25 m resolution in four looks. Its task is to demonstrate the reliability of ocean dynamics measurement. Although the SAR system operated for only 3 months due to the power system fault, during the whole flight, it worked 500 times, 5–10 min each time, and obtained high-quality image data of one hundred million square kilometers of earth's surface. It demonstrated that SAR can obtain high-resolution images from the space.

In November 1981, SIR-A SAR carried out by American "Columbia" space shuttle was successfully launched. It worked in L band, HH polarization, 47° of antenna look angle, 50 km swath width, 40 m range and azimuth resolution, optical data recording, and optical imaging processing. In the flight of two and a half days, it imaged 10 million square kilometers earth's surface, and obtained a large number of information. One of the most notable achievements was the discovery of the ancient river way of Nile River in the Sahara desert, which caused a shock in the international scientific community.

In October 1984, SIR-B SAR carried out by American "Challenger" space shuttle was successfully launched. It worked in L band, HH polarization, antenna look angle between 15° and 60°, 25 m resolution, variation range resolution between 14 and 46 m as antenna look angle, and the swath width between 20 and 50 km. Benefiting from the variable look angle, the radar can observe the same area from different angles and provide the first-hand space experiment data of the backscatter coefficients change in different look angles. It first achieves digital recording and digital processing of the data.

In December 1993, the United States successfully launched the SIR-C/X space shuttle SAR system. The system had three bands of X, C, and L. Among them, L and C bands had four polarization modes including HH, VV, HV, and VH. And they had variable swath width between 15 and 90 km. X band was jointly developed by Germany and Italy, with VV polarization and the observation swath width of 15–40 km. It was the first time used for human beings with multiband, multi-polarization, and multi-look angle SAR experiments in space.

In June 1983 and January 1984, the former Soviet Union launched two SAR satellites for the exploration of Venus. After that, in July 1987, the "Universe-1870" satellite was launched, which was equipped with a set of SAR with 25 m resolution. The radar was mainly used for imaging the perennial cloud covered areas, continuous observation of natural disaster area, and monitoring of sea surface pollution. In March 1991, Almaz-1 imaging radar satellite was launched, which operated in the S band, with variable look angle and 15 m resolution for two looks.

Since entering the twentieth century, the countries had gradually realized the important value of spaceborne SAR in the field of remote sensing and actively developed high-resolution spaceborne SAR system. The developing intensity of the spaceborne SAR system has been increased, and new technology has been used to improve the performance of spaceborne SAR.

In September 1991, European Space Agency (ESA) launched the earth resources satellite ERS-1. The system operated in C band, VV polarization, 100 km swath width, and the range and azimuth resolution of 30 m. In 1995, ESA launched ERS-2 satellite with similar performance. In March 1, 2002, ESA launched Envisat satellite equipped with advanced SAR (ASAR) system, the system operated in C band, multi-polarization, multi-work modes, swath wide between 100 and 400 km, resolution between 30 and 1000 m, and variable incidence angle between 15° and 45°.

In November 1995, Canada successfully launched a SAR satellite Radarsat-1. The system had a variable swath width of 45–500 km, range and azimuth resolution of 10–100 m, C band, HH polarization, and variable incidence angle between 20 and 59°. It first used the ScanSAR mode. With 500 km footprint, it could cover the Arctic area once a day, cover almost the whole of Canada, cover the United States and other northern latitudes once every 3 days, and cover global surface a time only needing 5 days. As the follow-up satellite of Radarsat-1, Canada successfully launched Radarsat-2 satellite. Its best resolution reached to 1 m, and had multi-polarization operating skill.

In February 1992, Japan launched the earth resources satellite JERS-1 for geological exploration. It had an observation swath width of 75 km, the resolution of 18 m, L band, HH polarization, and antenna incidence angle of 35°. Information gathering system (IGS), which is being developed in Japan, consists of four satellites including two optical satellites and two high-resolution SAR satellites. The radar has a resolution of 1–3 m and can cover global surface once a day. On April 24, 2006 Japan launched the Advanced Land Observation Satellite (ALOS) and equipped a polarimetry SAR with L band phase array antenna and resolution of about 10 m.

In 2002, Italy defense ministry and ESA jointly funded the development of the small satellite constellation (COSMO-SkyMed). It consisted of four low-orbit satellites. Each satellite was equipped with a multimode high-resolution SAR, worked in the X band and the ground resolution of 1 m. It was a ground observation system for military and civilian applications. On June 7, 2007, December 8, 2007, and October 24, 2008, respectively, three of them were launched. On November 6, 2010, the fourth satellite was launched.

On December 19, 2006, July 2, 2007, November 1, 2007, March 27, 2008, and July 22, 2008, German launched five SAR-Lupe satellites. They are distributed in three different orbits, composed of complete space-based radar reconnaissance systems. The best image resolution of SAR-Lupe can reach to 0.5 m.

In January 21, 2008, Israel launched the first radar imaging reconnaissance satellite "TECSAR". The orbit height was 550 km. It adopted the parabolic antenna. It had strip map, wide coverage Scan SAR, Mosaic and spotlight modes, and the best resolution could reach to 1 m.

On October 22, 2008, India launched a lunar exploration satellite "Chandrayaan-1", which equipped with a Mini SAR developed by the Applied Physics Laboratory of US Hopkins John College and the United States Naval Aviation tactical center. Its quality was 6.5 kg, and central frequency was 2.38 GHz. It could image whole permanent shadow on the moon and detect whether the surface of the moon contains water, ice, and mineral resources. On April 20, 2009, India launched X band radar imaging satellite (RISAT). Its orbit was 550 km, and best resolution reached to 1 m.

International spaceborne SAR reconnaissance systems are rapidly developing toward high resolution (1–2 m) and higher resolution (better than 1 m).

Imaging reconnaissance satellite has become an important means of reconnaissance in the United States. On December 2, 1988, American famous Lacrosse (Lacrosse 1) military reconnaissance satellite was launched into orbit. Its orbit inclination angle is 57°, perigee height is 680 km, apogee height is 690 km, and satellite weight is 15 tons. It was mainly used for small area high-resolution imaging mode and could accurately get the needed information for the local battlefield. Its spotlight imaging mode azimuth resolution can reach 1 m. After that, in March 1991, October 1997, August 2000, and April 2005, Lacrosse-2, Lacrosse-3, Lacrosse-4, and Lacrosse-5 were launched into space. These satellite radars have a bilateral look function, around the earth 12-14 circles each day. In a very poor weather condition, they can achieve the global reconnaissance with 0.3 m resolution, even though find the tanks and ground to air missiles hidden in the forest. In 2003, Lacrosse 2 & 3 & 4 was used in the Iraq war. Each satellite passes over Iraq two times a day, and a large amount of military intelligence was obtained for USA. During the Kosovo war, the two "Lacrosse" radar imaging satellites were used for combat reconnaissance before military strike and attack effect evaluation. They flied over the target area of Kosovo/Yugoslavia two times each day and can eastward and westward side-looking image the region. The resolution of the radar image is 0.9–0.3 m and can carry out basic bombing damage assessment in the night and all kinds of weather conditions.

The demand for high-resolution microwave imaging in military and civilian application is steadily increasing. For example, if a SAR image with a few meters geometric resolution is used to analyze the technical feasibility in military long-range reconnaissance, since a few meters size of the targets present as an open point on the radar image, it is difficult to application effectively. For most civil applications, such as the application of airborne SAR images to detect dangerous infrastructure, the requirements of the resolution are suitable for military applications. The contemporary civil spaceborne SAR resolution is meter scale. There are several decimeter-scale resolution explorations of airborne SAR experiment system. For example, the bandwidth of RAMSES multiband SAR is 1.2 GHz in X band and Ku band. The bandwidth of PAMIR is 1.8 GHz in X band, and its resolution is better than decimeter scale. Using sliding spotlight mode, it makes a good tradeoff between azimuth resolution and azimuth swath. In the future, with the enhancement of long-range SAR image resolution, the system bandwidth has to be increased, and the antenna scanning angle range is also increased.

According to European Synthetic Aperture Radar (EUSAR2010) report, German Fraunhofer high-frequency physical and Radar Technology Research Institute (FHR) carried out 5 cm resolution concept verification experiment in X band airborne SAR [5].

Figure 1.1 shows an X band SAR image with 3 GHz bandwidth. Its resolution is 5 cm $\times$ 2 cm (range $\times$ azimuth). The details of the picture change clear. The streetlights, pipeline coverage, private cars, and different vegetations can be identified. Even two rearview mirrors of the car can be seen.

At present, a series of SAR schemes, designed to fully explore the potential of SAR and develop the new mode SAR, have been studied out and implemented. The countries are studying the new mode spaceborne and airborne SAR systems. High-resolution and wide swath earth observation is still a hot topic at present.

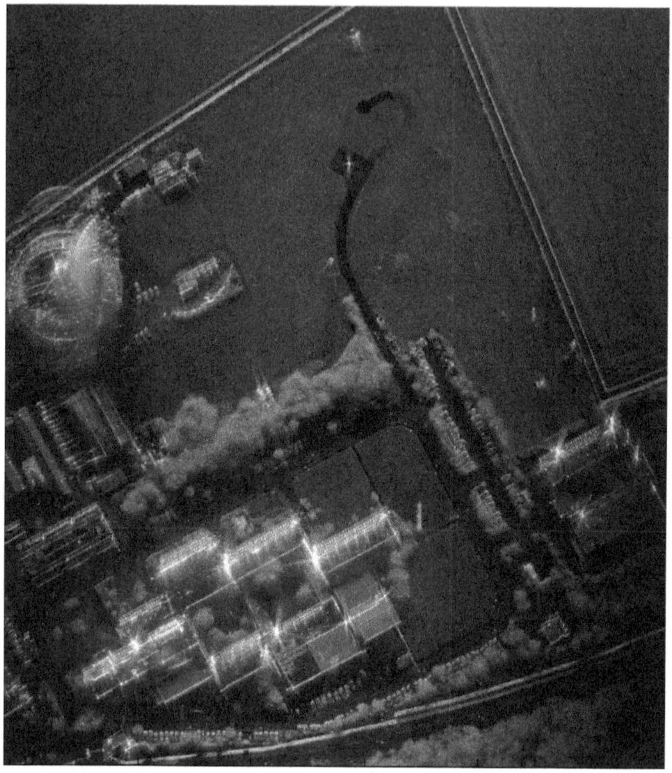

Fig. 1.1 X band SAR image with a resolution of 5 cm

1.2 High Range Resolution Microwave Imaging

The radar resolution depends on three factors: signal-to-noise ratio (SNR), signal form, and signal processing method. The radar transmitting waveform not only directly affects the potential system performance, such as resolution, imaging accuracy, and clutter suppression ability but also determines the signal processing method. On the condition of high SNR and optimum signal processing equipment, the range and azimuth resolution of SAR only depend on the signal form.

To improve the range resolution, wide bandwidth signal is required. The signal bandwidth is wider, and the resolution is higher [5]. The forms of high range resolution signal mainly include linear frequency modulation (FM) pulse signal waveform, stepped-frequency signal waveform, stepped-chirp waveform, impulse signal, binary phase encoding, discrete frequency encoding, linear FM continuous wave, and so on.

Linear frequency modulation pulse signal (chirp) is a widely used pulse compression signal, which obtains large time—bandwidth products by inner pulse linear frequency modulation. Its main advantage is insensitive to the Doppler frequency shift; even if echo signal has a larger Doppler frequency shift, the original matched filter still can play a role in the pulse compression, which will simplify signal processing. But its side lobe of matched filter output response is relatively high. To reduce the side lobe, it is necessary to lead in suitable weighting processing.

By transmitting stepped-frequency pulse trains, the equivalent wideband signal is obtained. It significantly reduces the instantaneous bandwidth of the system and the hardware requirements for the receiver. It is a kind of practical, convenient, and flexible high range resolution signal. The stepped-frequency signal can be considered as the discrete form of the wideband linear FM signal, which can reduce the A/D sampling rate of the receiver. The stepped-frequency pulse trains have high range resolution. But the waveform is very sensitive to radial velocity, and velocity compensation must be carried out.

The stepped-chirp waveform trains signal is a compromise between the linear FM signal and the stepped-frequency pulse signal. The waveform has the advantages of both stepped-frequency signal and linear FM signal. It can obtain a high range resolution image by digital signal processing without the increase of system instantaneous bandwidth. It is a kind of high-resolution signal form, which is easy to implement and flexible to design. The echo of the stepped-chirp signal SAR needs to go through range wideband synthesis and azimuth Doppler bandwidth synthesis, and eliminate the coupling between range and azimuth. For the targets with nonzero radial velocity, it is necessary to compensate the velocity when stepped-chirp sub-pulses waveform is synthesized to wideband echo signal.

Impulse signal has the ideal correlation characteristic, and the bandwidth is very wide. However, due to the limit of transmitting peak power, the signal form is very difficult to achieve.

Binary phase code has the advantages of easy digital processing, and its ambiguity function is thumbtack shape and suitable for both distance and speed high resolution. Its disadvantage is sensitive to the Doppler frequency shift. It is fit to the signal with narrow Doppler frequency shift, adopting low pulse repetition frequency (PRF) and large instantaneous wideband.

Continuous wave discrete frequency encoding signal coherence stacks the stepped-frequency waveform. It has no ambiguity Doppler response, independent of range ambiguity, and can improve the range and speed resolutions. However, the side lobe interference is serious in the target environment with strong clutter and large differential radar cross section (RCS).

Receiver of linear FM continuous wave radar adopts dechirp receiving. Its difference frequency bandwidth is narrow, and the high-resolution imaging with low sampling frequency can be achieved; the requirement of the video receiving channel, A/D sampling, and signal processing speed can be reduced greatly. FM continuous wave SAR does not require high peak power, which has the characteristics of low power consumption, lightweight, high maintainability, and low research and development costs. It draws more and more attention.

In this book, the time- and frequency-domain characteristics in range dimension of linear FM pulse signal, stepped-frequency signal, and stepped-chirp waveform trains signal were analyzed; the ambiguity function and distinguish of the signals were analyzed systematically and the SAR high-resolution imaging characteristics and methods of the signals were provided. The signal characteristics of linear FM continuous wave SAR were introduced.

1.3 Spotlight SAR and Azimuth Autofocusing

Spotlight model is a specialties operating model of synthetic aperture radar (SAR). It is an effective means of achieving high azimuth resolution. By rotation of the radar line of sight, the imaging area has always been under the radar beam irradiation, which can make spotlight SAR obtain longer synthetic aperture length than strip SAR and obtain a higher azimuth resolution.

In spotlight SAR operating, in addition to the same translation with Strip SAR, there is a rotation moving the antenna beam relative to the target area, and it is very similar with computer-aided X-ray tomography (CAT) in medicine. Therefore, the spotlight SAR imaging processing can be explained by means of the advantage of computer-aided tomography method.

Due to the continuous change of the radar incidence angle, the Doppler centroid of spotlight SAR varies with azimuth time, and frequency course of each point will possess the same starting and ending time. The azimuth bandwidth is related not only to the synthetic aperture time but also to the azimuth range of the imaging area. The azimuth bandwidth is composed of two parts including the instantaneous Doppler bandwidth and the Doppler bandwidth of point target echo. The instantaneous Doppler bandwidth of echo signal is constant; the Doppler centroid frequency is approximately linear FM signal. Doppler center frequency range B_d corresponds with the Doppler bandwidth of the center point target. The frequency difference between the target points and the center point of the scene is constant, and proportional to the distance from the point to the center of the scene.

The Doppler bandwidth of spotlight SAR is wide, and it requires high PRF. Considering the instantaneous Doppler bandwidth as the lower limit for PRF, the PRF sampling does not satisfy the Nyquist sampling law. It will cause azimuth aliasing of signal. However, using the azimuth downsampling processing, it can recover the aliased signal spectrum and get the signal spectrum before aliasing.

The spotlight SAR can improve the azimuth resolution by analyzing the Doppler frequency generated by the motion of the platform. However, the spotlight SAR still has the different characteristics with strip map mode including range migration, two-dimensional MTRC phenomenons, time-varying Doppler centroid, space varying phase course, coupling between range and azimuth, motion compensation, and so on.

Range Migration

Range Cell Migration (RCM) refers to the slant range changes between radar and target over a range resolution cell, which makes the echo signal from the same target and distributes to the different range cells. RCM includes range walking and range bending. Range migration makes the echo signal couple in azimuth and range, and two-dimensional independent criteria are not met, so echo signals cannot be decomposed into two one-dimensional processings directly.

Two-dimensional MTRC Phenomenon

Except for the relative translation on the plane velocity direction, considerable relative rotation between radar and target still exists in synthetic aperture time. The relative rotation between radar and target causes the distance course of target scattering point beyond a resolution cell at both range and azimuth simultaneously, that is a phenomenon of two-dimensional movement through resolution cells (MTRC). The MTRC phenomenon caused by the rotation is the reason that causes the image to defocus. Two-dimensional MTRC phenomenon has space varying characteristics. The variation law and the variation range depend on the geometrical position of the scattering points. This means that the imaging process for different scattering points in the target area requires different reference functions, and put forward strict requirements for the coherence and motion compensation of the radar system.

Time-Varying Doppler Centroid Frequency

Because of the continuous change of the Doppler centroid frequency with the radar incidence angle, the Doppler centroid frequency of the spotlight SAR varies with the azimuth time. Effective measures must be taken to remove the influence of the spectrum spread caused by the Doppler centroid frequency change.

Space Varying Phase Course

The relative motion between the radar and the measured target can be divided into translation and rotation. The phase error caused by the translation is space invariant. The phase error caused by the rotation varies with the geometric location of the scattering point. It has space varying characteristic. The space varying characteristic of phase course requires adopting different compensation phase factors for focusing different scattering points. This brings enormous difficulty to the imaging processing.

Motion Compensation

Spotlight SAR imaging has higher resolution and longer synthetic aperture time, and it puts forward extremely strict requirements for the motion compensation in imaging process and the coherence of the radar system.

The peculiar phenomenon of spotlight mode greatly increases the complexity of image processing.

In this book, the imaging algorithms of spotlight SAR were discussed in detail including polar format algorithm (PFA), range migration algorithm (RMA), chirp scaling algorithm (CSA), extended chirp scaling algorithm (ECSA), frequency scaling algorithm (FSA),two-step algorithm combining with azimuth deramp and chirp scaling algorithm on broadside and squint model, and so on. The principle and signal models of various algorithms were analyzed, and the computing procedure and special problems of imaging algorithms were given.

Sliding spotlight SAR is a recently developed work mode. It has the practical application in spaceborne SAR. Different with strip map and pure spotlight mode, spotlight focus point of sliding spotlight SAR is far from the center of the imaging scene. The azimuth imaging width is greater than that of spotlight mode, and the azimuth resolution is better than that of strip map mode.

In strip map mode, radar antenna beam is perpendicular to the flight track of the platform and points toward a fixed direction. Azimuth imaging width is not limited; the azimuth resolution is the half-length of antenna.

In spotlight mode, the beam center is always pointing to the center of the imaging scene in the imaging process. It increases the effective azimuth synthetic aperture length and improves the azimuth resolution. However, the resolution improvement is at the expense of the loss of irradiation length in azimuth dimension.

If controlling the antenna beam of sliding spotlight mode pointing to the ground irradiation center, it is the pure spotlight mode. If controlling the antenna beam pointing to the infinity point, it is the strip map mode. The strip map and pure spotlight modes can be seen as two special cases of sliding spotlight mode.

In this book, the imaging geometry model, the target irradiation time, the effective exposure area, the echo signal characteristics, the resolution, the echo-Doppler course, and the imaging algorithm of sliding spotlight SAR were analyzed and discussed more in detail.

Autofocusing technique is one of the key technologies in high-resolution SAR processing.

In SAR, the unsatisfactory trajectory of the platform, the irregular motion of the electromagnetic wave propagation medium such as atmospheric vortex, the phase jitter of the frequency source of the receiver, the approximation factors introduced by the imaging algorithm and data processing, etc. will cause the phase error of the SAR received signal. The compensation for the phase error effect in imaging is an indispensable part of the imaging process.

Autofocusing algorithm is a processing algorithm that can compensate the low-order phase error caused by the irregular motion of the SAR platform and atmospheric turbulence, and automatically accomplish image focusing. The basic principle is to estimate the phase error function from the echo data, and then compensate and obtain the well-focused image. In the high-resolution SAR system, autofocusing is one of the key techniques in the azimuth processing.

In this book, the classical autofocusing algorithm was discussed including power spectrum method (PSM), map-drift (MD) algorithm, multiple aperture map-drift (MAM) algorithm, shift-and-correlate (SAC) algorithm, phase difference (PD) algorithm, phase gradient algorithm (PGA), contrast optimization (CO) algorithm, and so on.

1.4 High-Resolution and Wide Swath SAR

A fundamental limitation in the system design of spaceborne SAR is the contradiction between swath width and azimuth resolution. The technical approaches to achieve high-resolution wide swath SAR include single phase center multiple azimuth beams (SPC-MAB), multiple phase center multiple azimuth beams (MPC-MAB), multiple range beams, Multiple-Input Multiple-Output (MIMO), and so on.

In SPC-MAB, the radar antenna forms N receiving beams in azimuth. They, respectively, cover the adjacent areas. The signal Doppler spectrums of the beams are adjacent to each other. After frequency-domain processing, the echo signals of N receiving beams can synthesize an azimuth echo signal with a wider Doppler bandwidth. The azimuth resolution obtained after the imaging processing improves about N times than that of the single beam azimuth resolution.

SPC-MAB has two operating modes including "Wide Transmitting Narrow Receiving" and "Narrow Transmitting Narrow Receiving".

The azimuth signal of the beams of SPC-MAB radar can be synthesized by interpolation and filtering. FFT is a fast computing tool of discrete Fourier transform and can be used to complete the interpolation, filtering, and other operations.

SPC-MAB technique simultaneously achieves high-resolution and wide swath by frequency-domain processing, while MPC-MAB technology simultaneously achieves high-resolution and wide swath by time-domain processing. The basic idea of MPC-MAB technique is to exchange for the reduction of time sampling rate with the increase of space sampling rate [6].

When adopting MPC-MAB technique, radar arranges multiple sub-antennas along azimuth. All of the sub-antennas have the same width, and cover the same area. The center sub-antenna transmits signal (or full aperture weighted broadened beam transmits signal). All of the sub-antennas receive the echo signal from the same area. There are N_B echo signal samples along azimuth for each transmitting pulse. The appropriate set of sub-antenna phase center space makes the echo signal samples independent of each other, allows PRF decreasing N_B times, and still guarantees correct sampling of echo signal in azimuth. By exchanging the reduction of time sampling rate with the increase of space sampling rate in azimuth, it allows decreasing the system PRF and broadening the swath width, meanwhile, guaranteeing certain azimuth resolution. Or at a certain PRF (corresponding to a certain swath width), it increases the equivalent sampling rate of the azimuth echo signal, allows the broadening the Doppler bandwidth, and improves the azimuth resolution.

The vacancy of phase course caused by the decrease of PRF just is filled up by the echo data for both sides of the multi-beam antennas. It just is the operating mechanism of MPC-MAB technology.

There are three kinds of errors that may be introduced by MPC-MAB technology: the first is the signal propagation path error, the second is the error caused by the receiving channel gain and non-equalizing phase shift, and the third is the error caused by the nonuniform sampling. These errors will cause the mismatch of azimuth matched filter. It leads to the paired echo after azimuth compression. The ambiguity interference will appear in the image, and the main wave amplitude will decrease. It leads to the decrease of image SNR. The errors caused by the disaccord of the receiving channels can be eliminated by calibration method. The errors caused by the azimuth nonuniform sampling can be compensated by interpolation method, filter bank method, etc. And the phase errors caused by discontinuous signal propagation path should be estimated and corrected before image processing.

Before the echo signal processing, the signals should be synthesized to one way. The echo data of all received beams should be corrected respectively. After phase error calculation, phase correction, data synthesis, and compression processing, the azimuth echo signal with a complete phase course is synthesized and is imaged by means of two-dimensional compression processing.

The idea that achieves high resolution and wide swath by azimuth multiple beam techniques was first proposed by England's A. Currie et al. in 1989 at the IEE SAR conference [7], Cooperating with C.D.Hall and M.A. Brown. They introduced range multiple beams and azimuth multiple beams techniques [8, 9]. A preliminary analysis of the azimuth resolution and ambiguity characteristics of SPC-MAB technology was carried out by Fujisaka Takahiko in Japan [10]. Until the end of the last century, high-resolution wide swath SAR research began to attract wide concerns. Successively, some researchers have proposed different system structures of high-resolution wide swath radar. In 1997, England's P.S. Cooper, A.F. Wons, and A.P. Gaskell proposed the scheme to achieve high-resolution SAR by multiple sub-band techniques [11]. Its main idea was to divide the antenna array into several sub-arrays in azimuth, achieve range high resolution by a wide transmitting signal bandwidth, and obtain two-dimensional high-resolution imaging without the expense of swath width. In 1999, Australia's G.D. Callaghan proposed to achieve wide swath by four-element antenna array. It can extend the swath about four times by the method of zero adjustment of the antenna, not affecting the azimuth resolution [12]. After that, Germany's M. Suess, B. Grafmueller, and R. Zahn proposed a new transmitting and receiving antennas independent structure to achieve high-resolution wide swath. Its transmitting and receiving antennas could be mounted on the same satellite or different satellites in the same constellation. The independent transmitting and receiving antennas were convenient for the optimized electric performance design, thereby reducing the system loss and improving the efficiency of [6, 13]. In 2002, England's E. Buckley further studied the method to realize the beam scanning and achieved the wide swath using transmitting and receiving independent antennas. A sub-antenna transmitted, multiple sub-antennas received, multiple transmit sub-pulses were inserted in a pulse repetition period, and to illuminate different regions respectively. It distinguishes the different sub-pulse echoes by space division, time division, frequency division, and code division, and achieved the same swath as the ScanSAR mode.

The technology realizing high-resolution wide swath with azimuth multi-beam technology had been equipped with the satellite and applied as an independent operating mode. The Canadian satellite Radarsat-2 launched on December 14, 2007, except to the existing functions of Radarsat-1 and Radarsat-2 also has a super fine imaging mode, full polarization, and ground moving target indication (GMTI) capability. In the superfine imaging mode, the azimuth multi-beam scheme is realized by setting two receivers in azimuth, and 3 m $\times$ 3 m resolution can be achieved.

From SAR imaging ambiguity view, range ambiguity limits swath width, and azimuth ambiguity limits azimuth resolution. The choice of PRF is seeking the optimal between the range and the azimuth ambiguity. It can realize high-resolution wide swath imaging by suppressing the ambiguity. Taking the azimuth ambiguity suppression as an example, the reduction of the azimuth ambiguity is to reduce the ambiguity signal energy folding into the azimuth processing Doppler bandwidth. This allows further broadening the Doppler bandwidth, obtaining a higher azimuth resolution, or reducing the PRF, and extending the swath. The method of ambiguity suppression can reduce the antenna side lobe energy or the antenna gain of ambiguity area by the design of the antenna pattern. Ambiguity suppression can be realized by signal processing and multi-beam technique.

Range ambiguity and azimuth ambiguity are overlapped with useful echo signal in time and frequency domains, but is separated in space domain. Controlling the antenna pattern zero point direction aligns to the ambiguity area to suppress ambiguity. In 1991, H.D. Griffiths and P. Mancini first proposed the method of zero point technology to solve the contradiction between the swath width and resolution [14]. And they studied two zero-forming adaptive algorithms including Howells–Applebaum algorithm which is easy to compute on satellite, and direct matrix transpose algorithm which processes on the ground receiving port and obtain antenna array element weights by computing the echo covariance matrix. In 1997, G.D. Callaghan and I.D. Longstaff proposed the method of zero point formed by the adaptive processing technology to suppress the ambiguity. The weight of antenna array is also required the range echo data [15]. In 1999, they also introduced the four-element antenna array to form zero point to suppress the azimuth and range ambiguity, proposed the algorithm for realizing zero point adaptive azimuth ambiguity suppression by Doppler centroid estimation, and pointed out the equivalent of between the MPC-MAB technology and the zero point technology. It should be pointed out that the MPC-MAB technology has strict requirements on the PRF and the distance between the sub-antenna phase center, but the zero point technology can get rid of the limit and realize the free control of zero point by changing the array weights. In 2006, R. Roman Jaime et al. studied the method to suppress Doppler ambiguity by means of forming the antenna pattern zero point using the array cell weights obtained from the optimal algorithm in the linear sparse array. Under the constraint that the useful echo signal gain was constant, the optimal weights were acquired by minimizing the energy of the ambiguity area.

In this book, the performances of SPC-MAB, MPC-MAB, and MIMO SAR systems were analyzed in detail.

1.5 SAR Super-Resolution Processing Technology

High resolution has been one of the important characteristics of radar imaging technology. As wide application of microwave imaging radar, the high resolution and good image quality of SAR image are got more and more attention.

In essence, the process of acquiring the mapping area SAR image is an inverse problem in mathematical physics. Its core is the estimation of parameters of the scattering source, that is, the estimation of spatial position and the backscattering coefficient of scattering source in the squint plane. Matched filter becomes the main data processing method of SAR imaging. It is an optimal filter for the detection of known signals from additive white noise. The traditional fast Fourier transform has the advantages of good robustness and high computational efficiency. However, due to the limited bandwidth of the radar system and synthetic aperture length, the collected phase course echo sampling data is also limited. For the finite data, the result of the Fourier transform is the convolution of the signal spectrum and the window spectrum, and the resolution of the Fourier transform is limited by the Rayleigh limit.

SAR super-resolution information processing technology adopts super-resolution imaging algorithm. That is, super-resolution imaging algorithm, which process microwave imaging radar data, uses a high-resolution spectral estimation method to replace Fourier spectrum estimation, extrapolation effective spectrum, adaptive weighted, and inverse problem-solving methods, and can break through the performance limitations of Fourier imaging method to improve image resolution further without increasing the signal bandwidth and the length of synthetic aperture.

The classical imaging algorithms, such as range-Doppler algorithm, $\omega - k$ algorithm, and chirp scaling algorithm, are highly dependent on the fast Fourier transform. They are very effective for frequency-domain processing. The common shortcoming of these methods is that the output SNR of the two-dimensional matched filter is approximate-optimal; they cannot achieve the minimized mean square error between the electromagnetic reflectivity ratio map and the reconstructed SAR image of the mapped plane. If the imaging process of microwave imaging radar is considered as a model-based estimation problem, the real image of the scene is described by a parametric model (such as autoregressive model); in the noise background, the minimized mean square error between estimated images is obtained from the observed signals and the real images. This will be a signal processing method which is different from the matched filter processing.

The development of super-resolution imaging algorithm passed through the following four primary stages: The first stage was the algorithm transplantation phase, which focused on the fast and effective super-resolution imaging algorithm. It could solve the problem of the large amount computation of modern 2D spectrum estimation. The second stage was specially tailored stage. The main research was suitable for the super-resolution microwave imaging algorithm, such as the adaptive side lobe suppression algorithm (ASR) which adopting the adaptive FIR filter to suppress the side lobe, spatially variant apodization (SVA) algorithm, and amplitude and phase estimation (APES) algorithm which can improve the accuracy of image amplitude estimation, etc. The third stage was the stage to improve the performance; the main research was the robust super-resolution imaging algorithm, such as relaxation method which is not sensitive to noise and MCRELAX method which can compensate the motion errors. The fourth stage was the application extending stage; the main research was the super-resolution algorithm for microwave imaging and combined with target recognition.

Modern spectrum estimation techniques and array super-resolution techniques were a class of power spectrum estimation methods based on the parametric model, which was developed for 1960s. It was able to give a much higher frequency resolution than the conventional periodic graph method. The substitution of Fourier spectrum estimate with modern spectral estimation was very attractive for SAR imaging. The classical spectrum estimation algorithm was used in the early SAR super-resolution technology. These nonlinear methods had the potential to improve the resolution and contrast and reduce the speckle. Using the adaptive disturbance zero setting, the linear prediction model, the signal clutter subspace decomposition, or the parametric sinusoidal signal course (point scattering) model, it could improve the resolution and reduce the side lobe false peak. By means of a non-coherent average of the signal course in the estimation of the power spectral density (PSD), the coherent speckle can be reduced. The contrast could be improved by the signal clutter subspace decomposition or singular value decomposition.

In 1977, Mancill used the maximum entropy (MEM) method as a super-resolution technique for radar imaging [16]. Nine years later, Friedlander applied AR autoregressive mode to the spaceborne SAR imaging. Its aim was to reduce the transmitting data rate between space and ground without reducing the system resolution.

Super-resolution microwave imaging algorithms include spectrum estimation algorithm, spectrum extrapolation method, adaptive weighting algorithm, regularization algorithm, point speed function deconvolution algorithm, and so on.

(1) **Super-resolution spectrum estimation algorithm**

Improving the radar image quality by modern spectrum estimation technique is one of the most active fields in the research of super-resolution microwave imaging. Spectrum estimation technique can be roughly divided into two categories: non-parametric method and parameter method.

Super-resolution nonparametric method can directly estimate the spectrum (or power spectrum density) from the data, which includes most classical spectrum estimation methods. The classical method is used to estimate the power spectrum density (PSD) with correlation diagram, window period diagram, and Blackman–Tukey (BT) spectrum estimation. The correlation graph method can calculate PSD. Using correlation sequence estimation, the periodic method can directly calculate PSD from data.

Parameter spectrum estimation is also known as modern spectrum estimation method including CAPON minimum variance method (MVM1) [17–19], amplitude phase spectrum estimation (APES) [20–22], relaxation method (RELAX) [23], CLEAN method, subspace orthogonal decomposition method (including Pisarenco method, mini-norm method, multiple signal classification (MUSIC) method [24, 25], and Prony method [26]), spatially rotation invariant (ESPRIT) method [27–29], autoregressive spectrum estimation (AR), parametric maximum likelihood (ML), etc.

Subspace orthogonal decomposition method decomposes the signal autocorrelation matrix, using the feature vector as a set of base to construct the linear space. When the signal is composed of multiple harmonic and white noise, the linear space can be decomposed into the sum of the signal subspace and the noise subspace, which are mutually orthogonal. Based on the orthogonal principle, a class of harmonic frequency estimation method is proposed including Pisarenco method, mini-norm method, multiple signal classification (MUSIC) method, Prony method, and so on. The subspace orthogonal decomposition method is characterized by a strong ability to suppress noise and a high accuracy to estimate the position of the scattering points, but cannot estimate the magnitude and type of the parameter of the scattering points.

Spatially rotation invariant method (ESPRIT) is a kind of subspace decomposition algorithm. It decomposes the received data and estimates the signal parameters by using the rotation invariant property of the signal subspace of the data covariance matrix. A pair of Toeplitz matrices is constructed from the observation data, and the relationship of the two matrices is determined by the poles of the model. The problem of estimating the model parameters is transformed into a problem of solving the generalized matrix eigenvalue and a problem of solving a Vandermonde-type linear equation group. This method has the advantages of high accuracy pole estimation, strong anti-noise ability, small computation amount, and no need to carry out spectrum peak search.

The maximum likelihood estimation (ML) method is a special case of the bias parameter estimation method and is the bias optimal estimation in the case of known white noise. It is a typically used estimation method. The maximum likelihood estimation (ML) method has better estimation performance than the subspace orthogonal decomposition method and the space rotation invariant method.

In this book, the methods of super-resolution modern spectrum estimation were studied including nonparametric method and parameter estimation, the practical application of various methods in SAR imaging was analyzed, and the algorithm procedure suitable for the super-resolution SAR imaging was given.

(2) **Super-resolution imaging algorithm based on spectrum bandwidth extrapolation**

Spectrum bandwidth extrapolation [30–33] is the data of the known band-limited signal in the given limited bandwidth by means of transcendental information to find out the datum out of the bandwidth and to improve the resolution.

Signal extrapolation is one of the basic problems in signal processing. For the continuous signal with limited band, because of the analytic nature of mathematics, it can be extrapolated by Taylor series method. However, Taylor series requires high-order derivative. When only discrete data can be measured, or accompanied by noise, higher order derivative may be not found, or its error is very great. In fact, it is very difficult to carry out the extrapolation for a continuous band-limited signal.

In 1961, the mathematician of the Bell laboratory presented the mathematical tool of extrapolation: prolate spheroidal function. Due to the orthogonality of the local and global prolate spheroidal functions, it was very easy to understand and analyze the signal extrapolation problem. Although the calculation and storage amount for prolate spheroidal function and its coefficient are very large, the signal extrapolation problem still did have the practical application. Until the 1970s, Gerchberg and Papoulis successively proposed the iterative extrapolation of continuous band-limited signal; in this way, the signal extrapolation can be widely applied.

Signal extrapolation is essentially "solving the whole by the known part". Under the strict constraints, it is possible. For the energy-limited continuous band-limited signal, as long as knowing a short signal in time-domain continuous interval, we can infer the whole signal in the time domain. However, many physical signals are not strictly limited energy continuous band-limited signal. They can only be understood as the approximate band-limited signal; it is difficult to carry out signal extrapolation. In addition, when observing band-limited signal, due to the discretization and the bandpass noise of the signal, the uniqueness and stability of signal extrapolation are proposed. Due to the pathological nature, all kinds of extrapolation problems need to be limited according to a variety of prior knowledge.

Because of the extrapolation of the limited target spectrum, the bandwidth is increased and the resolution is improved. Therefore, the extrapolation method can be used for super-resolution study.

In this book, we have studied the SAR super-resolution spectrum extrapolation imaging algorithm, analyzed the autoregressive model parameter estimation method, autoregressive linear prediction extrapolation method, and an improved two-dimensional autoregressive linear prediction extrapolation super-resolution imaging method. In addition, we have analyzed the minimum weighting norm constraint, adaptive weighting norm extrapolation (AWNE) method, and minimum weighting norm extrapolation (CAWNE) super-resolution imaging method using minimum variance spectrum estimation. The minimum variance spectrum estimation is used as constrained weights of weighted norm constraint. According to the weighting norm constraint, the phase course domain data are extrapolated to increase the effective bandwidth and improve the resolution.

(3) **Super-resolution adaptive weighting imaging algorithm**

Super-resolution adaptive weighting imaging algorithm is a nonlinear side lobe suppression method. A weighting network is introduced to adaptively deal with mismatch signal and suppress the side lobe without loss of the main lobe resolution. Super-resolution adaptive weighting imaging algorithm includes spatially variant apodization method (SVA) and adaptive side lobe reduction method (ASR) [34–38].

SVA calculates SAR image by point and makes that impulse response (IPR) is optimum and the side lobe is minimum. In 1994, Stankwitz first introduced the one-dimensional SVA, which was an extension of two-dimensional SVA. The two-dimensional SVA compares the uniform weighting images with arbitrary window function weighting images, point-by-point detects, and select the small amplitude as the output. SVA adaptively weights with raised cosine function for each output data point (pixel). It makes the output response of the side lobe minimum, and the calculation is simple, and hypothesis model is not required. This adaptive weighting maintains the same resolution as the rectangle weighting. SVA is a special case of single raised cosine weighting function. The weighted output energy is minimized, and the weighting function constraint changes between the uniform weighting function and the Hanning weighting function.

High-order expansion of SVA is adaptive side lobe reduction (ASR) method. DeGraaf introduced the adaptive side lobe reduction method (ASR), and used a more general raised cosine function. It used the cost function constraint to deal with the overdetermined parameter space.

In this book, we discuss the SAR SVA bandwidth extrapolation super-resolution imaging algorithm and the adaptive weighting super-resolution imaging algorithm. Using the bandwidth expansion effect, the new signal spectrum is processed by inverse weighting, the signal spectrum is broadened, and the resolution is improved.

(4) **Imaging system point spread function deconvolution super-resolution algorithm**

By using the imaging system point spread function (PSF) [39], super-resolution image is obtained by deconvolution. The reflectivity estimation of the scene can be regarded as the inverse problem. It recovers the image from the measured value by the inverse matrix operation. Ideally, the imaging system point spread function is needed. Pseudoinverse operation of point spread function is equivalent to the least square estimation of the scene reflection intensity. The inverse problem is morbid. In order to obtain a stable estimation valuation, it needs to use the regularization technique. In the actual microwave imaging radar system, using system internal and external calibrations, it can approximately accurately obtain system point spread function. By deconvoluting each received data, it can obtain high-resolution radar image.

(5) Super-resolution regularization imaging algorithm

Regularization method is an important method for microwave imaging super-resolution information processing [40]. In essence, the process of SAR acquiring the image of mapping area is a kind of inverse problem in mathematical physics. Its core is to estimate the spatial position in the squint plane and the backscattering coefficient of the scattering source. The regularization method derives from the solution of inverse problem in mathematics. It is based on the prior information to be solved, as many as possible, steadily restoring the information of the problem. So it can process super-resolution information by using the regularization method.

Frequency-domain imaging model of microwave imaging radar describes the relationship between backscattering coefficient and echo data of radar imaging scene. It is the discrete observation model of microwave imaging radar image. According to the observed quantity, by using all relevant image prior information, the regularization method obtains the real scene scattering coefficient as true as possible and improves resolution and image quality. This is an ill-posed problem. The key to solve this problem is to use the prior information to construct the constraint conditions and to transform the ill-posed problem into a properly posed problem. The regularization method is suitable for solving these problems. For different microwave imaging radar images, the regularization method needs to estimate the parameters according to the microwave imaging characteristics and choose different regularization parameters.

The super-resolution regularization imaging algorithm is suitable for uniform background and high radiation intensity images, such as the military targets on the surface of the water, desert background, grassland background and open background, the ground to air Inverse synthetic aperture radar (ISAR) imaging, and so on. For the distributed targets and complex background targets, the super-resolution regularization imaging algorithm requires a more detailed model to get better results.

In this book, the super-resolution regularization imaging algorithm is discussed including Tikhonov regularization, non-secondary regularization, variety difference method, and SAR super-resolution regularization processing method.

(6) Image super-resolution processing

At abroad, a lot of theoretical and practical research work on image super-resolution processing has been carried out. Using the not exactly same, the mutual complement of information between the multiple low-resolution images, the super-resolution technology of the multi-frame image has been developed rapidly. The super-resolution technique considers multiple low-resolution images as that is obtained by high-resolution images downsampling, and the translation and rotation of different parameters. Due to the sub-pixels displacement between the multiple low-resolution images, when low-resolution images are registered and all put into high-resolution mesh, the mutual complement of information between multiple low-resolution images fills the unknown true information of a single low-resolution image during iterative extrapolation. So it may improve the spatial resolution. France's SPOT, US's Earth Sat company, etc. have been trying to get high-resolution image from multiple low-resolution infrared images obtained from single satellite reconnaissance and multiple satellite imagings at the same time by using super-resolution information processing technology.

1.6 Configuration in the Book

This book is divided into two parts, which are high-resolution microwave imaging and super-resolution information processing. There are 17 chapters in this book.

The first chapter: Introduction. The development of SAR technology is discussed including high range resolution microwave imaging technology, spotlight SAR technology, autofocus SAR technology, high-resolution wide swath SAR, and microwave imaging super-resolution information processing technology.

The second chapter: The basic principle of microwave imaging. The principle and the theoretical models of microwave imaging radar are discussed. The pulse compression and synthetic aperture processing technology, microwave imaging radar focus depth, range migration, and the main technical parameters of spaceborne SAR are studied. The limitations of antenna area, the system quality factor, and the quality parameter of microwave imaging radar image are analyzed.

The third chapter: The principle of radar resolution. The optimal detection of radar signal, the mathematical model of the radar echo signal, the range ambiguity function and range resolution, velocity ambiguity function, and Doppler resolution are studied. The SAR resolution principle, the classification and the design principle of the waveform, the high-resolution waveform, and the processing are discussed.

The fourth chapter: Linear frequency modulation signal. The linear FM pulse signal of microwave imaging radar is studied. The spectrum, ambiguity function, pulse compression characteristic, and weighting processing of the linear FM pulse signal are discussed. Error analysis (phase error analysis, amplitude error analysis, quadrature modulation, and demodulation error analysis) of linear FM signal is done. Linear FM continuous wave SAR and its imaging algorithm are introduced.

The fifth chapter: Stepped-frequency SAR imaging. The high range resolution principle of stepped-frequency signal is discussed. The time-domain and frequency-domain characteristics of the synthesis wideband signal with stepped-frequency signal, the ambiguity function and resolution performance of stepped-frequency signal, and the effect and compensation of radial velocity on the target range image of stepped-frequency radar are analyzed. The design of the stepped-frequency SAR and the waveform parameters is studied.

The sixth chapter: Stepped-chirp SAR imaging. The principle of phase coherent synthesis wideband signal of inner pulse FM pulse trains waveform is discussed. The time-domain and frequency-domain characteristics of the inner pulse FM pulse trains, the ambiguity function and resolution performance of the inner pulse FM pulse trains, and the effect and compensation method of radial velocity on the synthetic range image of the inner pulse FM pulse trains are analyzed. The SAR system and imaging method of the inner pulse FM pulse trains are studied.

The seventh chapter: Spotlight SAR technology. The principle of spotlight SAR is introduced. The parameters of the spotlight mode SAR, such as spaceborne and airborne spotlight SAR azimuth resolution, synthetic aperture length, etc., are analyzed. The characteristics of the echo signal of the spotlight SAR and the special problems of the spotlight SAR imaging processing are studied including range migration, two-dimensional MTRC phenomenon, time-varying Doppler centroid, space varying phase course, and the coupling of range and azimuth, and motion compensation method of the spotlight SAR.

The eighth chapter: Spotlight SAR imaging algorithm. Spotlight SAR imaging algorithm is discussed. The polar format algorithm, range migration algorithm, chirp scaling imaging algorithm, extended chirp scaling algorithm, frequency scaling algorithm, and azimuth deramp + chirp scaling algorithm (broadside model and squint model two-step processing algorithm) are analyzed. The signal model and algorithm flow of the algorithms are presented.

The ninth chapter: Sliding spotlight SAR. The sliding spotlight SAR is discussed. The azimuth resolution, swath width, the echo signal model, and the echo-Doppler frequency course of the sliding spotlight SAR are analyzed. The PRF selection and beam pointing control technology of the sliding spotlight SAR is discussed. The imaging algorithm of sliding spotlight SAR is studied including extended chirp scaling (ECS) algorithm, wave number domain algorithm, baseband azimuth scaling algorithm, etc.

The tenth chapter: Single phase center multiple azimuth beams (SPC-MAB) SAR. The technology principle and azimuth signal characteristics of SPC-MAB SAR are discussed. Using the complex signal interpolation theory, the azimuth signal processing method of SPC-MAB SAR is studied. The azimuth ambiguities of "Wide Transmitting Narrow Receiving and Narrow Transmitting Narrow Receiving SPC-MAB SAR" are analyzed. The system model of SPC-MAB SAR is given.

The eleventh chapter: Multiple phases center multiple azimuth beams (MPC-MAB) SAR. The principle and echo signal mathematical model of MPC-MAB SAR are discussed. The ambiguity characteristic and systematic errors of MPC-MAB SAR are analyzed including the error introduced by the path difference of signal propagation, the phase error introduced by the amplitude and phase imbalance of the receiving channels and the azimuth nonuniform sampling, etc. The systematic model, azimuth phase correction, and signal processing method of MPC-MAB SAR are studied.

The twelfth chapter: Multiple-input multiple-output (MIMO) SAR. The technology principle of MIMO SAR and the virtual aperture technology is introduced. The high-resolution wide swath imaging, the ground moving target indicator, and the full polarimetric imaging of MIMO SAR working mode are discussed. The signal characteristics and the echo separation technology of MIMO SAR are analyzed including time-domain, frequency-domain, and space-domain separation technique. A detailed analysis is made on the mode, signal characteristics, the echo signal processing method, and the system performance of the frequency-domain diversity MIMO SAR. The new technology of MIMO SAR, such as the space-time encoding technology of the transmitter, the digital beam-forming technology, and the space-time encoding digital beam-forming technology of the receiver are studied. The distributed MIMO SAR system is studied. The signal characteristics and signal processing methods of the along-track distributed MIMO SAR and the cross-track distributed MIMO SAR are analyzed.

The thirteenth chapter: Autofocus algorithm for synthetic aperture radar. Autofocus algorithm for SAR are studied including sub-aperture correlation autofocus algorithm, multi-aperture correlation autofocus algorithm, phase difference autofocus algorithm, shift frequency correlation autofocus algorithm, phase gradient autofocus algorithm, image contrast optimization method, etc.

The fourteenth chapter: Super-resolution technology basics in microwave imaging. The classical Fourier spectrum analysis, the periodic graph, the average periodic graph, and the Blackman–Tukey (BT) spectrum estimation are discussed. The concept of SAR super-resolution technology is given. The super-resolution processing signal models of SAR, such as 2D frequency-domain model, dechirp processing model, $\omega - k$ domain model, etc., are analyzed.

The fifteenth chapter: Super-resolution spectrum estimation imaging method. Modern spectrum estimation super-resolution algorithm is analyzed in detail including minimum variance spectrum estimation method and its improvement, amplitude phase spectrum estimation (APES) algorithm, subspace decomposition method including multiple signal classification method, Pisarenko method, minimum modulus method and Prony method, signal parameter estimation method based on rotation invariant technique, relaxation spectrum estimation algorithms including RELAX method and CLEAN method, parametric maximum likelihood (MLE) method including frequency angle domain MLE, image domain MLE, autoregressive spectrum estimation method, etc. Experiments validation and analyses are carried out.

The sixteenth chapter: Super-resolution imaging algorithm based on spectrum extrapolation and adaptive weighting. The spectrum extrapolation imaging algorithm is analyzed and discussed including linear prediction extrapolation, forward and backward autoregressive linear prediction (ARLP) extrapolation, Tufts–Kumaresan ARLP extrapolation, and SAR super-resolution autoregressive linear prediction (ARLP) extrapolation. In adaptive minimum norm weighting extrapolation, the minimum weighting norm extrapolation super-resolution imaging algorithm and minimum norm iterative extrapolation (IWNE) algorithm are analyzed with minimum variance spectrum estimation.

Adaptive weighting imaging algorithm is mainly discussed in the analysis of adaptive weighting side lobe suppression method including spatially variant apodization (SVA) method and adaptive side lobe suppression algorithm. SVA mainly includes double variant apodization algorithm (DA), complex double variant apodization algorithm (CDA), and SVA. Adaptive side lobe suppression algorithm is discussed for one-dimensional and two-dimensional adaptive side lobe suppression methods. SAR adaptive weighting super-resolution imaging method is analyzed including SAR spatially variant apodization spectrum extrapolation super-resolution imaging algorithm and SAR adaptive weighting super-resolution imaging algorithm.

The seventeenth chapter: SAR super-resolution based on regularization method and deconvolution method. Regularization method summary, Tikhonov regularization, non-secondary regularization, total variant difference method, SAR regularization method, signal model of SAR regularization method, and SAR regularization processing method are analyzed. The deconvolution method using system point spread function is discussed. The image super-resolution information processing method is introduced.

References

1. Wehner DR (1995) High Resolution Radar. Artech House, Boston, London
2. Freeman A, Evans D, van Zyl JJ (1996) SAR Applications in the 21st Century, EUSAR'96, Königswinter Germany, pp 25–30
3. Zhang C (1989) Principle, system analysis and application of synthetic aperture radar. Science Press, Beijing
4. Zhang X (1998) Integrated ultra wide band range resolution technical papers. The Twentieth Research Institute of Ministry of Electronics Industry, internal data
5. Brenner AR (2010) Proof of concept for airborne SAR imaging with 5 cm resolution in X-band. EUSAR2010, pp 615–618
6. Suess M, Grafmueller B, Zahn R (2001) A novel high resolution, wide swath SAR system. IGARSS'2001 3:1013–1015
7. A Currie (1989) Wide-swath SAR imaging with multiple azimuth beams. IEE Colloquium on Synthetic Aperture Radar, 29 Nov 1989
8. Currie A, Brown MA (1992) Wide-swath SAR. IEE Proceedings-F 139(2):122–135
9. A. Currie, Hall CD (1990) A synthetic aperture radar technique for the simultaneous provision of high-resolution wide-swath coverage. Conference Proceedings of Military Microwaves'90, Microwave Exhibitions & Publishers Ltd., July 1990
10. Takahiko Fujisaka (2000) Azimuth high resolution and wide swath multi beam synthetic aperture radar. Airborne Radar 4:36–44
11. Cooper PS, Wons AF, Gaskell AP (1997) High resolution synthetic aperture radar using a multiple sub-band technique. Radar'97, No. 449, pp 263–167
12. Callaghan GD, Longstaff ID (1999) Wide-swath space-borne SAR using a quad-element array. IEE Proceedings-F 146(3):159–165

13. Suess M, Zubler M, Zahn R (2002) Performance investigation on the high resolution, wide swath SAR system. EUSAR'2002, pp 187–190
14. Griffiths HD, Mancini P (1991) Ambiguity suppression in SARs using adaptive array techniques. Proceedings of IGARSS'PI Symposium. pp 1015–1018
15. Callaghan GD, Longstaff ID (1997) Wide-swath space-borne SAR and range ambiguity. Proceedings of Radar 97. pp 248–252
16. Hua Y (1992) Estimating two-dimensional frequencies by matrix enhancement and matrix pencil. IEEE Trans Signal Proc 40:2267–2280
17. Jakobsson AJ, Marple SL, Stoica P (2000) Computationally efficient two-dimensional Capon spectral analysis. IEEE Trans Signal Process 48:2651–2661
18. Yadin E, Olmar D, Oron O, Nathansohn R (2008) SAR imaging using a modern 2D spectral estimation method. IEEE Radar Conference, RADAR'08, Rome, Italy
19. Li J, Stoica P, Wang Z (2004) Doubly constrained robust Capon beam former. IEEE Trans Signal Process 52:2407–2423
20. Li J, Stoica P (1996) An adaptive filtering approach to spectral estimation and SAR imaging. IEEE Trans Signal Proc 44:1469–1484
21. Jian L, Biao R et al (1999) A robust semi-parametric method for feature extraction and SAR image formation of targets consisting of trihedral and dihedrals. SPIE 3721:92–103
22. DeGraaf SR (1998) SAR imaging via modern 2-D spectral estimation methods. IEEE Trans Image Process 7:729–761
23. Li J, Stoica P (1996) An adaptive filtering approach to spectral estimation and SAR imaging. IEEE Trans Signal Process 44:1469–1484
24. Sacchini JJ, Steedly WM (1993) Two-dimensional prony modeling and parameter estimation. IEEE Trans Signal Process 41:3127–3136
25. Odendaal JW, Bernard E, Pistorius CWI (1994) Two-dimensional super resolution radar imaging using the MUSIC algorithm. IEEE Trans Antennas Propag 42:1386–1391
26. Barbarossa S, Marsili L, Mungari G (1996) SAR super-resolution imaging by signal subspace projection techniques. Proceedings of EUSAR'96, Konigswinter, Germany
27. Roy R, Kailath T (1986) ESPRIT-A subspace rotation approach to estimation of parameters of cissoids in noise. IEEE Trans Acoust Speech Signal Process ASSP-14:1340–1342
28. Wu Q, Wang G, Xu R (1999) ESPRIT super resolution ISAR imaging. Syst Eng Electron Technol 21:33–36
29. Feng D, Wang X, Chen Z et al (2005) Unitary ESPRIT super resolution ISAR imaging method. Electron J 33:2097–3000
30. Gupta Inder J, Beals Mark J, Moghaddar Ali (1994) High-Resolution Radar Imaging Using 2-D Linear Prediction. IEEE Trans Antennas Propag 42:31–37
31. Gabriel WF (1993) Improved range super resolution via bandwidth extrapolation. 1993 IEEE National Radar Conference, Lynnfield, MA, USA
32. Gupta IJ, Beals Mark J, Moghaddar Ali (1994) Data extrapolation for high resolution radar imaging. IEEE Trans Antennas Propag 42:1540–1545
33. Suwa Kei, Iwamoto Masafumi (2007) A two-dimensional bandwidth extrapolation technique for polarimetric synthetic aperture radar images. IEEE Trans Geosci Remote Sens 45:45–55
34. Stankwitz HC, Dallaire RJ, Fienup JR (1995) Non-linear apodization for side lobe control in SAR imagery. IEEE Trans Aerosp Electron Syst 31:267–278
35. Zhai W, Zhang Y (2007) Apply spatially variant apodization to SAR/INSAR image processing. 1st Asian and Pacific Conference on Synthetic Aperture Radar, 2007. APSAR 2007, Huangshan
36. Castillo-Rubio CF, Llorente-Romano S, Burgos-García M (2007) Spatially variant apodization for squinted synthetic aperture radar images. IEEE Trans Image Process 16:2023–2027
37. Brian Hendee Smith (2000) Generalization of spatially variant apodization to noninteger nyquist sampling rates. IEEE Trans Image Process 9:1088–1093

38. Xu X, Narayanan Ram M (2005) Enhanced resolution in SAR/ISAR imaging using iterative side lobe apodization. IEEE Trans Image Process 14:537–547
39. Blacknell D, Quegan S (1989) SAR super-resolution with a stochastic point spread function. IEE colloquium on Synthetic aperture radar, pp 1–9
40. Cetin M, Karl WC (2000) Super resolution and edge-preserving reconstruction of complex-valued synthetic aperture radar images. Proceedings of the 2000 IEEE International Conference on Image Processing, Vancouver, Canada

Chapter 2
The Basis of Microwave Imaging Radar

Synthetic aperture radar (SAR) obtains the two-dimensional high-resolution image by means of pulse compression technology in range dimension and synthetic aperture processing in azimuth dimension. The high range resolution is achieved by transmitting linear frequency modulation (LFM) signal and using pulse compression technology. Due to the movement of the platform, radar echo in azimuth has got the characteristic of LFM signal, and the azimuth resolution is realized by azimuth focusing. LFM signal and pulse compression technology are the basis of SAR imaging.

2.1 Principle of LFM Signal Pulse Compressing

There has been a lot of research on the theory of radar signal for better detection ability and resolution. It is found that ranging accuracy and range resolution are mainly determined by the structure of the signal frequency. As a result, a large bandwidth signal is needed to improve the ranging accuracy and range resolution, and a high power signal is needed to improve the ability of target detection. For the limitation of the radar peak transmit power, the high average transmission power can only be obtained by increasing the time width of the signals. LFM signal is a typical example of a large time–bandwidth product pulse compression signal, which is the earliest and most widely used pulse compression signal also.

The LFM signal with the normalized amplitude can be expressed by

$$x(t) = \mathrm{rect}\left(\frac{t}{T_{\mathrm{p}}}\right)\cos(\omega_0 t + \pi k t^2) \tag{2.1}$$

R. Yang et al., *High-Resolution Microwave Imaging*,
https://doi.org/10.1007/978-981-10-7138-6_2

where T_p is the pulse length, $\mathrm{rect}\left(\frac{t}{T_p}\right)$ is rectangular function, which is defined as

$$\mathrm{rect}\left(\frac{t}{T_p}\right)=\begin{cases}1 & \frac{T_p}{2}\leq t\leq\frac{T_p}{2}\\ 0 & 其他\end{cases}$$

The instantaneous angular frequency

$$\omega(t)=\frac{\mathrm{d}}{\mathrm{d}t}\left(\omega_0 t+\pi k t^2\right)=\omega_0+2\pi k t$$

where k is the chirp rate. During the pulse duration, the scope of the signal frequency is the bandwidth.

$$B=kT_p \tag{2.2}$$

The LFM signal is represented as a complex form

$$s(t)=\mathrm{rect}\left(\frac{t}{T_p}\right)\mathrm{e}^{j\left(\omega_0 t+\pi k t^2\right)} \tag{2.3}$$

The LFM pulse compression can be achieved by matched filter. According to the matched filter theory, the matched filter corrects the phase shift of each component in the signal spectrum and adds coherently at a certain time together. Due to the signal energy unchanged during this period, the matched filter makes the time width of LFM signals compressed greatly, and forms a high signal spike. The matched filter is

$$h(t)=s^*(t_0-t) \tag{2.4}$$

After the matched filter, the output signal is

$$s_0(t)=T_p\cdot\frac{\sin\left[\pi k T_p(t-t_0)\right]}{\pi k T_p(t-t_0)}\cdot\mathrm{e}^{j\omega_0(t-t_0)} \tag{2.5}$$

The peak of output signal appears at $t=t_0$, and the waveform is in the form of sinc function shown in Fig. 2.1.

Time resolution can be defined as the width of output pulse at peak fall to 4 dB

$$\rho_\tau=\frac{1}{B} \tag{2.6}$$

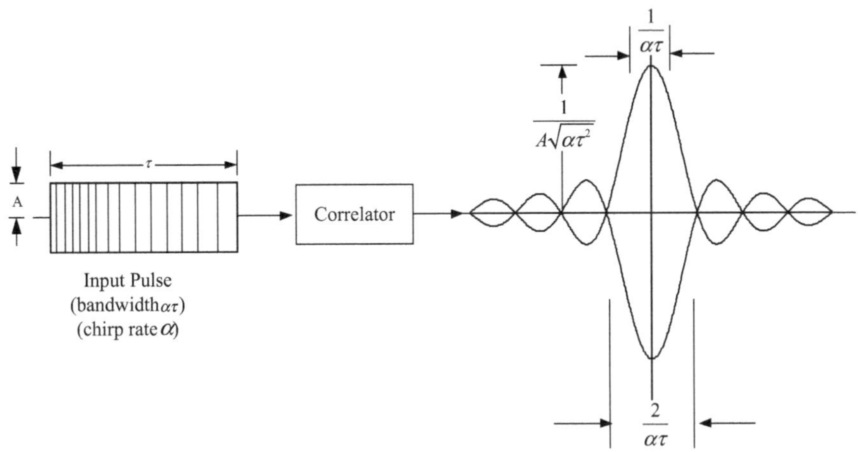

Fig. 2.1 The output waveform of LFM signal compressed by the matched filter

Time resolution is usually defined as the width of output pulse at peak fall to 3 dB

$$\rho_\tau = \frac{0.886}{B} \tag{2.7}$$

2.2 Principle of Synthetic Aperture Radar Imaging

The radar is mounted on the flight vehicle to make a straight line motion along with the platform. The direction of flight is defined as the azimuth dimension, and perpendicular to the flight direction is the range dimension. Schematic diagram of the side-looking SAR is shown in Fig. 2.2.

In Fig. 2.2, the range extent W_s of radar beam is called swath width, and the azimuth extent L_s of radar beam at the same slant range R is called synthetic aperture length, the time T_s of radar flying over synthetic aperture length is called synthetic aperture time.

The principle of synthetic aperture radar imaging can be explained from several points of view.

Doppler Sharpening: Using Doppler processing to sharpen the narrow azimuth beam of antenna.

Pulse Doppler: Using Doppler filter to distinguish the targets from different positions.

Synthetic Aperture: Using a small moving antenna synthesizes a large antenna aperture to form a narrow beam.

Matched Filter: Using Matched filter to optimize detection of signals.

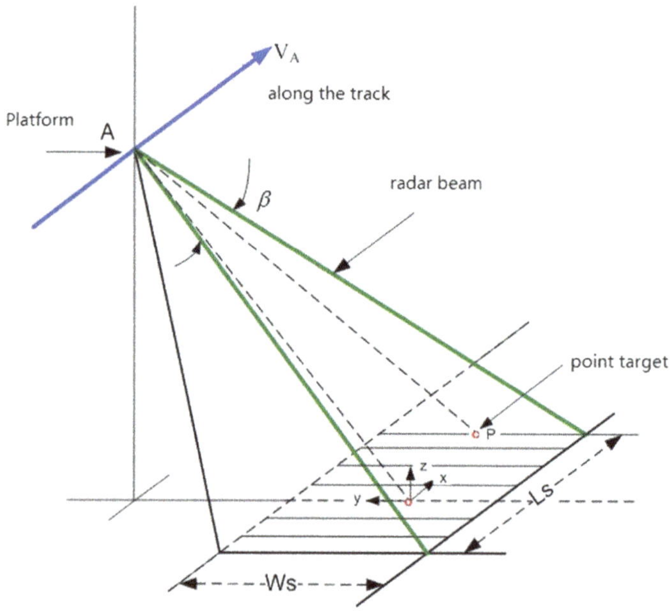

Fig. 2.2 Schematic diagram of the side-looking SAR

Holography: SAR is regarded as a kind of microwave holographic imaging technology from the point of optical holography.

Although the interpretations of the above points of view are different, the essence is the same.

(1) **Doppler Frequency analysis**

Geometry of SAR moving along track from S to S1 and passing over the target P is presented in Fig. 2.3. The closest slant range is R_0 when radar is at position S ($t = 0$). If the influence of earth autorotation can be ignored, at time $\boldsymbol{t}$, radar moves to the S1, and the distance between the radar and the target is $R(t)$.

Fig. 2.3 Geometry of SAR imaging in the slant plane

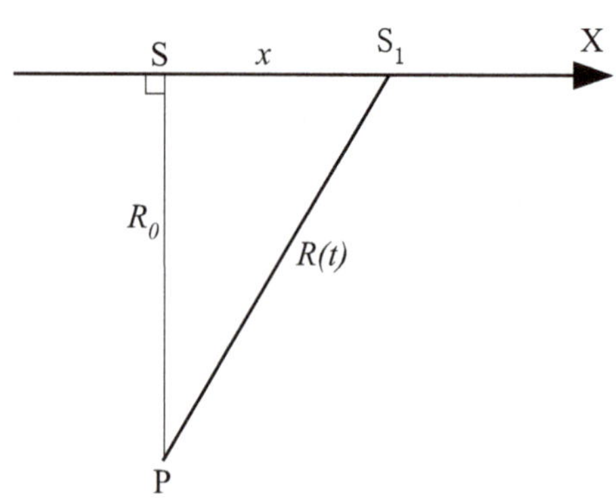

Radar echo is represented as a complex form

$$s_{\mathrm{r}}(t) = \mathrm{rect}\left[\frac{t - \frac{2R(t)}{c}}{T_{\mathrm{p}}}\right] e^{j\left\{\omega_0\left[t-\frac{2R(t)}{c}\right] + \pi k_{\mathrm{r}}\left[t-\frac{2R(t)}{c}\right]^2\right\}} \tag{2.8}$$

where c is the velocity of light, since $x \ll R_0$, the distance from $S1$ to P is approximated as

$$R(t) = \sqrt{R_0^2 + x^2} = \sqrt{R_0^2 + (v_{\mathrm{s}}t)^2} \approx R_0 + \frac{(v_{\mathrm{s}}t)^2}{2R_0} \tag{2.9}$$

where v_{s} is the velocity of flight vehicle, the phase $\phi(t)$ is

$$\varphi(t) \cong \omega_0\left(t - \frac{2R_0}{c}\right) + \pi k_{\mathrm{r}}\left(t - \frac{2R_0}{c}\right)^2 - \frac{2\pi}{\lambda R_0}(v_{\mathrm{s}}t)^2 - 2\pi k_{\mathrm{r}}\left(t - \frac{2R_0}{c}\right)\frac{(v_{\mathrm{s}}t)^2}{cR_0} \tag{2.10}$$

where the first and second terms compose a LFM signal in range, the third term is azimuth Doppler signal; the fourth term is coupling between range and azimuth. In the airborne SAR processing, the coupling phase term is small enough to satisfy the condition of two dimensions independent and separated between the range and azimuth, and can be ignored. However in the spaceborne SAR processing, this coupling phase term cannot be ignored. The main purpose of difference imaging algorithm is to compensate the effect of this coupling phase term. After compensated, the phase of echo signal is

$$\phi(t) \cong \omega_0\left(t - \frac{2R_0}{c}\right) + \pi k_{\mathrm{r}}\left(t - \frac{2R_0}{c}\right)^2 - \frac{2\pi}{\lambda R_0}(v_{\mathrm{s}}t)^2 \tag{2.11}$$

Azimuth instantaneous Doppler frequency in the third term is

$$f_{\mathrm{d}}(t) = \frac{1}{2\pi}\cdot\frac{\mathrm{d}}{\mathrm{d}t}\left[-\frac{2\pi}{\lambda R_0}(v_{\mathrm{s}}t)^2\right] = -\frac{2v_{\mathrm{s}}^2}{\lambda R_0}t \tag{2.12}$$

The echo in azimuth is a linear frequency modulated signal, and the chirp rate is

$$k_{\mathrm{a}} = -\frac{2v_{\mathrm{s}}^2}{\lambda R_0} \tag{2.13}$$

The difference of the instantaneous Doppler frequency between two targets apart Δx in azimuth is

$$\Delta f_{\mathrm{d}} = \frac{2v_{\mathrm{s}} \cdot \Delta x}{\lambda R_0} \tag{2.14}$$

The Doppler bandwidth and time duration of echo for each point target are

$$B_{\mathrm{a}} = \frac{2v_{\mathrm{s}} L_{\mathrm{s}}}{\lambda R_0} \tag{2.15}$$

$$T_{\mathrm{S}} = \frac{L_{\mathrm{s}}}{v_{\mathrm{s}}} \tag{2.16}$$

其中 $L_{\mathrm{s}} = \frac{\lambda R_0}{D_{\mathrm{a}}}$ 为合成孔径长度, D_{a}为天线方位向长度.

where $L_{\mathrm{s}} = \frac{\lambda R_0}{D_{\mathrm{a}}}$ is synthetic aperture length, D_{a} is antenna aperture length in azimuth.

In order to distinguish the two point targets, the echoes can be multiplied by the LFM signal with the same chirp rate of Eq. (2.13), and the output signals are two pulses modulated by different frequencies with pulse duration T_{S}. According to the different modulation frequency, the different bandpass filters can be used to separate the output signals, as shown in Fig. 2.4.

The frequency spectrum of the single frequency pulse signal with a time length of T_{S} is present in the form of sinc function, the spectrum width is

$$B = \frac{1}{T_{\mathrm{S}}} = \frac{v_{\mathrm{s}}}{L_{\mathrm{s}}} \tag{2.17}$$

The bandwidth of bandpass filter is B. If the difference between the instantaneous Doppler frequencies for two point targets is less than B, the two targets cannot be distinguished. The azimuth resolution is defined as

$$\rho_{\mathrm{a}} = \Delta x = \frac{\lambda R_0}{2L_{\mathrm{s}}} = \frac{D_{\mathrm{a}}}{2} \tag{2.18}$$

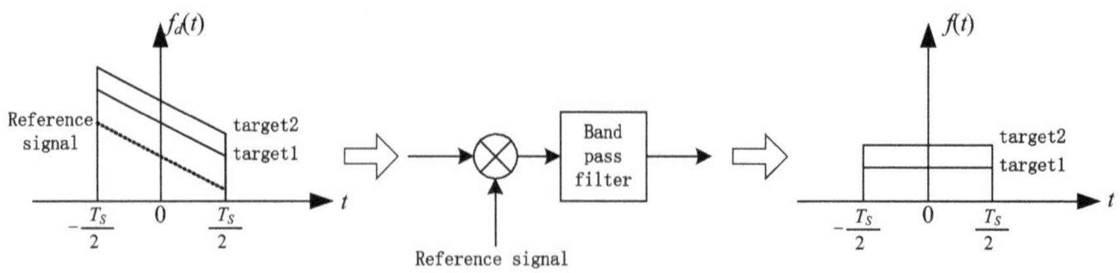

Fig. 2.4 Doppler frequency analysis processes

The azimuth resolution is half of the length of the antenna aperture, which is independent of the distance from the radar to the target and the working wavelength. This is one of the unique advantages of SAR. The reasonable antenna size design can provide a higher azimuth resolution.

(2) **Synthetic Aperture**

The concept of linear array antenna is used for reference in understanding synthetic aperture radar.

Figure 2.5 shows the schematic diagram of linear array antenna. The array element spacing is d, and off-axis angle is θ. Assume that each element is same, echo from each element in-phase stacked in the $\theta = 0$ direction reaches the maximum. The array pattern is

$$F(\theta) = \frac{\sin\left(\frac{\pi N d}{\lambda}\sin\theta\right)}{N\sin\left(\frac{\pi d}{\lambda}\sin\theta\right)} \tag{2.19}$$

when θ is very small, the array pattern is approximate to

$$F(\theta) = \frac{\sin\left(\frac{\pi N d}{\lambda}\sin\theta\right)}{\frac{\pi N d}{\lambda}\sin\theta} = \operatorname{sinc}\left(\frac{\pi N d}{\lambda}\sin\theta\right) \tag{2.20}$$

Synthetic aperture radar transmits signals periodically along with the movement of the platform along the track. Figure 2.6 shows the position of the radar transmitting signal that is arranged into a linear array in azimuth and each element is the real antenna. Array element spacing is $\Delta x = \frac{v_s}{\mathrm{PRF}}$, and array length is $L_s = \frac{\lambda R_0}{D_a}$.

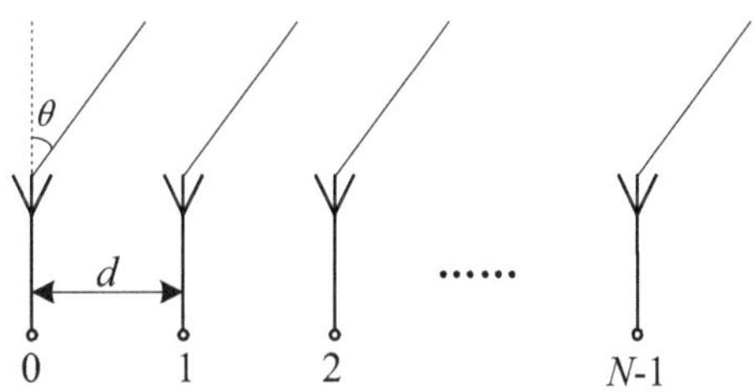

Fig. 2.5 Linear array antenna

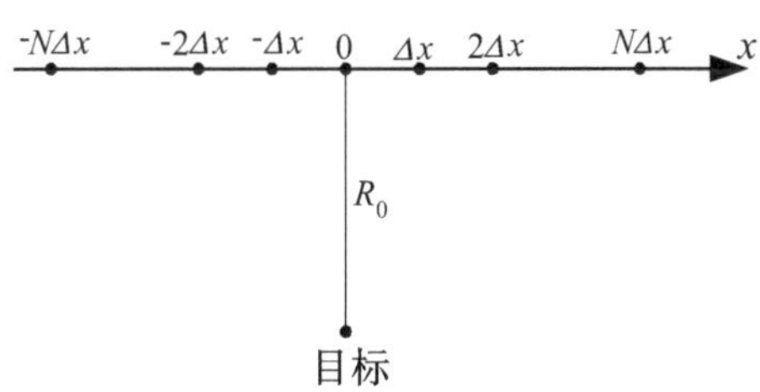

Fig. 2.6 Synthetic aperture principle of SAR

In terms of Eq. (2.8), the azimuth echo in each space position can be expressed as

$$s_a(n\Delta x) = e^{-j\frac{2\pi}{\lambda R_0}(n\Delta x)^2} \tag{2.21}$$

Coherent processing in azimuth needs to correct the phase of the echo signal for different azimuth positions. The phase compensation function is as

$$\phi(n\Delta x) = \frac{2\pi}{\lambda R_0}(n\Delta x)^2 \tag{2.22}$$

After phase correction, the echo in different azimuth positions can be stacked in same phase. Considering the echo round-trip phase shift between the target and radar, the beamwidth of synthetic antenna array is

$$\beta_{az} = \frac{\lambda}{2L_s} \tag{2.23}$$

So the azimuth resolution is defined as

$$\rho_a = \beta_{az} R_0 = \frac{\lambda R_0}{2L_s} = \frac{D_a}{2} \tag{2.24}$$

The azimuth resolution in Eq. (2.24) from the viewpoint of synthetic aperture is the same as Eq. (2.18).

(3) **Matched filter**

Equation (2.11) represents the phase of point target echo for the synthetic aperture radar. The third term for azimuth instantaneous Doppler frequency shows that the echo in azimuth can be viewed as LFM signal, and the duration is the synthetic aperture time T_s. The Doppler bandwidth is defined as the scope of the Doppler frequency changing over the synthetic aperture length, which can be expressed as

$$B_a = |k_a| \cdot T_s = |k_a| \cdot \frac{L_s}{v_s} = \frac{2v_s}{D_a}, \tag{2.25}$$

where L_a is the azimuth length of antenna. The echo in azimuth can be compressed by the matched filter with Doppler bandwidth B_a, the impulse response of matched filter is

$$h_a(t) = e^{j\frac{2\pi}{\lambda R_0}(v_s t)^2} \tag{2.26}$$

After compressing processing of matched filter, the azimuth resolution is

$$\rho_a = \frac{v_s}{B_a} = \frac{D_a}{2} \tag{2.27}$$

The azimuth resolution in Eq. (2.27) from the viewpoint of the matched filter remains the same with Eq. (2.18).

Synthetic aperture radar can also be understood from the point of view of microwave holography. The microwave holography draws lessons from the concept of optical holography, and has played a prominent role in proposing and improving the early synthetic aperture radar optical signal processing.

2.3 Theory Model of Synthetic Aperture Radar [1, 2]

Synthetic aperture radar is a kind of microwave imaging radar, and radar image can reflect microwave backscattering characteristics of the terrain. If the microwave backscattering coefficient is expressed as $\sigma_0(x, r)$, the synthetic aperture radar can be equivalent to a two-dimensional filtering network. Based on the network theory, if the impulse response function (IRF) of a linear network is $h(x, r)$, the output of the system is the radar image $\widehat{\sigma}_0(x, r)$.

$$\widehat{\sigma}_0(x, r) = \sigma_0(x, r) \otimes h(x, r) \tag{2.28}$$

Only if the IRF is δ function, radar image can be the correct reproduction of backscattering coefficient of the terrain. In order to measure the radar back scatter of targets, it is necessary that the IRF of SAR will close to δ function as far as possible.

The IRF of SAR is illustrated by two-dimensional pulse compression in Fig. 2.7.

The LFM signal transmitted by radar is

$$S(t) = \mathrm{Re}[\mu(t)\exp(j2\pi f_0 t)] = \mathrm{Re}\left[\mathrm{A}\ \exp\left(j \cdot \frac{1}{2} k_r t^2\right) \exp(j2\pi f_0 t)\right], \tag{2.29}$$

where $u(t)$ is complex amplitude, which is a function of time. A is the amplitude of LFM signal, and k_r is the chirp rate. The echo backscattered from a point target is

$$S_r(t) = \mathrm{Re}[k\mu(t - t_d)\exp(j2\pi f_0(t - t_d))] \tag{2.30}$$

Using Eq. (2.9) to express the delay time t_d, the following can be obtained:

$$t_d = \frac{2R}{c} = \frac{2R_0}{c} + \frac{(x - X_p)^2}{cR_0} = \frac{2R_0}{c}\left[1 + \frac{(x - X_p)^2}{2R_0^2}\right], \tag{2.31}$$

where $X_t = v_s t$.

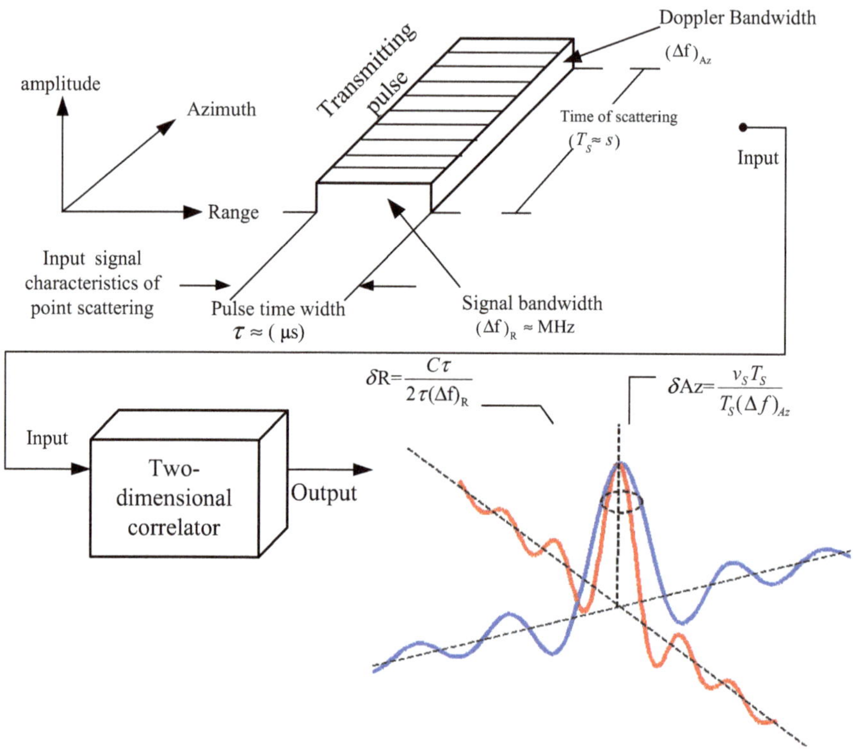

Fig. 2.7 Two-dimensional pulse compression of SAR

From Eq. (2.30) and Eq. (2.31), two kinds of time concepts are included in the signal complex amplitude and the delay. The former Eq. (2.30) reflects the change within the signal duration, and the latter Eq. (2.31) reflects the position change caused by the movement of the vehicle. Usually, in the signal duration the radar position change caused by the vehicle going forward is not worth mentioning, and can be ignored. So it can be defined that the time concept in the complex amplitude as "fast" time, and the time in the signal delay as "slow" time. Since the vehicle movement can be ignored in signal duration, by considering the signal duration as "fast" time, the "slow" time change can be viewed as a zero. It is to say the "fast" time and the "slow" time can be separated by processing under certain condition.

As we know, the signal included in complex amplitude changing with the "fast" time determines the range resolution, and the signal included in echo delay changing with "slow" time determines the azimuth resolution.

The echo signal is expressed as the two-variable function of the "fast" changing time t and the position of the vehicle x.

$$S_r(x,t) = \mathrm{Re}\left[k\mu\left(t - \frac{2R_0}{c}\left[1 + \frac{(x - X_p)^2}{2R_0^2}\right]\right)\exp\left(j2\pi f_0\left(t - \frac{2R_0}{c}\left[1 + \frac{(x - X_p)^2}{2R_0^2}\right]\right)\right)\right] \tag{2.32}$$

Similar to the derivation of Sect. 2.2, the echo signal is match filtered and self-correlation integrated after synchronization demodulation and the amplitude normalization, the output signal is the IRF $h(x,r)$ of SAR.

$$\begin{aligned} h(x,t) &= \mathrm{Re}\left\{\iint\left[e^{j\frac{1}{2}k_r(t'-t_d)^2}\cdot e^{-j\frac{2\pi}{\lambda R_0}\left(x'-X_p\right)^2}\right]\left[e^{-j\frac{1}{2}k_r(t+t'-t_d)^2}e^{-j\frac{2\pi}{\lambda R_0}\left(x+x'-X_p\right)^2}\right]\mathrm{d}x'\mathrm{d}t'\right\} \\ &= \mathrm{Re}\left\{\left[T\cdot e^{-j\frac{1}{2}k_r t^2}\cdot\frac{\sin\left[\pi\left(\frac{t}{\tau_r}\right)\right]}{\pi\left(\frac{t}{\tau_r}\right)}\right]\cdot\left[L_s e^{j\frac{2\pi}{\lambda R_0}x^2}\frac{\sin\left[\pi\left(\frac{x}{\rho_a}\right)\right]}{\pi\left(\frac{x}{\rho_a}\right)}\right]\right\} \end{aligned} \tag{2.33}$$

where T is pulse duration of LFM signal, τ_r is the main lobe width of the IRF which determines the range resolution. L_s is aperture length, ρ_a is the azimuth resolution.

The envelope of the IRF is

$$E\{h(x,t)\} = \left|\frac{\sin[\pi(t/\tau_r)]}{\pi(t/\tau_r)}\right|\cdot\left|\frac{\sin[\pi(x/\rho_a)]}{\pi(x/\rho_a)}\right| \tag{2.34}$$

Formula (2.34) shows that the envelope of IRF has the shape of sinc function both in azimuth and in range for the unweighted SAR. Due to the independence of the two variables, the IRF of SAR can be regarded as the product of two impulse responses in range dimension and azimuth dimension. That is

$$h(x,t) = h_r(x,t)\cdot h_a(x,t) \tag{2.35}$$

The theoretical model of SAR is represented in Fig. 2.8. SAR signal processing can be divided into two steps: pulse compression in range and beam sharpening in azimuth.

Based on the analysis above, the echo of SAR will be divided into the independent two-dimensional signal with the "fast" time in range and the "slow" time in azimuth while the position of vehicle can be ignored in the duration of transmitting pulse. The range processing and the azimuth processing can be independent deal with; However, there is the coupling between signals of range and azimuth and the coupling cannot be ignored in most cases, especially for spaceborne SAR, which will result in serious range migration.

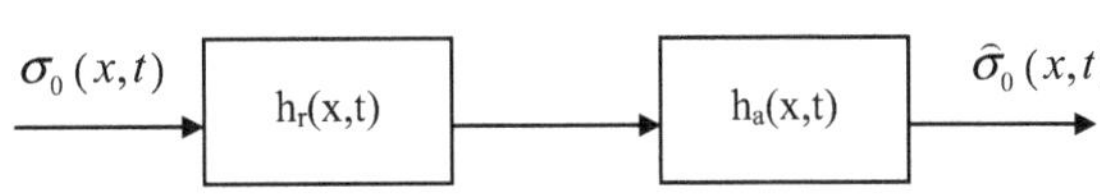

Fig. 2.8 Theoretical model of SAR

2.4 Pulse Compression Technology and Synthetic Aperture Processing

High resolution of SAR is achieved by pulse compression in range and synthetic aperture processing in azimuth. In general, the radar transmitting signal (range) and the synthetic aperture signal (azimuth) have the character of linear frequency modulation. In fact, the signal processing of SAR is pulse compression of LFM signal by matched filtering in range and azimuth dimensions, which provides fine delay time resolution in range and fine Doppler frequency resolution in azimuth.

Since the echo-Doppler signal can be approximated as a LFM signal, the process of focused SAR is similar to the pulse compression. The similarity comparison is shown in Table 2.1.

Because the echo-Doppler signal is generated by uniform speed and linear forward motion of the vehicle, so its parameters are different from that of the normal LFM signal. There are three main differences:

(1) Echo-Doppler signal duration (synthetic aperture time) is generally in the second order, and the normal LFM signal pulse duration is usually in the microsecond order.
(2) The echo-Doppler signal bandwidth is very small, generally a few hundred Hertz, while the normal LFM signal bandwidth is greater than MHz. Although the time–bandwidth product of the echo-Doppler signal in azimuth and the normal LFM signal in range is equivalent to the order of magnitude, the time duration and bandwidth are very different.
(3) The frequency modulation slope of the echo-Doppler signal varies inversely with the slant range, while the chirp rate the normal LFM signal is constant.

For convenient next analysis, some expression forms of azimuth resolution are as follows:

Using Doppler bandwidth B_d representation

Table 2.1 The similarity comparison between synthetic aperture processing and pulse compression

	Pulse compression	Synthetic aperture processing
Bandwidth	B	Doppler bandwidth $B_d = 2\ v_s\ /D_a$
Time duration	T_p	Synthetic aperture time $T_s = \lambda R/D_a \cdot v_s$
Chirp rate	$K = B/T_p$	$f_{dr} = -2\ v_s^2/(\lambda \cdot R)$
Time–bandwidth product	BT_p	$B_d \cdot T_s$
Velocity	Light velocity c	Carrier velocity v_s
Time resolution	$1/B$	$1/B_d$
Space resolution	$c/2B$	$v_s\ /B_d$

$$\rho_{\mathrm{a}} = \frac{v_s}{B_{\mathrm{d}}} \tag{2.36}$$

Using synthetic aperture L_{s} representation

$$\rho_{\mathrm{a}} = \frac{\lambda R}{2L_{\mathrm{s}}} \tag{2.37}$$

Using synthetic time T_{s} representation

$$\rho_{\mathrm{a}} = \frac{\lambda R}{2v_s T_{\mathrm{s}}} \tag{2.38}$$

Using real antenna length D_a representation,

$$\rho_{\mathrm{a}} = \frac{D_a}{2} \tag{2.39}$$

2.5 Focus Depth and Range Migration [3, 4]

The matched filtering can be also called as focusing process, which compensates fully the quadratic phase of the echo signal, and got the best azimuth resolution $D/2$. There is the focus depth for focus SAR.

The frequency modulation slope of SAR azimuth echo is inversely proportional to the slant range, so the matched filter has to change with the range for focus processing. In the actual processing, the matched filter parameter is not necessary to change with each range gate due to the complexity of hardware; it can use the same matched filter parameter to focus processing within a certain range.

The depth of focus is an important concept in SAR imaging processing. The depth of focus is the variation of the detection range that is adapted to the same matched filter in azimuth to satisfy the requirement of given resolution. The depth of focus determines the number of different reference signals that the processor

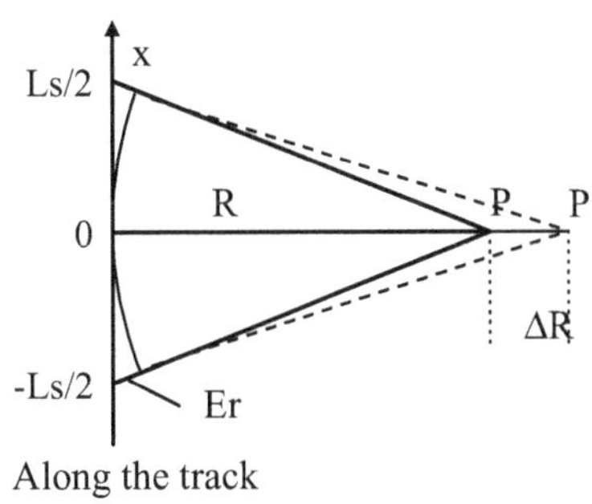

Fig. 2.9 The depth of focus

needs. If the depth of focus is large enough, the full range of action in the system requires only a reference signal; otherwise many different reference functions are needed.

Figure 2.9 shows the depth of focus. The target P is at the distance R far from the radar. When the aircraft flies from $-L_s/2$ to $L_s/2$ along the coordinates x, all echoes from the target P can be received. The wave front of the echo for P point target is a sphere (round is for plane). As shown in Fig. 2.9, the maximum phase error occurs at both ends of the point. Er can be expressed as

$$\mathrm{Er} = \sqrt{R^2 + \left(\frac{L_s}{2}\right)^2} - R \approx \frac{L_s^2}{8R} \tag{2.40}$$

Maximum phase error is

$$\frac{4\pi}{\lambda} \cdot \mathrm{Er} = \frac{\pi L_s^2}{2\lambda R} \tag{2.41}$$

If the azimuth resolution ρ_a is known, the synthetic aperture length L_s is

$$L_s = \frac{\lambda R}{2\rho_a} \tag{2.42}$$

Equation (2.41) can be rewritten as

$$\frac{4\pi}{\lambda} \cdot \mathrm{Er} = \frac{\pi\lambda R}{8\rho_a^2} \tag{2.43}$$

The Doppler phase $\frac{4\pi}{\lambda} \cdot \mathrm{Er}$ will be corrected to stack in the same phase during synthetic aperture focusing. Equation (2.43) shows that $\frac{4\pi}{\lambda} \cdot \mathrm{Er}$ is proportional to slant range. If the processor is focused on the target at the distance R, the focusing performance will get worse for target at the distance $R + \Delta R$.

The variation of the Doppler phase duo to ΔR is

$$\frac{4\pi}{\lambda} \Delta\mathrm{Er} = \frac{\pi\lambda}{8\rho_a^2} \Delta R \tag{2.44}$$

In order to focus processing approximately for the target at the distance R + ΔR, and the azimuth resolution cannot be decreased remarkable, the maximum allowable quadratic phase error will be π/4. That is

$$\frac{\pi\lambda \cdot \Delta R}{8\rho_a^2} \leq \frac{\pi}{4} \tag{2.45}$$

The variation of slant range ΔR is

$$\Delta R \leq \frac{2\rho_{\mathrm{a}}^2}{\lambda} \tag{2.46}$$

The depth of focus is

$$\Delta R_{\mathrm{m}} = \frac{2\rho_{\mathrm{a}}^2}{\lambda} \tag{2.47}$$

Equation (4.27) shows that the depth of focus depends on azimuth resolution and wavelength. The lower the resolution, the shorter the wavelength, the greater the depth of focus, and the simpler the focusing processor. Otherwise, the higher resolution will make the higher requirement on the maximum phase error, the depth of focus becomes smaller and the filter parameters will be updated faster.

It is assumed that the echo from the target at the distance R is received and stored within the same range gate. But that is not always the case. When the LFM signals are transmitted from one pulse to the next one along the track, the echoes of one static point target are not sampled always in the same range gates. Due to the coupling between the range and azimuth dimensions, the peak of range compression responses will appear at different range gates (Fig. 2.10), that the meaning of range migration. In order to synthetic aperture processing, it necessary to sample the echo at the right range gate and compensate the range migration. The range migration is divided into two parts: the range walk and the range curvature (Fig. 2.10).

The range migration is

$$R(s) \approx R_0 + R_0'(s - s_0) + R_0''(s - s_0)^2/2 \tag{2.48}$$

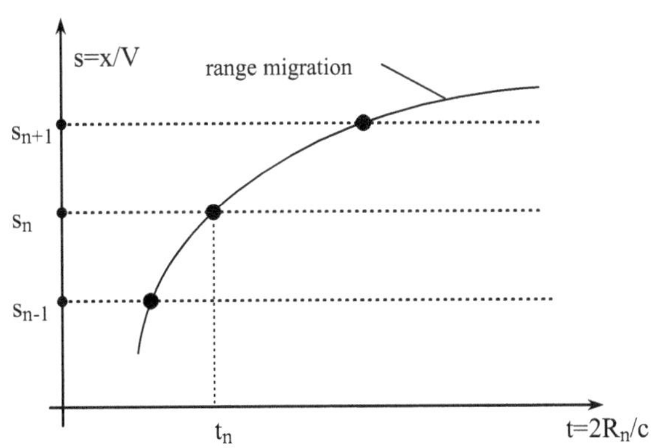

Fig. 2.10 Point target position after range compression in the (s, t) plane

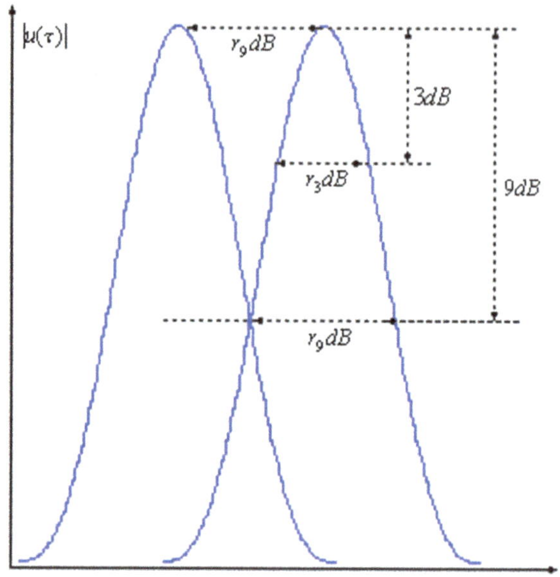

Fig. 2.11 Definition of resolution

where s is the "slow" time in azimuth, $R(s)$ is second-order Taylor expansion of the slant range from the radar to target in synthetic aperture time, and S_0 is the midpoint of the synthetic aperture time. The Doppler phase is

$$\phi(s) = -\frac{4\pi}{\lambda}R(s) \tag{2.49}$$

The Doppler centroid frequency is

$$f_{\mathrm{dc}} = \frac{\phi'(s)}{2\pi} = -\frac{2}{\lambda}R'(s_0) \tag{2.50}$$

The Doppler frequency rate is

$$f_{\mathrm{dr}} = \frac{\phi''(s)}{2\pi} = -\frac{2}{\lambda}R''(s_0) \tag{2.51}$$

So Eq. (2.48) can be rewritten as

$$R(s) = R_0 - \frac{\lambda f_{\mathrm{dc}}}{2}(s - s_0) - \frac{\lambda f_{\mathrm{dr}}}{4}(s - s_0)^2 \tag{2.52}$$

It is defined that the linear item is range walk and the quadratic term is range curve in Eq. (2.52). When the synthetic aperture length is comparable to the slant range, the range curve is particularly evident. The total range change is called as the range migration $\Delta R = R(s) - R_0$, which is inevitable in SAR processing, and it changes with system parameters.

When $s - s_0 = \pm T_s/2$, the range migration reaches the maximum for orthogonal side-looking SAR that is expressed as

$$\Delta R = \frac{\lambda f_{dr}}{4} \cdot \left(\frac{T_s}{2}\right)^2 = \frac{\lambda^2 R_0}{32\rho_a^2} \tag{2.53}$$

Generally, if the maximum range migration is less than $\rho_r/4$,

$$\left(\frac{\rho_a}{\lambda}\right)^2 \geq \frac{R_0}{8\rho_r} \tag{2.54}$$

The range migration does not have to be compensated. The migration is proportional to illumination time and slant range. If the migration is less than one range gate for short or medium range, this range migration effects can be ignored.

2.6 The Main Technical Parameters of SAR

Due to high altitude and wide coverage, the spaceborne SAR has been paid great attention by lot of countries in the world. The final output of SAR is a grayscale image of the terrain, and the image quality depends on the system specifications, such as resolution, side lobe ratio, swath, slant range, system sensitivity, ambiguity, radiometric resolution, radiation precision, dynamic range, etc. The main technical parameters of SAR system are introduced briefly as follows.

2.6.1 Resolution [5]

Resolution is defined as the ability of the imaging radar to distinguish the two closely spaced point targets. Figure 2.11 shows the magnitude of two point signals which present the same radar section at the receive output. It is the different possible definitions of the radar resolution.

The first definition: Nominal resolution

The nominal resolution is defined as the signal peak width of a single point target echo measured at 3 dB below its maximum value. This is the simplest definition of the radar resolution, which is only related to the echo of a single point target.

The second definition: Two-point target echo

Resolution is defined as the minimum interval required between two point targets with the same RCS in order to observe a trough between the two pulse compression peaks. It corresponds to the peak width of a single point target echo measured at 6 dB below its maximum value.

Suppose the amplitude of the two identical point targets echo is a, with a position shift equal to the 6 dB peak width, and random phase shift φ, the trough amplitude between the two signals is

$$\left|u(0)_{a,\varphi}\right| = \left|\frac{a}{2} + \frac{a}{2}\mathrm{e}^{j\varphi}\right| = a\sqrt{\frac{\cos(\varphi) + \cos^2(\varphi)}{2}}. \tag{2.55}$$

Equation (2.55) has the maximum value a only for $\varphi = 0$, this means that for any value of phase shift between both target echo, $\left|u(0)_{a,\varphi}\right| \leq a$. There will be a trough between both peaks if the two target position shift is greater than 6 dB peak width.

In most cases, the definition has a limited operational interest, because the trough is not deep enough to be observed. Therefore, the third definition is given as follows.

The third definition:

Resolution is defined as the minimum interval required between two point targets with the same RCS, in order to give a trough of more than 3 dB between the two pulse compression peaks. The definition gives a value about two times as great as the resolution at 3 dB of the first definition.

In order to establish a link between the first and the third definition, suppose that the phase shift of the matched filter outputs of is φ, when $\varphi = 0$, the two echoes are added. If the intersection of the two peaks is 9 dB below their maximum value, then the trough between the two peaks is 3 dB. Under any other relation phase between signals from the two targets, the trough is greater than 3 dB. Therefore, the third definition of resolution will be corresponding to 9 dB width of the peak at the output of the matched filter for the first definition.

2.6.1.1 Range Resolution [6, 7]

The range resolution of SAR is mainly determined by the bandwidth B of LFM signal, that is

$$\rho_{\mathrm{r}} = \frac{c}{2B} \tag{2.56}$$

The wider bandwidth, the higher the range resolution. In Eq. (2.56), ρ_{r} is the slant range resolution, the ground range resolution ρ_{g} can be obtained.

$$\rho_{\mathrm{g}} = \frac{\rho_{\mathrm{r}}}{\sin\eta}, \tag{2.57}$$

where η is incident angle.

2.6.1.2 Azimuth Resolution [8]

The azimuth resolution of SAR, independent to the slant range and wavelength, is half of the antenna length.

$$\rho_{\mathrm{a}} = \frac{D_{\mathrm{a}}}{2} \tag{2.58}$$

The azimuth Doppler bandwidth is $B_{\mathrm{a}} = 2v_{\mathrm{s}}/D_{\mathrm{a}}$. In order to satisfy the sampling theorem, the system pulse repetition frequency (f_{PRF}) must be satisfied

$$f_{\mathrm{PRF}} > B_{\mathrm{a}} = \frac{2v_{\mathrm{s}}}{D_{\mathrm{a}}} \quad \Rightarrow \quad \frac{1}{f_{\mathrm{PRF}}} < \frac{D_{\mathrm{a}}}{2v_{\mathrm{s}}} \tag{2.59}$$

Azimuth sampling interval is

$$\Delta x = \frac{v_{\mathrm{s}}}{f_{\mathrm{PRF}}} < \frac{D_{\mathrm{a}}}{2} \tag{2.60}$$

In order to satisfy the requirement of sampling theorem, the azimuth sampling interval must be less than azimuth resolution.

The azimuth resolution given by Eq. (2.60) is the best resolution of SAR. In fact, due to the weight of the antenna pattern, the azimuth Doppler spectrum does not have the shape of the rectangular spectrum, which makes the main lobe broadening and the resolution decreased. On the other hand, the processing bandwidth in the actual system does not necessarily select the whole Doppler bandwidth. Taking into account the different operating modes of SAR system, the signal processing bandwidth may only select a part of the Doppler bandwidth, which will also cause a decrease in the azimuth resolution.

The synthetic aperture length illuminated by the 3 dB main lobe of the uniform aperture antenna is

$$L_{\mathrm{S}} = K_{\mathrm{A}} \cdot \frac{\lambda R_0}{D_{\mathrm{a}}}, \tag{2.61}$$

where K_{A} is antenna pattern factor in azimuth, and $K_{\mathrm{A}} = 0.886$. The Doppler bandwidth is

$$B_{\mathrm{a}} = K_{\mathrm{A}} \cdot \frac{2v_{\mathrm{s}}}{D_{\mathrm{a}}} \tag{2.62}$$

If K_P is the radio between signal processing bandwidth B_P and Doppler bandwidth B_a, $K_P \leq 1$ B_P is

$$B_P = K_P \cdot B_a = K_P K_A \frac{2v_s}{D_a} \tag{2.63}$$

The broadening factor of azimuth compression output duo to antenna pattern weighting is expressed as K_a, $K_a > 1$,

$$\rho_a = K_a K_A \frac{v_s}{B_P} = \frac{K_a}{K_P} \cdot \frac{D_a}{2} \tag{2.64}$$

2.6.2 Peak Side Lobe Ratio and Integrated Side Lobe Ratio [5]

Peak-to-side lobe ratio (PSLR) is a power ration between the main peak and the side lobes located in an interval of 10 times the peak width, which is usually expressed in dB. Figure 2.12 shows the definition of PSLR.

Secondary side lobe ratio (SSLR) is a power ration between the main peak and the side lobes located in an interval between 10 times and 20 times the peak width, which is usually expressed in dB. Figure 2.13 shows the definition of SSLR.

Integrated side lobe ratio (ISLR) is defined as a power ratio between the main peak lobe and the total power of all the side lobes, which is usually expressed in dB. Figure 2.14 shows the definition of ISLR.

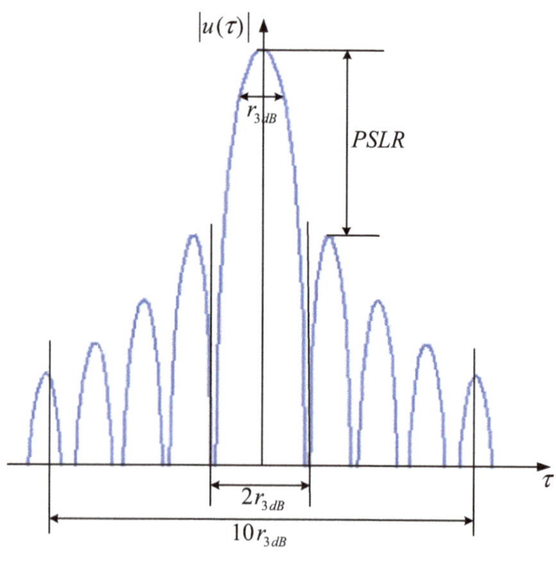

Fig. 2.12 Definition of PSLR

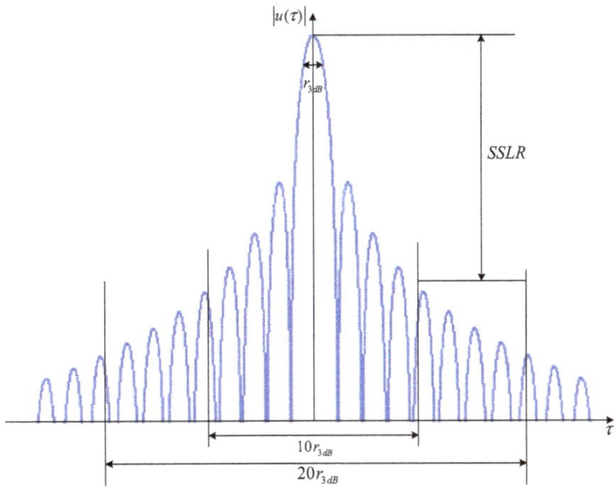

Fig. 2.13 Definition of SSLR

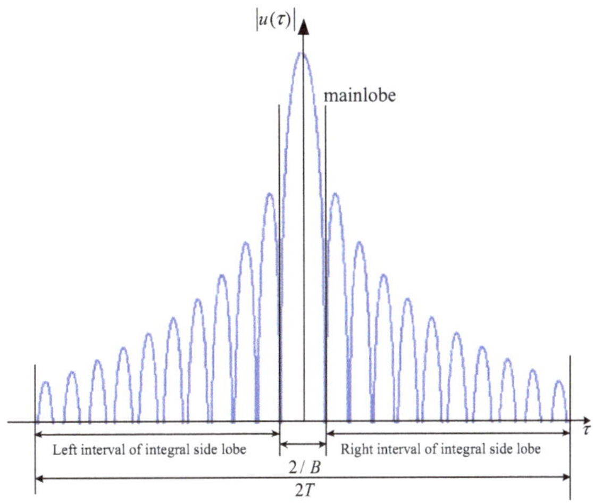

Fig. 2.14 Definition of ISLR

The mathematical formal definition of f_{ISLR} is

$$f_{\mathrm{ISLR}} = \frac{\int_{1/B}^{-1/B} |u(\tau)|^2 \mathrm{d}\tau}{\int_{-T}^{-1/B} |u(\tau)|^2 \mathrm{d}\tau + \int_{-1/B}^{T} |u(\tau)|^2 \mathrm{d}\tau}, \tag{2.65}$$

where $|u(\tau)|$ is the impulse response function (IRF) at the output of matched filter, T is the pulse length. The length of the impulse response function (IRF) is 2T. In practice, the far side lobe is too low to be measured, so it is necessary to introduce the second definition of ISLR. The side-lobe integration is limited to an interval of 20 times the main lobe width; Measurement of ISLR is illustrated in Fig. 2.15.

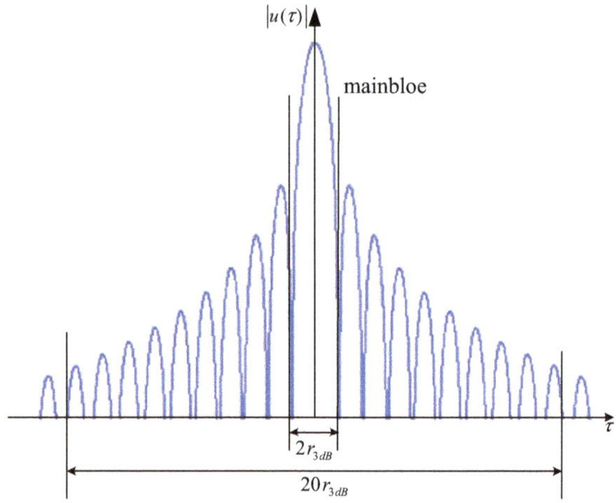

Fig. 2.15 Measurement of ISLR

2.6.3 Swath Width

The swath of SAR is the ground coverage area in range direction illuminated by antenna beam, which depends on the range beamwidth, the incident angle, the height of platform and PRF, etc.

The swath is defined as the coverage area illuminated by antenna in 3 dB beamwidth.

$$W_g = \frac{\beta_e R_0}{\cos\eta} = \frac{\lambda R_0}{H_a \cdot \cos\eta}, \tag{2.66}$$

where β_e is the range beamwidth of the antenna, η is the incident angle, H_a is the width of antenna. For the radar to operate correctly, the echo time in swath W_g should be less than pulse repetition time (PRT), the relationship is given by

$$W_g < \frac{c}{2 \cdot f_{PRF} \cdot \sin\eta} \tag{2.67}$$

2.6.4 System Sensitivity (Noise Equivalent Back Scattering Coefficient) [1]

Sensitivity of SAR System is an important parameter to measure the imaging capability for weak targets, and it is always be estimated by Noise Equivalent Sigma Zero(NESZ). That means that the power of echo scattered by the terrain, whose back scattering coefficient is NESZ, is equal to power of the SAR system noise.

Radar equation is the basis in radar system design, and the power of the transmission and the SNR of the end-to-end system can be estimated

$$\frac{S}{N} = \frac{P_t G^2 \lambda^2 \sigma}{(4\pi)^3 R^4 (k_0 T_0 B_n F_n)}, \tag{2.68}$$

where P_t is transmit peak power, G is antenna gain, σ is radar cross section (RCS), $k_0 = 1.38 \times 10^{-23}$ J/K is the Boltzmann constant, T_0 is absolute temperature, B_n is equivalent noise bandwidth that is approximately equal to system bandwidth B_r, F_n is noise figure.

The SNR of SAR image can be improved by pulse compression in range and azimuth, the signal processing gain is $G_{pr} \cdot G_{pa}$.

The processing gain in range G_{pr} is expressed as

$$G_{pr} = B_r \cdot T_p, \tag{2.69}$$

where T_p is the pulse length of the transmitting signal, and B_r is the band width of radar system. The azimuth focusing processing is equivalent to the coherent accumulation of pulses, and the coherent accumulation gain can be expressed as

$$G_{pa} = T_S \cdot f_{PRF}, \tag{2.70}$$

where T_S is the synthetic aperture time and f_{PRF} is the pulse repetition frequency.

In Eq. (2.68), σ is radar cross-section area of the minimum resolution cell that can be given by

$$\sigma = \sigma_0 \cdot \rho_g \rho_a, \tag{2.71}$$

where σ_0 is normalized backscattering coefficient, ρ_g is the ground range resolution, and ρ_a is azimuth resolution. Using Eqs. (2.69) and (2.71) in Eq. (2.68), the radar equation of SAR can be expressed as

$$\frac{S}{N} = \frac{P_{av} G^2 \lambda^3 \sigma_0 \rho_g}{2(4\pi)^3 R^3 v_s (k_0 T_0 F_n) L_{oss}}, \tag{2.72}$$

where L_{oss} is system loss, $P_{av} = P_t T_p \cdot PRF$ is average transmission power. According to Eq. (2.72), the SNR of radar output can be calculated under a certain average transmit power. Equation (2.72) can be rewritten as follows

$$P_{av} = \frac{2(4\pi)^3 R^3 v_s (k T_0 F_n) L_{oss}}{G^2 \lambda^3 \sigma_0 \rho_g} \cdot \frac{S}{N} \tag{2.73}$$

Equation (2.73) gives the average transmits power for satisfying the requirement of the SNR at output of radar system.

Noise equivalent back scattering coefficient (NEσ^0) is defined as

$$NE\sigma^0 = \frac{\sigma_0}{S/N} = \frac{2(4\pi)^3 R^3 v_s (k_0 T_0 F_n) L_{oss}}{P_{av} G^2 \lambda^3 \rho_g} \tag{2.74}$$

The SNR at output of radar system is the dB number that the actual back scattering coefficient σ^0 is higher than the NEσ^0.

2.6.5 Ambiguity [1, 8]

Since SAR operates in the pulse mode, ambiguity is inevitable as same as the pulse radar that means that the radar will receive useful echo signals and the ambiguity interference at the same time. The ambiguity signal is the echo signal from the outside of the image area, which is combined with the imaging signal to enter the radar receiver, and after signal processing it will degrade the image quality. Ambiguity in the spaceborne SAR is particularly prominent, and it is divided into the range ambiguity and the azimuth ambiguity.

Ambiguity ratio is defined as the ratio of the power between the ambiguity signal and the useful signal, and it is an index of the system's ambiguity interference degree.

2.6.5.1 Azimuth Ambiguity

Since the radar antenna pattern has got the side lobes, its Doppler spectrum is non-band limited, as the echo is sampled by a limited f_{PRF} in azimuth, it will cause the spectrum overlapping. It is the azimuth ambiguity. The formation process of the SAR azimuth signal is shown in Fig. 2.16.

The reason of the azimuth ambiguity is that the Doppler spectrum is sampled by the non-band limited PRF. According to the sampling theorem, the spectrum after sampled is extended along azimuth by the periodicity of PRF. Since the side lobes of the antenna azimuth pattern give rise to no strict non-band limited Doppler spectrum, the signal component outside effective processing bandwidth is folded into the signal processing bandwidth, and interferes the useful echo signal, that

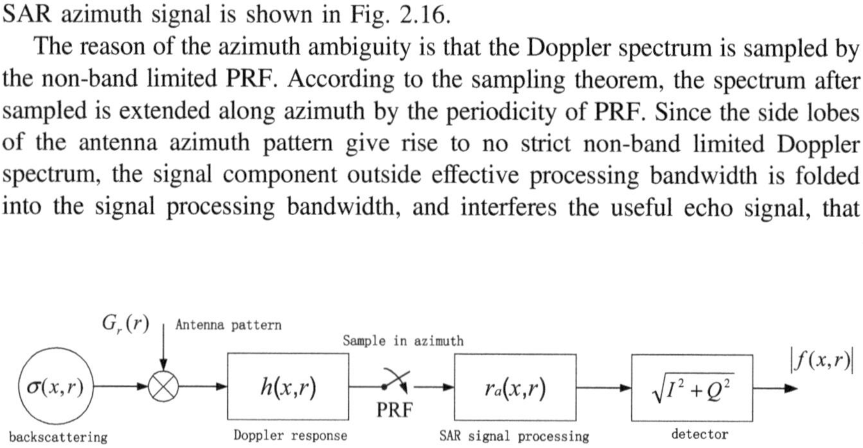

Fig. 2.16 The formation process of the SAR azimuth signal

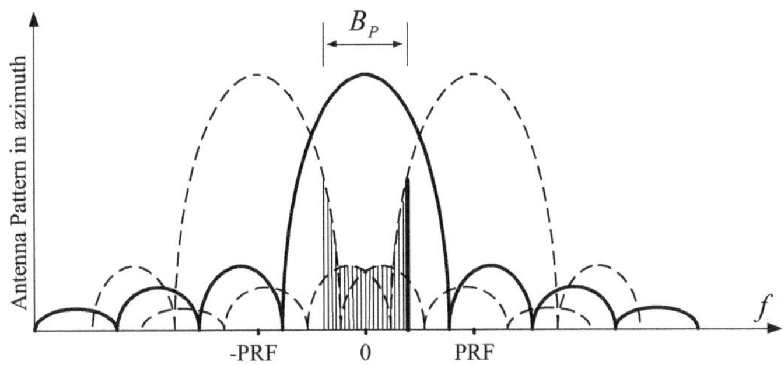

Fig. 2.17 A schematic diagram of azimuth ambiguity

causes the azimuth ambiguity. Figure 2.17 shows a schematic diagram of azimuth ambiguity, and B_{p} is Doppler processing bandwidth.

For convenient calculating, assuming that the terrain scattering characteristics of targets at different azimuth positions are uniform and the azimuth antenna pattern is the same in each range gate. The azimuth ambiguity-to-signal ratio (AASR) can be defined as

$$R_{\mathrm{AASR}} = \frac{\sum\limits_{\substack{m=-\infty \\ m \neq 0}}^{+\infty} \int_{-B_{\mathrm{p}}/2}^{+B_{\mathrm{p}}/2} G^2(f + m \cdot f_{\mathrm{PRF}})\mathrm{d}f}{\int_{-B_{\mathrm{p}}/2}^{+B_{\mathrm{p}}/2} G^2(f)\mathrm{d}f}, \tag{2.75}$$

where f is Doppler frequency, and $G(f)$ is azimuth antenna pattern.

In terms of Eq. (2.75), R_{AASR} can be improved by increasing PRF, but, in turn, increasing PRF will make the range ambiguity worse. Therefore, the choice of PRF is a tradeoff between the azimuth ambiguity and the range ambiguity.

2.6.5.2 Range Ambiguity

The generation mechanism of range ambiguity in SAR is the same as that of conventional pulse radar. Since the antenna has got the side-lobes in range, the echo signals from different ranges are received in the same sampling window, Fig. 2.18 shows a schematic diagram of range ambiguity. In a spaceborne SAR, since the range is far away, the platform flying velocity is fast and PRF is high, the echo is usually received after a number of PRTs as the pulse is transmitted. In this way, as soon as the radar receives the echo of the target in the swath (the second pulse), the echoes of the previous pulse (the first pulse) and the next pulse (the third pulse) may be received at the same time that is the origin of the range ambiguity. The echoes from the outside of the swath will interfere with the effective echo from the targets.

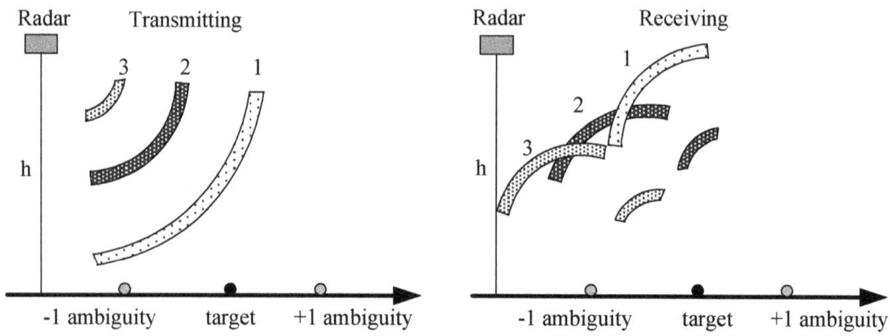

Fig. 2.18 The schematic diagram of range ambiguity

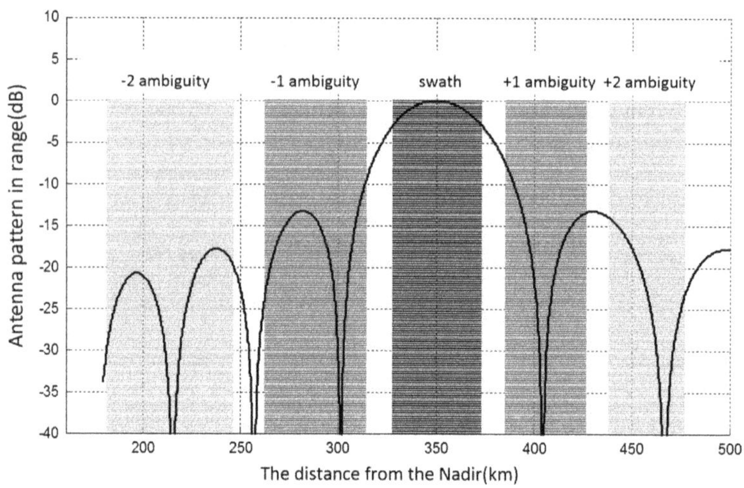

Fig. 2.19 The spatial relationship between swath and range ambiguity region

In order to further illustrate the range ambiguity, Fig. 2.19 shows the spatial position relationship between the swath and the ambiguity region. The swath is illuminated by the 3 dB beamwidth of the antenna pattern in range, and the first and the second ambiguity regions are given.

In order to evaluate the range ambiguity, the position of ambiguity points should be determined first. If the echo delay time of a target in the swath is t_i, the ambiguity signal can be in the following range:

$$R_{ij} = \frac{c}{2}(t_i + \frac{j}{f_{\text{PRF}}}) \quad j = \pm 1, \pm 2, \ldots, \pm n_{\text{h}} \tag{2.76}$$

If j is positive, the ambiguity signal comes from far range targets during the previous pulse. If j is negative, the ambiguity signal comes from near range targets during the next pulse. When j is equal to n_h, the ambiguity target is located at the horizon.

Figure 2.20 shows the geometry of the spaceborne SAR. The incident angle η_{ij} can be calculated from the looking angle γ_{ij}.for the ambiguity point

$$\eta_{ij} = \sin^{-1}(\frac{R_{\mathrm{e}} + h}{R_{\mathrm{e}}} \sin \gamma_{\mathrm{ij}}), \tag{2.77}$$

where R_{e} is the earth radius, h is the altitude of satellite. The looking angle γ_{ij} is given by

$$\gamma_{ij} = \cos^{-1}\left[\frac{R_{ij} + (R_{\mathrm{e}} + h)^2 - R_{\mathrm{e}}^2}{2R_{ij}(R_{\mathrm{e}} + h)}\right] \tag{2.78}$$

The range ambiguity-to-signal ratio (R_{RASR}) of the swath can be calculated by

$$R_{\mathrm{RASR}} = \frac{\sum_{i=1}^{N} S_{\mathrm{a}i}}{\sum_{i=1}^{N} S_i}, \tag{2.79}$$

where $S_{\mathrm{a}i}$ is the power of the ambiguity signal of the ith sample, and S_i is the effective echo signal power at the same sampling point. Equation (2.80) gives the expression.

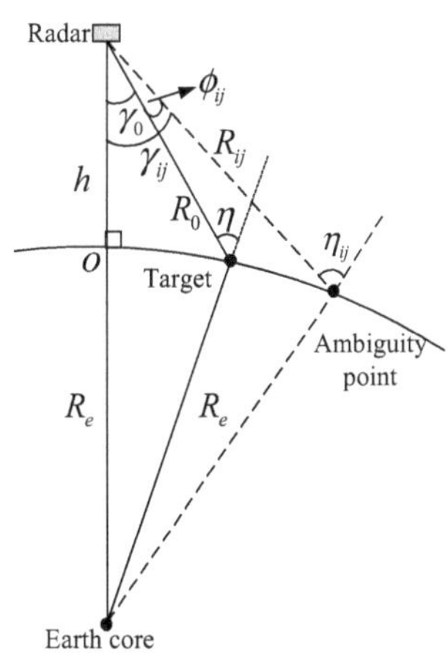

Fig. 2.20 The geometry of spaceborne SAR

$$\begin{cases} S_i = \dfrac{\sigma_{ij}^0 \cdot G_{ij}^2}{R_{ij}^3 \sin \eta_{ij}} & j=0 \\ S_{ai} = \sum\limits_{\substack{j=-n_h \\ j \neq 0}}^{n_h} \dfrac{\sigma_{ij}^0 \cdot G_{ij}^2}{R_{ij}^3 \sin \eta_{ij}} & j \neq 0 \end{cases} \tag{2.80}$$

G_{ij} is the antenna gain in the range dimension, σ_{ij}^0 is the normalized backscattering at the incident angle η_{ij}, which is usually approximated by a negative exponential function $\sigma_{ij}^0 = \mathrm{e}^{-\eta_{ij}/\eta_0}$.

From the above analysis, it can be found that choosing those SAR parameters is closely related to the azimuth ambiguity and the range ambiguity. Since the azimuth ambiguity is caused by the undersampling of the echo in the azimuth, it will effectively reduce the azimuth ambiguity by increasing the sampling PRF. In addition, the Doppler bandwidth can be decreased by changing the antenna azimuth length or designing the appropriate operating orbit to reduce the azimuth ambiguity, although the azimuth ambiguity is reduced, but it is also at the expense of azimuth resolution.

The range ambiguity-to-signal ratio(RASR) means the ratio of the range ambiguity power to the effective echo power in range. Reducing PRF can increase the PRT, and reduce the ambiguity signal into the data record window, thereby improving the RASR. In contrast, increasing PRF will degrade the RASR. In addition, in order to reduce the RASR of system, it should choose the appropriate antenna pattern to reduce the power of side lobe signal.

2.6.6 *Dynamic Range [5]*

The dynamic range of the microwave imaging radar can be measured at different points of the receiving channel.

At the receiver input, the dynamic range is the ratio of the maximum received signal and thermal noise. As soon as the maximum received signal is obtained, the saturation noise and other nonlinear signals must be limited under a given threshold at the same time. This threshold is chosen to avoid degrading the image contrast so much that weak echo region and the shadows no longer are distinguished.

On the image itself, at the output of the signal processing, the dynamic range is the ratio between the maximum point target echo and the noise. The received signal is the average superposition of all points on the instantaneous observation area (IFOV), which is defined by the transmission pulse width and azimuth beamwidth. Due to this averaging effect, the maximum signal power received corresponds to a mean backscatter coefficient that is smaller than the power of the maximum backscatter coefficient in the observed area.

In receiver, the amplifier gain should be set so that the maximum signal is passed without noticeable saturation. The main source of nonlinearity and saturation are the analog to digital conversion (ADC). Due to the Gauss distribution statistics of the echo, the saturation level of the ADC has to be set much higher than the level of the maximum received signal.

2.6.6.1 Noise Introduced by ADC Converter

The probability density function (PDF) of noise introduced by ADC is

$$p(x) = \frac{1}{\sigma\sqrt{2\pi}}\exp(-\frac{x^2}{2\sigma^2}) \tag{2.81}$$

σ is standard deviation of input signals for I and Q CHANNEL?

Signal quantization noise ratio (R_{SQNR}) is

$$R_{\mathrm{SQNR}} = \frac{\sigma^2}{\int_{-\infty}^{+\infty} (x - q(x))^2 p(x)\mathrm{d}x}, \tag{2.82}$$

where q is quantization function.

If the quantization order is Q and quantization bits is 2M, R_{SQNR} is

$$R_{\mathrm{SQNR}} = \left\{1 - \sqrt{\frac{2}{\pi}}\frac{Q}{\sigma}\sum_{m=-M+1}^{M-1}\exp\left(-\frac{mQ}{\sigma\sqrt{2}}\right)^2 + \frac{Q^2}{\sigma^2}\left(\left(M-\frac{1}{2}\right)^2 - \sum_{m=-M+1}^{M-1} m\cdot\mathrm{erf}\left(\frac{mQ}{\sigma\sqrt{2}}\right)\right)\right\}^{-1} \tag{2.83}$$

Figure 2.21 shows the relationship between quantization bits and SQNR, where the number of quantization bits is a function of σ/V_{sat}. ADC saturation level $V_{\mathrm{sat}} = MQ$. Table 2.2 summarizes the relationship among these parameters.

2.6.6.2 Dynamic Range of Input Signal

Due to the input signal fluctuation and the observed area is very broad for microwave imaging radar, the dynamic range of the receiver input signal is large.

Total dynamic range at the receiver input

Total dynamic range at the receiver input should be concerned with two parts: the dynamic range caused by range varied, and the echo variation caused by different backscattering coefficient into the observation area.

The dynamic range caused by range varied: according to the SAR range equation, the signal level of receiver is inverse proportion with R^3. For radar with range from 1 to 100 km, the variation of receiving signal levels will exceed 60 dB.

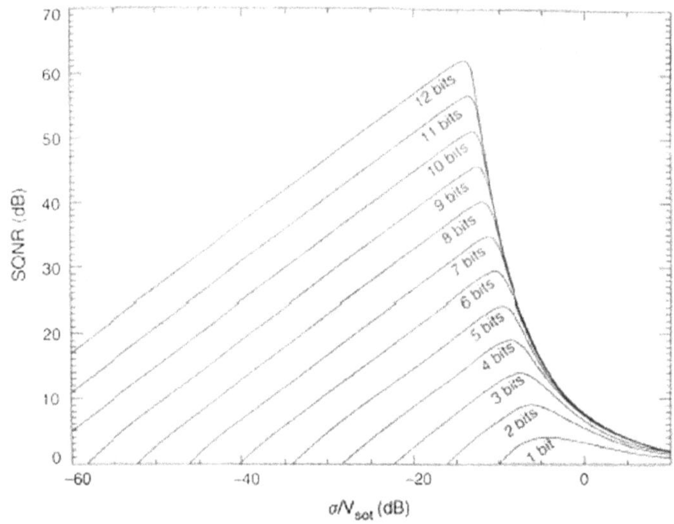

Fig. 2.21 The relationship curve between quantization bits and SQNR

Table 2.2 The relationship between the quantization bits, optimal Q/σ, optimal σ/V_{sat}, and optimal SQNR

Quantization bits	Optimal Q/σ	Optimal σ/V_{sat} (dB)	Optimal SQNR (dB)
1	1.596 E+0	−4.06	4.40
2	9.957 E−1	−5.98	9.25
3	5.860 E−1	−7.40	14.27
4	3.352 E−1	−8.57	19.38
5	1.881 E−1	−9.57	24.57
6	1.041 E−1	−10.45	29.83
7	5.687 E−2	−11.22	35.17
8	3.076 E−2	−11.90	40.57
9	1.650 E−2	−12.51	46.03
10	8.785 E−3	−13.06	51.55
11	4.650 E−3	−13.55	57.11
12	2.448 E−3	−14.00	62.71

The dynamic range means the change of different back scattering coefficients due to the different frequency band, different incident angle, different polarization, and so on. The strong backscattering coefficient may be 5 dBm^2/m^2, and a weak backscattering coefficient is −35 dBm^2/m^2.

For a pulse compressing radar, transmitting pulse duration is larger, observe area is wider, and the signals are always attenuated after echo round-trip propagation from radar to targets, which caused the echo signal change greatly.

The total dynamic range at the receiver input may be 70 dB.

Dynamic range in swath

The signal dynamic range in swath can be adjusted by automatic gain control (AGC) or time sensitive control (STC) to match with the dynamic range of ADC convert. The main factors affecting the dynamic range of the echo in the swath are included:

The backscattering coefficient changes from 20 to 35 dB.

The antenna gain in range dimension will change 2–6 dB from the center to the edge of the swath. And the signal attenuation in far range is different from that in near range that makes 0–8 dB dynamic range.

In general, the saturation level of the ADC converter is much higher than the maximum received signal in order to suppress the saturation noise. Therefore, it is needed to persist a certain interval between each code as the signal is encoded. The dynamic range of the ADC with 8–12 bits quantization will be 40–60 dB.

Instantaneous dynamic range

The receiver gain can be adapted to the average signal level received during a shorter period of time. This receiver parameter is the instantaneous dynamic range.

Since the variation of the signal envelope is spread over the entire pulse length and antenna azimuth beamwidth, the dynamic range of echo signals between two adjoining area is always small. For a part of the echo signal, the instantaneous dynamic range is much smaller than the whole dynamic range of the receiver.

The dynamic range of SAR image is usually 60–80 dB. Even if the point target is submerged in the echo of the IFOV, the ADC quantization does not eliminate the scattering information of the target. Similarly, the information of the small and strong scattering can still be retained after compression, but the thermal noise or clutter will be attenuated. Therefore, as long as the number of quantization bits is large enough, a suitable image dynamic range can be achieved.

2.7 Antenna Area Limitation and System Quality Factor

For a spaceborne SAR system, on the one hand, the resolution should be as high as possible to obtain the rich target information, and convenient for target detection and identification. On the other hand, the imaging swath should be as wide as possible to shorten the revisit period. However, there is the ambiguity in range and azimuth in SAR system, the high resolution and wide swath incompatible, and both cannot be improved at the same time. In order to obtain high resolution and wide swath as soon as possible, a minimum antenna area has to be limited in the system design.

2.7.1 Limitation Of Minimum Antenna Area [8, 9]

Figure 2.22 shows the geometric relationship of an orthogonal side-looking SAR, where D_a is the length of antenna, H_a is the width of antenna, V_s is the flying velocity of the satellite platform, W_g is the swath on the ground, and W_r is the swath in slant range.

$$W_g = \frac{\beta_e R_0}{\cos\eta} = K_E \frac{\lambda R_0}{H_a \cos\eta}, \tag{2.84}$$

where β_e is the 3 dB beamwidth of the antenna in range, K_E is the radiation pattern factor, which is relative to the antenna gain decrease dB at the edge of the swath. λ is wavelength, and R_0 is the slant range from the center of antenna beam to target.

The SAR system is operating with pulse model, ambiguities in range and azimuth have to be considered in system design. The basic requirement for the range ambiguity is that the whole echo in the swath must be received and sampled within a pulse repetition time (PRT), it is to say. The time at which the antenna receives the echo as the reflection of a transmitting pulse at nearest slant range point within the imaging swath, has to be late than the time at which the antenna receives the echo as the reflection of the previous transmitting pulse, at farthest slant range point within the imaging swath. This can be expressed as

$$\frac{2(R_f - R_n)}{c} = \frac{2W_r}{c} < f_{PRT} = \frac{1}{f_{PRT}}, \tag{2.85}$$

where c is the velocity of light, R_n is the nearest slant range, and R_f is the farthest slant range. In terms of Eq. (2.85), the upper bound of the allowable PRF is given by

$$f_{PRT} < \frac{c}{2W_r} = \frac{c}{2(R_f - R_n)} \tag{2.86}$$

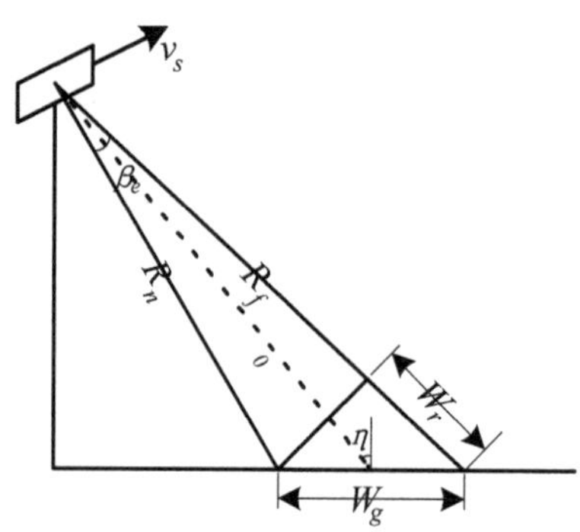

Fig. 2.22 Geometric relationship of radar imaging

The best azimuth resolution of orthogonal side-looking SAR is half of the antenna length.

$$\rho_a \geq \frac{D_a}{2} \tag{2.87}$$

The echo signal is sampled at a PRF in azimuth. In order to ensure that the echo spectrum do not overlap alias after sampled, the PRF must be greater than the Doppler bandwidth.

$$f_{PRT} > v_s/\rho_a = 2K_A v_s/D_a, \tag{2.88}$$

where K_A is the antenna pattern factor in azimuth, which is relative to the antenna gain decrease at the edge of the Doppler bandwidth.

From (2.85) to (2.88) just given above, the following can be obtained

$$W_r < \frac{c}{2 \cdot f_{PRT}} < \frac{c}{2v_s} \cdot \rho_a \quad \Rightarrow \quad \frac{W_r}{\rho_a} < \frac{c}{2v_s} \tag{2.89}$$

From Fig. 2.22, geometric relationship of radar imaging, W_r can be expressed as

$$\left.\begin{aligned} W_r &= W_g \sin\eta \\ W_g \cos\eta &= K_E \tfrac{\lambda R_0}{H_a} \end{aligned}\right\} \quad \Rightarrow \quad W_r = K_E \frac{\lambda R_0}{H_a} \cdot \tan\eta \tag{2.90}$$

Using Eq. (2.90) to rewrite Eq. (2.89), the following can be obtained

$$\frac{W_r}{\rho_a} = \frac{2K_A K_E \lambda R_0 \cdot \tan\eta}{D_a \cdot H_a} < \frac{c}{2v_s} \tag{2.91}$$

The limitation of antenna area is given as follows:

$$A_a = D_a \cdot H_a > \frac{4K_A K_E v_s \lambda R_0}{c} \cdot \tan\eta \tag{2.92}$$

This is restricting for the minimum antenna area in spaceborne synthetic aperture radar system design.

From (2.92), it can be seen that once the slant range R_0, incident angle η, wavelength λ, and the platform velocity v_s are confirmed, the limitation of minimum antenna area is ensured. For the steady velocity of the satellite, if the orbit altitude, the incident angle and wavelength of the radar are known, then the limitation of antenna area can be calculated. At present, the system performance has to be a tradeoff between the swath width and the azimuth resolution, and scanning imaging mode and spotlight imaging mode are two typical examples for it.

2.7.2 *Quality Factor [10–15]*

Swath width and azimuth resolution are two key parameters of spaceborne SAR performances. The SAR quality factor M_f can be defined as the ratio of swath width to azimuth resolution.

$$M_{\mathrm{f}} = \frac{W_{\mathrm{g}}}{\rho_{\mathrm{a}}} \tag{2.93}$$

The quality factor M_f depends on swath width in range and Doppler bandwidth of azimuth antenna radiation under the limitation of ambiguity ratio.

When the range ambiguity is considered

$$W_{\mathrm{g}} < \frac{c}{2 \cdot f_{\mathrm{PRT_{max}}} \cdot m_{\mathrm{R}} \cdot \sin \eta_{\max}}, \tag{2.94}$$

where W_g is swath, $f_{\mathrm{PRT_{max}}}$ is the maximum PRF under the meeting of the RASR, and $\eta_{\max}$ the maximum incident angle. m_R is related to the RASR, which is numerically equal to the ratio of PRT to the echo window.

When the azimuth ambiguity is considered

$$f_{\mathrm{PRT_{min}}} > m_{\mathrm{D}} \cdot B_{\mathrm{a}} \tag{2.95}$$

$$\rho_{\mathrm{a}} = \frac{v_{\mathrm{s}}/\xi}{B_{\mathrm{a}}} \tag{2.96}$$

$$\xi = \frac{R_{\mathrm{e}} + H}{R_{\mathrm{e}}}, \tag{2.97}$$

where B_a is the azimuth Doppler bandwidth, ρ_a is azimuth resolution, R_e is the earth radius, H is the orbit altitude, and v_s is the platform flying velocity. m_D is related to the AASR, which is numerically equal to the ratio of PRF to the Doppler bandwidth.

When range ambiguity and azimuth ambiguities are comprehensively considered, M_f can be represented by the expression as follows:

$$M_{\mathrm{f}} < \frac{c}{2v_0} \cdot \xi^{1.5} \frac{f_{\mathrm{PRT_{min}}}}{f_{\mathrm{PRT_{max}}}} \cdot \frac{1}{\sin \eta_{\max} \cdot m_{\mathrm{R}} m_{\mathrm{D}}}, \tag{2.98}$$

where $v_0 = \sqrt{GM/R_e}$, gravitational constant GM is equal to 3.986005×10^{14} m^3/s^2. The flying velocity of the satellite $v_s = v_0 \cdot \xi^{-0.5}$. Since .$\xi$ and η_{max} depend on the satellite–earth geometry model, m_R and m_D depend on the ambiguity ratio. The quality factor is mainly ensured by $c/2v_0$ as long as the orbit altitude, the maximum incident angle, and the ambiguity ratio are confirmed. For spaceborne SAR, $c/2v_0$ is equal to approximately 20,000, and for the airborne system, the value is generally about 500,000. Therefore, the quality factor of the spaceborne SAR will be restricted greatly.

Under the critical design conditions, the antenna is uniformly weighted, and Doppler bandwidth is corresponding to the 3 dB beamwidth of the antenna. When the AASR is −20 dB, m_D approximately is equal to 1.17. When the AASR is −25 dB, m_D approximately is equal to 1.26. If the swath is defined as the area illuminated by 1 dB beamwidth of the antenna, m_R is equal to 2 when the RASR is −22 dB. So system quality factor under critical design conditions is defined as

$$M_f = K_{amb} \cdot \xi^{1.5} / \sin \eta_{max}, \tag{2.99}$$

where $K_{amb} = \frac{c}{2v_0} \cdot \frac{f_{PRT_{min}}}{f_{PRT_{max}}} \cdot \frac{1}{m_R m_D}$, it depends on the AASR and RASR. When the AASR is −25 dB and the RASR is −22 dB, K_{amb} equals 6900. When the AASR is −20 dB and the RASR is −22 dB, K_{amb} equals 7300. Therefore under critical design conditions, the quality factor approximately is equal to 10^4 ($45° < \eta_{max} < 65°$). In general, it means that the ratio of swath width to azimuth resolution approximately equals 10^4 for spaceborne SAR with the maximum incidence angle from 45° to 65°.

The minimum effective area, length, and height of the antenna under the critical design conditions are given as follows.

The effective height of the antenna H_a on vertical to sight line direction should meet the requirement of the swath W_g.

$$\beta_e = K_E \lambda / H_a \tag{2.100}$$

$$W_g = K_E \frac{\lambda}{H_a} \cdot \frac{R_{max}}{\cos \eta_{max}}, \tag{2.101}$$

where β_e is the 1 dB beamwidth of the antenna, K_E is the factor of range radiation pattern, and R_{max} is the slant range at the maximum incident angle.

The effective length of the antenna D_a can be derived by the following formulas.

$$B_a = \frac{2v_s}{\lambda} \beta_a \tag{2.102}$$

$$\beta_a = K_A \frac{\lambda}{D_a} \tag{2.103}$$

$$\rho_a = \frac{v_s}{\xi B_a} = \frac{\lambda}{2\xi \beta_a} = \frac{D_a}{2\xi K_A}, \tag{2.104}$$

where β_a is the beamwidth of the Doppler bandwidth, K_A is the factor of azimuth radiation pattern. The minimum effective area of the antenna is given by

$$A = D_a H_a = \frac{2K_E K_A \xi \lambda R_{max}}{M_f \cos \eta_{max}} \tag{2.105}$$

Using Eq. (2.98) to express Eq. (2.105)

$$A > 4K_E K_A m_R m_D \cdot \frac{v_0}{c} \cdot \xi^{-0.5} \cdot \frac{f_{PRT_{max}}}{f_{PRT_{min}}} \cdot \lambda R_{max} \tan \eta_{max} \tag{2.106}$$

Under the critical design condition, if the swath width is defined as the area illuminated by the 1 dB beamwidth of the antenna in range, K_E is equal to 0.56. If the Doppler bandwidth is defined as the bandwidth by the 3 dB beamwidth of the antenna in azimuth, K_A is equal to 0.9. And then the minimum effective area, length, and height of the antenna under the critical design are given as follows.

$$A = \frac{\xi^{-0.5} \lambda R_{max} \tan \eta_{max}}{K_{amb}} \tag{2.107}$$

$$D_a = 1.8 \xi \rho_a \tag{2.108}$$

$$H_a = \frac{0.55 \xi^{-1.5} \lambda R_{max} \tan \eta_{max}}{K_{amb} \rho_a} \tag{2.109}$$

In terms of Eq. (2.107), the antenna area can be estimated under the condition of the ambiguity, which can be called the least effective area of the antenna in the design of the SAR system. But Eq. (2.92) gives the minimum area estimation without considering the ambiguity signal, which can be called the minimum area of the antenna. Comparison of Eq. (2.107) with Eq. (2.92), the difference between them is a m_R m_D factor that is just used to limit the ratio of ambiguity to signal.

From the previous analysis, it can be known that swath width and azimuth resolution of spaceborne SAR contradict each other. Once the system geometry and ambiguity requirements are specified, the upper limitation of the system quality factor is ensured if one wants to obtain wide swath and high azimuth resolution at the same time, considering the new ways. One idea is to obtain a wider Doppler bandwidth than the conventional radar for the enhancement of azimuth resolution without increasing the PRF, meanwhile the echo is sampled properly in range.

References

1. Zhang C (1989) Synthetic aperture radar principle, system analysis and application. Science Press, Beijing
2. John J (1978) Kovaly, synthetic aperture radar. Artech House, Boston
3. Liu Y (1999) Radar imaging technology. Harbin Institute of Technology Press

4. Lin M, Ke Y (1984) Radar signal theory. National Defense Industry Press
5. Lacomme P, Hardange J-P, Marchais J-C, Normant E (2001) Air and space borne radar system: an introduction. William Andrew Publishing, LLC, New York
6. Rihaczek AW (1973) Principle of high resolution radar. McGraw-Hill, New York
7. Wehner DR (1995) High resolution radar. Artech House, Boston
8. Clurlander JC (1991) Synthetic Aperture radar system and processing, California Institute of Technology, JPL, Pasadena, California. Wiley, New York
9. Freeman A, Johnson WTK et al (2000) The "Myth" of the minimum SAR antenna area constraint. IEEE Trans Geosci Remote Sens 38(1):320–324
10. Aguttes JP, Lapierre L (1999) Prose: a low cost P band transponder radar satellite based on the SAIL concept. IGARSS'99
11. Aguttes JP, Sombrin J, Conde E (1996) Radar "SAIL" satellite concept. Acta Astronaut 39(9–12):847–853
12. Aguttes JP, Sombrin J et al (2000) Radar "SAIL" satellite concept and design. Acta Astronaut 46(9):565–576
13. Aguttes JP (2001) New designs or modes for flexible space borne SAR. IGARSS'2001 1:674–676
14. Aguttes JP (1999) Short antenna space borne SAR concepts. Aerosp Sci Technol 6:367–378
15. Aguttes JP (2001) Radically new design of SAR satellite: short vertical antenna approach. IEEE Trans Aero Electr Syst 37(1):50–64

Chapter 3
Radar Resolution Theory

Radar resolution is the ability of a radar system to distinguish between two and more adjacent targets in a multi-target environment, which can be described by one of the target's position parameters (range, azimuth, and elevation) or motion parameters (velocity and acceleration) to distinguish target. The resolution is a quantify express of distinguishing ability for targets. Under the condition of the high SNR and the optimum processor, radar distinguishes targets according to the range and velocity can be attributed to the distinguishing problem of time delay and frequency shift.

For static targets, the delay difference between adjoining targets can represent the range difference from target to radar. So the range distinguish can be translated into the time (range–delay) distinguish. However, for the targets which the radial velocity is not zero, it is necessary to analyze the effect of the radial velocity, set up the mathematical model of echo signal, and evaluate the distinguish performance.

3.1 Optimal Detection of Radar Signals

The primary function of radar is to extract useful information from the echo of mixed background noise. When the background noise is Gauss distribution, the signal-to-noise ratio (SNR) decides the ability to detect the target in the noise background. The optimal linear filter based on the maximum SNR is a matched filter.

Suppose that the input of the linear time-invariant filter is mixed with the signal and the noise

$$x(t) = s_i(t) + n_i(t), \tag{3.1}$$

R. Yang et al., *High-Resolution Microwave Imaging*,
https://doi.org/10.1007/978-981-10-7138-6_3

where the energy of signal is $E = \int_{-\infty}^{\infty} |s_i(t)|^2 \mathrm{d}\tau$, and the noise is zero mean white noise with the bilateral power spectrum density $\frac{N_0}{2}$. The impulse response of the filter is $h(t)$, and the output of the matched filter can be expressed by

$$y(t) = x(t) * h(t) = s_i(t) * h(t) + n_i(t) * h(t) = s_o(t) + n_o(t) \tag{3.2}$$

When the peak value of the output signal component appears at t_0, the peak power of the signal is

$$E_0 = |x_o(t_0)|^2 = \left| \int_{-\infty}^{\infty} s_i(t_0 - \tau)h(\tau)\mathrm{d}\tau \right|^2 \tag{3.3}$$

The average power of output noise is

$$\mathrm{E}\left[|n_o(t)|^2\right] = \mathrm{E}\left[\left| \int_{-\infty}^{\infty} n_i(t - \tau)h(\tau)\mathrm{d}\tau \right|^2\right] = 2N_0 \int_{-\infty}^{\infty} |h(\tau)|^2 \mathrm{d}\tau \tag{3.4}$$

The SNR at the filter output can be defined as

$$\rho = \frac{E_0}{\mathrm{E}\left[|n_o(t)|^2\right]} = \frac{\left|\int_{-\infty}^{\infty} s_i(t_0 - \tau)h(\tau)\mathrm{d}\tau\right|^2}{2N_0 \int_{-\infty}^{\infty} |h(\tau)|^2 \mathrm{d}\tau} \tag{3.5}$$

In order to obtain the filter impulse response $h(t)$ for maximum SNR using the Schwartz inequality, the SNR can be rewritten as

$$\rho \leq \frac{\int_{-\infty}^{\infty} |s_i(t_0 - \tau)|^2 \mathrm{d}\tau \cdot \int_{-\infty}^{\infty} |h(\tau)|^2 \mathrm{d}\tau}{2N_0 \int_{-\infty}^{\infty} |h(\tau)|^2 \mathrm{d}\tau} = \frac{\int_{-\infty}^{\infty} |s_i(t_0 - \tau)|^2 \mathrm{d}\tau}{2N_0} = \frac{E}{2N_0} \tag{3.6}$$

The equation occurs only in

$$h(t) = cs^*(t_0 - t), \tag{3.7}$$

where c is the arbitrary constant.

From formula (3.6), the SNR of the matched filter is equal to the ratio of the signal energy to the input noise power spectral density $\frac{N_0}{2}$, and is independent of radar signal waveform. The impulse response of the matched filter is uniquely determined by the input signal and is the conjugate image of the signal. For a physically implemented filter, the input signal must be completed before the time t_0.

The Fourier transformation is used to Eq. (3.7), it can get the matched filter in the frequency domain:

$$H(f) = cS^*(f)\exp(-j2\pi f t_0), \tag{3.8}$$

where $S(f)$ is the spectrum of the input signal. Equation (3.8) shows that the amplitude–frequency characteristic of the matched filter is identical with the input signal frequency spectrum, and the phase frequency characteristic is opposite to the input signal phase spectrum.

3.2 Mathematical Model of Radar Echo [2, 3]

3.2.1 *Doppler Effect Caused by Radial Velocity*

Assuming the initial range between the radar and the target is R_0, the target flies move away from the radar at a constant radial velocity v. The radar transmits the signal with single frequency f_0 at zero and T_0 time, where $T_0 = 1/f_0$. The signal is given by

$$s_t(t) = g(t)\exp\{j[\omega_0 t + \theta(t) + \varphi_0]\}, \tag{3.9}$$

where $g(t)$ is the amplitude modulation function, $\theta(t)$ is the phase change caused by frequency modulation, and φ_0 is the initial phase.

The change of the slant range between the transmitting and receiving signal at zero time is shown in Fig. 3.1.

From Fig. 3.1, the distance that the signal travels from radar to the moving target is R_1, and $R_1 = c\Delta t_1$, $R_1 = R_0 + v\Delta t_1$, where Δt_1 is the one-way time from radar to target, and the round-trip time can be expressed as

$$t_1 = 2\Delta t_1 = \frac{2R_0}{c - v} \tag{3.10}$$

The change of the slant range of the radar transmitting and receiving signal at T_0 time is shown in Fig. 3.2. The distance that the signal travels from radar to the moving target is R_2, and $R_2 = R_0 + vT_0 + v\Delta t_2 = c\Delta t_2$ where Δt_2 is the one-way time from radar to target.

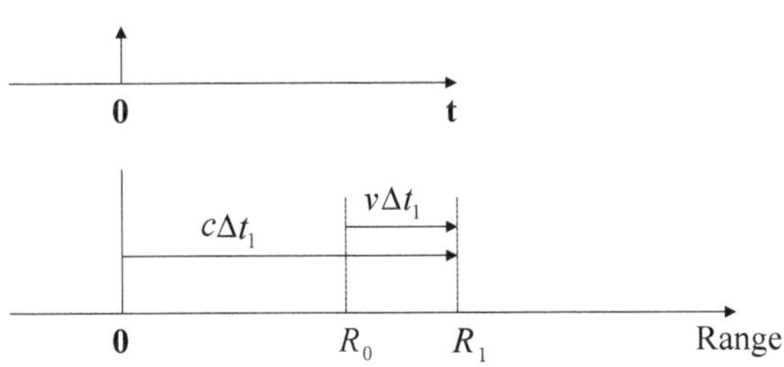

Fig. 3.1 The slant range of radar transmitting and receiving signals at zero time

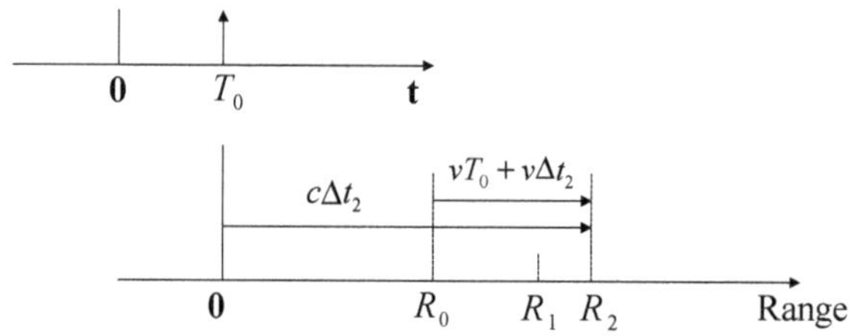

Fig. 3.2 The slant range of radar transmitting and receiving signals at T_0 time

The time from zero to receive time is

$$t_2 = T_0 + 2\Delta t_2 = \frac{(c+v)T_0 + 2R_0}{c-v} \tag{3.11}$$

Since the transmitting interval of two signals is the reciprocal of carrier frequency, the echo signal frequency can be calculated by

$$f_r = \frac{1}{t_2 - t_1} = \frac{c-v}{c+v} f_0 \tag{3.12}$$

The Doppler frequency is

$$f_d = f_r - f_0 = \frac{-2vf_0}{c+v} \approx \frac{-2v}{c} f_0, \quad |v| \ll c \tag{3.13}$$

Equation (3.13) shows that as long as the target moves away from the radar, the Doppler frequency shift is negative, and the received signal frequency is less than the frequency of the transmitted signal. On the contrary, as long as the target moves toward the radar, the Doppler frequency shift is positive and the received signal frequency is larger than the transmitted signal frequency.

It can be seen that the radial velocity between the radar and the target causes the change of the echo frequency, and the Doppler shift is proportional to the radial velocity and inversely proportional to the radar wavelength. The effect of radial velocity on echo delay will be discussed as follows.

3.2.2 *Effect of Radial Velocity on Range Time Delay*

Suppose the target moves away from the radar with a constant velocity, the distance between radar and target varies along with time, which is defined as $R(t) = R_0 + vt$. Radar transmitted signal at zero time, the echo signal scattered from target is expressed as

$$s_r(t) = \alpha s_t(t - \tau_R(t)), \tag{3.14}$$

where α is the target scattering coefficient and the round-trip time $\tau_R(t)$ is given by

$$\tau_R(t) = \frac{2(R_0 + vt)}{c + v} \tag{3.15}$$

The echo time axis can be expressed as

$$t - \tau_R(t) = \left(\frac{c - v}{c + v}\right)\left(t - \frac{2R_0}{c - v}\right), \tag{3.16}$$

where $\frac{c-v}{c+v}$ is the stretch factor of time axis caused by radial velocity. Using Eq. (3.16) to Eq. (3.9), the echo can be given by

$$\begin{aligned} s_r(t) = {} & \alpha \cdot g\left\{\left(\frac{c - v}{c + v}\right)\left(t - \frac{2R_0}{c - v}\right)\right\} \\ & \times \exp\left\{j\omega_0\left(\frac{c - v}{c + v}\right)\left(t - \frac{2R_0}{c - v}\right) + j\theta\left[\left(\frac{c - v}{c + v}\right)\left(t - \frac{2R_0}{c - v}\right)\right] + j\varphi_0\right\}, \end{aligned} \tag{3.17}$$

where $\frac{c-v}{c+v}$ in modulation function $g(t)$ and $\theta(t)$ is approximately equal to 1, usually since the carrier frequency is large, that in (3.7) the terms concerned ω_0 cannot ignore the stretch of time axis caused by radial velocity. It will be rewritten as

$$\omega_0\left(\frac{c - v}{c + v}\right)\left(t - \frac{2R_0}{c - v}\right) = (\omega_0 + \omega_d)(t - \tau_R), \tag{3.18}$$

where $\omega_d = 2\pi f_d$

$$\tau_R = \frac{2R_0}{c - v} \approx \frac{2R_0}{c}, \quad |v| \ll c.$$

If the approximation condition is established, the echo will be expressed as

$$s_r(t) = \alpha \cdot g(t - \tau_R)\exp\{j[(\omega_0 + \omega_d)(t - \tau_R) + \theta(t - \tau_R) + \varphi_0]\} \tag{3.19}$$

It is generally assumed that the target motion only causes the echo delay and Doppler shift, and the mathematical model in Eq. (3.19) will be used for waveform selection.

3.3 Ambiguity Function

According to the echo mathematical model in Eq. (3.19), it can get the expression of any radial velocity at any position in space. For a basic reference target, the echo will be given by

$$s_{r1}(t) = g(t - \tau_R)\exp[j2\pi(f_0 + f_d)(t - \tau_R)] \tag{3.20}$$

The echo of the adjacent target with the time delay τ and Doppler frequency shift ξ relative to the reference target is

$$s_{r2}(t) = g(t - \tau_R - \tau)\exp[j2\pi(f_0 + f_d + \xi)(t - \tau_R - \tau)] \tag{3.21}$$

Thus, the mean variance of echo signals for the two targets can be written as

$$\begin{aligned} |\varepsilon|^2 &= \int_{-\infty}^{+\infty} |s_{r1}(t) - s_{r2}(t)|^2 \mathrm{d}t \\ &= 2\left\{E - \mathrm{Re}\left[e^{j2\pi(f_0 + f_d)\tau} \int_{-\infty}^{+\infty} g(t)g^*(t+\tau)e^{j2\pi\xi t}\mathrm{d}t\right]\right\} \end{aligned} \tag{3.22}$$

Time–frequency complex autocorrelation function of signal complex envelope is defined as

$$\chi(\tau, \xi) = \int_{-\infty}^{+\infty} g(t)g^*(t+\tau)e^{j2\pi\xi t}\mathrm{d}t \tag{3.23}$$

Equation (3.22) can be rewritten as

$$|\varepsilon|^2 = 2E - |\chi(\tau, \xi)|\cos[2\pi(f_0 + f_d)\tau + \arg\chi(\tau, \xi)] \tag{3.24}$$

It can be seen that $|\chi(\tau, \xi)|$ is the only factor that confirms the joint resolution of range and velocity for the adjacent target. The value of $|\chi(\tau, \xi)|$ is larger, the variance is smaller, to distinguish the target is more difficult, and the ambiguity is greater. Generally, $|\chi(\tau, \xi)|$ will be called the uncertain function, and $|\chi(\tau, \xi)|^2$ will be called the fuzzy function, and sometimes the two will be called the fuzzy function. If the fuzzy function is normalized to 1 $|\chi(0, 0)|^2 = 1$, $|\chi(\tau, \xi)|^2$ represents the interference power of the matched filter output, and it can be used to measure the parameters of distinguishability at different ranges and velocities in a multi-target environment.

It can be seen from Eq. (3.23) that the ambiguity function only depends on the waveform of the transmitted signal and the characteristics of the filter. Under the condition of optimal signal processor, the ambiguity function can answer the questions, such as which waveform should be transmitted? Which processing filter should be adopted? What abilities for examples resolution, ambiguity, measurement accuracy, clutter-suppressing performance, etc. the system will have? The ambiguity function is the main tool for signal analysis and system design. It is an important measure for analyses and compares optimize degree of signal processing. The ambiguity function plays an important role not only in target distinguish theory but also in accuracy analysis.

The study of various radar waveforms, in essence, is studying the ambiguity function of the signal. The ambiguity function expresses the basic characteristics of signal. The optimal signal processing of SAR is actually how to reproduce the ambiguity of transmitted waveform. The volume under the surface of the ambiguity figure depends on the total energy of the signal, and the energy is distributed along the surface height of ambiguity figure. If the ambiguity figure is sharper, the side lobe is lower; the energy will be concentrated as well. The area of ambiguity figure expresses the range and velocity resolutions of radar, the smaller the area, the higher the resolution. According to the theory of matched filter, the output waveform is proportional to the correlation between input signal and the transmitting copy signal. If the input signal is matched with the matched filter, the output is the autocorrelation function of the transmitted signal. Therefore, in the absence of noise and interference, the output of the radar receiver matched filter is the reproduction of radar signal ambiguity figure but the peak point is no longer (0, 0) [1].

3.3.1 Range Ambiguity Function and Range Resolution

If the Doppler frequency shift ξ in Eq. (3.23) equals zero, then

$$|\chi(\tau,0)| = \left| \int_{-\infty}^{+\infty} g(t)g^*(t+\tau)\mathrm{d}t \right| = |R_{gg}(\tau)|, \tag{3.25}$$

where $R_{gg}(\tau)$ is the autocorrelation function of complex envelope of signals, which can be renamed as time ambiguity function, which quantitatively reflects the delay distinguishing ability of the signals. According to the Parseval theorem, Eq. (3.25) can also be expressed as

$$|\chi(\tau,0)| = \left| \int_{-\infty}^{+\infty} |G(f)|^2 e^{j2\pi f\tau}\mathrm{d}f \right|, \tag{3.26}$$

where $G(f)$ is the Fourier transformation of $g(t)$. Equation (3.26) shows that the time ambiguity function is inverse Fourier transform of the square of signal complex envelope spectrum modular.

According to the principle of matched filtering, the time ambiguity function can be seen as the complex envelope of the response function of the signal passing through matched filter. The main lobe width of the matched filter output response (the half power width of the main lobe) is usually defined as the nominal resolution. Roughly, the half power width is the reliably distinguished interval between the two main lobes of targets with the same RCS.

If $|\chi(\tau,0)| = 1/\sqrt{2}$ in Eq. (3.26), the nominal resolution will be

$$\Delta\tau = \frac{1}{\Delta f} \approx \frac{1}{\beta} \tag{3.27}$$

where Δf is the bandwidth of signal. β is the root-mean-square (rms) bandwidth:

$$\beta = 2\pi\left[\int_{-\infty}^{+\infty} f^2|A(f)|^2 \mathrm{d}f\right]^{\frac{1}{2}} \tag{3.28}$$

The time resolution can be converted into range resolution from the relationship between echo delay and range:

$$\Delta r = \frac{c\Delta\tau}{2} = \frac{c}{2\Delta f} \tag{3.29}$$

It can be seen that the nominal range resolution depends only on the signal bandwidth. The wider the signal bandwidth, the higher the resolution.

As studying the radar system resolution, only paying attention to the nominal resolution of adjacent targets is not enough, since it only reflects the main lobe width of the matched filter output, without considering the influence of side lobe interference on the target resolution. In general, it can reflect the ability of a radar system to distinguish the target and is still widely used as the performance index of radar system.

As the time delay is larger, in order to consider the main lobe and side lobe distinguished overall, the time distinguish constant T_{res} has to be defined. The time ambiguity function is shown in Fig. 3.3.

The time distinguish constant T_{res} is defined as

$$T_{\text{res}} = \frac{\int_{-\infty}^{+\infty} |\chi(\tau,0)|^2 \mathrm{d}\tau}{|\chi(0,0)|^2} \tag{3.30}$$

It is a parameter which can measure the range resolution ability.

It is shown in Fig. 3.3 that the smaller the time distinguish constant T_{res}, the higher the range resolution. The time distinguish constant T_{res} is the lower limit of time resolution. Therefore, only the time interval of target range is greater than T_{res}, and the target can be distinguished.

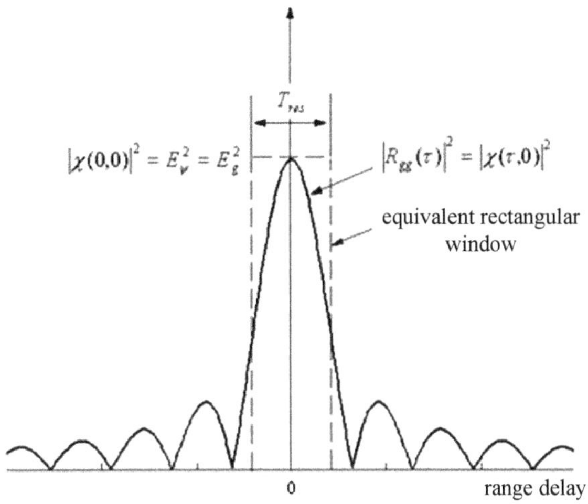

Fig. 3.3 Time ambiguity figure and equivalent rectangular window

Using Fourier transforms $g(t) \leftrightarrow G(f)$ and Parseval's theorem, Eq. (3.30) can be rewritten as

$$T_{\text{res}} = \frac{\int_{-\infty}^{+\infty} |R_{\text{gg}}(\tau)|^2 \mathrm{d}\tau}{E_g^2} = \frac{\int_{-\infty}^{+\infty} |G(\omega)|^4 \mathrm{d}\omega}{2\pi E_g^2} \tag{3.31}$$

As the signal energy is fixed, in theory the broadband signal can be designed for smaller the integral value of square of the autocorrelation function and higher resolution.

The reciprocal of the time distinguish constant can be defined as the effective bandwidth of the signal, sometimes referred to as the spectral duration width, which represents the sum of all the spectral components:

$$B_{\text{eff}} = \frac{1}{T_{\text{res}}} (\text{Hz}) \tag{3.32}$$

In summary, the range resolution depends on the spectral structure of the signal. In order to improve the range resolution, the signal must occupy a large bandwidth.

3.3.2 Velocity Ambiguity Function and Doppler Resolution

Using the time–frequency complex autocorrelation function in Eq. (3.23) to express velocity ambiguity function,

$$|\chi(0,\xi)| = \left| \int_{-\infty}^{+\infty} |g(t)|^2 e^{j2\pi\xi t} dt \right| \tag{3.33}$$

In the same way as the range resolution, the nominal Doppler resolution is defined as the 3 dB width of the main lobe to express the resolution of target with similar velocity. As a larger velocity difference of target, in order to considerate effect of the main lobe and side lobe to velocity resolution overall, the frequency distinguish constant Ω_{res} is defined as

$$\Omega_{\text{res}} = \frac{\int_{-\infty}^{+\infty} |\chi(0,\xi)|^2 d\xi}{|\chi(0,0)|^2} = \frac{\int_{-\infty}^{+\infty} |R_{gg}(\xi)| d\xi}{|R_{gg}(0)|^2} \tag{3.34}$$

In terms of Parseval's theorem, Ω_{res} is rewritten as

$$\Omega_{\text{res}} = \frac{2\pi \int_{-\infty}^{+\infty} |g(t)|^4 dt}{\left[\int_{-\infty}^{+\infty} |g(t)|^2 dt\right]^2} = \frac{2\pi \int_{-\infty}^{+\infty} |g(t)|^4 dt}{E_g^2} \tag{3.35}$$

It can be seen, as the signal energy is fixed, in theory the time width of signal can be designed arbitrarily; let autocorrelation function on frequency domain become as narrow as possible, and get small frequency distinguish constant.

Figure 3.4 shows the velocity ambiguity function and the equivalent rectangular window.

The effective duration time width T_e is defined as the reciprocal of the frequency distinguish constant Ω_{res}:

$$T_e = \frac{2\pi}{\Omega_{\text{res}}} \tag{3.36}$$

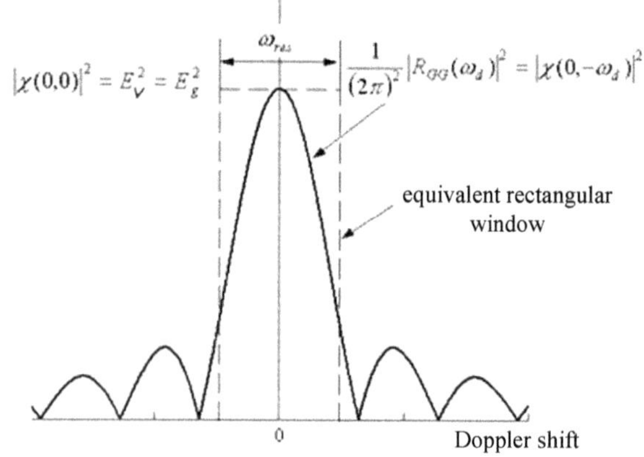

Fig. 3.4 The velocity ambiguity function and the equivalent rectangular window

It can be seen that the velocity resolution depends on the temporal structure of the signal. In order to improve the velocity resolution, the signal must occupy large time duration.

The range resolution depends mainly on the spectral structure of the signal under the condition of the optimal processing and a certain SNR. In order to increase the ranging precision and resolution, a large bandwidth is necessary. The velocity measurement accuracy and velocity resolution depend on the time structure of the signal, and in order to increase the velocity precision and velocity resolution, a large time duration is necessary. In addition, in order to increase the detection probability of radar system, a large amount of energy of the signal is necessary. In conclusion, the waveform with large time-bandwidth product is important for high resolution imaging SAR.

3.4 SAR Resolution Principles [5, 6]

Synthetic aperture radar (SAR) is a kind of two-dimensional microwave imaging radar, whose basic task is to obtain the image information of the earth's surface. In order to ensure the SAR image to reflect the microwave scattering characteristics of the terrain real, the range impulse response and azimuth impulse response of SAR system are required to be as close as to δ function. SAR obtains the two-dimensional high-resolution image by means of pulse compression technology in range dimension and synthetic aperture processing in azimuth dimension. This section focuses on the two-dimensional resolution of SAR.

From the conclusion of Sect. 3.3, the range resolution depends mainly on the spectral structure of the signal, inversely proportional to the signal bandwidth. The normal range resolution of SAR is given by

$$\rho_r = \frac{c}{2B}, \tag{3.37}$$

where B is the bandwidth of the transmitted signal. The range resolution depends only on the effective bandwidth of the transmitted signal, independent on the waveform.

Figure 3.5 shows the impulse response of SAR in two dimensions. The factors that affect the radar resolution will be discussed.

The main lobe width of SAR impulse response in azimuth dimension is

$$\rho_a = \frac{v_a}{\Delta f_d}, \tag{3.38}$$

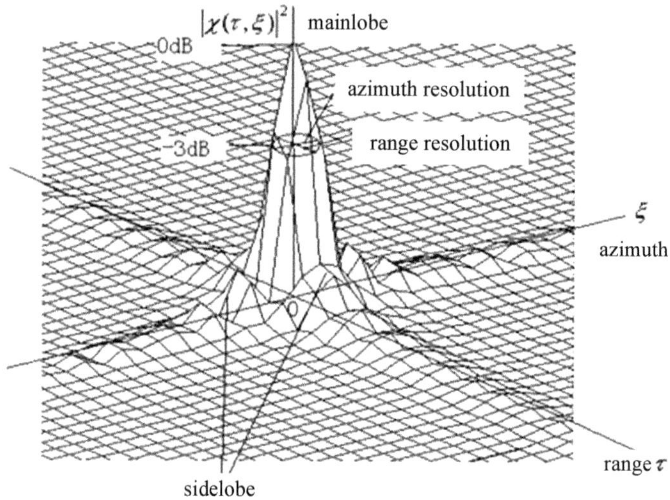

Fig. 3.5 The impulse response of SAR in two dimensions

where v_a is the velocity of the platform in the azimuth. Δf_d is the Doppler bandwidth in synthetic aperture time and is given by

$$\Delta f_d = \frac{2\beta v_a}{\lambda}, \tag{3.39}$$

where β is the beamwidth of the antenna in azimuth and λ is the wavelength.

Equation (3.38) shows that the azimuth resolution is inversely proportional to the Doppler bandwidth. The wider the Doppler bandwidth, the higher the azimuth resolution. Using Eq. (3.39) to express Eq. (3.38), we can get

$$\rho_a = \frac{D_a}{2} \tag{3.40}$$

It can be shown that the azimuth resolution of SAR depends only on the antenna azimuth size and is independent of the wavelength and range, which is the important feature of SAR azimuth resolution.

As the azimuth size of the antenna is selected, the SAR can only increase the synthetic aperture time to get higher azimuth resolution by means of the spotlight mode operating. However, the improvement of the range resolution depends on the large time–bandwidth product of the transmitted signal. On the one hand, the signal has a large bandwidth to satisfy the high-resolution requirement, and on the other hand, the time duration of the energy demand is satisfied for SNR. Under the limitation of the peak power of the system, it is very important to design the signal with large time–bandwidth product.

According to the physically means of ambiguity function in Sect. 3.3, the two-dimensional impulse response of SAR is equivalent to the ambiguity function of signal; $|\chi(\tau,\xi)|^2/|\chi(0,0)|^2$ can be used to measure the joint resolution of the range velocity. As $|\chi(\tau,\xi)|^2/|\chi(0,0)|^2 = 3/4$, the ellipse cut out from the curved surface of ambiguity function will provide the normal range resolution and normal velocity resolution.

The resolution of radar system depends on the SNR, the waveform, and signal processing method. The transmitting waveform not only directly affects the performance, such as resolution, imaging precision, and clutter suppression ability, but also decides the signal processing method. Under the condition of a certain SNR and the optimal processing, SAR resolution in range and azimuth dimensions depends only on the waveform. To design a practical and ideal ultra-wideband waveform becomes an important substance for the SAR system design.

3.5 Waveform Classification and Design Principles [1, 4]

3.5.1 *Waveform Classification*

There are many kinds of radar signal waveforms, which can be classified by different methods, and the classification by means of signal modulation is commonly used. The different radar systems transmit signals with the different modulations and parameters. However, it is the most reasonable method to classify signals by means of ambiguity figure, so the radar signal can be classified into four types, such as positive edge shape (single rectangle or Gauss type single frequency pulse), cutting edge shape (pulse linear or nonlinear frequency modulated pulse), thumbtack (pure noise signal is the ideal thumbtack, and phase encoding pulse signal and noise FM signal is the signal of non-ideal thumbtack), and nail bed (Correlated pulse signal). Comparison of four types of signal forms and characteristics is shown in Table 3.1.

3.5.2 *Waveform Design Principles*

How to make the ambiguity function optimal matching to the radar operating environment and information requirement is the basic method of waveform synthesis and design. Radar waveform design can be summed up in two methods, one is proposed by Sussman, which obtains the optimal waveform by means of least squares optimal synthesis of ambiguity function for a given function $|\chi(\tau,\xi)|$ and argument $\arg(\tau,\xi)$. This method not only faces the difficulties in the mathematical derivation but also often the optimum waveform is very complex modulation signal, which is technically difficult to achieve. Up to now, this method has not got any

Table 3.1 Radar signal classification according to the ambiguity function

Classification	A	B1	B2	C
TB Time–bandwidth product	TB = 1	TB $\geq$ 1	TB > 1	TB > 1
The shape of ambiguity figure	Positive cutting edge	Thumbtack	Cutting edge	Nail bed
Normal resolution unit	1	1/TB	1	1/TB
Ambiguity	/	/	Range-Doppler ambiguity	Spike
Side lobe	Low	High	Low	Low
Common signal waveform	Single frequency pulse	1. Nonlinear frequency modulation signal 2. Pseudo-random phase-coded pulse signal 3. Frequency-coded pulse signal 4. Pseudo-random phase, frequency, stagger pulse signal	1. Linear frequency modulation signal 2. Stepped-frequency signal 3. Frequency modulated burst signal	Uniform spacing coherent burst signal
The characteristics of ambiguity function	The ambiguity volume is concentrated on the ridge. The range and Doppler side lobe depend on the shape of the envelope, but the side lobe is still only a tiny fraction of the ambiguity volume	The shape of ambiguity function is approximate to thumbtack. The larger TB product, more close to the thumbtack shape	The ambiguity function can be viewed as that of a class cut, and the energy outside of the main lobe is very small. It has good range resolution and velocity resolution for areas out of the edge	Due to the periodicity, the ambiguity function presents a regular arrangement on the $\tau - \xi$ plane. The central ambiguity region is the same as the B1 signal. But most of the ambiguity volume is concentrated in the repeated isolated spikes

(continued)

Table 3.1 (continued)

Classification	A	B1	B2	C
Advantages and disadvantages	The range and velocity high resolution cannot be provided at the same time. The range and range resolution affect each other	The range resolution, velocity resolution, and measurement accuracy can be provided at the same time. In the target environment with strong clutter and large RCS area, the base side lobe interference is serious, and the original matched filter is no longer applicable when Doppler detuning	When the detuning of Doppler is not greater than the signal bandwidth B, the matched filter will still play the role of pulse compression, but coupling between the range and the Doppler occurs	A maximum unambiguous area around the center main peak eliminates the basic side lobe interference as signal B1. But there is serious measurement ambiguity and peak interference
Application consideration	Signal generation and processing are the most simple. It is suitable for radar with low precision and low resolution, and near range	Generation and processing of non-LFM signal are simple. The pseudo-random phase-coded pulse signal is suitable for the target with narrow Doppler bandwidth. The shape of ambiguity figure of pseudo-random coded stagger burst is close to thumbtack, but generation and processing of signal are very difficult	In order to eliminate the range-Doppler coupling, the LFM signal with different orientations of the blade must be transmitted alternately. A weighted method is used to suppress the side lobe but the main lobe is broadened and the SNR is degraded	In order to eliminate the ambiguity, the amplitude and phase weighting or phase, frequency, interval stagger will adopt and convert the ambiguity into quasi-thumbtack. Signal generation and processing simple waveform parameter flexible can be controlled

more application results. The other method, proposed by Rihaczek, is simple waveform selection. According to the target environment and radar tactical requirements, the suitable signal type is selected first, and then the difficulty and cost of the implementation of the signal waveform and parameters are taken into account for satisfying the requirements of the system performance. In this case the waveform. Is not the "optimal waveform" under the specific conditions, but the designer clearly known the differences between this waveform and "optimal waveform", and the practical value for the further optimum? For different applications of radar, the optimal ideal waveform for various target environments and operating requirements does not exist. Under a given target environment and tactical conditions, there may be several suitable types of signal. As long as the appropriate parameters are selected, the performance specifications of the radar system can be guaranteed.

The waveform design is to get the signal with large time–bandwidth product according to the given $\chi(\tau, \xi)$. Since the waveform synthesis directly from the ambiguity function will face mathematics difficulty seriously, it is easy to use the simple waveform selection route. At first, select the suitable signal type, and then to make the ambiguity figure match with the target environment. Finally, selecting the specific waveform and parameters is to satisfy the requirements of the system performance.

Since the change of each parameter of radar signal is not independent of each other, and there is the mutual constraints between each parameter; the changing mode and scope of radar signal are limited, while selecting and designing signal parameters. It is necessary to analyze the mutual constraints between parameters of different types of radar signal in detail. In practice, some requirement of parameter index is often contradictory, such as between range and velocity measurement; between accuracy, resolution, and multi-valued ambiguity; between measuring range and velocity unambiguously; between optimal waveform generation and processing complexity; and between implementation and cost. So a compromise scheme is often used in the design of radar signals. Usually, there is no optimal waveform to make all specification best and suitable for all applications. In general, several different waveforms are used, and each waveform can be suitable for the specific application.

As designing radar signal, the suitable signal type is selected first according to the target environment and radar tactical requirements, and at the same time, giving consideration for the difficulty and cost of the implementation of the signal waveform to choice suitable waveform and parameter of signal, and satisfying the requirements of the system performance. Although the designed waveform is not "the best waveform", its advantages, weakness, and the difference from the optimal waveform can be easy understood and will help designer to find effective measures to improve the design.

References

1. Fei Y, Su G, M Hong et al (2002) The generating technology of wideband radar signals. National Defense Industry Press, Beijing
2. Maoyong L, Youan K (1984) Radar signal theory. National Defense Industry Press, Beijing
3. Peyton ZP (1998) Radar principles. Wiley, Hoboken
4. Rihaczek AW, Hershkowitz SJ (1996) Radar resolution and complex-image analysis. Artech House, Boston, London
5. Zhang C (1989) Synthetic aperture radar principle, system analysis and application. Science Press, Beijing
6. Immoreev IJ (Russia), Taylor JD (UAS) (2000) Ultra-wideband (UWB) radars: today and future. In: Progress in electromagnetics research symposium, July 5–14, 2000, Cambridge, MA, USA. http://www.uwb.org/Conference/Proceedings.htm

Chapter 4
Linear Frequency Modulation Pulse Signal

Linear Frequency Modulation (LFM) pulse signal (chirp) is a widely used pulse compression signal, which obtains large time–bandwidth products by inner-pulse linear frequency modulation. Its main advantage is insensitive to the Doppler frequency shift, even if echo signal has a larger Doppler frequency shift, the original matched filter still can play a role in the pulse compression, which will simplify signal processing. But, its side lobes of matched filter output response are relatively high. In order to reduce the side lobes, proper weighting processing must be introduced. In SAR, LFM signal is also the most commonly used signal form. The process of its generation and compression is essentially consistent with the principle of azimuth synthetic aperture.

4.1 Linear Frequency Modulation Pulse Signal

4.1.1 Spectrum of Linear Frequency Modulation Pulse Signal

Assuming that the central frequency of the LFM signal is f_0, the bandwidth is B, the pulse duration is T, and the chirp rate is k. The complex expression of LFM signal can be written as

$$s(t) = u(t)\exp(j2\pi f_0 t) = \mathrm{rect}\left(\frac{t}{T}\right)\exp\left\{j2\pi\left(f_0 t + \frac{k}{2}t^2\right)\right\} \tag{4.1}$$

where

$$u(t) = \mathrm{rect}\left(\frac{t}{T}\right)\exp\left(j\pi k t^2\right) \tag{4.2}$$

R. Yang et al., *High-Resolution Microwave Imaging*,
https://doi.org/10.1007/978-981-10-7138-6_4

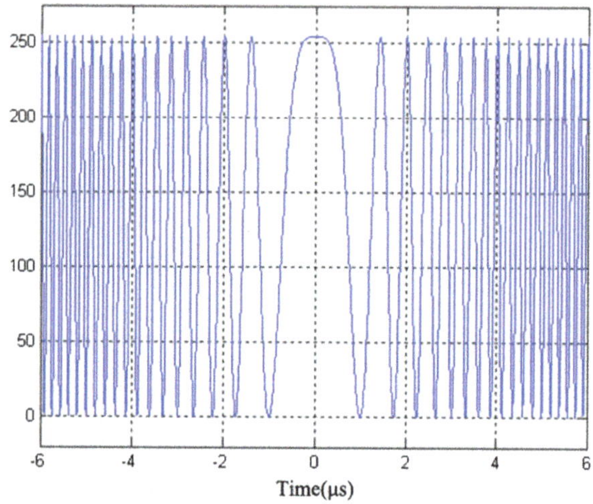

Fig. 4.1 Sketch of LFM baseband complex signal

is the complex envelope of the signal, also known as baseband complex signal. Its waveforms are shown in Fig. 4.1.

The instantaneous frequency of the signal can be obtained as

$$f = \frac{\mathrm{d}}{\mathrm{d}t}\left[f_0 t + \frac{k}{2}t^2\right] = f_0 + kt \tag{4.3}$$

where $k = B/T$ is frequency modulation slope.

The zero center frequency spectrum of the signal can be obtained by Fourier transformation for (4.2)

$$U(f) = \int\limits_{-\frac{T}{2}}^{\frac{T}{2}} \exp\left(j\pi kt^2\right)\exp(-j2\pi ft)\mathrm{d}t = \exp\left(-\frac{j\pi f^2}{k}\right)\int\limits_{-\frac{T}{2}}^{\frac{T}{2}} \exp\left(j\pi k\left(t - \frac{f}{k}\right)^2\right)\mathrm{d}t \tag{4.4}$$

The variable substitution is performed $x = \sqrt{2k}\left(t - \frac{f}{k}\right)$, and the Fresnel integral formula is used

$$C(X) = \int\limits_0^X \cos\left(\frac{\pi x^2}{2}\right)\mathrm{d}x \tag{4.5}$$

$$S(X) = \int\limits_0^X \sin\left(\frac{\pi x^2}{2}\right)\mathrm{d}x \tag{4.6}$$

The signal spectral expression can be written as

$$U(f) = \frac{1}{\sqrt{2k}} \exp\left(-\frac{j\pi f^2}{k}\right)\{[C(X_1) + C(X_2)] + j[S(X_1) + S(X_2)]\} \tag{4.7}$$

where

$$X_1 = \sqrt{2BT}\left(\frac{1}{2} - \frac{f}{B}\right) \quad X_2 = \sqrt{2BT}\left(\frac{1}{2} + \frac{f}{B}\right) \tag{4.8}$$

The amplitude spectrum is

$$|U(f)| = \frac{1}{\sqrt{2k}}\sqrt{\left\{[C(X_1) + C(X_2)]^2 + [S(X_1) + S(X_2)]^2\right\}} \tag{4.9}$$

The phase spectrum is

$$\varphi(f) = -\frac{\pi}{k}f^2 + \arctan\left[\frac{S(X_1) + S(X_2)}{C(X_1) + C(X_2)}\right] \tag{4.10}$$

When the time–bandwidth product of the signal is large, X_1 and X_2 will be very large. According to the properties of the Fresnel integral, as shown in Fig. 4.2, the Fresnel integral value tends to fluctuate around a fixed value and the ripple wave is small.

At the time, the amplitude spectrum of the signal $|U(f)|$ is close to the constant in $-\frac{B}{2} \leq f \leq \frac{B}{2}$, but there is Fresnel fluctuation at the edge $\pm\frac{B}{2}$. When $|f| > \frac{B}{2}$, the Fresnel integral value is offset to zero. Therefore, when the *TB* product is large, the amplitude spectrum has an approximate rectangular spectrum. The first term in

Fig. 4.2 Fresnel integral

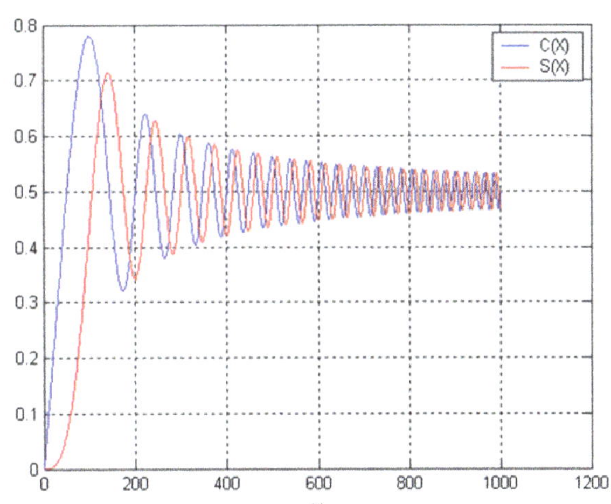

the signal phase spectrum is called the square rate term, which is the quadratic phase term. The second term is called the residual phase term and tends to be constant $\frac{\pi}{4}$ when the TB product is large.

Numerical integration proves the following conclusion. When $TB \geq 10$, 95% of the energy is included in $\left[-\frac{B}{2}, \frac{B}{2}\right]$. When $TB \geq 100$, the percentage rose to 99%. Figures 4.3 and 4.4 show the amplitude and phase spectrum of the LFM signal.

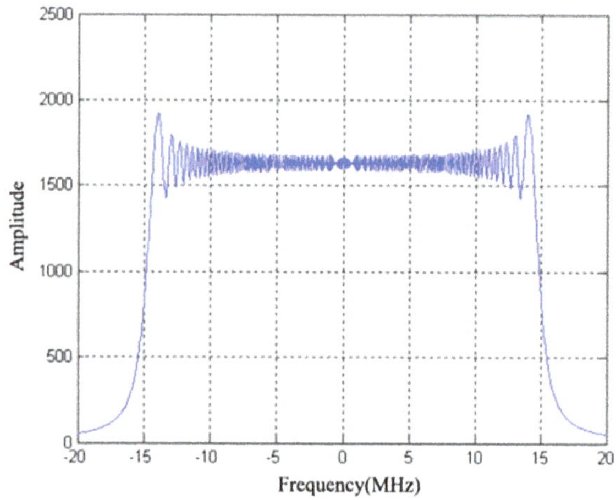

Fig. 4.3 Amplitude spectrum of LFM

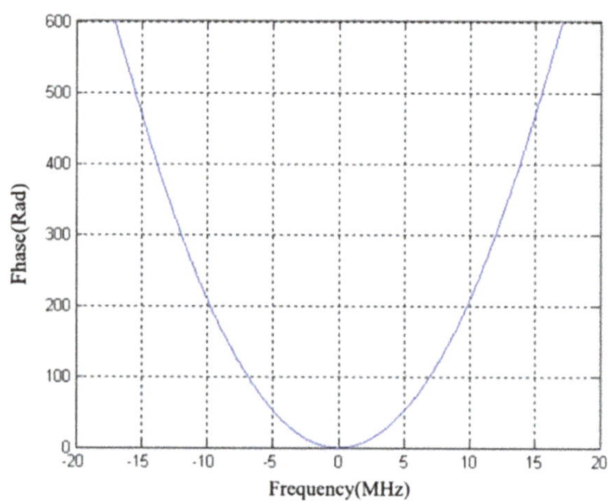

Fig. 4.4 Phase spectrum of LFM

In high-resolution radar applications, the TB product of the signal is very large, so the amplitude spectrum and phase spectrum of LFM can be approximated as

$$|U(f)| = \frac{1}{\sqrt{k}}\mathrm{rect}\left(\frac{f}{B}\right) \tag{4.11}$$

$$\varphi(f) = -\frac{\pi f^2}{k} + \frac{\pi}{4} \tag{4.12}$$

The entire signal presented in the paper is large time–bandwidth product signal. The following derivation will directly use the Eqs. (4.11) and (4.12) to represent the spectrum of the LFM signal.

4.1.2 Ambiguity Function of Linear Frequency Modulation Signal [1]

First, the ambiguity function of the single carrier rectangular pulse signal is deduced, and the ambiguity function of the LFM signal is obtained by means of the properties of the ambiguity function. The normalized complex envelope of the single carrier frequency rectangular pulse signal can be written as

$$u(t) = \mathrm{rect}\left(\frac{t}{T}\right) \tag{4.13}$$

where T is the pulse width. Through substituting the above formula into (3.23), we can be obtain

$$\chi(\tau,\xi) = \int_{-\infty}^{\infty} u(t)u^*(t+\tau)\exp(j2\pi\xi t)\mathrm{d}t = \int_a^b \exp(j2\pi\xi t)\mathrm{d}t \tag{4.14}$$

The integral limits can be discussed as follows:

(1) When $0<\tau<T$, $a=0$; $b=-\tau+T$,

$$\chi(\tau,\xi) = \int_0^{-\tau+T} \exp(j2\pi\xi t)\mathrm{d}t = \exp(j\pi\xi(T-\tau))\cdot\frac{\sin\pi\xi(T-\tau)}{\pi\xi(T-\tau)}\cdot(T-\tau) \tag{4.15}$$

(2) When $-T < \tau < 0$, $a = -\tau$; $b = T$,

$$\chi(\tau,\xi) = \int_{-\tau}^{T} \exp(j2\pi\xi t)\mathrm{d}t = \exp(j\pi\xi(T-\tau)) \cdot \frac{\sin\pi\xi(T+\tau)}{\pi\xi(T+\tau)} \cdot (T+\tau) \quad (4.16)$$

(3) When $|\tau| > T$, $u(t)u^*(t+\tau) = 0$

$$\chi(\tau,\xi) = 0 \quad (4.17)$$

The above three formulas can be combined as

$$\chi(\tau,\xi) = \mathrm{rect}\left(\frac{\tau}{2T}\right) \cdot \exp\left(j\pi\xi(T-\tau)\right) \cdot \frac{\sin\pi\xi(T-|\tau|)}{\pi\xi(T-|\tau|)} \cdot (T-|\tau|) \quad (4.18)$$

Its model is

$$|\chi(\tau,\xi)| = \mathrm{rect}\left(\frac{\tau}{2T}\right) \cdot \left|\frac{\sin\pi\xi(T-|\tau|)}{\pi\xi(T-|\tau|)}\right| \cdot (T-|\tau|) \quad (4.19)$$

Its three-dimensional ambiguity figure is shown in Fig. 4.5. Let $\xi = 0$, the range ambiguity function of the single carrier frequency rectangular pulse signal is obtained. It is the envelope of the autocorrelation function, as shown in Fig. 4.6. Let

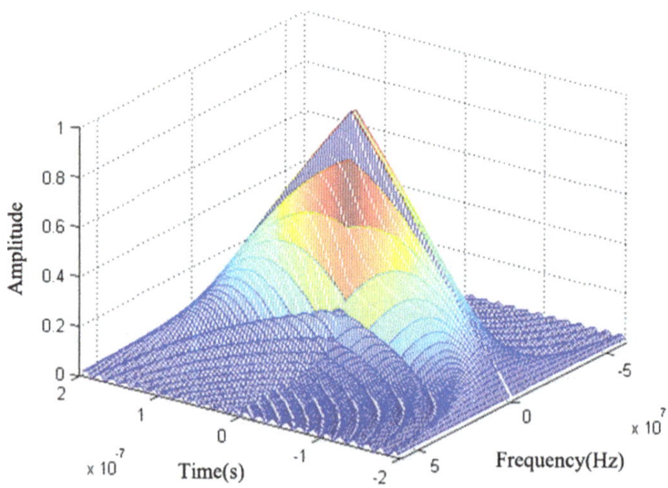

Fig. 4.5 The ambiguity function of the single carrier frequency rectangular pulse signal

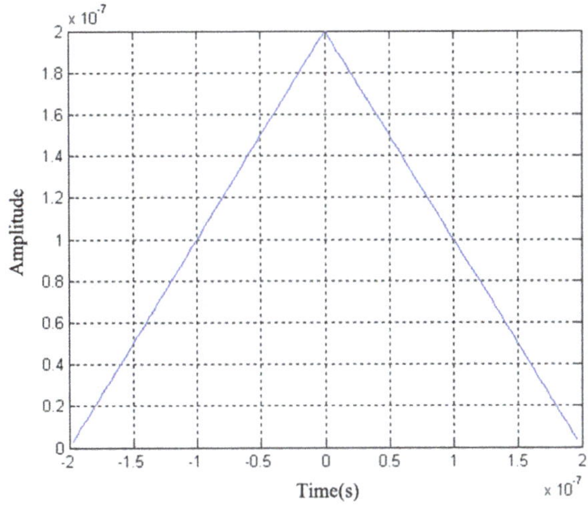

Fig. 4.6 The range ambiguity function of the single carrier frequency rectangular pulse signal

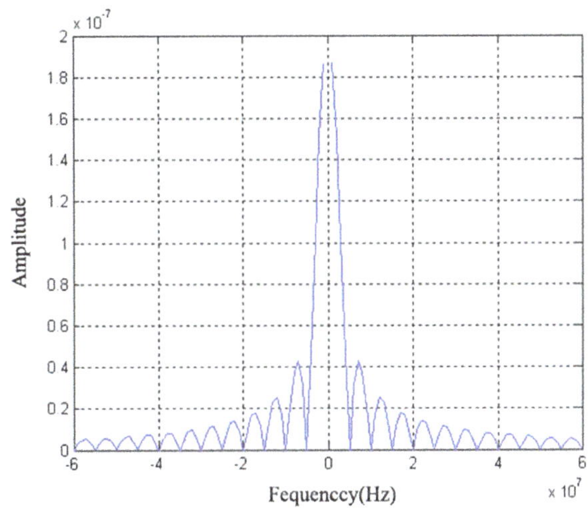

Fig. 4.7 The velocity ambiguity function of the single carrier frequency rectangular pulse signal

$\tau = 0$, the velocity ambiguity function of the single carrier frequency rectangular pulse signal is obtained, as shown in Fig. 4.7. It can be seen that the signal cannot obtain high resolution of range and velocity simultaneously. The pulse compression signal is required to improve its performance.

The ambiguity function has the following properties, if $s(t) = u(t)\exp(j\pi ct^2)$, the ambiguity functions of $s(t)$ can be written as

$$\chi_s(\tau,\xi) = \int\limits_{-\infty}^{\infty} u(t)u^*(t+\tau)\exp\left(j\pi ct^2\right)\exp\left(-j\pi c(t+\tau)^2\right)\exp(j2\pi\xi t)\mathrm{d}t \\ = \exp\left(-j\pi c\tau^2\right)\cdot\chi_u(\tau,\xi-c\tau) \tag{4.20}$$

Let the complex envelope of single carrier frequency rectangular pulse signal be $u_1(t)$, then the complex envelope $u(t)$ of LFM pulse signal can be expressed as

$$u(t) = \mathrm{rect}\left(\frac{t}{T}\right)\exp\left(j\pi kt^2\right) = u_1(t)\exp\left(j\pi kt^2\right) \tag{4.21}$$

By using the property of (4.20), the ambiguity function of LFM signal is obtained as follows:

$$\chi(\tau,\xi) = \mathrm{rect}\left(\frac{\tau}{2T}\right)\cdot\exp\left\{j\pi\left[(\xi-k\tau)(T-\tau)-k\tau^2\right]\right\}\cdot\frac{\sin\pi(\xi-k\tau)(T-|\tau|)}{\pi(\xi-k\tau)(T-|\tau|)} \\ \cdot(T-|\tau|) \tag{4.22}$$

Its model is

$$|\chi(\tau,\xi)| = \mathrm{rect}\left(\frac{\tau}{2T}\right)\cdot\left|\frac{\sin\pi(\xi-k\tau)(T-|\tau|)}{\pi(\xi-k\tau)(T-|\tau|)}\right|\cdot(T-|\tau|) \tag{4.23}$$

Figure 4.8 shows the three-dimensional ambiguity figure, it can be seen that it has the shape of the nail plate. When$\xi=0$, the range ambiguity function $|\chi(\tau,0)|$ of LFM pulse signal is obtained. It is the envelope of the autocorrelation function, as shown in Fig. 4.9.

$$|\chi(\tau,0)| = \mathrm{rect}\left(\frac{\tau}{2T}\right)\cdot\left|\frac{\sin\pi k\tau(T-|\tau|)}{\pi k\tau(T-|\tau|)}\right|\cdot(T-|\tau|) \tag{4.24}$$

When $\tau=0$, the velocity ambiguity function $|\chi(0,\xi)|$ of LFM pulse signal is obtained as shown in Fig. 4.10.

$$|\chi(0,\xi)| = \left|\frac{\sin\pi\xi T}{\pi\xi T}\right|\cdot T \tag{4.25}$$

It can be seen that both the range ambiguity and the velocity ambiguity have sinc function shape, and the first zero point of $|\chi(\tau,0)|$ is located in $\frac{1}{kT}$ point,the first zero point of $|\chi(0,\xi)|$ is located in $\frac{1}{T}$ point. When T is larger, the better range resolution and velocity resolution can be obtained simultaneously, and the farther action distance as well.

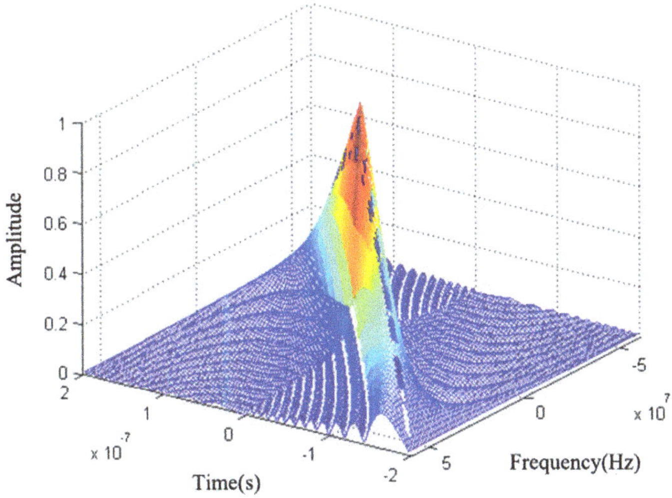

Fig. 4.8 The ambiguity function of LFM pulse signal

Fig. 4.9 The range ambiguity function of LFM pulse signal $|\chi(\tau, 0)|$

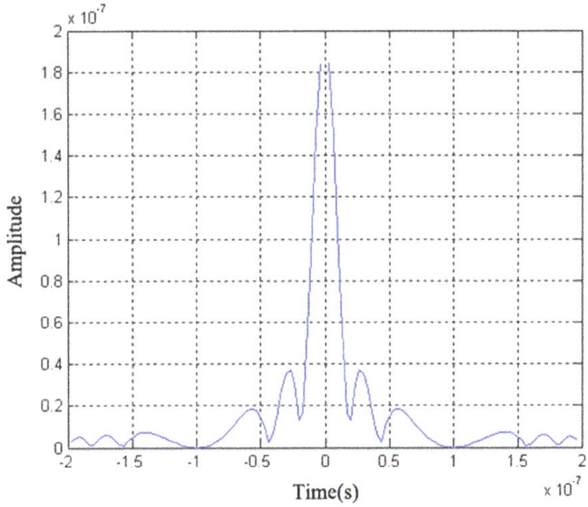

However, the side lobe of sinc function is high, about 13.2 dB, and the method of weighted processing is required to suppress the side lobe.

The −3 dB LFM pulse ambiguity figure and single carrier pulse ambiguity figure are combined together as shown in Fig. 4.11. It can be seen that the ambiguity figure of the former is obtained from the latter by inclining at an angle $\theta = \text{arctg}(k)$.

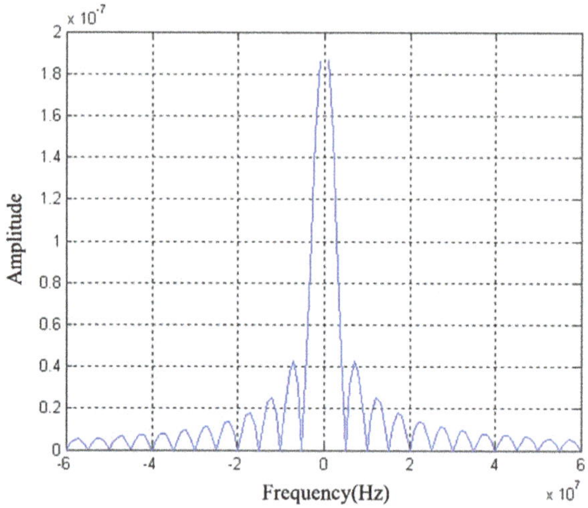

Fig. 4.10 The velocity ambiguity function of LFM pulse signal $|\chi(0,\xi)|$

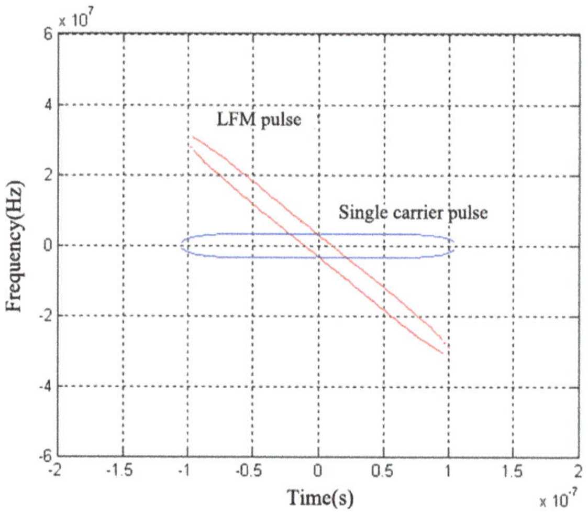

Fig. 4.11 The ambiguity figure of two signals

The k value is greater; the inclining angle is greater too. When k is positive, the inclining angle is counterclockwise; when k is negative, the inclining angle is clockwise.

The benefits of ambiguity figure inclining are [1]:

(1) When the target velocity is known, it can be a very high range accuracy; when the distance of the target is known, it can have high-velocity accuracy.
(2) In a multi-target environment, a very high range resolution can be obtained when the target is at the same velocity. A very high-velocity resolution can be obtained when the target is at the same distance.

But it also brings the following drawbacks:

(1) For the targets whose distance and velocity are not known, only the joint value can be accurately measured, and the distance and velocity of the target cannot be known exactly.
(2) The multiple targets in the vicinity of the oblique blade completely cannot be distinguished.

The resolution parameters of LFM signal can be further obtained. The delay resolution constant A_τ or continuous spectrum width W_e which express the range resolution can be written as

$$A_\tau = \frac{1}{W_e} = \frac{\left[\int_{-\frac{B}{2}}^{\frac{B}{2}} \left|\frac{1}{\sqrt{k}}\right|^2 df\right]^2}{\int_{-\frac{B}{2}}^{\frac{B}{2}} \left|\frac{1}{\sqrt{k}}\right|^4 df} = \frac{1}{B} \tag{4.26}$$

It can be seen that continuous spectrum width is equal to the frequency offset, which is the signal bandwidth B.

The Doppler resolution constant A_ξ or continuous time width T_e express the velocity resolution, and it can be written as

$$A_\xi = \frac{1}{T_e} = \frac{\left[\int_{-\frac{T}{2}}^{\frac{T}{2}} \left|\mathrm{rect}\left(\frac{t}{T}\right)\right|^2 dt\right]^2}{\int_{-\frac{T}{2}}^{\frac{T}{2}} \left|\mathrm{rect}\left(\frac{t}{T}\right)\right|^4 dt} = \frac{1}{T} \tag{4.27}$$

It can be seen that continuous time width is equal to the pulse width T.

4.1.3 Pulse Compression Characteristics and Weighted Processing of LFM Signals [1]

The high-frequency bandpass system can be transformed into an equivalent low-pass system. From the formulas (4.11) and (4.12), it can be seen that the matched filter of the baseband complex signal should have the following frequency characteristics (ignoring the constant phase term $\frac{\pi}{4}$)

$$H(f) = \mathrm{rect}\left(\frac{f}{B}\right)\exp\left(j\frac{\pi f^2}{k}\right) \tag{4.28}$$

If the baseband complex signal $u(t)$ contains Doppler shift ξ, then

$$u(t) = \mathrm{rect}\left(\frac{t}{T}\right) \exp\left(j\pi k t^2\right) \exp(j2\pi\xi t) \tag{4.29}$$

Its frequency spectrum is

$$U(f) = \int_{-\frac{T}{2}}^{\frac{T}{2}} u(t) \exp(-j2\pi f t) \mathrm{d}t \tag{4.30}$$

The output signal spectrum of the matched filter is

$$\begin{aligned} Y(f) &= U(f) \cdot H(f) \\ &= \mathrm{rect}\left(\frac{f}{B}\right) \exp\left(j\frac{\pi f^2}{k}\right) \cdot \int_{-\frac{T}{2}}^{\frac{T}{2}} \exp\left(j\pi k t^2\right) \exp(j2\pi\xi t) \exp(-j2\pi f t) \mathrm{d}t \end{aligned} \tag{4.31}$$

The waveform of the output signal is

$$\begin{aligned} y(t) &= \int_{-\infty}^{\infty} Y(f) \exp(j2\pi f t) \mathrm{d}f \\ &= \sqrt{kT^2} \cdot \frac{\sin[\pi(\xi + kt)T]}{\pi(\xi + kt)T} \cdot \exp\left(-j\pi k t^2\right) \exp\left(j\frac{\pi}{4}\right) \end{aligned} \tag{4.32}$$

$$= \sqrt{D} \cdot \frac{\sin[\pi(\xi + kt)T]}{\pi(\xi + kt)T} \cdot \exp\left(-j\pi k t^2\right) \exp\left(j\frac{\pi}{4}\right) \tag{4.33}$$

It can be seen that the output signal contains a negative slope linear frequency modulation. When the Doppler shift $\xi = 0$, the output signal envelope has a sinc function type. Its -4 dB main lobe width is $\frac{1}{B}$ and first side lobe height is -13.2 dB. $\sqrt{D} = \mathrm{BT}$ is called the compression ratio. It indicates that the maximum of output signal envelope is increased by $\sqrt{D}$ times of the input signal and the pulse width is reduced by $\frac{1}{\sqrt{D}}$ of the input signal.

When the Doppler shift $\xi \neq 0$, the output envelope will produce displacement, resulting in ranging error. And the amplitude of the output envelope will decrease, and its pulse width will increase, resulting in a decrease in signal-to-noise ratio and range resolution. This is the range-Doppler coupling.

From the above analysis, it can be seen that the output signal envelope of the matched filter has a sinc envelope. It is a compression waveform with side lobe. Its first side lobe level is of -13.2 dB, the pulse pressure results are shown in Fig. 4.12. In a multi-target environment, the side lobe of the strong signal

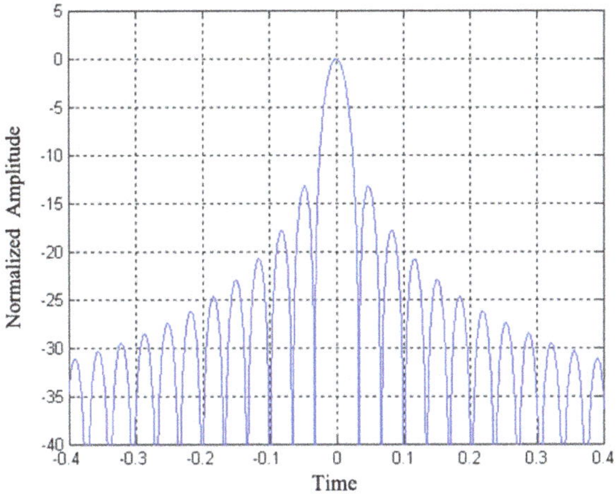

Fig. 4.12 The match filter output result of LFM signal

submerges the main lobe of the weak signal so that it cannot be distinguished. In order to improve the ability to distinguish multiple targets, it is needed to introduce weighting technology to suppress the side lobes, however at same time, it also causes loss of signal-to-noise ratio and the widening of the main lobe. Weighting can be carried out in amplitude or phase in the frequency domain or in amplitude or phase in the time domain. The general expression of the weighting function can be written as

$$H(f) = K + (1 - K)\cos^n\left(\frac{\pi f}{B}\right) \tag{4.34}$$

The different values of K and n can be different weighting function, with different performance indicators as shown in Table 4.1. Figure 4.13 shows the Taylor-weighting LFM-matched filter output waveform. The widening of the main lobe and the decline of side lobe can be seen in the figure.

4.2 Error Analysis of LFM Signal [2]

The error of LFM is divided into the frequency domain error which is introduced by not ideal for the amplitude–frequency characteristic of the system, and the time domain amplitude modulation and phase modulation error contained in the LFM signal itself. This section will focus on the amplitude and phase error analysis in the frequency domain, which can put forward the system allowed requirements for the amplitude and phase distortion range. At the same time, the orthogonal modulation error and frequency error are also analyzed in this section.

Table 4.1 Several commonly used weighting functions

Weighting functions	Maximum side lobe level	Loss of signal-to-noise ratio	−3 dB main lobe broadening factor	Side lobe decay rate dB/octave
Rectangle function $\mathrm{rect}\left(\frac{f}{B}\right)$	−13.6	0	1.00	4
Taylor weighting, $\bar{n} = 8$	−40	1.14	1.41	6
Hamming Weighted $0.08 + 0.92\cos^2\left(\frac{\pi f}{B}\right)$	−42.56	1.34	1.47	6
3: 1 cone ratio $0.33 + 0.67\cos^2\left(\frac{\pi f}{B}\right)$	−25.7	0.52	1.21	6
Hamming and 3:1 cone ratio series	−17.3	2.12	2.1	6
$\cos^2\left(\frac{\pi f}{B}\right)$	−31.7	1.76	1.62	18
$\cos^3\left(\frac{\pi f}{B}\right)$	−39	2.38	1.87	24

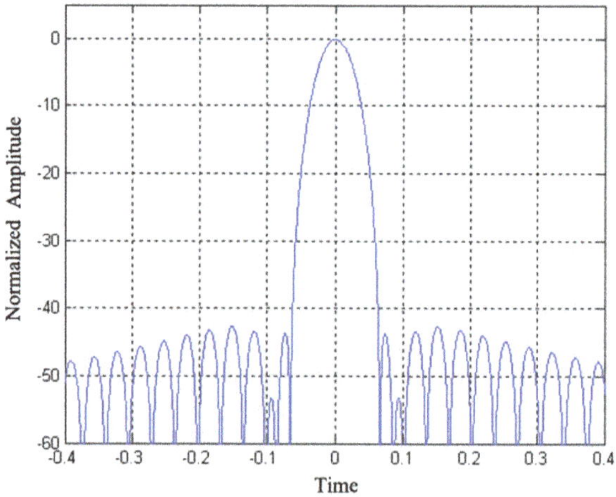

Fig. 4.13 The match filter output result after Hamming weighting

4.2.1 *Phase Error Analysis*

Deterministic phase errors can be divided into periodic and non-periodic. The non-periodic phase error can be roughly divided into first-order error, second-order error, and high-order error. The analysis of the phase difference of the LFM signal is carried out in the frequency domain. Assume that there is the phase error $\exp\{j\phi(f)\}$ in the frequency domain. The following will be discussed separately:

(1) **Periodic phase error**

Periodic phase error $\phi(f)$ can be developed into Fourier series:

$$\varphi(f) = a_0 - \sum_{n=1}^{N} a_n \sin(nc2\pi f) \tag{4.35}$$

In order to simplify the analysis, only the first two items are analyzed. The constant items a_0 do not affect on the compression, so only take the first harmonic.

$$\exp\{j\varphi(f)\} = \exp\{ja_1 \sin(c2\pi f)\} \tag{4.36}$$

The spectrum of the LFM signal containing the periodic phase error can be expressed as

$$U^e(f) = U(f) \cdot \exp\{ja_1 \sin(c2\pi f)\} \tag{4.37}$$

The output spectrum of the matched filter is

$$Y^e(f) = U^e(f) \cdot H(f) = Y(f) \cdot \exp\{ja_1 \sin(c2\pi f)\} \tag{4.38}$$

The output signal $y^e(t)$ of inverse Fourier transform is

$$y^e(t) = \int_{-\infty}^{\infty} Y(f) \cdot \exp\{ja_1 \sin(c2\pi f)\} \exp(j2\pi ft) df \tag{4.39}$$

Using the Bessel function, it can be developed into

$$\exp\{ja_1 \sin(c2\pi f)\} = J_0(a_1) + \sum_{m=1}^{\infty} J_m(a_1)[\exp\{jmc2\pi f\} + (-1)^m \exp\{-jmc2\pi f\}] \tag{4.40}$$

(3.39) can be transformed into

$$y^e(t) = J_0(a_1)y(t) + \sum_{m=1}^{\infty} J_m(a_1)[y(t+mc) + (-1)^m y(t-mc)] \tag{4.41}$$

It is shown that the phase distortion term will produce an infinite "paired echoes" in the matched filter output. When $a_1 < 0.5$ radian, it can be approximated as

$$y^e(t) = y(t) + \frac{a_1}{2}[y(t+c) - y(t-c)] \tag{4.42}$$

Then only a pair of echoes remaining. Their values are same, but in the opposite direction, the echo appears in the position $\pm c$. c has a time dimension. When $c = \frac{1}{B}$, the error signal fluctuates only one cycle within the signal bandwidth and the paired echo appears at the first zero crossing of the compressed main lobe. When $c < \frac{1}{B}$, it can be considered that the error becomes a non-periodic signal.

The above analysis is about a sine signal, which is an odd function with odd symmetry. If the error is cosine signal, let $\exp\{j\phi(f)\} = \exp\{ja_1 \sin(c2\pi f + \frac{\pi}{2})\}$. After a simple derivation, when $a_1 < 0.5$ radian, It can be obtained as

$$y^e(t) = y(t) + \frac{a_1}{2}[y(t+c) + y(t-c)] \tag{4.43}$$

The above equation shows that the paired echo caused by the even symmetry error about the signal center frequency is even symmetrical, and the paired echo caused by the odd symmetry error is odd symmetrical.

The simulation results are given in Figs. 4.14 and 4.15. Figure 4.14 shows the pulse compression results including periodic phase error (odd symmetry). It can be seen that because the paired echo has odd symmetry, adding to the main signal, the results are asymmetry. Figure 4.15 shows the pulse compression results including periodic phase error (even symmetry). It can be seen that adding to the main signal, the results are symmetry.

(2) **Non-periodic phase error**

The analysis of the non-periodic phase error can be used a series expansion method. Here it will expand $\varphi(f)$ into a Legendre polynomial, so that each

Fig. 4.14 The pulse compression results including periodic phase error (odd symmetry)

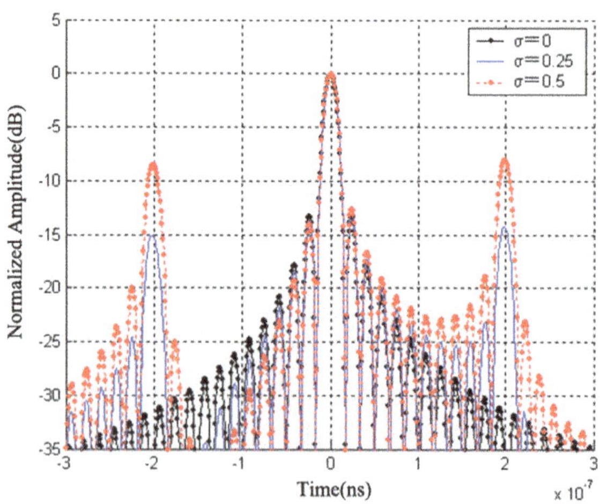

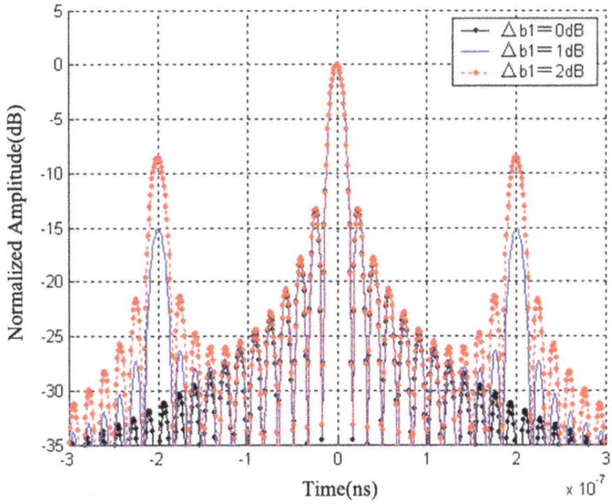

Fig. 4.15 The pulse compression results including periodic phase error (even symmetry)

component is orthogonal, to achieve the best fit. $\varphi(f) = \sum_{n=0}^{\infty} b_n P_n(f)$, where $P_n(f)$ is a normalized Legendre polynomial. Its first three items are

$$P_0(f) = \frac{1}{\sqrt{B}} \quad P_1(f) = \sqrt{\frac{3}{B}} \cdot \frac{2f}{B} \tag{4.44}$$

$$P_2(f) = \sqrt{\frac{5}{B}} \cdot \left(\frac{6f^2}{B^2} - \frac{1}{2}\right) \quad P_3(f) = \sqrt{\frac{7}{B}} \cdot \left(\frac{20f^3}{B^3} - \frac{3f}{B}\right) \tag{4.45}$$

According to the orthogonally of Legendre polynomials, its variance is $\sigma_n = \frac{b_n}{\sqrt{B}}$. Thus, the Legendre polynomial can be expressed as

$$\varphi(f) = \sum_{n=0}^{\infty} \sqrt{B}\sigma_n P_n(f) \tag{4.46}$$

Obviously, the fixed phase error does not affect the compression. Analysis the impact of first-order term, then first-order phase error term can be expressed as

$$\exp\{j\varphi(f)\} = \exp\left\{j\sigma_1 \cdot \frac{2\sqrt{3}f}{B}\right\} \tag{4.47}$$

The output signal of the matched filter is

$$y^e(t) = \mathrm{IFT}\left[Y(f)\exp\left\{j\sigma_1 \cdot \frac{2\sqrt{3}f}{B}\right\}\right] = y\left(t - \frac{1}{\pi} \cdot \sigma_1 \cdot \frac{\sqrt{3}}{B}\right) \tag{4.48}$$

It can be seen that the first-order phase error only causes the translation of the target position. It does not cause distortion in the case of single-point target, but will cause geometric distortion of the SAR image in the case of multiple point targets.

Analysis the impact of second-order term, ignoring the constant term, only second-order terms are considered.

$$\varphi_2(f) = \sqrt{B}\sigma_2 \cdot \sqrt{\frac{5}{B}} \cdot \left(\frac{6f^2}{B^2}\right) \tag{4.49}$$

Let $\frac{\pi}{2}a_1^2 = \sigma_2 \cdot \frac{6\sqrt{5}}{B^2}$, using Fresnel integral, we can obtain the following:

$$y^e(t) = \sum_{n=1}^{N} \mathrm{IFT}\left[Y(f)\exp\left\{j\frac{\pi}{2}a_1^2 f^2\right\}\right] = I_0 \cdot \{C(X_0) - C(X_1) + j[S(X_0) - S(X_1)]\} \tag{4.50}$$

where $I_0 = \exp\left\{-j\frac{\pi}{2} \cdot \frac{4(t-\tau_0)^2}{a_1^3}\right\}$ is a constant term

$$X_0 = a_1 \cdot \frac{B}{2} + \frac{2(t-\tau_0)}{a_1} \quad X_1 = -a_1 \cdot \frac{B}{2} + \frac{2(t-\tau_0)}{a_1} \tag{4.51}$$

The envelope of the matched output under the influence of the second-order phase error can be obtained by the equation, as shown in Fig. 4.16. It can be seen that the second-order phase error will cause the main lobe of the compression waveform to broaden, the resolution to decrease, the main lobe peak of the compression waveform to decrease, and the side lobe level to increase (mainly the peak side lobe level).

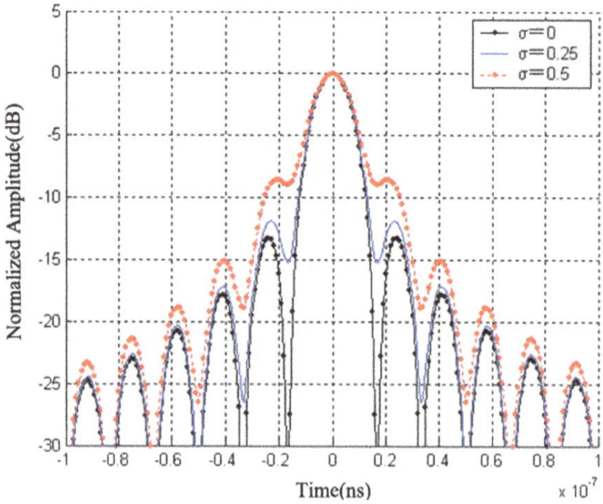

Fig. 4.16 Matched output results under the influence of the second-order phase error

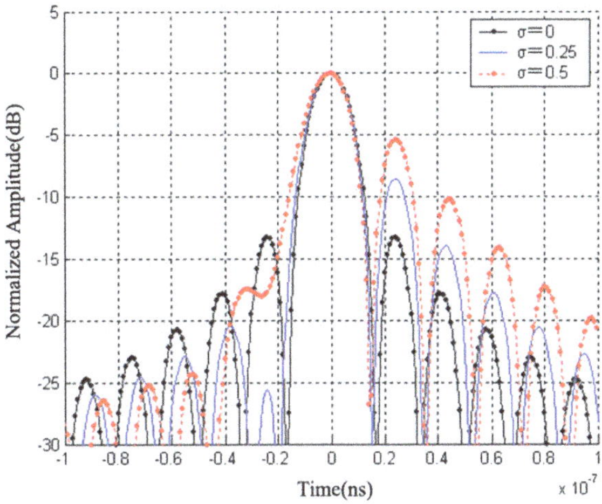

Fig. 4.17 Matched output results under the influence of the third-order phase error

Based on the above analysis method, the influence of the third-order phase errors can be analyzed.

$$\varphi_3(f) = \sqrt{B}\sigma_3 P_3(f) = \sigma_3\sqrt{7}\cdot\left(\frac{20f^3}{B^3} - \frac{3f}{B}\right) \tag{4.52}$$

Through the simulation, as shown in Fig. 4.17, it can be seen that the third-order phase error will cause the asymmetric distortion of the compression waveform, the right side lobe rises, the left side lobe declines, the main lobe will be widened to the left. In SAR, the rise in side lobe level will result in "ghost" of the image. In fact, the shape of the third-order phase error is similar to the one-cycle sine phase error of $c = \frac{1}{B}$, and the distortion of the matched output is also similar.

The matched output result with fourth-order phase error is shown in Fig. 4.18.

More than fourth-order phase errors have oscillating function graph. It can be seen as an approximation of the periodic phase error. It will mainly make the integral side lobe level rise; the impact on the main lobe can be ignored. Odd-order phase errors are odd symmetry, so they cause the side lobe asymmetry. Even-order phase errors are even symmetry, so the side lobes with even-order phase errors are symmetry. This is consistent with the results obtained above. The results are shown in Fig. 4.18.

4.2.2 *Amplitude Error Analysis*

The analysis of amplitude errors can also use the pair the echo theory. Similar to the phase error, the analyzed amplitude errors in frequency domain has the form of harmonic oscillation. It can be expressed as

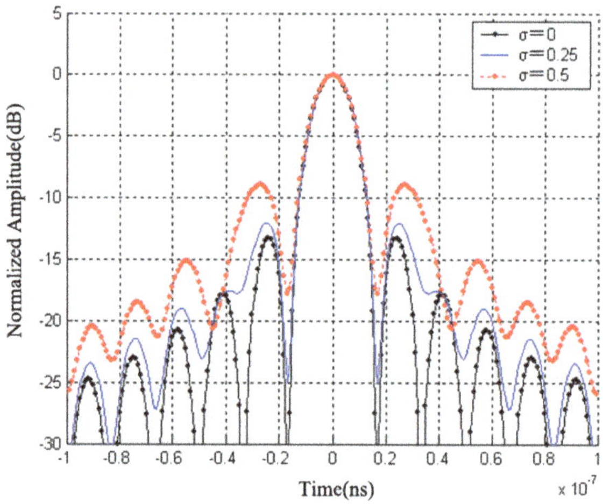

Fig. 4.18 Matched output results with fourth-order phase error

$$A(f) = b_1 \cos(c2\pi f) = \frac{b_1}{2}[\exp(jc2\pi f) + \exp(-jc2\pi f)] \tag{4.53}$$

The spectrum of the LFM signal with amplitude error can be written as

$$U^e(f) = U(f)[1 + A(f)] \tag{4.54}$$

The output spectrum of the matched filter can be written as

$$Y^e(f) = U^e(f) \cdot H(f) = Y(f) \cdot [1 + A(f)] \tag{4.55}$$

The output signal $y^e(t)$ can be obtained by inverse Fourier transform

$$y^e(t) = \int_{-\infty}^{\infty} Y(f) \cdot \left[1 + \frac{b_1}{2}[\exp(jc2\pi f) + \exp(-jc2\pi f)]\right] \exp(j2\pi ft)\mathrm{d}f \tag{4.56}$$

$$= y(t) + \frac{b_1}{2}[y(t+c) + y(t-c)] \tag{4.57}$$

The above formula proves that, similar to the effect of phase error, the amplitude error will also produce "paired echo" around the main lobe. The amplitude of the error fluctuation b_1 determines the amplitude of the pair of echoes, and the fluctuation frequency c determines the position where it appears. And regardless of the amplitude error on the signal, center frequency is odd symmetry or even symmetry, the paired echoes have same phase. Figure 4.19 shows the matched output results with amplitude error.

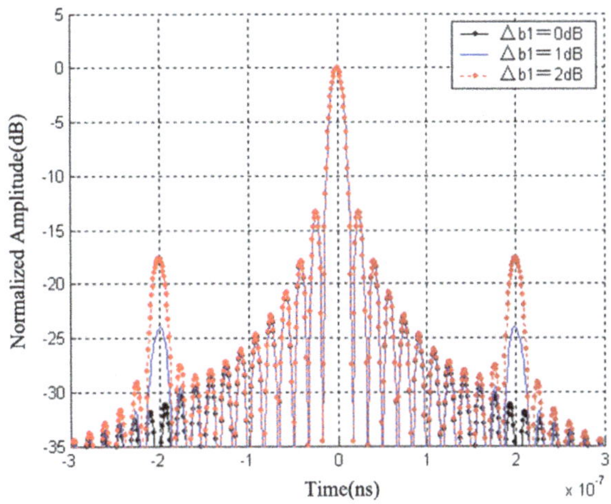

Fig. 4.19 Matched output results with amplitude error

4.2.3 *Analysis of Orthogonal Modulation and Demodulation Errors*

In practical applications, orthogonal modulation techniques are used to reduce the sampling frequency required by the digital baseband signals that are generated; the orthogonal demodulation technique preserves the phase information of the echo signal while reducing the data forming sampling rate. Two technologies are widely used in radar. So, it is necessary to study their impact on the pulse compression.

First, the error analysis of the orthogonal modulation is performed. The principle of orthogonal modulation is shown in Eq. (4.58). The I, Q channel baseband signals are, respectively, mixed with the orthogonal local oscillator signal, and then superimposed to obtain the modulated output intermediate frequency signal. Using the formula, it can be expressed as

$$\cos\left(\pi kt^2\right)\cos(2\pi f_0 t) - \sin\left(\pi kt^2\right)\sin(2\pi f_0 t) = \cos\left(2\pi f_0 t + \pi kt^2\right) \tag{4.58}$$

where

$$I = \cos\left(\pi kt^2\right) \quad Q = \sin\left(\pi kt^2\right) \tag{4.59}$$

are baseband signals. The analysis of the high-frequency narrow-band signal can be equivalent to the analysis of the signal complex envelope. The orthogonal modulation process analysis can also be equivalent to the analysis that the baseband signal I, Q synthesis baseband complex signal $U(t)$

$$\cos\left(\pi kt^2\right) + j\cdot\sin\left(\pi kt^2\right) = U(t) \tag{4.60}$$

(1) If I, Q channels contain a fixed phase error φ_{d}, then the phases of the two channels deviate from the orthogonal, there is

$$U^{e}(t) = \cos\left(\pi kt^{2} + \varphi_{\mathrm{d}}\right) + j\cdot\sin\left(\pi kt^{2}\right) \tag{4.61}$$

$$\begin{aligned} &= \cos\left(\pi kt^{2}\right)\cos(\varphi_{\mathrm{d}}) - \sin\left(\pi kt^{2}\right)\sin(\varphi_{\mathrm{d}}) + j\cdot\sin\left(\pi kt^{2}\right) \\ &= \cos\left(\pi kt^{2}\right)\left\{1 - 2\sin^{2}\left(\frac{\varphi_{\mathrm{d}}}{2}\right)\right\} - \sin\left(\pi kt^{2}\right)\sin(\varphi_{\mathrm{d}}) + j\cdot\sin\left(\pi kt^{2}\right) \\ &= U(t) - 2\sin\left(\frac{\varphi_{\mathrm{d}}}{2}\right)\sin\left(\frac{\varphi_{\mathrm{d}}}{2} - \pi kt^{2}\right) \end{aligned} \tag{4.62}$$

$$= U(t) - 2\sin\left(\frac{\varphi_{\mathrm{d}}}{2}\right)\left\{\frac{\exp\left(\frac{\pi}{2} - \frac{\varphi_{\mathrm{d}}}{2} + \pi kt^{2}\right) + \exp\left(-\frac{\pi}{2} + \frac{\varphi_{\mathrm{d}}}{2} - \pi kt^{2}\right)}{2}\right\} \tag{4.63}$$

$$= U(t) - \sin\left(\frac{\varphi_{\mathrm{d}}}{2}\right)\exp\left(\frac{\pi}{2} - \frac{\varphi_{\mathrm{d}}}{2}\right)\cdot U(t) - \sin\left(\frac{\varphi_{\mathrm{d}}}{2}\right)\exp\left(-\frac{\pi}{2} + \frac{\varphi_{\mathrm{d}}}{2}\right)\cdot U^{*}(t) \tag{4.64}$$

From Eq. (4.64), it can be seen that the baseband I, Q signals are not orthogonal which leads to the error in the orthogonally modulated signal. The error is divided into two parts: The first part is the original signal attached to a smaller amplitude and phase error $\sin\left(\frac{\varphi_{\mathrm{d}}}{2}\right)\exp\left(\frac{\pi}{2} - \frac{\varphi_{\mathrm{d}}}{2}\right)\cdot U(t)$. The other part is the image frequency components of the original signal $\sin\left(\frac{\varphi_{\mathrm{d}}}{2}\right)\exp\left(-\frac{\pi}{2} + \frac{\varphi_{\mathrm{d}}}{2}\right)\cdot U^{*}(t)$. Figure 4.20 shows the modulation of the signal envelope after orthogonal modulation. Figure 4.21 shows the image frequency components of single sideband LFM signal after orthogonal modulation caused by phase non-orthogonal.

Fig. 4.20 The envelope of modulation signal when I/Q phase is not orthogonal

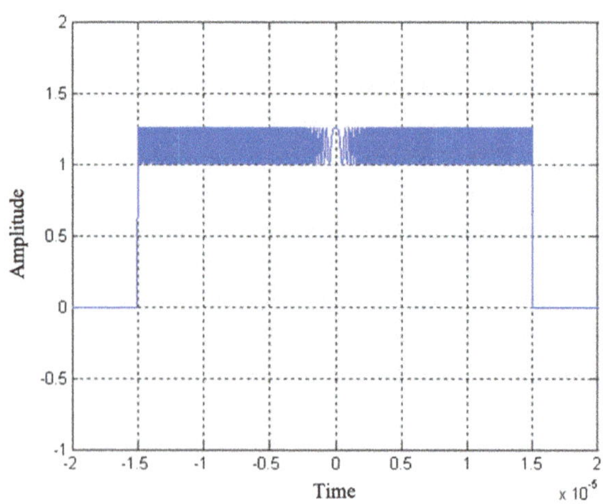

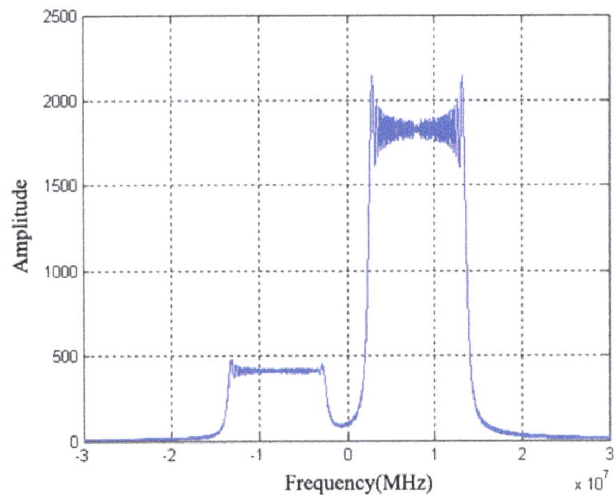

Fig. 4.21 The image frequency component caused by phase non-orthogonal

When φ_d is smaller, $0 < \varphi_d < \frac{\pi}{4}$ the amplitude of the error increases as φ_d increases.

Analyzing from the matched filtering, the first error is the fixed amplitude and phase error, the fixed phase error has no effect on pulse compression, the fixed amplitude error will result in the main lobe widening and side lobe descanting. The second error is the image frequency error. It is completely mismatched with the matched filter of the original signal $U(t)$; the energy cannot be gathered, and scattered throughout the entire signal bandwidth. So it can be judged that the influence of the image frequency component on the compressed main lobe is very small, and the main effect is to the integral side lobe. Because φ_d is usually very small, so the influence on the integral side lobes will be very small too.

It is analyzed in the following by the simulation, the simulation parameters are $B = 60$ MHz, $T = 30$ μs. The simulation results are plotted in Figs. 4.22 and 4.23.

The comparison of main lobe width, peak side lobe and integral side lobe when φ_d respectively is 0, $\frac{\pi}{8}$, $\frac{\pi}{4}$ is shown in Table 4.2.

It can be seen from the above table, the impact on the integral side lobe and the main lobe is very small when $0 < \varphi_d < \frac{\pi}{4}$. In the actual circuit, φ_d is generally about a few degrees, so the impact on pulse compression can be ignored.

(2) If the amplitudes of I, Q channels are inconsistent, the phase difference is ΔA dB, the linear amount of imbalance amplitude is α, which can be computed by

$$\alpha = 10^{(\Delta A/20)} - 1 \tag{4.65}$$

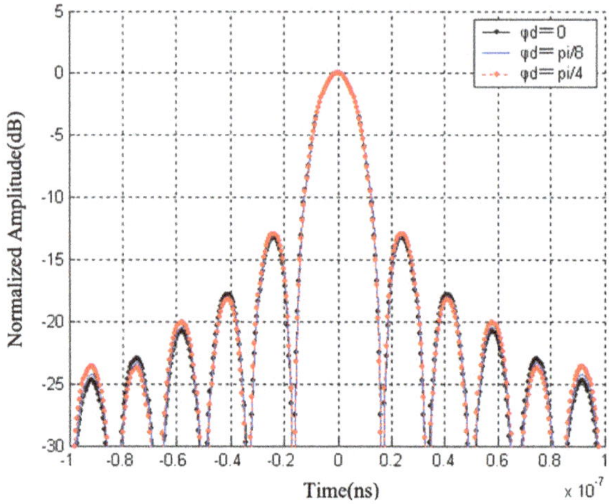

Fig. 4.22 The matched output results when I/Q channels are not orthogonal (main lobe)

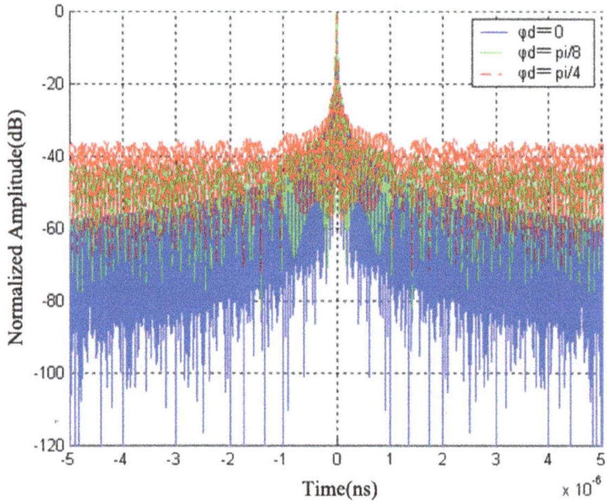

Fig. 4.23 The matched output results when I/Q channels are not orthogonal (side lobe)

Table 4.2 The comparison of pulse compression when the baseband I/Q phase is not orthogonal

Phase errors (rad)	Pulse compression parameters		
	Main lobe width ML (ns)	Peak side lobe PSL (dB)	Integral side lobe ISL (dB)
0	14.7	−13.2	−9.78
$\frac{\pi}{8}$	14.7	−13.1	−9.52
$\frac{\pi}{4}$	14.7	−12.9	−9.13

So the orthogonal modulated signal with error is

$$U^e(t) = (1+\alpha)\cos\left(\pi kt^2\right) + j\cdot \sin\left(\pi kt^2\right) \tag{4.66}$$

$$\begin{aligned} &= \cos\left(\pi kt^2\right) + \alpha\cos\left(\pi kt^2\right) + j\cdot\sin\left(\pi kt^2\right) \\ &= U(t) + \alpha\cos\left(\pi kt^2\right) = U(t) + \alpha\cdot\frac{U(t)+U^*(t)}{2} \\ &= U(t) + \frac{\alpha}{2}\cdot U(t) + \frac{\alpha}{2}\cdot U^*(t) \end{aligned} \tag{4.67}$$

As shown in the Eq. (4.67), the inconsistent amplitudes of the baseband I, Q channels after orthogonal modulation cause the errors which are also divided into two parts. The first part is the impact on the original signal amplitude, and the other part is the image frequency component. The first part has no effect on the matched filter results. The second part, as analyzed above, has little effect on the main lobe and will increase the integral side lobes. It is analyzed by the simulation too. The simulation parameters and results are shown in Figs. 4.24 and 4.25.

The comparison of main lobe width, peak side lobe, and integral side lobe when the inconsistent radio ΔA, respectively, is 0, 2, 4 dB is shown in Table 4.3. It can be seen that the amplitude inconsistency also affects the integral side lobe of the pulse pressure, which has a very small effect on the main lobe. Usually, the amplitude inconsistent degree is less than 2 dB, so the impact on the pulse compression can also be ignored.

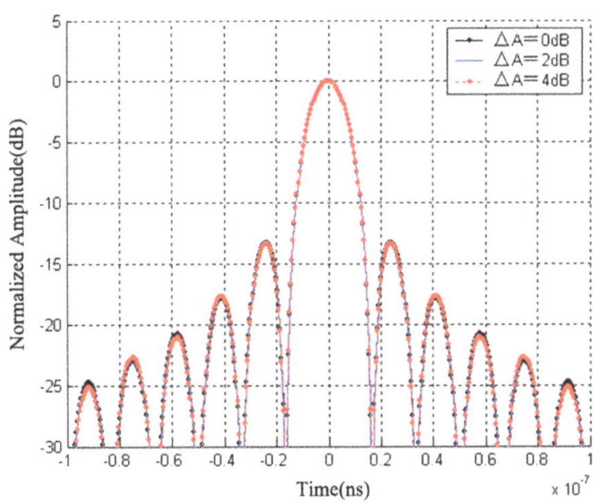

Fig. 4.24 The matched output results when I/Q channels are inconsistent (main lobe)

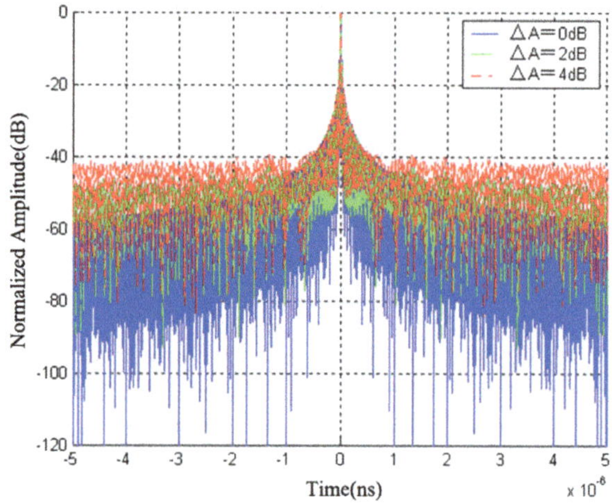

Fig. 4.25 The matched output results when I/Q channels are inconsistent (side lobe)

Table 4.3 The comparison of pulse compression when the baseband I/Q amplitude are inconsistent

Inconsistent radio (dB)	Pulse compression parameters		
	Main lobe width ML (ns)	Peak side lobe PSL (dB)	Integral side lobe ISL (dB)
0	14.7	−13.2	−9.78
2	14.7	−13.2	−9.70
4	14.7	−13.2	−9.63

(3) I or Q channels contain the direct current (DC) component

When I channel contains a DC signal, that is $I = \cos(\pi kt^2) + b$, then the signal after orthogonal modulation can be presented as

$$\left\{\cos\left(\pi kt^2\right) + b\right\}\cos(2\pi f_0 t) - \sin\left(\pi kt^2\right)\sin(2\pi f_0 t) \tag{4.68}$$

$$\begin{aligned} &= \cos\left(\pi kt^2\right)\cdot\cos(2\pi f_0 t) + b\cdot\cos(2\pi f_0 t) - \sin\left(\pi kt^2\right)\sin(2\pi f_0 t) \\ &= \cos\left(2\pi f_0 t + \pi kt^2\right) + b\cdot\cos(2\pi f_0 t) \end{aligned} \tag{4.69}$$

From Eq. (4.69), it can be seen that the DC component of the baseband is converted into carrier frequency leakage as after orthogonal modulation. And the envelope of the signal will also be affected. The carrier frequency leakage signal energy cannot be gathered through the matched filter of LFM signal, so the impact on the main lobe and integral side lobe is very small. The scattered energy will raise the integral side lobe. The results of the simulation analysis are shown in Figs. 4.26 and 4.27. The parameters are same with the above simulation.

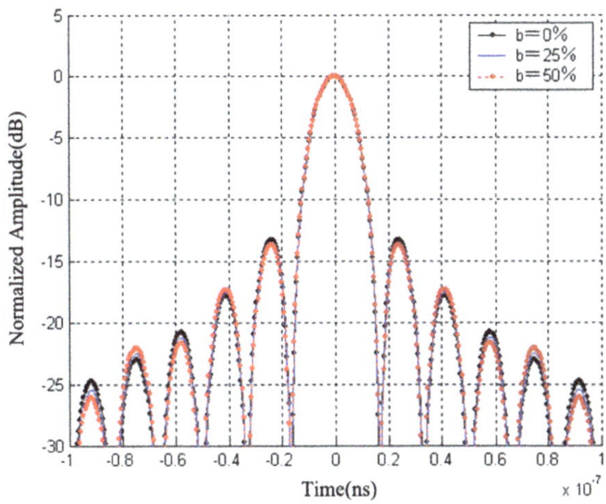

Fig. 4.26 The matched output results when the baseband signal contains the DC component (main lobe)

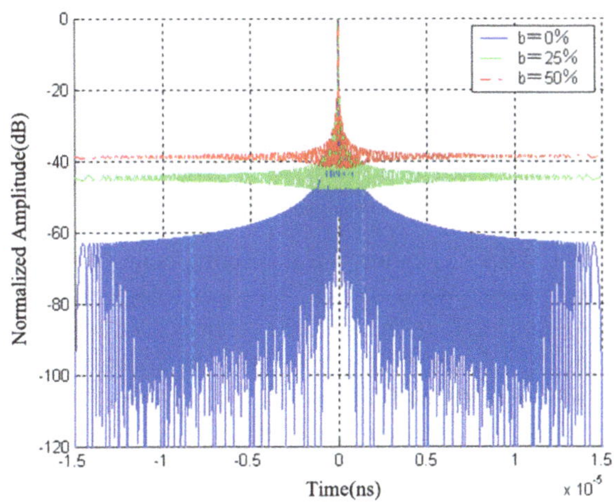

Fig. 4.27 The matched output results when the baseband signal contains the DC component (side lobe)

The comparison of main lobe width, peak side lobe, and integral side lobe when the ratio of the DC component b to the total signal amplitude in the baseband signal, respectively, is 0, 25, 50% is shown in Table 4.4.

In the table, it can be seen that, when the baseband signal contains a smaller DC component, it primarily affects the integral side lobe after the pulse compression, has a very small effect on the main lobe and the peak side lobes. In the actual circuit, the baseband signal DC component can be done <10%, so the impact on the pulse pressure will be very small.

In summary, through the above analysis, in the orthogonal modulation process, when the baseband I/Q signal phases are not orthogonal, amplitudes are inconsistent; it will cause the image frequency component and the amplitude changes of the original signal. When the baseband I or Q contains a DC component, it will cause

Table 4.4 The comparison of pulse compression when the baseband I/Q signals contain the DC component

Percentage of DC component (%)	Pulse compression parameters		
	Main lobe width ML (ns)	Peak side lobe PSL (dB)	Integral side lobe ISL (dB)
0	14.7	−13.2	−9.78
25	14.7	−13.4	−9.77
50	14.8	−13.6	−9.53

the carrier frequency leakage. These errors will mainly affect the integral side lobe of the pulse compression waveform and have little effect on the main lobe and the peak side lobes. When the error is small, the effect on the pulse compression results can be ignored.

Orthogonal demodulation is the inverse process of orthogonal modulation, the properties of the errors such as amplitude and phase inconsistent and the effect on the pulse compression results are the same as those of the orthogonal modulation.

4.3 Linear Frequency Modulation Continuous Wave Synthetic Aperture Radar (LFMCWSAR)

LFMCWSAR is a microwave imaging radar combined with Frequency Modulation Continuous Wave (FMCW) technology and Synthetic Aperture Radar (SAR) technology. As the receiver uses dechirp reception, the echo signal is mixed with the reference signal, resulting in a smaller difference frequency bandwidth, which can achieve high-resolution imaging at a lower sampling ratio, greatly reducing the data amount in the system, and reduce the requirements for video reception channels, A/D sampling, and signal processing speed.

FMCWSAR continuously transmits the signal in a repetition interval, and does not need a high peak power. SAR adopting the continuous wave system has attracted more and more attention because of its low power consumption, light weight, strong maintainability, and relatively low cost for development.

4.3.1 Frequency Modulation Continuous Wave (FMCW) Signal [3]

The frequency of the FMCW signal is a function of time, and the two widely used modulation forms are sawtooth and triangular wave modulation. If the transmitted signal is a sawtooth wave frequency modulation signal, the target echo is the duplicate wave of the transmitted waveform and its round trip echo delay is $\tau = 2R/$

c, where R is the range from radar to target, c is the speed of light, Δf is the scan frequency bandwidth, and PRI is the repetition period, that is, the time of scan frequency. For a static target, the instantaneous frequency difference of the received signal and the transmitted signal is a single frequency signal proportional to the range, and the difference frequency signal contains the range information.

$$f_b = \frac{\Delta f 2R}{PRIc} \tag{4.70}$$

The frequency of the difference frequency signal is proportional to the range from radar to target. To detect targets at different range, FFT transforms are usually used. When the target has a radial velocity relative to the radar, the received signal frequency will move up or down along the frequency axis, where the difference frequency signal contains the range information and the velocity information. For moving targets, range and time delay vary continuously via time. The range and velocity of the moving target can also be calculated using a two-dimensional FFT. The "fast time dimension" (the time in each sawtooth wave) gives the range information, and the "slow time dimension" (sawtooth wave number) gives the target speed information. The classical FMCW radar not only estimates the range to the target, the radial velocity of the target, but also the radial acceleration of the target.

In order to carry out the digital processing of the radar signal, the start and end of the sampled FMCW radar data are period for the transmitted signal sweep frequency. In order to avoid the signal frequency jump in the beginning, the signal sampling is delayed for a time, which is dependent on the maximum delay of the target echo.

4.3.2 Analysis of Frequency Modulation Continuous Wave SAR Signal [4, 5]

FMCWSAR operates in the orthogonal side strip map mode, the aircraft flying height is h, the platform speed is v, the antenna azimuth beam width is β, and the vertical distance from the scene center line to the route is $R_0 = \sqrt{h^2 + y^2}$.

Figure 4.28 shows the frequency relation of LFMCWSAR. Figure 4.28a shows the instantaneous frequency variation of transmission and reception. The dashed line is the instantaneous frequency of the transmitted signal, the solid line is the instantaneous frequency of the received signal, and the point line is the instantaneous frequency of the reference range echo. The transmitted signal is a sawtooth wave frequency modulation signal, the target echo is the delay signal of the transmitted signal, and the reference range is the vertical distance from the scene center line of the mapping swath to the route. T_{echo} is the round trip propagation delay corresponding to the mapping swath. PRI is the repetition period and T_{ad} is

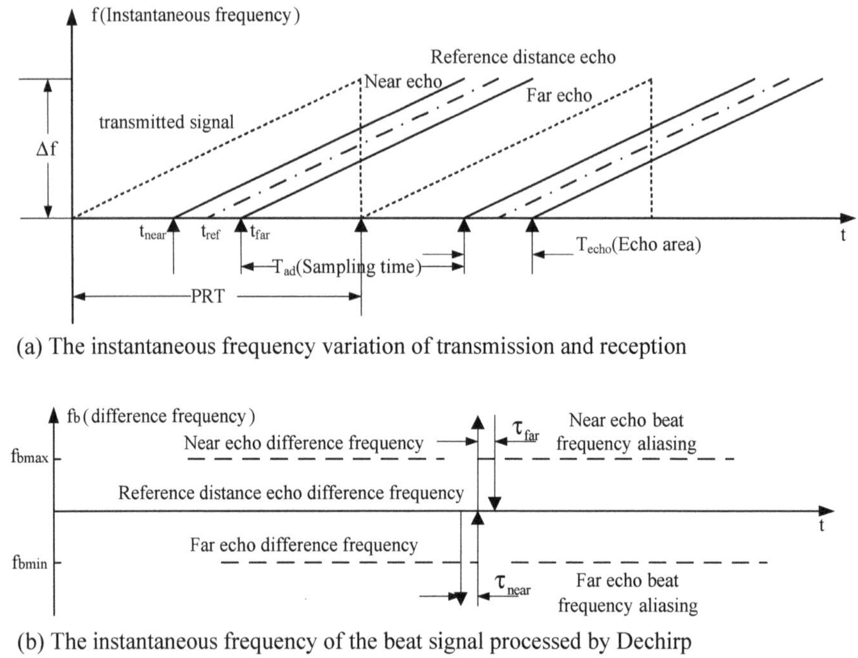

Fig. 4.28 The frequency relation of LFMCWSAR

the fast time length of the echo data at range AD sampling. When the sawtooth wave modulation is adopted, the time range of the echo data AD sampling should be selected after the echo reflected from furthest target arrives at the receiver, and before the echo reflected from nearest target arrives at the receiver.

Figure 4.28 shows the instantaneous frequency of the beat signal processed by dechirp. The maximum beat frequency of near echo is f_{bmax}. The minimum beat frequency of far echo is f_{bmin}. The beat echo in reference distance is DC. Except for the echo signal at the reference distance, the beat echo signal spectrums of the other target alias entirely. The duration times of the beat signal spectrum aliasing of the nearest and furthest targets is τ_{near} and τ_{far}, respectively, $T_{\text{echo}} = \tau_{\text{near}} + \tau_{\text{far}}$.

The transmitted signal of LFMCWSAR is

$$S_t(t) = \exp(j2\pi f_c t + j\pi k_r t^2) \tag{4.71}$$

where f_c is the carrier frequency, $k_r = \Delta f/\text{PRT}$ is the FM slope of LFM signal.

The echo at the slant range R can be expressed as

$$S_R = \exp\left[j2\pi f_c\left(t - \frac{2R}{c}\right) + j\pi k_r\left(t - \frac{2R}{c}\right)^2\right] \tag{4.72}$$

The reference signal processed by dechirp is

$$S_{\text{ref}}(t) = \exp\left[j2\pi f_{\text{c}}\left(t - \frac{2R_{\text{ref}}}{c}\right) + j\pi k_{\text{r}}\left(t - \frac{2R_{\text{ref}}}{c}\right)^2\right] \tag{4.73}$$

The difference frequency signal obtained by the complex conjugate multiplication between the received signal and the reference signal can be expressed as

$$\begin{aligned} S_{\text{rb}} = {} & A\text{rect}\left(\frac{t - 2R(t, t_{\text{m}})/c}{T_{\text{ad}}}\right)\exp\left(-\text{j}\frac{4\pi}{\lambda}R(t, t_{\text{m}})\right) \\ & \exp\left(-j\frac{4\pi}{c}k_r(R(t, t_{\text{m}}) - R_{\text{ref}})\right)\left(t - \frac{2R_{\text{ref}}}{c}\right) \\ & \exp\left(j\frac{4\pi}{c}k_r(R(t, t_{\text{m}}) - R_{\text{ref}})^2\right) \end{aligned} \tag{4.74}$$

where t is the range "fast" time, t_{m} is the azimuth to "slow" time. $R\ (t,\ t_{\text{m}})$ is the instantaneous distance from the target to the radar. R_{ref} is the reference distance.

In Eq. (4.74), the first exponential term is azimuth phase history, which determines the azimuth focus. The second exponential term is the phase corresponding to the Doppler shift term. Since it is first-order term of range fast time, it will cause the movement of the target echo envelope and affect the focusing effect. The third exponential term is the RVP term (residual video phase), which affects the azimuth focus.

After dechirp the echo signal of the LFMCW radar is sampled. The sampling is divided into range "fast" time sampling and azimuth "slow" time sampling, forming a two-dimensional data matrix.

In range dimension, LFMCWSAR has many restrictions; the minimum sawtooth wavelength is limited by the target maximum range. Usually, the sawtooth wavelength is 3–9 times larger than the maximum signal delay. It results in a relatively low sawtooth frequency modulation slope, which limits the maximum no ambiguity Doppler frequency (range velocity). The frequency modulation slope of LFM signal is $k_{\text{r}} = \Delta f/\text{PRT}$. The range resolution depends on the LFM width $\rho_r = c/2\Delta f$. The time range of the echo data AD sampling should be selected after the echo reflected from furthest target arrives at the receiver, and before the echo reflected from nearest target arrives at the receiver. AD sampling ratio should be greater than $2k_{\text{r}}R_{\text{swth}}/c = k_{\text{r}}T_{\text{echo}}$. Regardless of the transmitted LFM signal bandwidth, it is only related to the frequency modulation slope and the mapping swath width.

The value range of azimuth "slow" time sampling t_{m} is $T_{\text{s}}/2 \leq t_{\text{m}} \leq -T_{\text{s}}/2$. T_{s} is the synthetic aperture time.

(1) **The influence of the Doppler Effect**

In the LFMCWSAR, the Doppler frequency change caused by the radar motion cannot be neglected during the signal duration since the radar continuously transmits and receives the echo. Doppler effect will bring the second-order coupling of between the range and azimuth signal, resulting in the azimuth echo nonuniform sampling and echo envelope time shift, which will affect the image focusing in range dimension and azimuth dimension. The effect of Doppler frequency shift should be compensated.

The second-order coupling between the azimuth signal and range signal consists of the cross-coupling between the first- and second-order information in range dimension and the first-order information in azimuth dimension. It will cause the defocus of the range signal, and the target position offset in azimuth after range compression. Since there is a "fast" time offset along the range for the sampling time of each azimuth echo, for a point target, the sampling interval of the azimuth sampling point exhibits a nonuniform sampling change. If the signal is not interpolated in azimuth, the defocus phenomenon will appear after the azimuth compression, if the error caused by the Doppler Effect is not compensated, it will seriously affect the image quality, resulting in image degradation.

In the imaging process, it must eliminate the error caused by the Doppler effect and refocus the echo energy of the target. In LFMCWSAR, the azimuth "slow" time echo sampling signals are added a "fast" time echo delay in each echo cycle. The accurate compensation can be selected in the azimuth Doppler domain. In the echo time domain, the "fast" time offset is added for each "slow" time sampling period, which corresponds to a fixed frequency offset that changes with the range sampling point in the azimuth Doppler domain.

The original echo is transformed into the azimuth Doppler domain, and the error caused by the Doppler effect can be compensated by the complex multiplication operation. After the compensation, the point target can be refocused in range and azimuth, and the impact of defocus can be eliminated.

(2) **Range migration**

The coupling between range and azimuth causes the range migration. During a scan frequency cycle, the motion of the LFMCWSAR antenna phase center includes: the movement between the start time of scan frequency cycle and the arrival time of nearest point echo, the maximum slant range change amount $\Delta r_1 = 2R_{\text{near}}/cv$; the motion during the acquisition of echo data $\Delta r_2 = T_{\text{ad}}v$; the maximum slant range change amount $\Delta r_3 = (\tau_{\text{near}} + \tau_{\text{far}})v = T_{\text{echo}}v$ between the end time of echo data collection and the start time of the next repetition period.

For the target whose instantaneous slant range is $R(\tau, t_n, x_0, r_0)$, in the scan frequency cycle, the antenna continuous movement causes the Doppler shift, and the range cell migration caused by it is

$$\Delta R_{cm} = -\frac{c}{2k_r} f_d = -\frac{f_c v}{k_r} \frac{v t_n - x_0}{r(t_n, x_0, r_0)} \tag{4.75}$$

The range cell migration and carrier frequency caused by the Doppler effect are proportional to the sine of the instantaneous squint angle and inversely proportional to the frequency modulation slope. And the range cell migration correction can be implemented in the frequency domain.

(3) **Residual Video Phase (RVP) error**

Residual Video Phase error (RVP) is the unique phenomenon when the receiver adopts the dechirp. In most cases, the impact of RVP on imaging is relatively small, and in the same system bandwidth and resolution, the longer duration of the LFM signal and the smaller frequency modulation slope, the smaller effect of RVP on imaging. The maximum RVP of LFMCWSAR is

$$|\phi_{RVP}|_{max} = \frac{4\pi k_r}{c^2} \left(\sqrt{(R_0 + R_{swath}/2)^2 + L_S^2} - R_{ref} \right)^2 \tag{4.76}$$

$$|\phi_{RVP}|_{max} = \frac{4\pi k_r (R_0 - R_{ref})^2}{c^2} \left(1 + \frac{R_0}{R_0 - R_{ref}} \left(\sqrt{\left(1 + \frac{R_{swath}}{2R_0} \right)^2 + \frac{\lambda^2}{D^2}} - 1 \right) \right)^2 \tag{4.77}$$

RVP is related to the range resolution, the distance between the target and the mapping swath center, the width of the mapping swath and so on. The distance between the target and the mapping swath center is farther, the RVP is greater. The resolution is higher, the RVP is larger. If the mapping swath is wide, it should be considered to compensate for RVP.

In the frequency domain, the compensation function of RVP can be expressed as

$$H_{RVPC}(\omega) = \exp\left(-j \frac{\omega^2}{4\pi k_r} \right) \tag{4.78}$$

The echo data is multiplied by the compensation function of RVP in the range frequency domain after the range Fourier transforms. RVP is compensated by the RVP filter. And the echo data which have been compensated the RVP is obtained after the inverse Fourier transform.

4.3.3 Imaging Process of LFMCWSAR [3–6]

The imaging algorithms of strip map mode LFMCWSAR mainly include Range-Doppler Algorithm (RDA), Frequency Scaling Algorithm (FSA), Improved Frequency Scaling Algorithm (NFSA, DCDFSA), and so on.

Range-Doppler Algorithm (RDA) separates the range dimension and azimuth dimension to process. The range migration is corrected using the interpolation operation in the range-Doppler domain. Using the Fresnel approximation, it is neglected the second-order term and the higher order of the range-azimuth coupling phase. It is ignored the range curvature of the scene and the space-variant range effect.

In Frequency Scaling Algorithm (FSA), the range migration is divided into the consistent range migration and the space-variant range migration. The frequency scaling operation is used to correct the space-variant range migration. And then the consistent range migration is corrected in the range-Doppler domain. The second-order term and the higher order of the range-azimuth coupling phase are processed by the mapping swath center distance approximation method in FSA. The space-variant range term of the second and third-order range-azimuth coupling is ignored. No interpolation operation is required, only complex multiplication and FFT operations are required.

Figure 4.29 shows the flowchart of range-Doppler imaging algorithm for the strip map mode LFMCWSAR.

AD convert and cache: Complete the digitized sampling of the received dechirp signal and reduce the sampling quantization data rate.

Data pre-filter: Using FIR low-pass filtering and downsampling, reduce the system data rate, avoid azimuth Doppler spectrum aliasing, and reduce the processing burden.

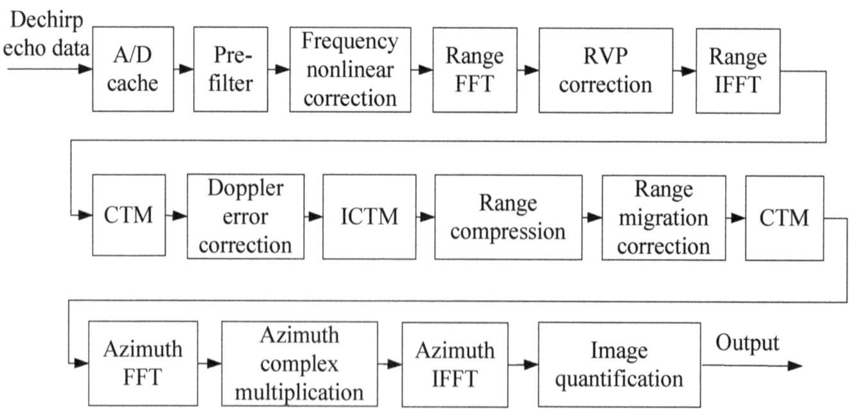

Fig. 4.29 The flowchart of range-Doppler imaging algorithm for the strip map mode LFMCWSAR

Frequency nonlinear correction: The repetition scan frequency period of LFMCWSAR is very long. For such a long time, it is difficult to ensure that the transmitted broadband LFM signal keeps the frequency completely linear. Frequency nonlinearity will cause range spectrum spreading, and affect range resolution and target positioning. It is possible to estimate the nonlinearity of the echo signal frequency directly from the received dechirp data and compensate it.

RVP correction:Echo data is multiplied by the residual video phase error compensation function in the range frequency domain after the range Fourier transforms. The RVP filter is used to compensate for the residual video phase error, and then the echo signal compensated residual video phase error is obtained after the inverse Fourier transform.

CTM: Corner Turning Memory. It converts the echo data from the range dimension to the azimuth dimension to process.

Doppler effect error correction: The echo data is transformed into the azimuth Doppler domain, and the error caused by the Doppler effect is compensated by the complex multiplication operation, so that the point target can be refocused in both range and azimuth dimensions.

ICTM: Inverse Corner Turning Memory. It converts the echo data from the azimuth dimension to the range dimension.

Range compression:The LFMCW echo signal is match filtered to obtain the narrow pulse of the range compression, and determine the instantaneous slant range between the target and all the synthetic aperture elements.

Range migration correction: It includes range walk and range curvature correction.

CTM: Corner Turning Memory. It converts the echo data from the range dimension to the azimuth dimension to process.

Azimuth migration correction: The echo data are complex multiplied by the azimuth reference function to remove the azimuth phase modulation of the echo signal, complete the azimuth compression, and determine the azimuth position of the target.

The two-dimension data after the range and azimuth compression further complete the Doppler effect error correction, the RVP correction, and the range migration correction. It is outputted after the image quantization. The strip map mode LFMCWSAR imaging processing is completed.

References

1. Lin M, Ke Y (1984) Radar signal theory. National Defense Industry Press, Beijing
2. Liang Y, Wang H, Xing M, Bao Z (2008) The analysis of FMCW SAR signal and image study. J Electron Inf Technol 30(5):1017–10021
3. deWit JJM, Meta A, Hoogeboom P (2006) Modified range-Doppler processing for FMCW synthetic aperture radar. IEEE Geosci Remote Sens Lett 3(1):83–87
4. Zhang J, Mao E (2005) Research on imaging processing of linear frequency modulation continuous wave SAR. Mod Radar 27(4):42–45

5. Dawei Z, Fang W, Yanfei W (2008) Analysis of the Doppler Effect in FMCW-SAR and its influence on the image of the system. J Electron Inf Technol 30(3):559–563
6. Ma B (2010) Research on LFMCW SAR imaging algorithm and real-time processing. Institute of Electronics, Chinese Academy of Sciences, Doctoral Thesis

Chapter 5
Stepped-Frequency Waveform and SAR Imaging

Modern radar is developing towards high sensitivity, strong anti-interference ability, high flexible and good altitude performance, and the high range resolution is one of the key technologies. The high range resolution requires a large system bandwidth, and the increase of instantaneous bandwidth will improve the hardware requirements. Under the restriction of the existing hardware, the stepped-frequency waveform is a practical, convenient and flexible signal form with high range resolution.

Stepped-frequency waveform achieves large bandwidth by sequentially changing carrier frequency over pulses in one burst, which significantly reduces the instantaneous bandwidth and the receiver hardware requirements, and avoids some practical design problems of pulse compression waveform. In addition, it can skip those frequencies of external interference with anti-jamming [1], such as FM radio and mobile communications.

5.1 The Characteristics of Stepped-Frequency Waveform in Time Domain and Frequency Domain

Any signal can be described as a function of time or frequency. The spectrum can be obtained from the time domain waveform by Fourier transform that expresses the relationship among amplitude, phase and frequency. The measurements of the target echo are equivalent in both time domain and frequency domain, which can be used to determine the reflectivity of the target. The reflectivity refers to the amplitude and phase of the echo under a given set of radar parameters and incident angles.

In fact, high resolution imaging means that the radar can accurately measure the impulse response of the target. When the radar transmits impulse signal $\delta(t)$ towards the target whose linear transfer function is $h(t)$, the impulse response of the

R. Yang et al., *High-Resolution Microwave Imaging*,
https://doi.org/10.1007/978-981-10-7138-6_5

linear system is expressed as $h(t) = \int_{-\infty}^{+\infty} h(\tau)\delta(t-\tau)\,\mathrm{d}\tau$. In the practical measurement, it is impossible to obtain the ideal impulse function [2] in the whole frequency domain, but only in a certain frequency scope. The frequency scope can be defined by

$$W(f) = \begin{cases} 1 & |f \pm f_0| < B_w/2 \\ 0 & \text{else} \end{cases} \tag{5.1}$$

The target response in time domain is

$$\begin{aligned} h(t) &= \int_{-\infty}^{+\infty} H(f)W(f)\mathrm{e}^{j2\pi ft}\mathrm{d}f \\ &= 2B\frac{\sin(\pi B_w t)}{\pi B_w t}\cos(2\pi f_0 t) \end{aligned} \tag{5.2}$$

The distance between the peak of the main lobe and the first zero is $1/B_w$, and then the normal range resolution is $\Delta r = c/2B_w$. The range resolution is inversely proportional to the signal bandwidth. As the signal is weighted in frequency domain, the main lobe of target response is expanded in time domain, and the resolution is degraded. Since the frequency scope $W(f)$ contains infinite number of frequency components, only a number of discrete frequencies can be measured in the actual system, so the frequency scope has to be discretized. The target is usually measured in some discrete narrow bands in turn. Although the resolution of narrowband signal is low, but the coherent synthesis of all narrow bands can provide the wide frequency scope $W(f)$ for the specified range resolution.

The most direct measurement of the impulse response of a radar target is to transmit an impulse $\delta(t)$, Which can be decomposed into infinite sine wave signals with different frequencies and same amplitude,and superposed upon all this signals. When an infinite number of sine waves are transmitted at the same time, then the infinite echoes added are equivalent to the impulse $\delta(t)$. It is impractical to transmit infinite sine wave signals simultaneously by infinite transmitters. If the impulse response $h(t)$ of the target is time invariant, then the infinite sinusoidal signals with different frequencies can be transmitted in a time sequence, so the physical hardware can be further extended to realization. Sinusoidal signal in mathematics is infinite long signal in time domain, so that the signal can't be transmitted in different time. Each Sinusoidal signal can be truncated by the time window, and a series of infinite truncated sine pulses in time domain can be obtained.

Because of the infinite frequency, it is still impossible realize physically, so the frequency dispersion is needed in deed. The frequency in a certain scope can be discretized to a series of pulses with equal frequency interval. By means of the truncation in time domain and sampling in frequency domain for the impulse $\delta(t)$ signal, the stepped-frequency signal can be realized in engineering. After the coherent processing, the measurement results can be equivalent to that of the impulse $\delta(t)$, which means wideband signal can be synthesized from the narrow-band sub-pulses [3].

Let the center frequency is f_0, and the bandwidth is B_w. The infinite long sinusoidal signal can be truncated by the time window with width T_w. Because of each time window contains infinite frequency components, so the frequency must be discrete to a burst of single frequencies with the interval Δf. Suppose that the number of discrete sub-pulses is N ($N = B_w/\Delta f$), the duration of sub-pulse is T. If the N sub-pulses are transmitted in sequence with the pulse repetition period T_r, the transmitting sub-pulses are expressed as

$$s_i(t) = \begin{cases} u(t)A_i e^{j(2\pi f_i t + \theta_i)} & iT_r - \frac{T}{2} \le t \le iT_r + \frac{T}{2} \\ 0 & \text{else} \end{cases} \tag{5.3}$$

where $u(t)$ is the envelope of sub-pulse, A_i and θ_i are the initial amplitude and phase respectively, T_r is the pulse repetition period, T is the duration of sub-pulse. f_i is the carrier frequency of each sub-pulse, it is defined as

$$f_i = f_0 + i\Delta f, \quad i = 0, 1, 2, \ldots, N-1 \tag{5.4}$$

f_0 is the initial carrier frequency, Δf is step size of carrier frequency, i is the index of sub-pulses. Equations (5.3) and (5.4) are the expression of stepped-frequency waveform.

The simulation is provided to verify the feasibility of stepped-frequency coherent synthesis of wideband signals.

The main parameters are given as follows: The initial frequency f_0 is 2 GHz, the bandwidth B_w is 16 MHz, the step size of carrier frequency Δf is 1 MHz, the number of sub-pulses N is 16, the pulse duration T is 1 μs, and the pulse repetition period T_r is 10 μs. The amplitude spectrum of each sub-pulse of stepped-frequency waveform is shown in Fig. 5.1.

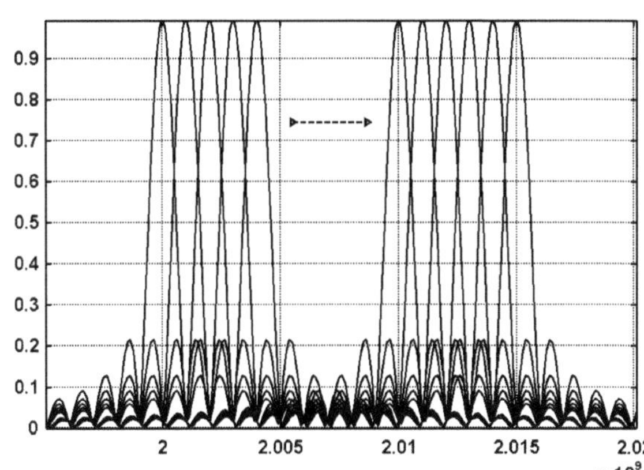

Fig. 5.1 The amplitude spectrum of each sub-pulse of stepped-frequency waveform

It can be seen from Fig. 5.1 that the spectrum of each sub-pulse is the sinc function, due to the sine wave is truncated by a time window, and the peak main lobe width is the reciprocal of time window $B = 1/T$, as a result, the time domain windowing causes the main lobe broadening in the frequency domain. Only frequency interval of two adjacent sub-pulses is $\Delta f \leq 1/T$, the synthetic spectrum from N sub-pulses can be approximated as a flat rectangular wideband spectrum. If the frequency interval of two adjacent sub-pulses equals to $\Delta f = 1/T$, the synthetic spectrum is illustrated by Fig. 5.2.

It can be seen that the 16 MHz spectrum is approximately synthesized by 16 sub-pulses with 1 MHz bandwidth of stepped-frequency. The simulation verifies feasibility of stepped-frequency coherent synthesis of wideband signals.

In summary, the stepped-frequency signal can be regarded as the discrete form of wideband LFM signal, and the equivalent wideband signal can be obtained under the condition of reducing the sampling rate of A/D. The principle of high resolution range profile of stepped-frequency will be discussed in the following.

5.2 The Ambiguity Function of Stepped-Frequency Signal [4]

Since Woodward put forward the famous radar ambiguity theory in 1962, ambiguity function has become an effective tool for signal analysis and waveform design. Under the premise of the optimal signal processing, the ambiguity function can reveal the potential performance of the system, such as resolution, ambiguity, precision and clutter-suppression. The ambiguity function may be the most complete description of the natural characteristics, which reflects the range-Doppler position of the ambiguity response, and explains the range and the Doppler resolution. In the following, the ambiguity function of the stepped-frequency signal is deduced and used to analyze the performance of the waveform.

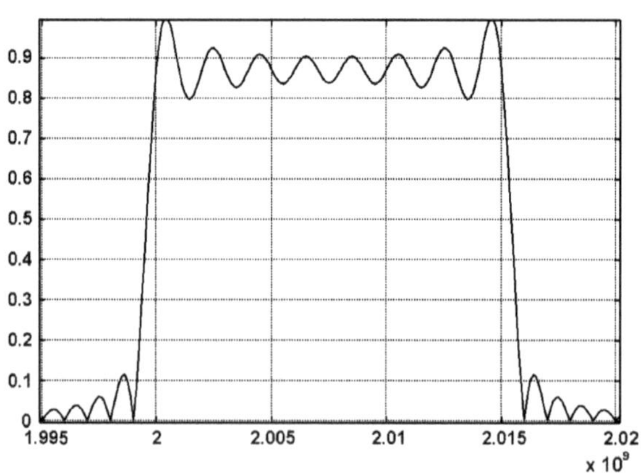

Fig. 5.2 The synthetic amplitude spectrum of N sub-pulses

5.2.1 The Ambiguity Function and Resolution Performance

The transmitting stepped-frequency signal is given by

$$s(t) = \sum_{n=0}^{N-1} u(t - nTr)e^{j2\pi f_n t} \tag{5.5}$$

where

$$u(t - nTr) = \begin{cases} 1/\sqrt{T} & nT_r \le t \le nT_r + T \\ 0 & \text{else} \end{cases}$$

$$f_n = f_0 + n\Delta f, \quad n = 0, 1, 2, \ldots, N - 1$$

$u(t)$ is the envelope of sub-pulse, f_0 is the initial carrier frequency, Δf is step size of carrier frequency, n is the index of sub-pulses, f_n is the carrier frequency of each sub-pulse (Fig. 5.3).

The transmitting signal in Eq. (5.5) can be rewritten as

$$s(t) = \sum_{n=0}^{N-1} u(t - nTr)\exp(j2\pi n\Delta f t)\exp(j2\pi f_0 t), \quad nTr < t < nTr + T \tag{5.6}$$

The ambiguity function is defined as

$$\begin{aligned}\chi(\tau, \xi) = &\int_{-\infty}^{+\infty} \sum_{n=0}^{N-1} u(t - nTr)\exp(j2\pi n\Delta f t) \\ &\cdot \sum_{m=0}^{N-1} u^*(t - mTr + \tau)\exp^*[j2\pi m\Delta f(t + \tau)]\exp(j2\pi\xi t)\mathrm{d}t\end{aligned} \tag{5.7}$$

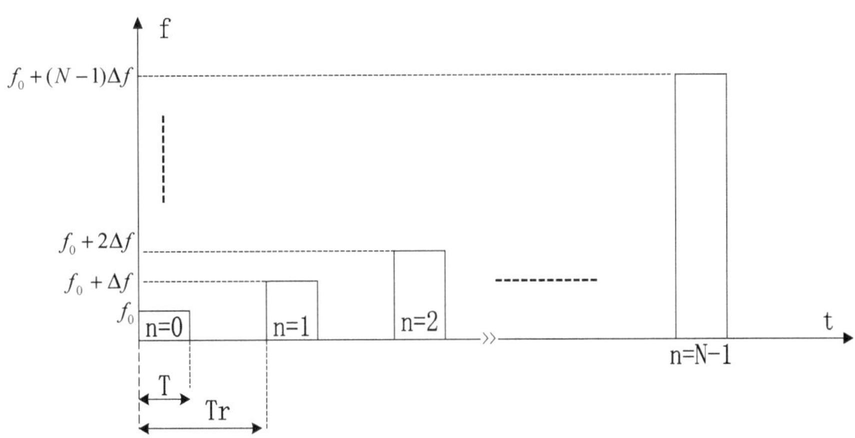

Fig. 5.3 Time-frequency diagram of stepped-frequency signal

The Eq. (5.7) can be further simplified as

$$\begin{aligned} |\chi(\tau,\xi)| = &\left|\sum_{n=0}^{N-1}\sum_{n=0}^{N-1}\exp(-j2\pi m\Delta f\tau)\exp[j2\pi(\xi-(m-n)\Delta f)nTr]\right. \\ &\left.\cdot\chi_0(\tau-(m-n)Tr,\xi-(m-n)\Delta f)\right|,\quad |\tau-(m-n)Tr|<T \end{aligned} \tag{5.8}$$

where $|\chi_0(\tau,\xi)|$ is the ambiguity function of single frequency pulse signal with rectangular envelope, it is given by

$$\chi_0(\tau,\xi) = e^{j\pi\xi(T-\tau)}\frac{\sin\pi\xi(T-|\tau|)}{\pi\xi(T-|\tau|)}\cdot\frac{T-|\tau|}{T},\quad |\tau|<T \tag{5.9}$$

Let $p = m - n$, then the Eq. (5.9) can be rewritten as

$$\sum_{n=0}^{N-1}\sum_{m=0}^{N-1}f(x,y) = \sum_{p=-(N-1)}^{0}\sum_{m=0}^{N-1-|p|}f(x,y) + \sum_{p=1}^{N-1}\sum_{n=0}^{N-1-|p|}f(x,y) \tag{5.10}$$

The Eq. (5.10) is expressed as

$$\begin{aligned} |\chi(\tau,\xi)| = &\left|\sum_{p=-(N-1)}^{0}\sum_{m=0}^{N-1-|p|}e^{-j2\pi(\xi-p\Delta f)pT_r}e^{j2\pi m[(\xi-p\Delta f)Tr-\Delta f\tau]}\chi_0[\tau-pT_r,\xi-p\Delta f]\right. \\ &\left.+\sum_{p=1}^{N-1}\sum_{n=0}^{N-1-|p|}e^{-j2\pi p\Delta f\tau}e^{j2\pi n[(\xi-p\Delta f)Tr-\Delta f\tau]}\chi_0[\tau-pT_r,\xi-p\Delta f]\right|,\quad |\tau-pTr|<T \end{aligned} \tag{5.11}$$

It can be seen from the Eq. (5.10) that the ambiguity function of stepped-frequency can be regarded as a weighted superposition of $N^2\chi_0(\tau,\xi)$ with the time shift step $(m-n)T_r$ and frequency shift step $(m-n)\Delta f$.

In terms of the formula of finite series,

$$\sum_{i=0}^{N-1}e^{-j2\pi i\beta\alpha} = e^{-j(N-1)\beta\alpha}\frac{\sin N\pi\beta\alpha}{\sin\beta\alpha} \tag{5.12}$$

Using Eq. (5.12) to express Eq. (5.7), the ambiguity function of stepped-frequency is defined as

$$\chi(\tau,\xi) = \sum_{p=-(N-1)}^{N-1}e^{j\pi[p^2\Delta fT_r-p\Delta f\tau-(N-1)p\Delta fT_r]}e^{j\pi(N-1)(\xi T_r-\Delta f\tau)}e^{j\pi\xi(T-\tau)}A(\tau,\xi)B(\tau,\xi) \tag{5.13}$$

where

$$A(\tau,\xi) = \frac{\sin\pi(\xi - p\Delta f)(T - |\tau - pT_r|)}{\pi(\xi - p\Delta f)(T - |\tau - pT_r|)} \cdot \frac{T - |\tau - pT_r|}{T}, \quad |\tau - pT_r| < T$$

$$B(\tau,\xi) = \frac{\sin\pi(N - |p|)[(\xi - p\Delta f)T_r - \Delta f\tau]}{\sin\pi[(\xi - p\Delta f)T_r - \Delta f\tau]}, \quad |\tau - pT_r| < T$$

The Eq. (5.13) is the expression of ambiguity function of stepped-frequency signal.

For any p, the ambiguity function is

$$\begin{aligned}|\chi_p(\tau,\xi)| = & \left|\frac{\sin\pi(\xi - p\Delta f)(T - |\tau - pT_r|)}{\pi(\xi - p\Delta f)(T - |\tau - pT_r|)}\right| \cdot \left|\frac{T - |\tau - pT_r|}{T}\right| \\ & \cdot \left|\frac{\sin\pi(N - |p|)[(\xi - p\Delta f)T_r - \Delta f\tau]}{\sin\pi[(\xi - p\Delta f)T_r - \Delta f\tau]}\right|, \quad |\tau - pT_r| < T\end{aligned} \tag{5.14}$$

If $\xi = 0$, the ambiguity function is simplified as:

$$\begin{aligned}|\chi_p(\tau,0)| = & \left|\frac{\sin\pi p\Delta f(T - |\tau - pT_r|)}{\pi p\Delta f(T - |\tau - pT_r|)}\right| \cdot \left|\frac{T - |\tau - pT_r|}{T}\right| \\ & \cdot \left|\frac{\sin\pi(N - |p|)\Delta f(\tau + pT_r)}{\sin\pi\Delta f(\tau + pT_r)}\right|, \quad |\tau - pT_r| < T\end{aligned} \tag{5.15}$$

It can be known from above analysis, for any p, the peak position of range ambiguity centers around pT_r, the time delay is limited between $-T + pT_r$ and $pT_r + T$. When the ambiguity function crosses zero for the first time, the time delay can be used to define the normal range resolution.

$$\rho_\tau = \min\left[\frac{1}{p\Delta f}, \frac{1}{(N - |p|)\Delta f}\right] \tag{5.16}$$

If $\tau = 0$, the ambiguity function of frequency is simplified as:

$$\begin{aligned}|\chi_p(0,\xi)| = & \left|\frac{\sin\pi(T - |pT_r|)(\xi - p\Delta f)}{\pi(T - |pT_r|)(\xi - p\Delta f)}\right| \cdot \left|\frac{T - |pT_r|}{T}\right| \\ & \cdot \left|\frac{\sin\pi(N - |p|)T_r(\xi - p\Delta f)}{\sin\pi T_r(\xi - p\Delta f)]}\right|, \quad |\xi - p\Delta f| < \frac{\Delta f}{T_r}T\end{aligned} \tag{5.17}$$

The peak position of Doppler ambiguity centers around $p\Delta f$ for any p, the Doppler frequency changes between $\Delta f/T_r \cdot T + p\Delta f$ and $pT_r + Tp\Delta f + \Delta f/T_r \cdot T$. When $T_r > T$ and $-(N-1) \le p \le N-1$, then $(N - |p|)T_r > T - |pT_r|$, the normal resolution of Doppler frequency main lobe is defined as

$$\rho_{\xi} = \frac{1}{(N - |p|)T_r} \tag{5.18}$$

Conclusion:

(1) Due to the periodicity of the signal, the ambiguity function presents periodic arrangement in the time delay and Doppler shift plane, the time delay ambiguity interval is equal to the pulse repetition period, and the Doppler ambiguity interval is equal to the frequency step increment.
(2) For any response with nail-bed shape ambiguity, the range is limited between $-T + pT_r$ and $pT_r + T$, and the normal range resolution is minimum between $1/p\Delta f$ and $1/(N - |p|)\Delta f$. The Doppler frequency changes between $\Delta f/T_r \cdot T + p\Delta f$ and $pT_r + Tp\Delta f + \Delta f/T_r \cdot T$, and the normal resolution of Doppler frequency main lobe is defined as $1/(N - |p|)T_r$.

5.2.2 *Ambiguity Function of Central Ambiguity Region*

In the following, the ambiguity function of the central ambiguity region is used as an example to illustrate the resolution performance of stepped-frequency signal. For the central ambiguity band, if $p = 0$, the ambiguity functions is simplified as:

$$|\chi_c(\tau,\xi)| = \left|\frac{\sin \pi N(T_r\xi - \Delta f\tau)}{\sin \pi(T_r\xi - \Delta f\tau)}\right| \cdot \left|\frac{\sin \pi\xi(T - |\tau|)}{\pi\xi(T - |\tau|)}\right| \cdot \left|\frac{T - |\tau|}{T}\right|, \quad |\tau| < T \tag{5.19}$$

If $\xi = 0$, the ambiguity function is simplified as:

$$|\chi_c(\tau,0)| = \left|\frac{\sin \pi N\Delta f\tau}{\sin \pi\Delta f\tau}\right| \cdot \left|\frac{T - |\tau|}{T}\right|, \quad |\tau| < T \tag{5.20}$$

Eq (5.20) shows that the ambiguity function in the central ambiguity region has the sinc shape, which decides the range resolution. The peak position of range ambiguity is $\tau = 0$, and the peak value is 1. When $|\tau| = 1/N\Delta f$, the ambiguity function crosses zero for the first time. The main-lobe width is $2/N\Delta f$, and the normal resolution is $\rho_\tau = 1/N\Delta f$.

If $\tau = 0$, then Eq. (5.19), the velocity ambiguity function can be rewritten as

$$|\chi_c(0,\xi)| = \left|\frac{\sin \pi NT_r\xi}{\sin \pi T_r\xi}\right| \cdot \left|\frac{\sin \pi\xi T}{\pi\xi T}\right|, \quad |\xi| < \frac{\Delta f}{T_r}T \tag{5.21}$$

The peak position of velocity ambiguity function is $\xi = 0$. Because of $T_r > T$, $1/NT_r < 1/T$ the first zero point of velocity ambiguity function is $|\xi| = 1/NT_r$, so the normal velocity resolution is $\rho_\xi = 1/NT_r$, which is defined as main lobe width with half power position.

Simulation parameters: step size of frequency $\Delta f = 20\,\mathrm{Hz}$, number of sub-pulses $N = 5$, synthetized bandwidth $B = 100\,\mathrm{Hz}$, duration of sub-pulses $T = 0.01$, pulse period time $T_r = 0.05$. The ambiguity function figure is shown in Fig. 5.4; time-frequency domain projection figure is shown in Fig. 5.5, range and Doppler ambiguity figure at peak, profile of central ambiguity region are all shown in Figs. 5.6 and 5.7.

Figure 5.4 shows that range ambiguity of stepped-frequency has nail-bed shape, and the ambiguity function presents the regularly arranged spikes in the time-delay and frequency-shift plane. It can be seen from the time-frequency domain projection in Fig. 5.5 that the time delay $2NT_r$ is 0.5 s, frequency shift $2N\Delta f$ is 200 Hz,

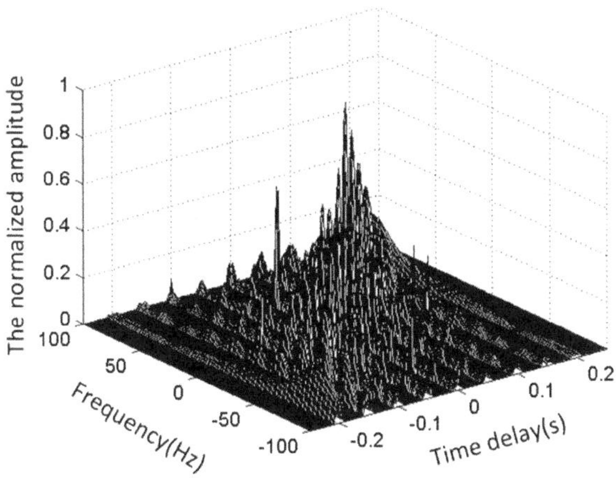

Fig. 5.4 The ambiguity function of stepped-frequency waveform

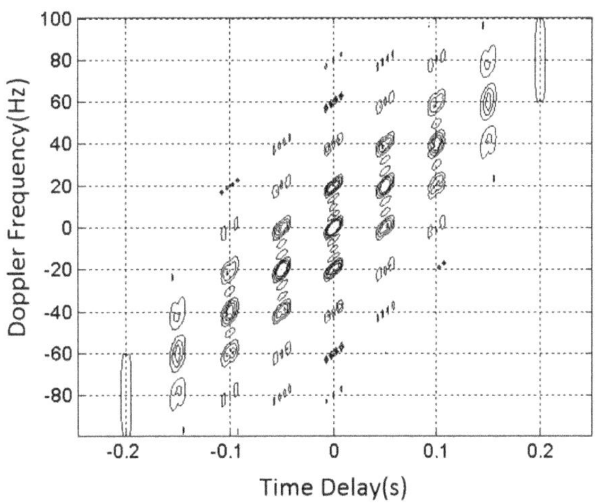

Fig. 5.5 Time-frequency domain projection of ambiguity function

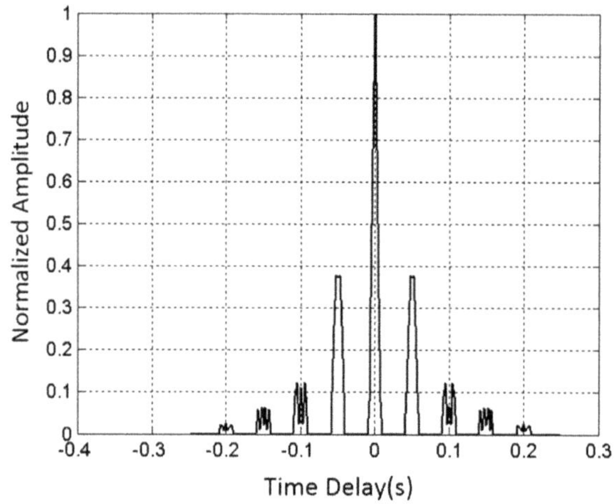

Fig. 5.6 Doppler ambiguity at peak

Fig. 5.7 Range ambiguity at peak

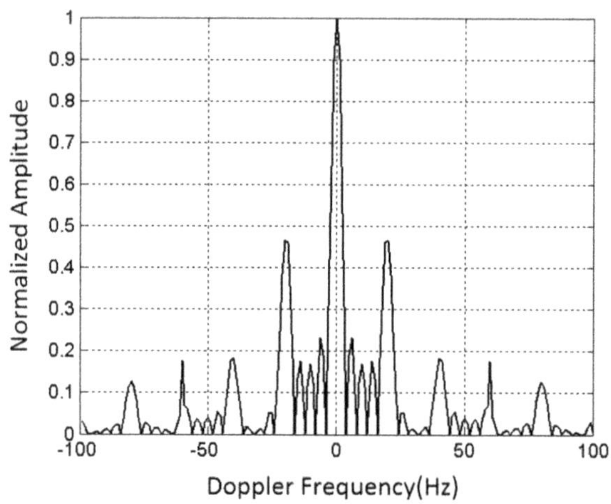

and the chirp rate $\Delta f/T_r$ is 400 Hz/s. The ambiguity function figure of stepped-frequency can be looked as that the ambiguity function figure of single frequency pulses burst rotated according to LFM slope $\Delta f/T_r$. Therefore, stepped-frequency waveform has the range-Doppler coupling.

The range ambiguity figure at peak shows that the time-delay ambiguity interval in different ambiguity centers just right equals to pulse repetition period $T_r = 0.05\,\text{s}$, and the Doppler ambiguity figure at peak shows that the Doppler-frequency ambiguity interval in different ambiguity centers equals to stepped-frequency increment $\Delta f = 20\,\text{Hz}$. Figures 5.8 and 5.9 present the range profile and Doppler profile in the central ambiguity region.

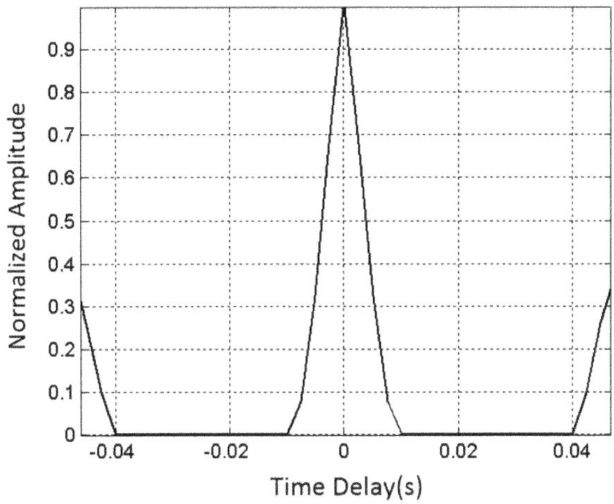

Fig. 5.8 The profile along the time delay axis in the central ambiguity region

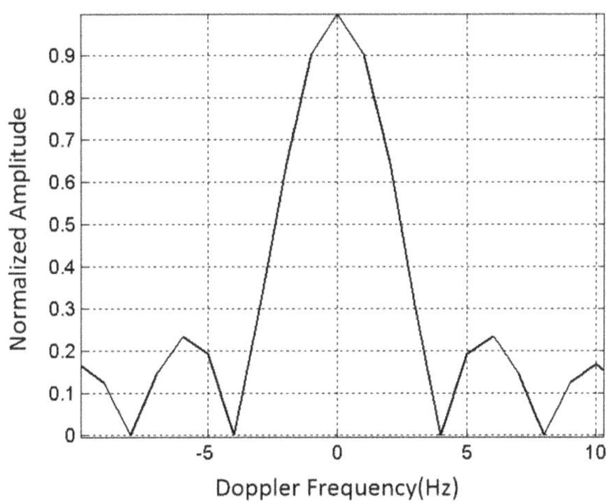

Fig. 5.9 The profile along the frequency shift axis in the central ambiguity region

Figure 5.8 shows that the 3 dB main lobe width of the profile along the time delay axis in the central ambiguity region is 0.01 s, that equals to the theoretical time resolution $1/N\Delta f$. Figure 5.9 shows that the 3 dB main lobe width of the profile along the frequency shift axis in the central ambiguity region is 4 Hz that equals to the theoretical velocity resolution $1/NT_r$.

5.3 The High Range Resolution Principle of Stepped-Frequency Waveform

5.3.1 The Profile in Range Dimension Obtained by IFFT Transforms [5]

The radar transmits a stepped-frequency signal with N sub-pulses expressed by Eq. (5.3). Let the complex scattering coefficient of stationary point target is $\rho(t) = Be^{j\phi}$, and the distance between the radar and the target is R_0, then the echo of each sub-pulse scattered from the target is

$$\begin{aligned} r_i(t) &= \rho(t) \otimes s_i(t) = \int\limits_{-\infty}^{+\infty} \rho(\tau)s_i(t-\tau)\mathrm{d}\tau \\ &= A_i B \mathrm{e}^{j(2\pi f_i t + \theta_i)} \int\limits_{-\infty}^{+\infty} \rho(\tau)u(t-\tau)\mathrm{e}^{-j2\pi f_i \tau}\mathrm{d}\tau \end{aligned} \tag{5.22}$$

The reference signal is

$$z_i(t) = \begin{cases} \mathrm{e}^{j(2\pi f_i t + \theta_i)} & iT_r - \frac{T}{2} \le t \le iT_r - \frac{T}{2} + T_r \\ 0 & \text{else} \end{cases} \tag{5.23}$$

The echo in Eq. (5.22) mixed with reference signal is

$$m_i(t) = r_i(t) \cdot z_i^*(t) = C_i \int\limits_{-\infty}^{+\infty} \rho(\tau)u(t-\tau)\mathrm{e}^{-j2\pi f_i \tau}\mathrm{d}\tau \tag{5.24}$$

where C_i is the amplitude of $m_i(t)$, and assuming C_i equal 1 for any sub-pulse. Let $q(t,\tau) = \rho(\tau)u(t-\tau)$, it presents the range profile scattered from target after the pulse complex envelope weighting. The output of mixer can be expressed as

$$m_i(t) = \int\limits_{-\infty}^{+\infty} q(t,\tau)\mathrm{e}^{-j2\pi f_i \tau}\mathrm{d}\tau \tag{5.25}$$

It can be seen from Eq. (5.25) that the output signal $m_i(t)$ after quadrature demodulation is the Fourier transform of the range profile scattered from target $q(t,\tau)$essential. Then

$$q(t,\tau) = \int\limits_{-\infty}^{+\infty} m_i(t)\mathrm{e}^{j2\pi f_i \tau}\mathrm{d}f = \sum_{i=0}^{N-1} m_i(t)\mathrm{e}^{j2\pi f_i \tau}\Delta f \tag{5.26}$$

Let $\tau = \tau_0 + n\Delta\tau$, $\tau_0 = M/\Delta f < \tau_{\min}$ M is integer, and $\Delta\tau = 1/(N\Delta f)$.
For $f_i = f_0 + i\Delta f$, $i = 0, 1, 2, \ldots, N-1$ then

$$\begin{aligned} q(t,\tau) &= \sum_{i=0}^{N-1} m_i(t) e^{j2\pi(f_0 + i\Delta f)(\tau_0 + n\Delta\tau)} \Delta f \\ &= e^{j2\pi f_0(\tau_0 + n\Delta\tau)} \sum_{i=0}^{N-1} m_i(t) e^{j2\pi i\Delta f(\tau_0 + n\Delta\tau)} \Delta f \\ &= e^{j2\pi f_0(\tau_0 + n\Delta\tau)} \sum_{i=0}^{N-1} m_i(t) e^{j2\pi i\Delta f n\Delta\tau} \Delta f \\ &= e^{j2\pi f_0(\tau_0 + n\Delta\tau)} \sum_{i=0}^{N-1} m_i(t) e^{j\frac{2\pi}{N}in} \Delta f \end{aligned} \tag{5.27}$$

The amplitude of complex range profile is

$$|q(t,\tau)| = \left| \sum_{i=0}^{N-1} m_i(t) e^{j\frac{2\pi}{N}in} \Delta f \right| \tag{5.28}$$

From the analysis above it can be seen that the range profile of the target illustrated by stepped-frequency essential is the inverse Fourier transform of the output signal after the sub-pulse mixer, and the time resolution $\Delta\tau$ is $1/(N\Delta f)$.

5.3.2 The Range Resolution of Stepped-Frequency Waveform

For the stepped-frequency waveform with N sub-pulses, the complex signal of each sub-pulse after quadrature demodulation is shown in Eq. (5.6). Assuming that the target scattering coefficient is a constant, the signal can be further represented to

$$m_i(t) = C_i e^{-j2\pi f_i \tau(t)}, \quad i = 0, 1, \ldots, N-1 \tag{5.29}$$

The time delay is

$$\tau(t) = \frac{2(R_0 - vt)}{c} \tag{5.30}$$

where the carrier frequency $f_i = f_0 + i\Delta f$, f_0 is the initial carrier frequency, Δf is step size of carrier frequency, the signal synthetic bandwidth is $B_w = N\Delta f$. R_0 is initial distance from the radar to the target, v is the radial velocity of the target, and c is the velocity of light.

For each sub pulse echo of the stepped frequency signal, the signal output by from the quadrature mixer represents the sampling of the target reflectivity in the frequency domain, all the complex samplings in each pulse burst can be IDFT processed to form a one-dimensional range profile of the target [6] that is

$$H(l) = \sum_{i=0}^{N-1} m_i(t) \mathrm{e}^{j(2\pi/N)li} \tag{5.31}$$

where l is the radial position of the target, $l = 0, 1, 2, \ldots, N-1$. For all l, assuming C_i equals 1. When the distance is R and the velocity is zero, the normal synthetic range profile is expressed as

$$H(l) = \mathrm{e}^{-j2\pi f_0 \frac{2R_0}{c}} \frac{\sin \pi y}{\sin \frac{\pi y}{N}} \mathrm{e}^{j\frac{N-1}{2}\frac{2\pi y}{N}} \tag{5.32}$$

The amplitude of the synthetic range profile is

$$|H(l)| = \left| \frac{\sin \pi y}{\sin\left(\frac{\pi y}{N}\right)} \right| \tag{5.33}$$

Where $y = \frac{-2NR\Delta f}{c} + l$

For isolated point target, when $y = 0, \pm N, \pm 2N, \ldots$, the value of target response is maximum, so all target positions associated with the peak distance l_0 can be expressed as

$$R = \frac{cl_0}{2N\Delta f}, \frac{c(l_0 \mp N)}{2N\Delta f}, \frac{c(l_0 \mp N)}{2N\Delta f}, \ldots \tag{5.34}$$

The Eq. (5.34) shows that as the distance from the radar to the target is increasing or decreasing by step size $c/2\Delta f$, all the peak positions of the range profile of point target appear in the l_0 range sampling cell. So it is clear the unambiguity distance from radar to target is $\Delta R_{\mathrm{ua}} = c/2\Delta f$. If the distance from radar to target is greater than $R_0 + \Delta R_{\mathrm{ua}}$, the range profile is wrapped. Therefore, the stepped- frequency signal only can obtain the high resolution range profile of relative position target, but it cannot got the range profile of absolute position target for any range scope. In addition, from the range profile of a point target located at l_0 it can be seen that the stepped- frequency signal of N sub-pulses generates N equal distance gaps, which are uniformly distributed in an unambiguity cell. The range resolution of the synthetic range profile is given by

$$\Delta r = \frac{\Delta R_{\mathrm{ua}}}{N} = \frac{c}{2N\Delta f} \tag{5.35}$$

The resolution of the N sub-pulse echoes of the stationary point target after IFFT is given by K. Ruttenberg in the literature [7].

In the practical range processing, commonly N is taken integral power of 2 in order to employ IFFT algorithm which is a fast algorithm of IDFT. In order to show the range profile clearly in detail, usually the method of adding supplement zero is adopted to increase the number of IFFT point to improve the sampling resolution of the range profile and make the range profile more in detail. Let the number of IFFT $M = 2^P N$, $p = 0, 1, 2, \ldots$. If $i \geq N$, $m_i(t) = 0$ then the normalized complex range profile of N sub-pulse echoes after the IFFT processing is:

$$H(l) = \sum_{i=0}^{N-1} \exp\left\{ j\frac{2\pi}{M}\left[\left(l - \frac{R_0}{\Delta r_m} + \frac{vT_r f_0/\Delta f}{\Delta r_m}\right) i + \frac{vT_r i^2}{\Delta r_m}\right]\right\} \tag{5.36}$$

where l is the position along the radial range, $l = 0, 1, \ldots, M-1$ and the range sampling resolution Δr_m is $c/(2M\Delta f)$. For a stationary target, the amplitude of the complex range profile can be expressed as

$$|H(l)| = \left| \frac{\sin \frac{\pi N}{M}\left(l - \frac{R_0}{\Delta r_m}\right)}{\sin \frac{\pi}{M}\left(l - \frac{R_0}{\Delta r_m}\right)} \right|, \quad l = 0, 1, \ldots, M-1 \tag{5.37}$$

Let $l_p = R_0/\Delta r_m$, $t_l = l - l_p$. If $t_l = kM$, $k = 0, \pm 1, \pm 2, \ldots$ then the peak of $|H(l)|$ is related to R_0. The normalized compression waveform has a sinc function shape near the peak. When $t_l = kM \pm M/N$, $|H(l)|$ crosses zero for the first time. Usually the distance between the peak and first zero is looked as the main lobe width, which approximately equals to half power width. So the range resolution of stepped-frequency of N sub-pulses is

$$\Delta r = \frac{M}{N}\Delta r_m = \frac{M}{N}\frac{c}{2M\Delta f} = \frac{c}{2N\Delta f} \tag{5.38}$$

The range resolution Δr only depends on the number N of sub-pulses and the step size of the frequency Δf, it is depends on the total bandwidth of signal, and independent on the number M of IFFT.

The following simulation is to verify the range resolution, and the parameters are same as the Sect. 5.3. The distance from the radar to the target is 10 km, and the number of stepped-frequency sub-pulses is 16. The range profiles after 32 points and 256 points IFFT are shown in the Figs. 5.10 and 5.11. Comparison between the two figures shows that along with increase the points number of IFFT, the range sampling resolution will improved and the range profile will reveal more in detail. In Fig. 5.10 the width of the main lobe will occupy two range cells, however in Fig. 5.11 that will occupy 16 range cells. By (5.38) the range resolution is

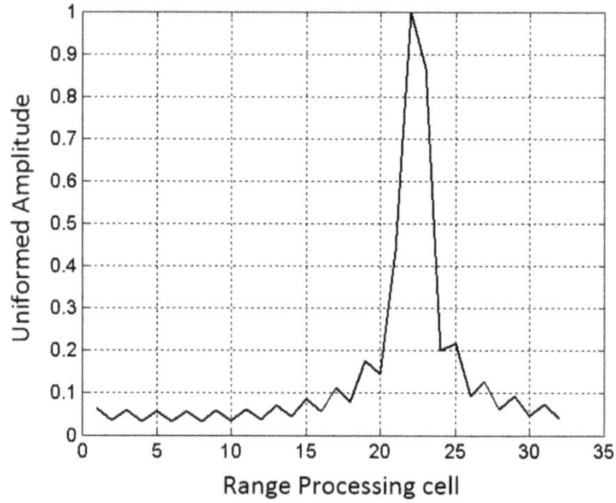

Fig. 5.10 The range profile after 32 point IFFT

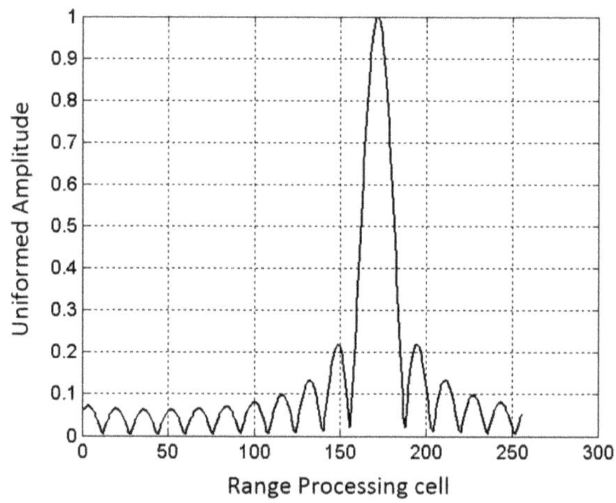

Fig. 5.11 The range profile after 256 point IFFT

$$\Delta r = \frac{M}{N}\Delta r_m = \frac{c}{2N\Delta f} = 9.375\,\text{m}$$

The simulation prove that the increment of IFFT number only can improve the range sampling resolution to reveal some in details, but it cannot change the range distinguishing ability.

According to the properties of Fourier transform, the M-point IFFT is corresponding to the echo sampled by the interval $1/M\Delta f$ in the frequency domain, and the range profile is extended by M sampling cells as a period in time domain. Since the position l in radial range is limited in a certain scope, only $l = l_p - k_pM \in [0, M-1]$, the target range response appears in the range profile correctly. Therefore, at the initial range R_0, the position of the stationary target in the range profile will be a position of range sampling cell which centering around l_p and be extended by the k_p cycle with period M.

5.4 The Effect of Radial Velocity on Synthetic Range Profile of Stepped-Frequency and the Velocity Compensation [8–13]

5.4.1 *The Effect of Radial Velocity on Synthetic Range Profile of Stepped-Frequency*

For each transmitting burst, the signal out of the quadrature mixer represents the sampling of the target reflectivity in the frequency domain, and all these complex samplings can be processed to form one dimension range profile of target by IDFT. The normalized complex range profile is expressed as

$$H(l) = \sum_{i=0}^{N-1} \mathrm{e}^{j\frac{2\pi}{N}\left[\left(l - \frac{R_0}{\Delta r} + \frac{vT_r f_0/\Delta f}{\Delta r}\right)i + \frac{vT_r i^2}{\Delta r}\right]} \tag{5.39}$$

where l is the position in radial range, $l = 0, 1, \ldots, N-1$. The range resolution is $\Delta r = c/2N\Delta f$. Since stepped-frequency waveform can be looked as the discretized LFM signal, so according to the instantaneous frequency algorithm of LFM signal, the instantaneous frequency of the range profile can be got:

$$f(i) = \frac{1}{N}\left[l - \frac{R_0}{\Delta r} + \frac{vT_r f_0}{\Delta f \Delta r} + \frac{2vT_r i}{\Delta r}\right] \tag{5.40}$$

N-point IFFT of is equivalent to sample by the interval $1/N\Delta f$ in frequency domain, and the range profile is extended by M range cells as a period in time domain.

$$l(i) = l - \frac{R_0}{\Delta r} + \frac{vT_r f_0/\Delta f}{\Delta r} + \frac{2vT_r i}{\Delta r}, \quad i = 0, 1, \ldots N-1 \tag{5.41}$$

Equation (5.41) represents the relationship between the range profile and the frequency step number. The first term l, is the position in the radial range. The second term is the peak position of the stationary target range profile at the initial distance R_0. The third term is Doppler frequency shift $2vf_0/c$, if it isn't compensated, in the range processing, it mistake the Doppler frequency shift as a frequency shift cause by range. It makes the range profile move away $vT_r f_0/(\Delta f \Delta r)$ range cells from the stationary target position. The fourth term is coupling between the velocity v and the index i of frequency step number, if it isn't compensated, during N sub-pulses, it will broaden the main lobe width of the target response into $2vNT_r/\Delta r$ range cells. Therefore, the migration factor L and the broadening factor P of the peak of the range profile are defined as

$$L = \frac{vNT_r}{\Delta r}\frac{f_0}{B} \tag{5.42}$$

$$P = \frac{2vNT_r}{\Delta r} \tag{5.43}$$

If the number of the FFT point is M, then the migration factor L and the broadening factor P are given by

$$L = \frac{M}{N}\frac{vNT_r}{\Delta r}\frac{f_0}{B} \tag{5.44}$$

$$P = \frac{M}{N}\frac{2vNT_r}{\Delta r} \tag{5.45}$$

when $M = N$, the migration and broadening factors are same as that in Eqs. (5.42) and (5.43). The simulation will verify the correctness of this text theoretical deduction.

Simulation

The main parameters are given as follows.

$$f_0 = 2\,\text{GHz},\ \Delta f = 1\,\text{MHz},\ N = 512,\ R_0 = 5\,\text{km},\ f\text{PRF} = 4\,\text{kHz}$$

The 1st experiment: The migration factor without the main lobe broadening of the range profile

(1) Target moving toward the radar.

Let five targets are moving toward the radar at 0, 10, 20, 30 and 40 m/s velocity. The range profile of N sub-pulse echoes can be obtained by 2N-point IFFT and shown in Fig. 5.12. Where v is the velocity, and l_p is the peak position. The comparison of the peak position with Eq. (5.42) at different velocities is shown in Table 5.1.

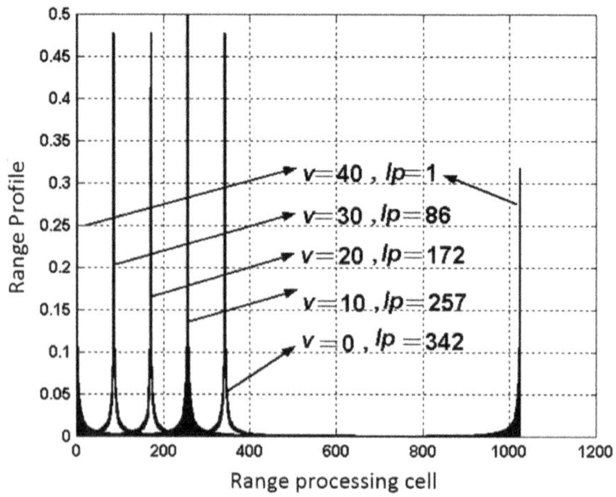

Fig. 5.12 The range profiles of the targets moving toward radar

Table 5.1 Comparison of peak position and theoretical migration at different positive velocities

Radial velocity (m/s)	0	10	20	30	40
Peak position	0	−85	−170	−256	−341
Theoretical migration	0	−85.3333	−170.6667	−256.0000	−341.3333
Relative deviation	0	0.3333	0.6667	0.0000	0.3333

Figure 5.12 shows that the peak of the stationary target is at the 342nd range cell. Along with the increase of the velocity, the distance between the moving target and the stationary target is increased. When velocity is 40 m/s, the range profile is wrapped from the 1st cell back to the 1024th cell. The simulation results are consistent with the theoretical values, and the error is less than a range cell.

(2) Target moving away from the radar.

Let five targets are moving away from the radar at 0, −20, −40, −60 and −80 m/s velocity. The range profile of N sub-pulse echoes can be obtained by $2N$-point IFFT and shown in Fig. 5.13. The comparison of the peak position with Eq. (5.42) at different velocities is shown in Table 5.2.

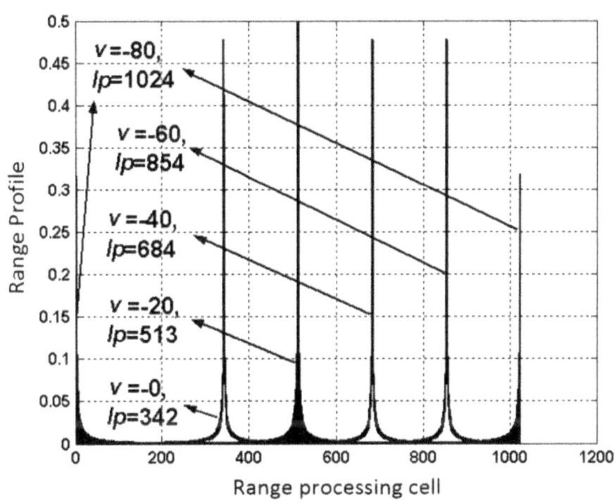

Fig. 5.13 The range profiles of the targets moving away radar

Table 5.2 Comparison of peak position and theoretical migration at different negative velocities

Radial velocity (m/s)	0	−20	−40	−60	−80
Peak position	0	171	342	512	683
Theoretical migration	0	170.3333	342.0000	511.6667	682.3333
Relative deviation	0	0.6667	0.0000	0.3333	0.6667

Figure 5.13 shows that when velocity is −80 m/s, the range profile will be wrapped. According to theory calculating, when velocity is −80 m/s, the peak of the range profile should be migrated 683 cells from 342nd cell of the stationary position. That means the peak position is the summation of the 342 and 683, and 1025th range cell will be wrapped back to the 1th cell by 1024 point IFFT.

It is proved that it is coincided between theory derivation and experiment.

The 2nd experiment: The broadening factor without the migration of the range profile.

(1) Target moving toward the radar.

The parameters are same as these in the 1st experiment. The range profile by N-point IFFT is shown in Fig. 5.14, and the comparison of the peak position with Eq. (5.43) at different velocities is shown in Table 5.3.

(2) Target moving away from the radar.

The parameters are same as these in the 1st experiment. The range profile by N-point IFFT is shown in Fig. 5.15, and the comparison of the peak position with Eq. (5.43) at different velocities is shown in Table 5.4.

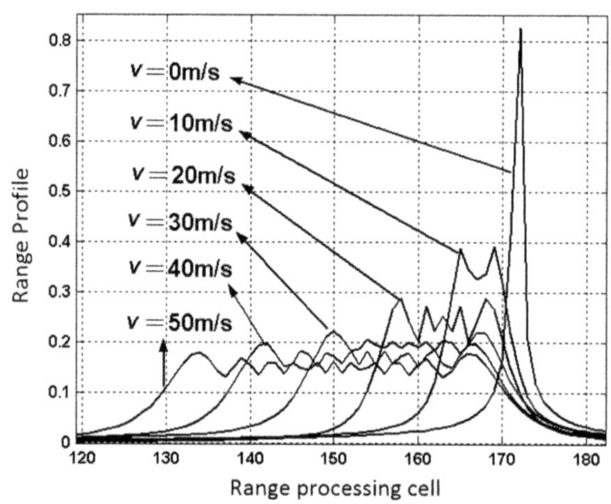

Fig. 5.14 The main lobe broadening of the targets moving towards radar

Table 5.3 Comparison of the main lobe broadening at different positive velocities

Radial velocity (m/s)	0	10	20	30	40	50
Main lobe width	1	7	15	24	31	39
Theoretical main lobe width	1	8.7381	17.4763	26.2144	34.9525	43.6907
Relative deviation	0	−1.7381	−2.4763	−2.2144	−3.9525	−4.6907

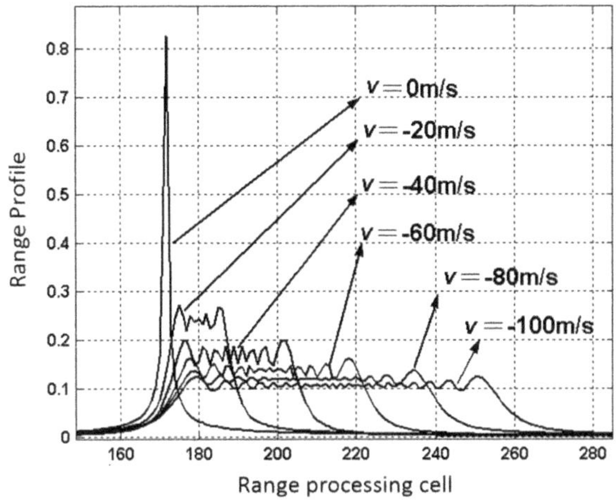

Fig. 5.15 The main lobe broadening of the targets moving away radar

Table 5.4 Comparison of the main lobe width broadening at different negative velocities

Radial velocity (m/s)	0	−20	−40	−60	−80	−100
Main lobe width	1	15	31	49	66	82
Theoretical main lobe width	1	17.4763	34.9525	52.4288	69.9051	86.3813
Relative deviation	0	−2.4763	−3.9525	−3.4288	−3.9051	−4.3813

The simulation shows that the radial velocity makes the matched filter mismatched, the main lobe widened, the range resolution degraded, and the SNR greatly reduced.

The 3rd simulation: The main lobe broadening factor of the range profile.

The broadening factor of the range profile is given in Ref. [6].

$$P_1 = \frac{vNT_r}{\Delta r} \tag{5.46}$$

The broadening factor of the range profile in this chapter is

$$P_2 = 2\frac{vNT_r}{\Delta r} \tag{5.47}$$

The following experiments will be performed to verify the Eq. (5.47).

The parameters of radar are same as these in the 1st experiment. The five targets move away from the radar at −50, −100, −150, −200 and −250 m/s velocity. The main lobes of the range profiles are shown in Fig. 5.16. The comparison of the simulation result with Eqs. (5.46) and (5.47) is shown in Table 5.5.

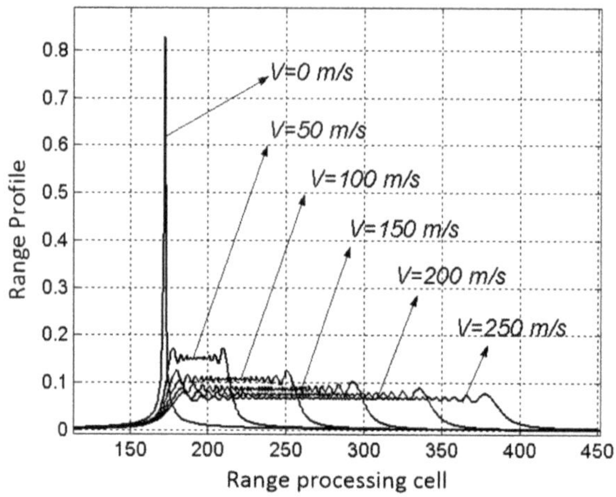

Fig. 5.16 The main lobe of the range profile at different radial velocities

Table 5.5 Comparison of theoretical, measured and theoretical main lobe width

Radial velocity (m/s)	−50	−100	−150	−200	−250
Measured main lobe width P_m	42	84	126	169	216
Theoretical main lobe width in Eq. (5.47) P_2	43.6907	87.3813	131.0720	174.7627	218.4533
Theoretical main lobe width in Eq. (5.46) P_1	21.8453	43.6907	65.5360	87.3813	109.2267
Deviation between Eq. (5.47) and measured value $\lvert P_{实际} - P_2 \rvert$	1.6907	3.3813	5.0720	5.7627	2.4533
Deviation between Eq. (5.46) and measured value $\lvert P_{实际} - P_1 \rvert$	20.1547	40.3093	60.4640	81.6187	106.7733

In Fig. 5.17 the No. 1 curve represents the theoretical broadening of the range profile in Eq. (5.46), the No. 2 curve is the measured main lobe width in simulation, the No. 3 curve represents the theoretical value given in Ref. [14], and the No. 4 curve represents the deviation between the measured main lobe widths with theoretical value in Eq. (5.46). The comparison of the main lobe width of the targets with different velocities is shown in Table 5.5.

The simulation shows that the broadening of the target range profile caused by the radial velocity is consistent with the simulation results, and the maximum deviation is less than 6 processing cells. The No. 3 curve given in Ref. [14] is seriously deviated from the measured results represented by the No. 2 curve, but the No. 1 curve is basically the same as the No. 2 curve, which proves the correctness of the results.

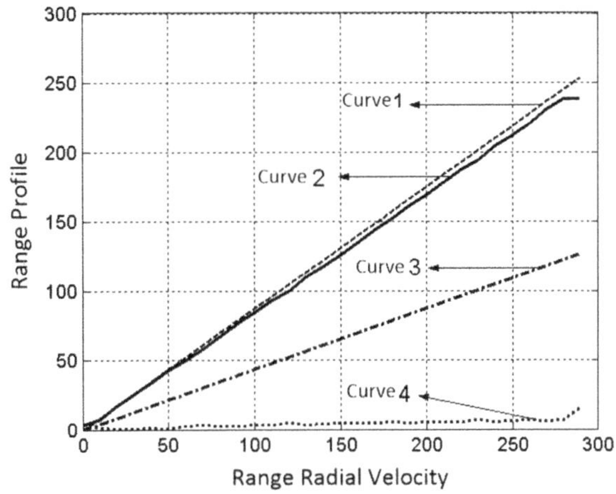

Fig. 5.17 Comparison of the main lobe broadening measured with theoretical value

The 4th experiment: The broadening factor and the migration factor of the range profile.

(1) Target moving toward the radar.

The parameters are same as these in the 1st experiment. The range profile by *N*-point IFFT is shown in Fig. 5.18. Duo to the main lobe broadening of the target moving, with 40 m/s as a critical velocity, the wrap of the range profile will appear, the velocity scope that can be displayed correctly in the range profile should be decreased.

(2) Target moving away the radar.

The parameters are same as these in the 1st experiment. The range profile by *N*-point IFFT is shown in Fig. 5.19. Duo to the main lobe broadening, the range profile of the target moving at −80 m/s as a critical velocity has wrapped.

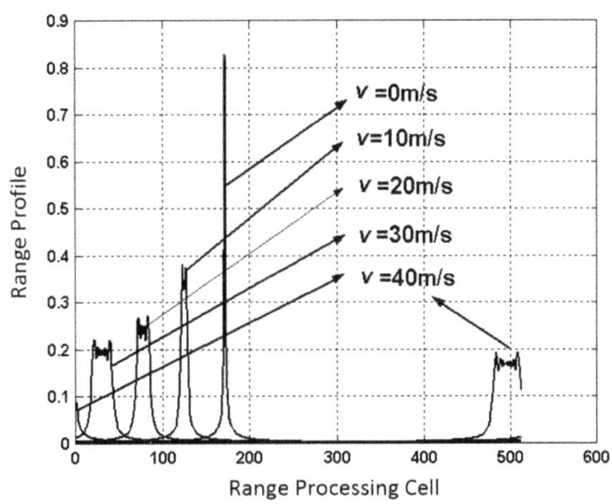

Fig. 5.18 The range profiles at positive redial velocity

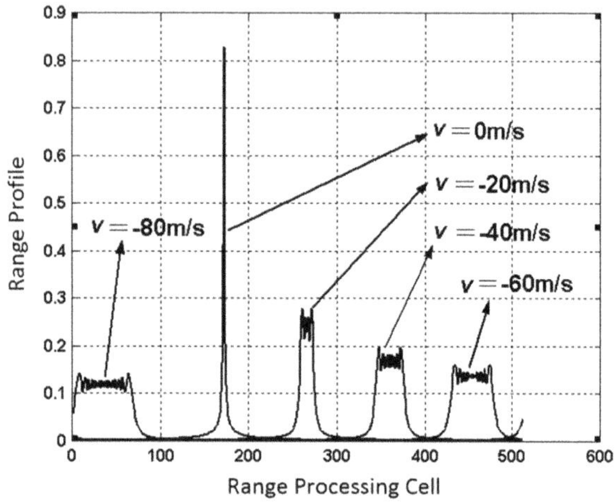

Fig. 5.19 The range profiles at negative redial velocity

From the above four groups of experiments, the following conclusions can be drawn:

(1) The migration and broadening of the range profile caused by the radial motion are proportional to the radial velocity. When the target moves towards the radar, the range profile is migrated and broadened to the direction of the decreasing distance delay. When the target moves away from the radar, the range profile is migrated and broadened to the direction of the increasing distance delay. Therefore, the radial velocity must be compensated.
(2) The range profile only can display the moving target in a certain scope of radial velocity. Once the velocity exceeds this scope, the wrap of the range profile will appear in the false targets or is called the ghost shadow that will degrade the image quality. If the range profile is used to estimate the velocity of the target, it will bring serious error and cannot get good velocity compensation effect.
(3) In the stepped-frequency parameters design, it is necessary to consider the effect of the target velocity on the peak value shift and the main lobe broadened. Appropriate waveform parameters should be selected for target velocity detection and synthetic aperture radar imaging.

5.4.2 The Target Velocity Estimation and Compensation of Stepped-Frequency Radar

The stepped-frequency waveform can be used to image the stationary targets in with high range resolution. If the target is moving, the radial velocity will lead to the migration of the range profile, the widening of main lobe and the loss of the SNR. In order to detect the moving target, it is necessary to design the reasonable parameters of the waveform according to the radial velocity scope of the target, and then the radial velocity can be estimated correctly.

The effect of the radial velocity on the target range profile shows that according to the determined parameters, the position and the main lobe width of the range profile can be predicted in advance without the velocity compensation. In turn, the radial velocity can be estimated according to the position and the main lobe width of the range profile, and then the velocity estimation is used to compensate the velocity. The method of velocity estimation is given as follows:

$$v_{e1} = L\frac{\Delta r \cdot B}{NT_r f_0} \tag{5.48}$$

$$v_{e2} = p\frac{\Delta r}{2NT_r} \tag{5.49}$$

$$v_e = \frac{v_{e1} + v_{e2}}{2} \tag{5.50}$$

The velocity compensation factor is

$$\mu(i) = \exp\left[-j2\pi f_i \frac{2v_e \cdot iT_r}{c}\right] \tag{5.51}$$

Each sub-pulse echo is multiplied by the corresponding velocity compensation factor $\mu(i)$, which is used to eliminate the migration and broadening of the target range profile caused by the radial velocity. Then the compensated echo burst is transformed by IFFT to obtain the high resolution range profile.

The simulation experiment of velocity estimation and velocity compensation is presented in the following. The parameters of stepped-frequency waveform are

$$f_0 = 2\,\text{GHz}, \Delta f = 1\,\text{MHz}, N = 512, R_0 = 5\,\text{km}, f\text{PRF} = 3\,\text{kHz}$$

Let the number of IFFT point M equals to N. The radial velocity is 100 m/s. Figure 5.20 presents the range profile with velocity compensation. The migration and the main lobe broadening make the resolution degraded and SNR decreased.

① The migration of the peak position is

$$L_e = 172 - 143 - 29$$

The peak position is migrated 29 sampling cells.

② The broadening of the main lobe is

$$P_e = 9$$

The main lobe width is 9 sampling cells.

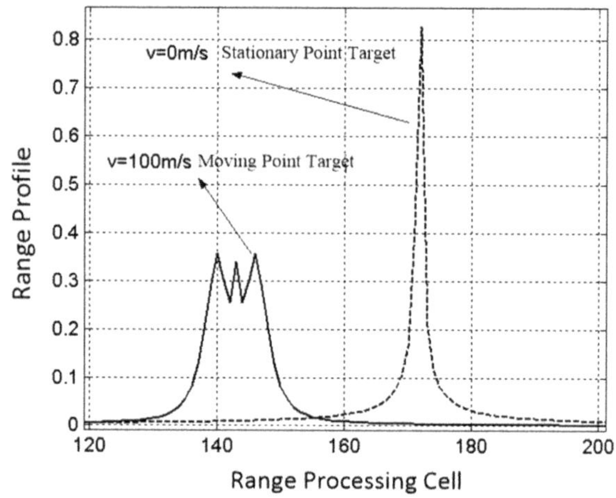

Fig. 5.20 The range profile without velocity compensation

Using Eqs. (5.48) and (5.49) to estimate the radial velocity

$$v_{e1} = f_0 \frac{2N \times f_{\mathrm{PRF}}}{c \times L_e} = 77.2476\,\mathrm{m/s}$$
$$v_{e2} = 2N\Delta f \frac{2N \times f_{\mathrm{PRF}}}{c \times P_e} = 127.4414\,\mathrm{m/s}$$

The joint target velocity estimation is $v_e = (v_{e1} + v_{e2})/2 = 102.3445\,\mathrm{m/s}$. The error between the estimation the real value is 2.3445 m/s. The estimation v_e is used to compensate the radial velocity; the range profile is shown in Fig. 5.21.

Comparison of Figs. 5.21 with 5.20 shows that after velocity compensation the response peak of the moving target is overlapped with that of the stationary target, and the main lobe width of the moving target is the same as that of the stationary

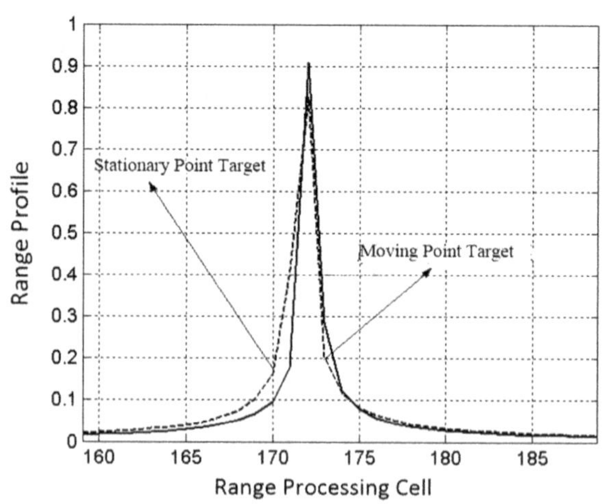

Fig. 5.21 The range profile after velocity compensation

target. Therefore the peak position and resolution of the moving target range profile are restored, and the velocity compensation is effective. Experimental results prove that the method is feasible to estimate the target velocity by using the joint migration factor and the broadening factor.

In this section, the effect of the radial velocity on the target range profile is discussed, and the expressions of the migration factor and the broadening factor of the range profile for different range sampling resolution are deduced, the expression of broadening factor is corrected. In addition, the velocity estimation and compensation method is presented and verified by simulations. The simulation experimental results are accordance with the theoretical calculation. The beneficial basis for the design of stepped-frequency waveform, velocity estimation and stepped-frequency SAR processing are provided in this section.

5.5 Stepped-Frequency SAR Imaging [8, 15–19]

SAR obtains the high azimuth resolution by means of synthetic aperture processing in azimuth dimension, and uses the stepped-frequency waveform can to achieve high range resolution by means of synthesizing large bandwidth signal via the step changing of the carrier frequency. If the stepped-frequency waveform is used in SAR, it can obtain the two-dimensional high resolution. Therefore, in this section two-dimensional high resolution of the stepped-frequency SAR will be discussed, the imaging method, waveform parameter design principle and signal processing steps are analyzed. The two-dimensional high resolution SAR imaging simulation of point targets is represented.

From above analysis it can be seen that the stepped-frequency waveform has the ability of higher range resolution, but it is very sensitive to the radial velocity. Since the range-Doppler coupling caused by the radial velocity is more complex, it makes the target energy dispersing to adjacent range cell, bring the resolution degraded, range accuracy reduced and the SNR decreased. So the radial velocity should be compensated. In fact, the SAR is just uses the relative motion between the radar and the target to improve the SAR azimuth resolution, the radial velocity component of the vehicle relative to the target has a great influence on the processing of the stepped frequency signal. In order to deal with this contradict, in the text, the influence of radial velocity on the range profile of stepped-frequency is discussed, the waveform design principle and signal processing steps of stepped-frequency SAR are represented, the simulation and results are provided.

5.5.1 Stepped-Frequency SAR Imaging Principle

The vehicle mounted the radar is flying forward along a special track with a constant velocity. Radar transmits and receives the signal at the pulse repetition frequency. The amplitude and phase information of the received signal are stored and superimposed with the previous echo signal, the equivalent linear array antenna will be formed along with radar moving forward. In the following, the stepped-frequency SAR imaging processing method will be discussed.

Schematic diagram of the side looking SAR is shown in Fig. 5.22. The radar moves at the velocity V_a along the track X. The vertical distance from the point target P to the radar air route is R_c, and the coordinate of target P is $(R_c, 0)$. The angle between the direction of the vehicle flying and the line from the radar to the target is θ. T_s is the synthetic aperture time, L_s is the synthetic aperture length. When the initial coordinate of the radar is $A(0, -L_s/2)$, the target at $(R_c, 0)$ is just illuminated by the beam of antenna.

When the radar is moving at the coordinate origin $Q(0,0)$, the target P is just at the center of the beam. When the radar moves at the $C(0, +L_s/2)$, the target just moves out of the beam of antenna. Taking the oblique plane, including the vehicle flying and the point target P, as reference plane, the coordinate system along the forward direction X and the oblique distance R is established and shown in Fig. 5.23.

Set up the radar transmits M stepped-frequency bursts within a synthetic aperture time, and each stepped-frequency burst consists of N sub-pulses. The pulse repetition period of each sub-pulse is T_r, and the sub-pulse time duration is T.

The transmitting waveform is shown in Fig. 5.24, and the signal is expressed as follows.

$$x_{ki} = A_{ki}\cos(2\pi f_i t + \theta_{ki}), \quad (i+kN)T_r \le t \le (i+kN)T_r + T \tag{5.52}$$

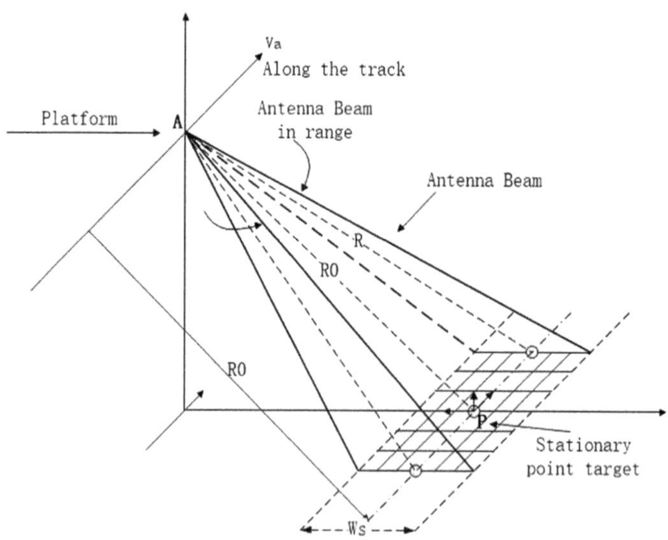

Fig. 5.22 Schematic diagram of the side looking SAR

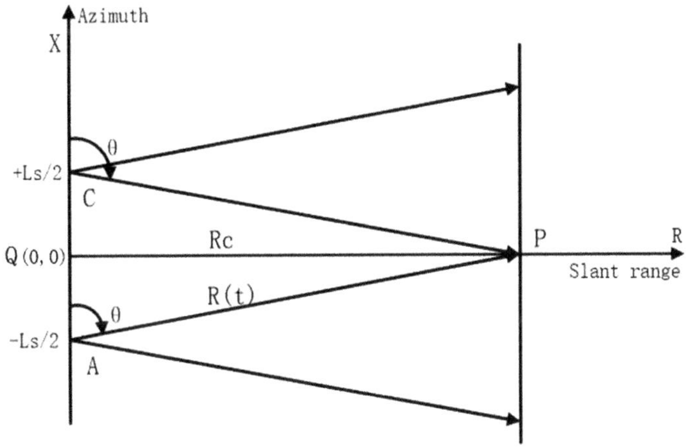

Fig. 5.23 The antenna pattern in the oblique plane

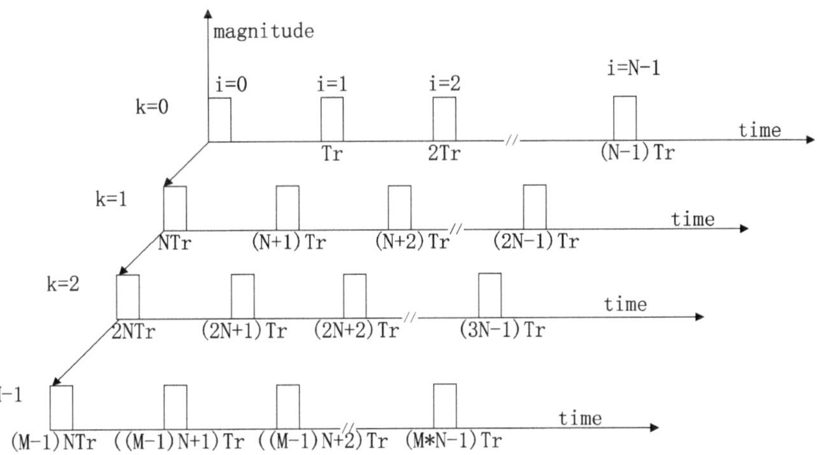

Fig. 5.24 Stepped-frequency burst waveform

where $k = -M/2, \ldots, M/2$ the index of bursts is, $i = -N/2, \ldots, N/2$ is the index of sub-pulses in one burst. A_{ki} is the initial amplitude, and θ_{ki} is the initial phase.

The received signal is

$$y_{ki} = B_{ki} \cos\{2\pi f_i[t - \tau(t)] + \theta_{ki}\}, \quad (i + kN)T_r \le t \le (i + kN)T_r + T \qquad (5.53)$$

where $\tau(t) = 2R(t)/c = 2\sqrt{R_c^2 + (V_a t)^2} \Big/ c$

Generally, the slant range is much larger than the synthetic aperture length, so Fresnel approximation can be used to Eq. (5.53), the round-trip delay between radar and target can be rewritten as

$$\tau(t) = 2\left[R_c/c + (V_a t)^2 \Big/ 2R_c c\right] \tag{5.54}$$

The reference signal is

$$z_{ki} = A_{ki}\cos(2\pi f_i t + \theta_{ki}), \quad (i + kN)T_r \leq t \leq (i + kN)T_r + T_r \tag{5.55}$$

The time relation between the transmitted signal and the reference signal is illustrated in Fig. 5.25.

The echo in Eq. (5.53) mixed with reference signal is

$$m_{ki} = C_{ki}\cos[-2\pi f_i \tau(t)], \quad (i + kN)T_r + \tau(t) \leq t \leq (i + kN)T_r + T + \tau(t) \tag{5.56}$$

Using Eq. (5.54) to express the round-trip delay of the Eq. (5.56), the phase is

$$\begin{aligned}\varphi_{ki} = -2\pi f_i \tau(t) = {} & -2\pi f_c \frac{2R_c}{c} - 2\pi i\Delta f \frac{2R_c}{c} - 2\pi f_c \frac{(V_a i T_r)^2}{cR_c} - 2\pi i\Delta f \frac{(V_a i T_r)^2}{cR_c} \\ & - 2\pi f_c \frac{(V_a T_r)^2 2kNi}{cR_c} - 2\pi i\Delta f \frac{(V_a T_r)^2 2kNi}{cR_c} - 2\pi i\Delta f \frac{(V_a kNT_r)^2}{cR_c} - 2\pi f_c \frac{(V_a kNT_r)^2}{cR_c}\end{aligned} \tag{5.57}$$

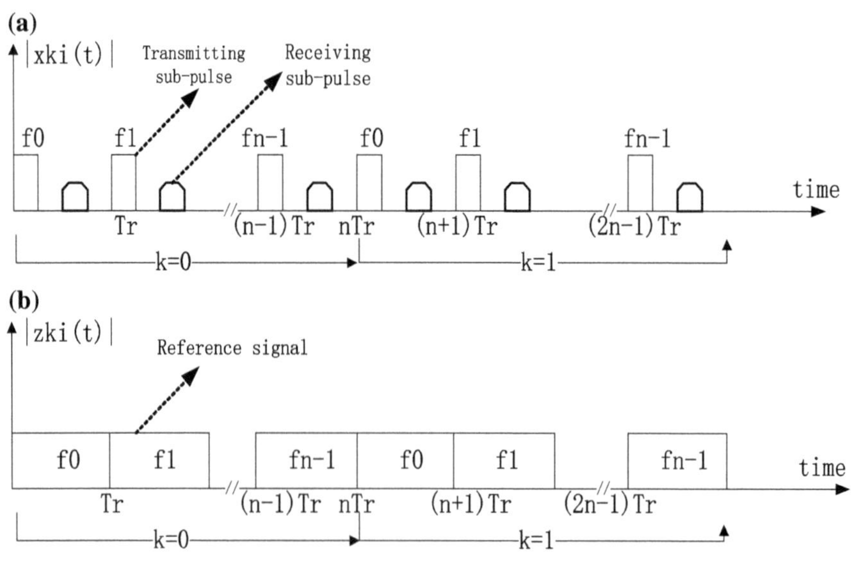

Fig. 5.25 The time relation between the transmitted signal and the reference signal

In the phase expression, the first term is constant, that is no meaning to for pulse compression in range and azimuth dimensions. The second term is necessary for pulse compression in range, and the third term and fourth term are the phase's duo to target motion and stepped-frequency within each burst, that results in the peak shift and main lobe broadening of the range profile; however that can be limited by means of waveform design. The fifth term, the sixth term and the seventh term are the coupling between the index of bursts and sub-pulses, that make the range profile shifted and main lobe broadened further, moreover, the mismatching of azimuth matched filter results in serious distortion of SAR image. The eighth term is the quadratic term of the index of bursts, which is necessary for pulse compression in azimuth.

For stepped-frequency SAR, in order to ensure the range resolution is not affected by the radial velocity component between radar and target, it is necessary to select reasonable waveform parameters or velocity compensation. In order to ensure azimuth pulse compression, the range migration of the SAR system and the range Doppler coupling problem must be considered. For the airborne SAR, the compensation factor can be obtained from the combination of the inertial navigation, which provides the aircraft velocity, altitude, pitching angle and azimuth angle, stepped-frequency source and timing parameters unite confirm the velocity compensation factor and the coupling compensation factor, to eliminate multiple phase of range-Doppler and the range-azimuth coupling. For the distributing target, it can be selected the interval of slant range according to the range resolution and using the compensation factor to eliminate from the third term to the seventh phase term in (5.57) for the different range echo data approximately. It also can be selected reasonable stepped-frequency parameter to limit the peak shift and main lobe broadening of the range profile within a permit scope. In this way the compensation factor only is as a coupling compensation factor:

$$\mu = \exp\left[2\pi f_c \frac{(V_a T_r)^2 2kNi}{c\tilde{R}_c} + 2\pi i\Delta f \frac{(V_a T_r)^2 2kNi}{c\tilde{R}_c} + 2\pi i\Delta f \frac{(V_a kNT_r)^2}{c\tilde{R}_c}\right] \quad (5.58)$$

Take the single point target as example in Fig. 5.23, after velocity compensation the phase in Eq. (5.57) is simplified as

$$\varphi_{ki} = -2\pi f_c \frac{2R_c}{c} - 2\pi i\Delta f \frac{2R_c}{c} - 2\pi f_c \frac{(V_a kNT_r)^2}{cR_c} \quad (5.59)$$

That is necessary phase for SAR two-dimensional processing. In the following, the design of stepped-frequency waveform for SAR is discussed, and to verify the waveform parameters the simulation experiments are carry out.

According to the above analysis and conclusion, the steps of stepped-frequency SAR imaging are as follows:

Step 1: The coupling phase compensation. Compensate the fifth term, the sixth term and the seventh term phase of each sub-pulse.

Step 2: Processing in range dimension. IFFT processing in range will get high resolution of range profile.
Step 3: Data rearrangement.
Step 4: Processing in azimuth. Pulse compression in azimuth and to get the high resolution SAR image.

5.5.2 Stepped-Frequency SAR System and Waveform Parameter Design

The block diagram of stepped-frequency SAR system is shown in Fig. 5.26.

The IF waveform generator forms the single frequency IF signal, which is mixed with the stepped-frequency signal $i\Delta f$ produced from the reference frequency source, and is converted up to the RF by the local oscillator f_0. Then the RF signal is amplified through the power amplifier and sent to the circulator. Finally, it is transmitted from the antenna.

The RF echo received from antenna is amplified and mixed with down-converter to IF signal, and then the IF signal is amplified and mixed with IF coherent signal. The signal after Quadrature detection is send to the low pass filter to get VF signal,

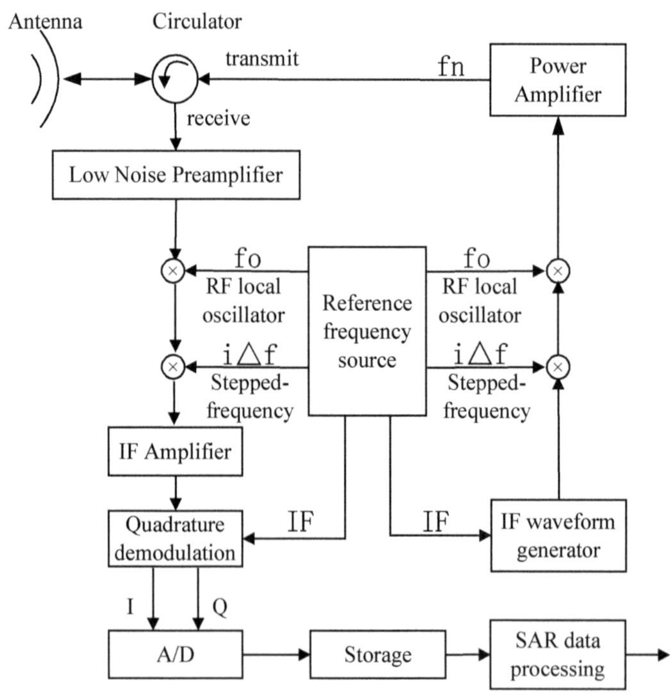

Fig. 5.26 The block diagram of stepped-frequency SAR system

the dual I/Q (In-phase and Quadrature) base band signal are sampled by A/D and saved. Finally, the high resolution SAR image is obtained through the real time or post image processing.

In the stepped-frequency waveform design, some parameters, such as the center frequency of transmitting signal, the velocity of the vehicle, the vertical distance from target to the vehicle track and the maximum slant range, the specified range resolution and azimuth resolution, are known firstly, and then the length of the antenna, signal bandwidth, PRF, etc., should confirmed. The key is to determine the number of the stepped-frequency, sub-pulse duration and stepped-frequency increment.

The selected method of waveform parameter is as follow:

(1) Pulse Repetition Frequency (PRF)
At first, the maximum radial velocity $V_{\max}$ and the minimum radial velocity $V_{\min}$ between radar and target can be determined from the geometric relationship between SAR and imaging target. Figure 5.23 shows that the radial velocity component of the moving vehicle that is in the direction of the radar and target on line is $V_a \cos\theta$, it is the absolute maximum at the starting point A and the ending point C of the synthetic aperture, and is zero at the coordinate origin. From the Refs. [20] and [21], the PRF should satisfy.

$$\frac{2\beta V_a}{\lambda} \leq f_{\mathrm{PRF}} \leq \frac{c}{2R_{\max}} \tag{5.60}$$

(2) The stepped-frequency number N and stepped-frequency increment Δf
From Sect. 5.5.1 it is known that the migration factors L and the broadening factor P of the peak of the range profile are:

$$L + P = \frac{2VNT_r}{c}(f_c + 2B) \tag{5.61}$$

In order to guarantee the requirement of high resolution SAR imaging, the range migration and the broadening are not required to limited within a range resolution cell, and then the stepped-frequency number should be satisfied.

$$N < \frac{c \times f_{\mathrm{PRF}}}{2V_{\max}(f_c + 2B)} \tag{5.62}$$

For the IFFT in range processing, the stepped-frequency number N generally take 2 integer powers.

The stepped-frequency increment is

$$\Delta f = B/N \tag{5.63}$$

(3) The sub-pulse duration T

The sub-pulse duration is usually taken as the reciprocal of the stepped-frequency increment.

$$T \leq \frac{1}{\Delta f} \tag{5.64}$$

(4) The burst number M

From pulse repetition frequency PRF, synthetic aperture time T_s and frequency stepped number N, the transmitting burst number within a synthetic aperture time is

$$M = f_{\mathrm{PRF}} \times T_s/N \tag{5.65}$$

The stepped-frequency waveform design method is verified by following simulation experiments.

The specified parameters are giving as follows.
The velocity of the vehicle V_a is 100 m/s, the vertical distance from the target to track R_c is 10 km, the center frequency f_c is 2 GHz, and the range resolution is 0.3 m.
The stepped-frequency waveform parameters are designed as follows.
Pulse Repetition Frequency: $f_{\mathrm{PRF}} = 15\,\mathrm{kHz}$
Synthetic Aperture Time: $T_s = 3.75\,\mathrm{s}$
Radial velocity: $V_{\max} = 3.75\,\mathrm{m/s}$, $V_{\min} = 0\,\mathrm{m/s}$
The number of the stepped-frequency sub-pulses: $N = 256$
The stepped-frequency increment: $\Delta f = 2\,\mathrm{MHz}$
The sub-pulse duration: $T = 0.5\ \mu\mathrm{s}$
The number of bursts: $M = 256$

The ambiguity function of the stepped-frequency signal is used to verify the rationality of the waveform design. The waveform parameters are substituted into the ambiguity function expression of the stepped-frequency signal in Eq. (5.13), and the ambiguity figure is shown in Fig. 5.27, the profile along the time-delay axis and Doppler frequency axis are shown in Figs. 5.28 and 5.29.

The ambiguity figure shows that the main lobe in the center region is very narrow, and the ambiguity figure is similar to thumbtack. Most of the ambiguity volume moves to the ambiguity region far from the origin, and the signal has a higher range resolution and velocity resolution. It can be seen that the waveform parameters match with the target environment and information requirements.

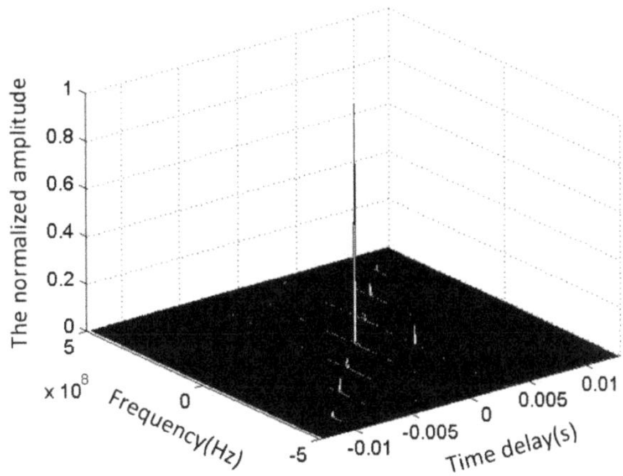

Fig. 5.27 The ambiguity function of stepped-frequency waveform

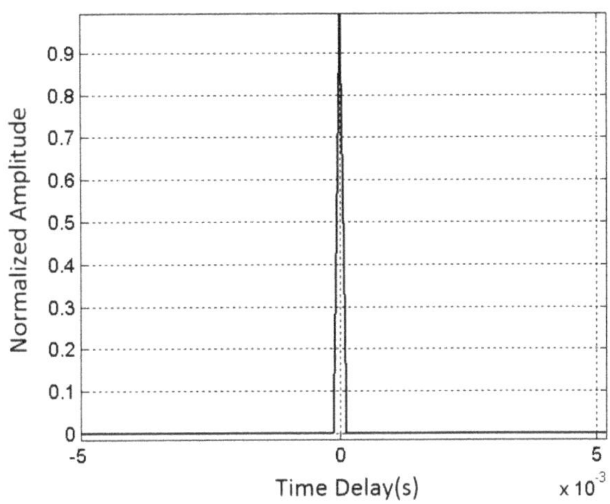

Fig. 5.28 The profile along the time delay axis in the central ambiguity region

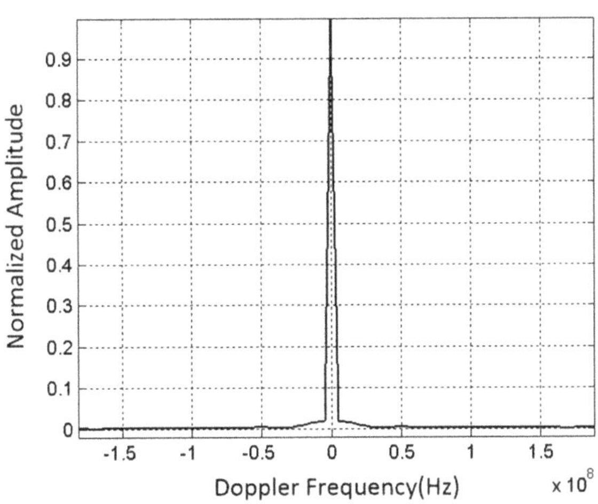

Fig. 5.29 The profile along the Doppler frequency axis in the central ambiguity region

5.5.3 Simulation

Simulation of single and multiple point targets will be performed to verify the rationality of the waveform parameter selection and the SAR imaging method.

Single point target simulation

The point target image in the 3D side view without coupling compensation is shown in Fig. 5.30, and the top view of the image is shown in Fig. 5.31. Furthermore, the profiles in range and azimuth dimensions at the target response peak are shown in Fig. 5.32.

The side lobe of the range response is very high, and there is a severe coupling between the azimuth and the range. Along with increase of beam departure from the center position, the azimuth matched filtering mismatching degree deepens, and the effect of pulse compression become worse. The main lobe width of the peak of the range response without compensation is 0.5859, that is approximately equal to 2 times the system required resolution, and the first side lobe in azimuth is −5.1927 dB, the azimuth matching filter has been severely mismatched and the main lobe has been diverged, So the compensation factors in Eq. (5.58) has to be used to compensate the coupling between the azimuth and the range.

Coupling compensation factor simulation

The coupling between the azimuth and the range in echo of each sub-pulse in Fig. 5.30 can be eliminated by means of multiplying the coupling compensation factor. The point target image in the 3D side view with coupling compensation is shown in Fig. 5.33, and the top view of the image is shown in Fig. 5.34. The profiles in range and azimuth dimensions at the target response peak are shown in Fig. 5.32. The output response has fine sinc function envelope, which eliminates the coupling basically. It can be seen from the range profile in Fig. 5.35 that the side lobe decreases obviously, and the first side lobe is about to be −13.3917 dB that is close to −13.3 dB of the impulse response of the LFM signal. Furthermore, the

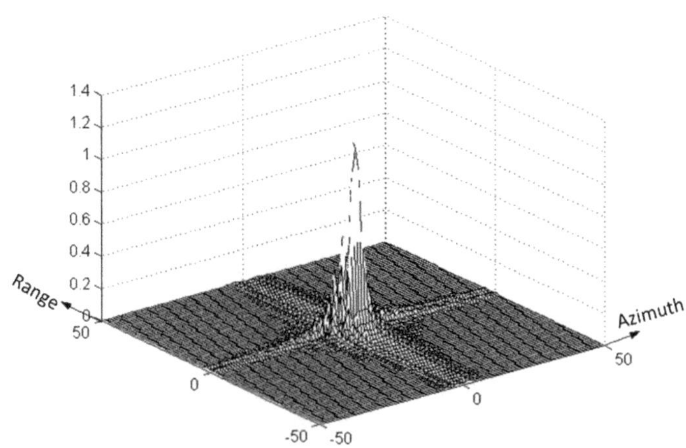

Fig. 5.30 The point target image without any compensation

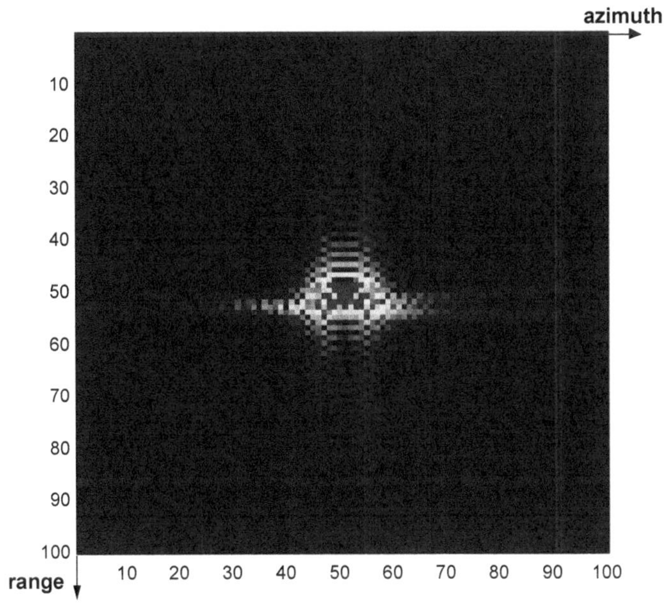

Fig. 5.31 The top view of the point target image

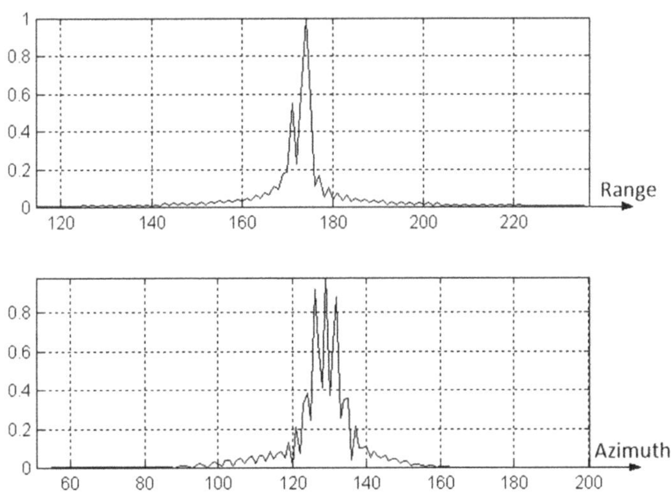

Fig. 5.32 The profile at the pulse response peak

3 dB width of the main lobe is 0.2930 that is consistent with the theoretical range resolution, and the first side lobe in azimuth is −13.9794 dB, the azimuth resolution is 0.8533 that is slightly less than 1 m required by the system. Therefore, the coupling compensation makes the range and azimuth resolution recovered, and the compensation effect is ideal.

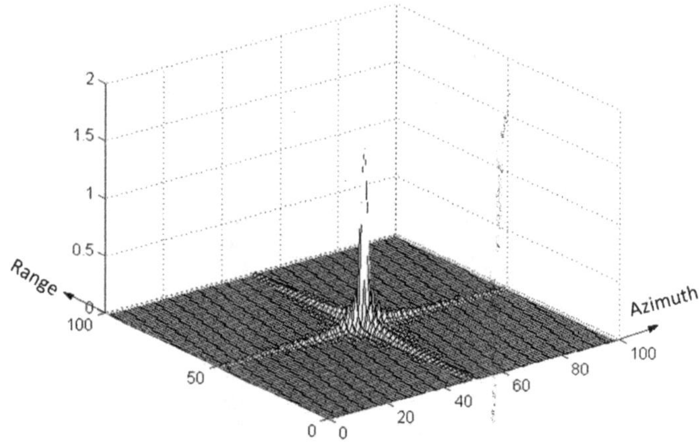

Fig. 5.33 The point target image decoupled and compensated

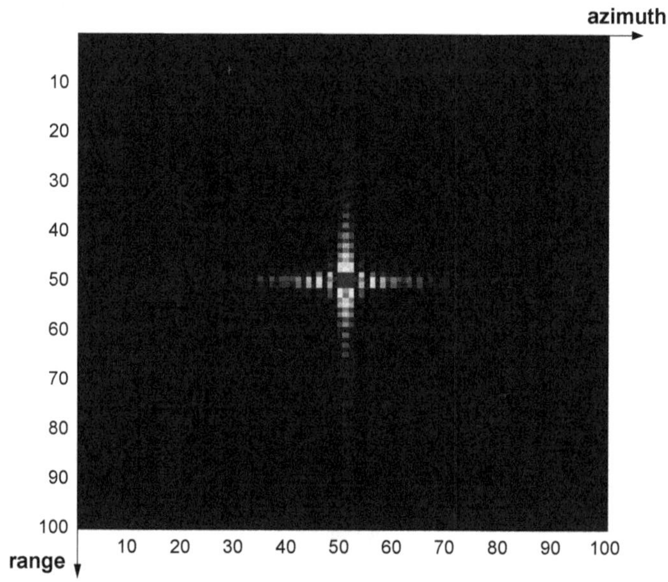

Fig. 5.34 The top view of the point target image

Multiple point targets simulation

The five point targets, spacing between 1 m, are evenly distributed in the range dimension. The image in the side view is shown in Fig. 5.36, and the profile in range and azimuth at peak are shown in Fig. 5.37.

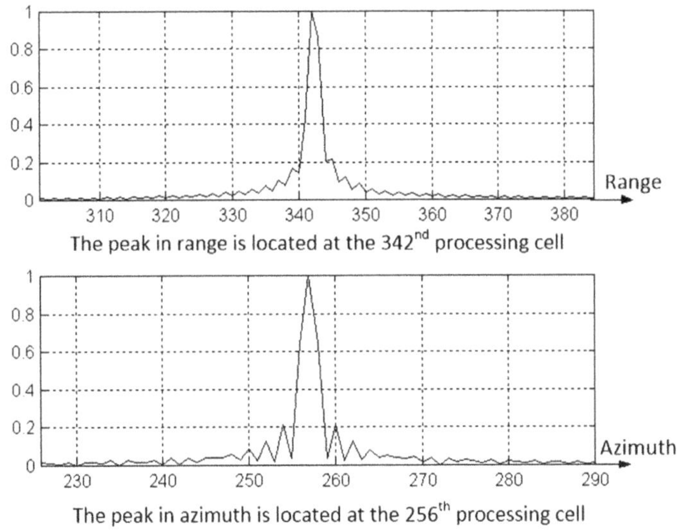

Fig. 5.35 The profile at the pulse response peak

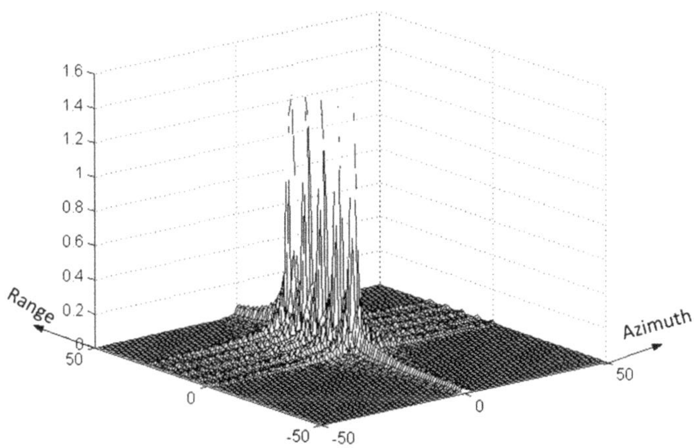

Fig. 5.36 The 3D side view image of five point targets decoupled and compensated

The Fig. 5.36 shows that the five response peaks can be distinguished, and it can be seen clearly that the five point targets have been completely separated in the profile. The main lobe width of each peak in the range profile is same as the range resolution of the single point target, and the first side lobe is −12.7127 dB. The azimuth resolution calculated from the azimuth profile is still 0.8533 m.

Based on the analysis of stepped-frequency processing and SAR imaging principle, the design principle of stepped-frequency waveform and the stepped-frequency SAR imaging method are proposed. By means of point target

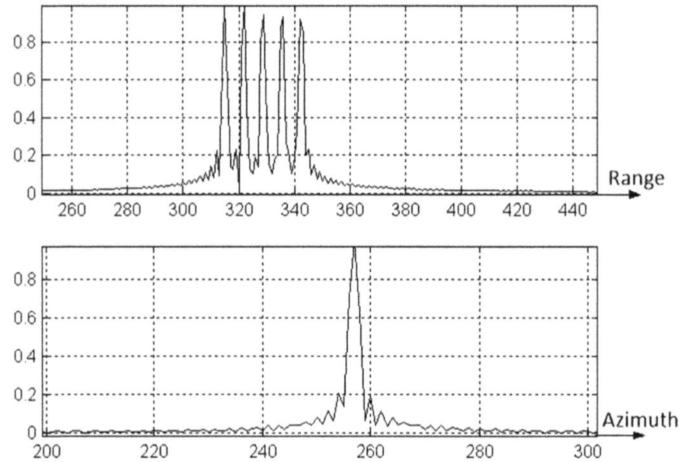

Fig. 5.37 The profile in azimuth and range

simulation, a high-resolution SAR image is obtained. The research shows that it is feasible to combine the synthetic aperture technology in space domain with the stepped-frequency in time domain to obtain high resolution SAR images [8].

References

1. Lord RT, Inggs MR (1996) High resolution VHF SAR processing using synthetic range profiling. In: Proceedings of IEEE geoscience remote sensing symposium, IGARSS'96, June 1996, vol 1. Lincoln, Nebraska, pp 454–456
2. Xiangqi Wu (1996) Signal, system and signal processing. Beijing Publishing House of Electronics Industry, Beijing
3. Running Z (1994) The Ultra-wideband and low side-lobe pulse compression signal design and processing technique. Ph.D. thesis, Beijing Institute of Technology, Beijing
4. Lin M, Ke Y (1984) Radar signal theory. National Defense Industry Press, China
5. Peiqing C (1998) Digital signal processing. Tsinghua University Press, Beijing
6. Wehner DR (1995) High resolution radar. Artech House, Boston, London
7. Ruttenberg K et al (1968) High range resolution by means of pulse to pulse frequency shifting. EASCON'68 Record, pp 47–51
8. Li H (2002) The ultra-wideband signal waveform and its SAR imaging. Ph.D. thesis, Graduate University of Chinese Academy of Sciences
9. Haiying Li, Ruliang Yang (2003) Analysis of radial velocity effect on range profile of stepped-frequency waveform. J Electron Inf 25(5):592–596
10. Peebles PZ (1998) Radar principles. Wiley, New Jersey
11. Freeman A, Evans D, van Zyl JJ (2006) SAR applications in the 21st century, EUSAR'96, Königswinter, Germany, pp 25–30
12. Haiying L, Ruliang Y (2002) Stepped-frequency waveform design for SAR imaging and radial velocity compensation. In: IEEE radar conference, Hyatt Regency, Long Beach, CA, pp 488–492, 22–25 April 2002

13. An Jianping, Zhang Running, Bai Kezhuang (2000) A nove waveform applied to SAR system and its doppler phase compensation technique. J Syst Eng Electron 22(1):59–62
14. Ma Y-E (1996) Velocity compensation in stepped frequency radar. Master thesis, Monterey CA, Naval Postgraduate School, Dec 1996
15. Haiying Li, Ruliang Yang (2003) Stepped-frequency synthetic aperture radar imaging. Chin J Electron 31(3):349–352
16. Inggs MR, Van Zyl MW, Knight A (1992) A simulation of synthetic range profile radar. In: IEEE 1992 international radar conference
17. Kovaly JJ (1978) Synthetic aperture radar. Artech House Inc, Norwood
18. Ming Li, Qizheng Hu (2002) Disaster and environmental monitoring system based on small satellite constellation. In: Proceedings of symposium on microwave remote sensing technology, pp 42–48, Dec 2002
19. Fei Yuanchun, Guangchuan Su, Mi Hong et al (2002) The generating technology of wideband radar signals. National Defense Industry Press, Beijing
20. Zhang Chenbo (1989) Synthetic aperture radar principle, system analysis and application. Science Press, Beijing
21. Soares PA (1996) Step frequency waveform design and analysis using the ambiguity function. Master thesis, Monterey CA, Naval Postgraduate School, June 1996
22. Xiushe Z (1998) Paper collection of ultra-wideband range resolution technique no. 20, vol 6. Research Institute of CETC

Chapter 6
Stepped-Chirp Burst Waveform and SAR Imaging

The stepped-frequency signal is a kind of flexible high-range resolution waveform, it is sensitive to the radial velocity, and the couple between range and Doppler is very complicated. The target energy is distributed to the adjacent range cells, which results in the degradation of the range resolution and the loss of SNR. At present, the main purpose of Ultra-Wideband (UWB) SAR using stepped-frequency is to increase the dynamic range of the receiver by reducing the instantaneous bandwidth of the system for the weak target detection in the strong clutter environment. However, as soon as the range resolution is higher, the disadvantages of stepped-frequency waveform cannot be ignored. Conventional SAR generally uses an LFM signal, the outstanding advantage of this waveform is that the matched filter is not sensitive to the Doppler shift, but the increase of the instantaneous system bandwidth is bound to increase the difficulty and cost of the radar [1–3].

The stepped-chirp burst waveform is a compromise between the LFM (Chirp) signal and the stepped-frequency signal; therefore, it has the advantages of both. In the condition of not increasing the instantaneous system bandwidth, the stepped-chirp burst waveform can obtain the high-resolution synthetic range profile by digital signal processing, so it is widely used in UWB radar system [4].

6.1 The Principle of Coherent Wideband Signal Synthesis of Stepped-Chirp Burst

6.1.1 The Characteristics of Wideband Signal in Time Domain and Frequency Domain [5, 6]

The characteristics of wideband signal in time domain and frequency domain are discussed first. Assuming that the central frequency of the wideband LFM signal is f_c, the bandwidth is B_w, and $B_w/f_c \geq 1\%$. The pulse duration is T_w, the pulse repetition time is T_r, and the chirp rate is r. The transmitting signal is expressed as

R. Yang et al., *High-Resolution Microwave Imaging*,
https://doi.org/10.1007/978-981-10-7138-6_6

$$s(t) = \mathrm{rect}\left(\frac{t}{T_w}\right)\exp j(2\pi f_c t + \pi r t^2) \tag{6.1}$$

If the scattering coefficient of the target is constant and the weight of the antenna is uniform, the normalized amplitude of echo scattered from the static point target at the range R_0 is given by

$$r(t) = \mathrm{rect}\left(\frac{t-\tau_0}{T_w}\right)\exp j\left[2\pi f_c(t-\tau_0) + \pi r(t-\tau_0)^2\right] \tag{6.2}$$

where the time delay is $\tau_0 = 2R_0/c$.

The reference signal is

$$s(t) = \mathrm{rect}\left(\frac{t}{T_w}\right)\exp j(2\pi f_c t) \tag{6.3}$$

The echo in Eq. (6.2) mixed with reference signal is sent to low-pass filter, the output of the filter is

$$m(t) = \mathrm{rect}\left(\frac{t-\tau_0}{T_w}\right)\exp j\left[-2\pi f_c\tau_0 + \pi r(t-\tau_0)^2\right] \tag{6.4}$$

The output signal after pulse compression via the matched filter is

$$y(t) = \mathrm{rect}\left(\frac{t-\tau_0}{2T_w}\right)\frac{\sin \pi r(t-\tau_0)(T_w - |t-\tau_0|)}{\pi r(t-\tau_0)} \tag{6.5}$$

Usually, we are interested only in $y(t)$ around the peak of the pulse compression response. When $T_w \gg |t-\tau_0|$, the $y(t)$ around the peak is

$$y(t) = \mathrm{rect}\left(\frac{t-\tau_0}{2T_w}\right)T_w\frac{\sin \pi B_w(t-\tau_0)}{\pi B_w(t-\tau_0)} \tag{6.6}$$

It is shown that the signal in Eq. (6.6) has a sinc function shape. The time resolution is $1/B_w$, and the range resolution is $c/2B_w$.

When the time–bandwidth product of LFM signal is much greater than 1, the energy is concentrated in the scope of $\left[-\frac{B_w}{2}, +\frac{B_w}{2}\right]$, the amplitude spectrum is close to rectangular, and the residual phase is approximately equal to $\pi/4$. The frequency spectrum of LFM signal envelope is

$$U(f) = \frac{1}{\sqrt{r}}\mathrm{rect}\left(\frac{f}{B_w}\right)\exp j\left(-\frac{\pi}{k}f^2 + \frac{\pi}{4}\right) \tag{6.7}$$

The frequency spectrum of baseband echo after matched filter is

$$Y(f) = M(f)H(f) = \frac{1}{r}\mathrm{rect}\left(\frac{f}{B_w}\right)\exp j(-2\pi f\tau_0) \tag{6.8}$$

6.1.2 The Characteristics of Stepped-Chirp Burst in Time Domain and Frequency Domain [7]

The wideband signal in Eq. (6.1) can be synthesized by N sub-pulses, and the bandwidth of each sub-pulse is B_w/N, the pulse duration is T_w/N.

The carrier frequencies of N sub-pulses are defined as

$$f_n = f_c + \Delta f_n, \quad \Delta f_n = \left(n + \frac{1-N}{2}\right)B, \quad n = 0, 1, \ldots, N-1 \tag{6.9}$$

The time shifts of N sub-pulses are defined as

$$\Delta t_n = \left(n + \frac{1-N}{2}\right)T, \quad n = 0, 1, \ldots, N-1 \tag{6.10}$$

The chirp rate is

If N is supposed to be 4, the relation between the frequency and the time of wideband LFM signal and stepped-chirp burst are illustrated, respectively, in Figs. 6.1 and 6.2.

Figure 6.1 shows that the wideband LFM signal can be looked as N sub-pulses coherent synthesis, it can be expressed as

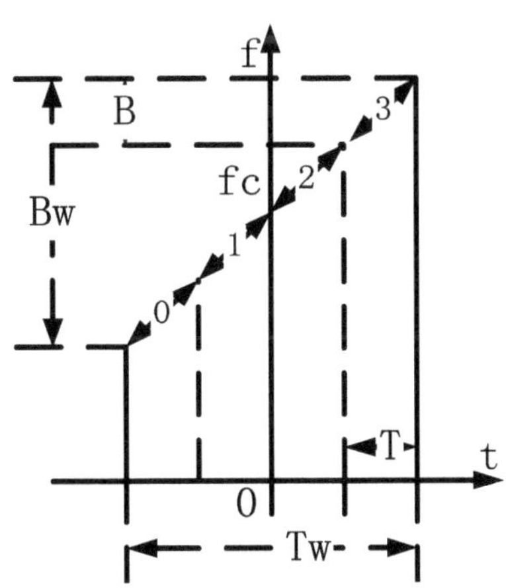

Fig. 6.1 Time-frequency diagram of wideband LFM signal

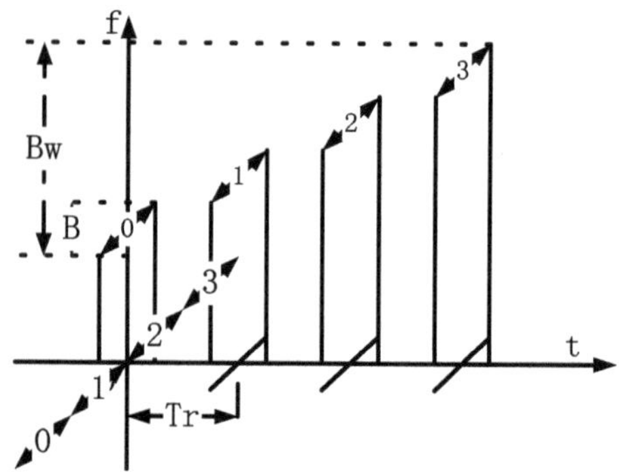

Fig. 6.2 Time-frequency diagram of stepped-chirp burst signal

$$s(t) = \sum_{n=0}^{N-1} s_n(t) = \sum_{n=0}^{N-1} \text{rect}\left(\frac{t - \Delta t_n}{T}\right) \exp j(2\pi f_c t + \pi r t^2) \tag{6.12}$$

The echo is

$$r(t) = \sum_{n=0}^{N-1} \text{rect}\left(\frac{t - \Delta t_n - \tau_0}{T}\right) \exp j\left[2\pi f_c(t - \tau_0) + \pi r(t - \tau_0)^2\right] \tag{6.13}$$

The output after echo demodulated orthogonally is given by

$$m(t) = \sum_{n=0}^{N-1} m_n^0(t) = \sum_{n=0}^{N-1} \text{rect}\left(\frac{t - \Delta t_n - \tau_0}{T}\right) \exp j\left[-2\pi f_c \tau_0 + \pi r(t - \tau_0)^2\right] \tag{6.14}$$

The matched filter is

$$h(t) = \sum_{n=0}^{N-1} h_n^0(t) = \sum_{n=0}^{N-1} \text{rect}\left(\frac{-t - \Delta t_n}{T}\right) \exp\left[-j\pi r(-t)^2\right], \quad t_0 = \tau_0 \tag{6.15}$$

The stepped-chirp radar transmits the sub-pulses and receives their echoes one by one in pulse repetition interval T_r, the relation between time and frequency is shown in Fig. 6.2. The transmitting signal is

$$s(t) = \sum_{n=0}^{N-1} s_{n0}(t) = \sum_{n=0}^{N-1} \text{rect}\left(\frac{t - nT_r}{T}\right) \exp j\left[2\pi f_n(t - nT_r) + \pi r(t - nT_r)^2\right] \tag{6.16}$$

The received signal is

$$r(t)=\sum_{n=0}^{N-1}\text{rect}\left(\frac{t-nT_r-\tau_0}{T}\right)\exp j\left[2\pi f_n(t-nT_r-\tau_0)+\pi r(t-nT_r-\tau_0)^2\right] \tag{6.17}$$

The output after echo demodulated orthogonally is

$$m(t)=\sum_{n=0}^{N-1}m_{n0}^0(t)=\sum_{n=0}^{N-1}\text{rect}\left(\frac{t-nT_r-\tau_0}{T}\right)\exp j\left[-2\pi f_n\tau_0+\pi r(t-nT_r-\tau_0)^2\right] \tag{6.18}$$

The matched filter is

$$h(t)=\sum_{n=0}^{N-1}h_{n0}^0(t)=\sum_{n=0}^{N-1}\text{rect}\left(\frac{-t+nT_r}{T}\right)\exp\left[-j\pi r(-t+nT_r)^2\right] \tag{6.19}$$

According to Eq. (6.14), the ideal output of the sub-pulses mixed with the reference is

$$m_n^0(t)=\text{rect}\left(\frac{t-\Delta t_n-\tau_0}{T}\right)\exp j\left[-2\pi f_c\tau_0+\pi r(t-\tau_0)^2\right] \tag{6.20}$$

According to Eq. (6.18), the actual output of the sub-pulses mixed with the reference is

$$m_{n0}^0(t)=\text{rect}\left(\frac{t-nT_r-\tau_0}{T}\right)\exp j\left[-2\pi f_n\tau_0+\pi r(t-nT_r-\tau_0)^2\right] \tag{6.21}$$

If the Eq. (6.21) is shifted by $nT_r-\Delta t_n$ in time domain, the following is obtained

$$m_{n0}^0(t+nT_r-\Delta t_n)=\text{rect}\left(\frac{t-\Delta t_n-\tau_0}{T}\right)\exp j\left[-2\pi f_n\tau_0+\pi r(t-\Delta t_n-\tau_0)^2\right] \tag{6.22}$$

where

$$f_n=f_c+\Delta f_n,\quad \Delta f_n=\left(n+\frac{1-N}{2}\right)B$$

$$(t-\Delta t_n-\tau_0)^2=(t-\tau_0)^2-2\Delta t_n(t-\tau_0)+\Delta t_n^2$$

The relation between Eqs. (6.20) and (6.21) is given by

$$m_n^0(t) = m_{n0}^0(t + nT_r - \Delta t_n)\exp(j2\pi\Delta f_n\tau_0)\exp[j2\pi r\Delta t_n(t - \tau_0)]\exp(-j\pi r\Delta t_n^2) \tag{6.23}$$

Because $\Delta f_n = r\Delta t_n$, Eq. (6.23) can be further simplified into

$$m_n^0(t) = m_{n0}^0(t + nT_r - \Delta t_n)\exp(j2\pi\Delta f_n t)\exp(-j\pi r\Delta t_n^2) \tag{6.24}$$

It can be seen that the steps of the wideband signal synthesis by the baseband sub-pulse $m_{n0}^0(t)$ are given as follows:

The first step: The baseband echoes $m_{n0}^0(t)$ of the sub-pulses in one burst can be shifted by $nT_r - \Delta t_n$, which make the time delay signal back to the original position of wideband signal echo $m_{n0}^0(t + nT_r - \Delta t_n)$.

The second step: The signal shifted in time domain $m_{n0}^0(t + nT_r - \Delta t_n)$ is shifted in frequency domain by $\exp j(2\pi\Delta f_n t)$, that makes the carrier frequency of sub-pulses back to f_c.

The third step: using $\exp(-j\pi r\Delta t_n^2)$ to phase compensate $m_{n0}^0(t + nT_r - \Delta t_n)$ $\exp j(2\pi\Delta f_n t)$, and form the continuous spectrum. Pass through above three steps; it can obtain the ideal output $m_n^0(t)$ of the sub-pulses mixed with the reference.

The fourth step: The echoes $m_n^0(t)$ of N sub-pulses are synthesized coherently to the wideband base echo $m(t)$ in (6.14).

The synthetic wideband base signal $m(t)$ is compressed by the matched filter in Eq. (6.15); the matched filter is given by

$$y(t) = m(t) \otimes h(t) = \sum_{n=0}^{N-1} m_n^0(t) \otimes \sum_{n=0}^{N-1} h_n^0(t) \tag{6.25}$$

In terms of the Fourier transform and the matched filtering principle, the filter is rewritten as

$$Y(f) = \sum_{n=0}^{N-1} M_n^0(f) \cdot \sum_{n=0}^{N-1} H_n^0(f) = \sum_{n=0}^{N-1} M_n^0(f) \cdot H_n^0(f) \tag{6.26}$$

$$y(t) = \mathrm{IDFT}[Y(f)] = \sum_{n=0}^{N-1} m_n^0(t) \otimes h_n^0(t) \tag{6.27}$$

Discussion:

(1) When $-T_w + \tau_0 \le t \le \tau_0$, the output signal of the sub-pulse after pulse compression via the matched filter is

$$
\begin{aligned}
y_n(t) &= \int\limits_{-\frac{T_W}{2}+\tau_0}^{\frac{T_W}{2}+t} m_n^0(\tau)h_n(t-\tau)\mathrm{d}\tau \\
&= \int\limits_{t_n+\tau_0-\frac{T}{2}}^{-t_n+\frac{T}{2}+t} m_n^0(\tau)h_n(t-\tau)\mathrm{d}\tau \\
&= \frac{1}{j2\pi r(t-\tau_0)}\left\{\exp\left[j2\pi r(t-\tau_0)\left(\frac{t+T-\tau_0}{2}-\Delta t_n\right)\right]\right. \\
&\quad \left. -\exp\left[-j2\pi r(t-\tau_0)\left(\frac{t+T-\tau_0}{2}-\Delta t_n\right)\right]\right\}
\end{aligned} \tag{6.28}
$$

The signal synthesized coherently by N sub-pulses is

$$
y(t) = \sum_{n=0}^{N-1} y_n(t) = \frac{\sin \pi NB(t-\tau_0)}{\sin \pi B(t-\tau_0)} \frac{\sin \pi r(t-\tau_0)(T+t-\tau_0)}{\pi r(t-\tau_0)}, \quad -T_w+\tau_0 \le t \le \tau_0 \tag{6.29}
$$

When $\tau_0 \le t \le T_w + \tau_0$, the output signal of each sub-pulse compressed by matched filter is

$$
\begin{aligned}
y_n(t) &= \int\limits_{-\frac{T_W}{2}+t}^{\frac{T_W}{2}+\tau_0} m_n^0(\tau)h_n(t-\tau)\mathrm{d}\tau \\
&= \int\limits_{-t_n+t-\frac{T}{2}}^{t_n+\frac{T}{2}+\tau_0} m_n^0(\tau)h_n(t-\tau)\mathrm{d}\tau \\
&= \frac{1}{j2\pi r(t-\tau_0)}\left\{\exp\left[j2\pi r(t-\tau_0)\left(\frac{T+\tau_0-t}{2}+\Delta t_n\right)\right]\right. \\
&\quad \left. -\exp\left[-j2\pi r(t-\tau_0)\left(\frac{T+\tau_0-t}{2}+\Delta t_n\right)\right]\right\}
\end{aligned} \tag{6.30}
$$

The signal synthesized coherently by N sub-pulses compressed is

$$
y(t) = \sum_{n=0}^{N-1} y_n(t) = \frac{\sin \pi NB(t-\tau_0)}{\sin \pi B(t-\tau_0)} \frac{\sin \pi r(t-\tau_0)(T-t+\tau_0)}{\pi r(t-\tau_0)}, \quad \tau_0 \le t \le T_w+\tau_0 \tag{6.31}
$$

It can be seen from the above two steps that the signal synthesized coherently by N sub-pulses is expressed as

$$y(t) = \mathrm{rect}\left(\frac{t-\tau_0}{2T_w}\right) T_w \frac{\sin \pi B_w (t-\tau_0)}{\pi B_w (t-\tau_0)} \tag{6.32}$$

The Eq. (6.32) is same as Eq. (6.6), so the method of synthesizing wideband signal by the stepped-chirp burst is feasible.

The baseband spectrum of any narrowband sub-pulse is

$$M_n(f) = \frac{1}{\sqrt{r}} \mathrm{rect}\left(\frac{f-f_n}{B}\right) \exp j(-2\pi f_c \tau_0) \quad n = 0, 1, \ldots, N-1 \tag{6.33}$$

The matched filter is $H_n(f) = M_n^*(f) \exp(-j2\pi f \tau_0)$

According to the principle of matched filtering, the frequency spectrum of after sub-pulse compressed is

$$Y_n(f) = M_n(f) \cdot H_n(f) = \frac{1}{r} \mathrm{rect}\left(\frac{f-f_n}{B}\right) \exp(-j2\pi f \tau_0) \tag{6.34}$$

The synthetic spectrum of N sub-pulses compressed is

$$Y(f) = \sum_{n=0}^{N-1} Y_n(f) = \frac{1}{r} \mathrm{rect}\left(\frac{f}{B_w}\right) \exp(-j2\pi f \tau_0) \tag{6.35}$$

The synthetic spectrum of stepped-chirp burst in Eq. (6.35) is same as that of the wideband spectrum in Eq. (6.8), so the stepped-chirp burst waveform can obtain a large equivalent bandwidth without increasing the instantaneous bandwidth of the receiver.

6.1.3 Simulation

The main parameters of wideband signal: the center frequency is 1.25 GHz, the bandwidth is 400 MHz, the pulse duration is 40 μs, and the PRF is 1600 Hz.

Using 4 stepped-chirp signals, which the bandwidth of each sub-pulse is 100 MHz and the sub-pulse duration is 10 μs. To synthesize the wideband signal with 400 MHz bandwidth.

The first experiment: Single point target simulation.

Here R_c the range from radar to target is 4.8 km. The amplitude spectrum of the wideband echo is shown in Fig. 6.3, and the output after pulse compression is shown in Fig. 6.4. If the stepped-chirp waveform is used, the amplitude spectrum of

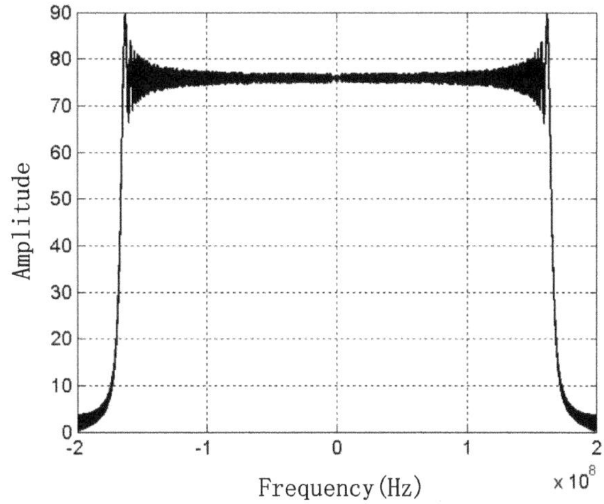

Fig. 6.3 The amplitude spectrum of wideband echo

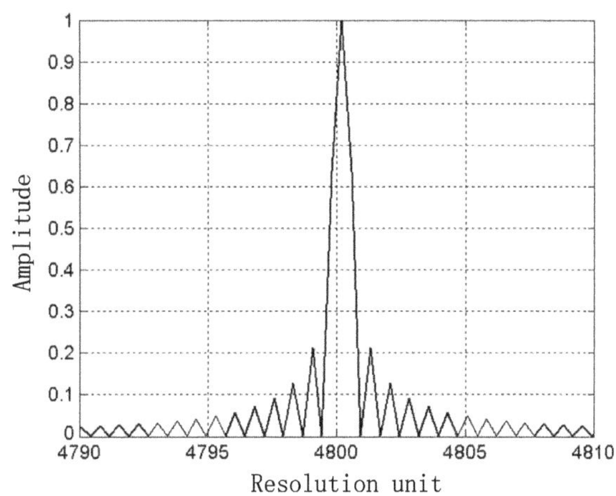

Fig. 6.4 The pulse compressed output of wideband echo

the sub-pulse echo is shown in Fig. 6.5, and the output after pulse compression of synthetic wideband echo is shown in Fig. 6.6.

The comparison between Figs. 6.6 and 6.4 show that the coherent synthetic spectrum of sub-pulse echoes is exactly same as the spectrum of wideband echo for the stationary point target. The matched filtering results of two kinds signal have the shape of sinc function, and the peak main lobe width equals to 1 range resolution unit. It can be seen that the wideband signal can be synthesized by a burst of narrowband pulses, and both signals have the same effective bandwidth and range resolution.

The second experiment: Multiple point targets simulation.

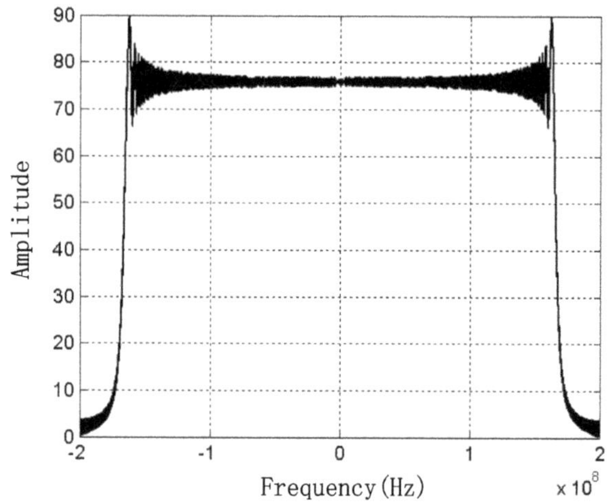

Fig. 6.5 The synthetic amplitude spectrum of the sub-pulses

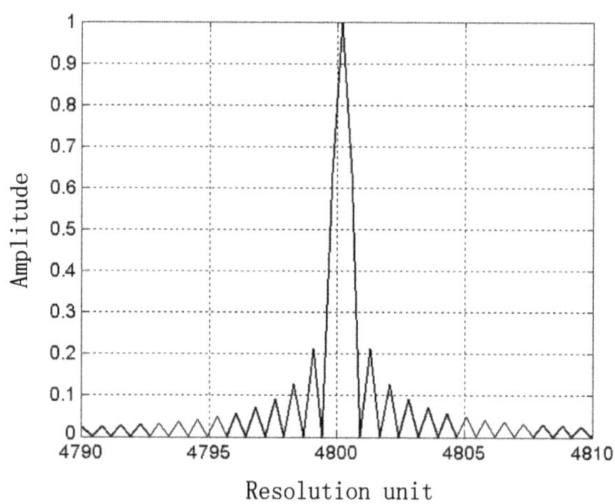

Fig. 6.6 The output of pulse compression of synthetic wideband signal

Assume that the distances between the radar and four stationary targets are 9999, 10,001, 10,002, and 10,003 m each other. The four sub-pulses with 100 MHz bandwidth is used to synthesize a 400 MHz wideband signal, the coherent synthesized spectrum of sub-pulse echoes mixed with the reference is shown in Fig. 6.7, the output signal after pulse compression via matched filtering is shown in Fig. 6.8.

The comparison between Figs. 6.7 and 6.2 show that the coherent synthetic spectrum of sub-pulse echoes is exactly same as the spectrum of wideband echo. The stepped-chirp burst waveform has the same resolution as the equivalent synthetic wideband signal and can be used for radar imaging with high-range resolution completely.

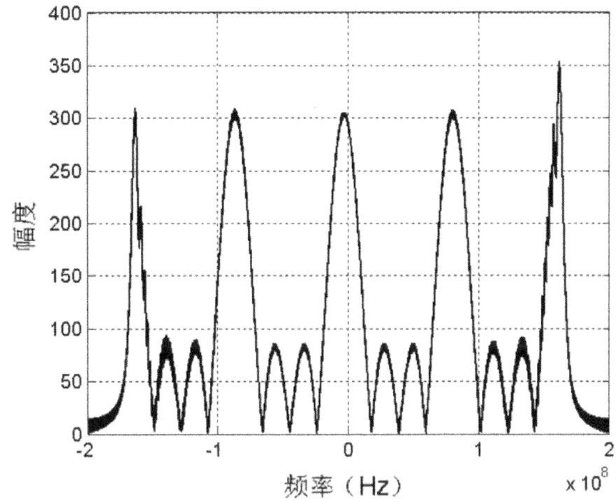

Fig. 6.7 The coherent synthetic spectrum of sub-pulse echoes

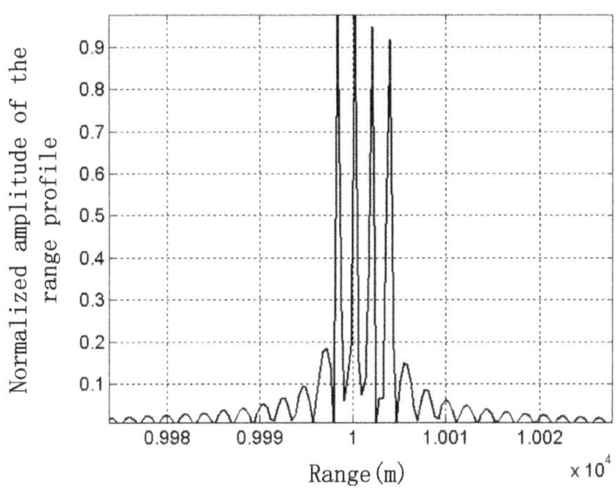

Fig. 6.8 The output signal after pulse compression via matched filtering

6.2 The Ambiguity Function of Stepped-Chirp Burst Signal

6.2.1 The Ambiguity Function and Resolution Performance

The transmitted stepped-chirp burst signal is given by

$$s(t) = \sum_{n=0}^{N-1} u(t - nTr)\mathrm{e}^{j\pi rt^2}\mathrm{e}^{j2\pi f_n t} \tag{6.36}$$

Where

$$u(t-nTr)=\begin{cases}1/\sqrt{T} & nT_r \le t \le nT_r+T \\ 0 & 其他\end{cases}$$

$$f_n = f_0 + n\Delta f, \quad n = 0, 1, 2, \ldots, N-1$$

$u(t)$ Is the envelope of sub-pulse. f_n Is the carrier frequency of each sub-pulse, f_0 is the initial carrier frequency, Δf is step size of carrier frequency, n is the index of sub-pulses (Fig. 6.9).

The transmitting signal can be rewritten as

$$s(t) = \sum_{n=0}^{N-1} u(t-nTr)\exp(j2\pi n\Delta ft)\exp(j\pi rt^2)\exp(j2\pi f_0 t), \quad nTr < t < nTr + T \tag{6.37}$$

The ambiguity function is defined as

$$\begin{aligned}\chi(\tau,\xi) = \int_{-\infty}^{+\infty} & \sum_{n=0}^{N-1} u(t-nTr)e^{j2\pi n\Delta ft}e^{j\pi rt^2} \\ & \times \sum_{m=0}^{N-1} u^*(t-mTr+\tau)e^{-j2\pi m\Delta f(t+\tau)}e^{-j\pi r(t+\tau)^2}e^{j2\pi\xi t}dt\end{aligned} \tag{6.38}$$

The ambiguity function can be expressed as

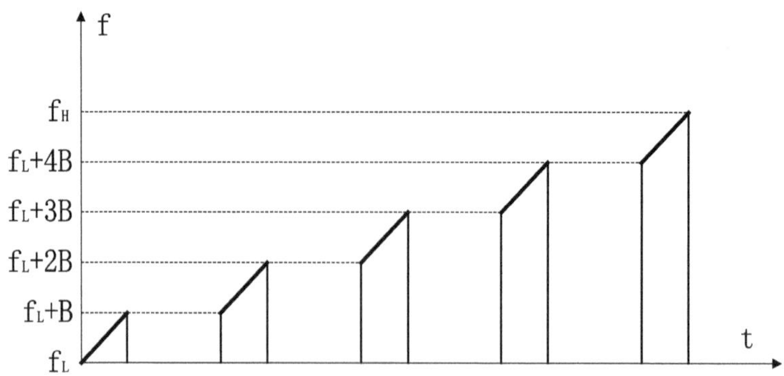

Fig. 6.9 Time-frequency diagram of stepped-chirp burst signal

$$\chi(\tau,\xi) = \mathrm{e}^{-j\pi r\tau^2} \sum_{n=0}^{N-1}\sum_{m=0}^{N-1} \mathrm{e}^{-j2\pi m\Delta f\tau} \mathrm{e}^{j2\pi(\xi-(m-n)\Delta f - r\tau)nTr} \times \int_{-\infty}^{+\infty} u(t)u^*(t-(m-n)Tr+\tau)\mathrm{e}^{j2\pi(\xi-(m-n)\Delta f - r\tau)t}\mathrm{d}t \tag{6.39}$$

The expression of ambiguity function is

$$|\chi(\tau,\xi)| = \left|\sum_{n=0}^{N-1}\sum_{m=0}^{N-1} \mathrm{e}^{-j2\pi m\Delta f\tau}\mathrm{e}^{-j\pi r\tau^2}\mathrm{e}^{j2\pi(\xi-(m-n)\Delta f - r\tau)nTr} \times \chi_0(\tau-(m-n)Tr, \xi-(m-n)\Delta f - r\tau)\right|, \quad |\tau-(m-n)Tr| < T \tag{6.40}$$

where $|\chi_0(\tau,\xi)|$ is the ambiguity function of single frequency pulse signal with rectangular envelope.

Let $p = m - n$, then Eq. (6.40) can be rewritten as

$$|\chi_0(\tau,\xi)| = \left| \sum_{p=-(N-1)}^{0} \sum_{m=0}^{N-1-|p|} \mathrm{e}^{-j2\pi(\xi-p\Delta f - r\tau)pT_r}\mathrm{e}^{j2\pi m[(\xi-p\Delta f - r\tau)Tr-\Delta f\tau]}\chi_0[\tau-pT_r, \xi-p\Delta f-r\tau] + \sum_{p=1}^{N-1}\sum_{n=0}^{N-1-|p|} \mathrm{e}^{-j2\pi p\Delta f\tau}\mathrm{e}^{j2\pi n[(\xi-p\Delta f-r\tau)Tr-\Delta f\tau]}\chi_0[\tau-pT_r, \xi-p\Delta f-r\tau]\right|, \quad |\tau-pTr| < T \tag{6.41}$$

In terms of the formula of finite series, using Eq. (6.10) to express Eq. (6.41), the ambiguity function of stepped-chirp burst is defined as

$$|\chi(\tau,\xi)| = \left|\sum_{p=-(N-1)}^{N-1} \mathrm{e}^{j\pi(N-p-1)(\xi-r\tau-p\Delta f)T_r}\mathrm{e}^{-j\pi(N+p-1)\Delta f\tau}\mathrm{e}^{j\pi\xi(T-\tau+pT_r)}A(\tau,\xi)B(\tau,\xi)\right| \tag{6.42}$$

There

$$A(\tau,\xi) = \frac{\sin\pi(\xi-p\Delta f-r\tau)(T-|\tau-pT_r|)}{\pi(\xi-p\Delta f-r\tau)(T-|\tau-pT_r|)}\cdot\frac{T-|\tau-pT_r|}{T}, \quad |\tau-pT_r| < T$$
$$B(\tau,\xi) = \frac{\sin\pi(N-|p|)[(\xi-p\Delta f-r\tau)T_r-\Delta f\tau]}{\sin\pi[(\xi-p\Delta f-r\tau)T_r-\Delta f\tau]}, \quad |\tau-pT_r| < T$$

For any p, the ambiguity function is

$$\begin{aligned}|\chi_p(\tau,\xi)| = {} & \left|\frac{\sin \pi(\xi - p\Delta f - r\tau)(T - |\tau - pT_r|)}{\pi(\xi - p\Delta f - r\tau)(T - |\tau - pT_r|)}\right| \cdot \left|\frac{T - |\tau - pT_r|}{T}\right| \\ & \times \left|\frac{\sin \pi(N - |p|)[(\xi - p\Delta f - r\tau)T_r - \Delta f\tau]}{\sin \pi[(\xi - p\Delta f - r\tau)T_r - \Delta f\tau]}\right|, \quad |\tau - pT_r| < T\end{aligned} \tag{6.43}$$

6.2.2 Ambiguity Function of Central Ambiguity Region

In the following, the ambiguity function of the central ambiguity region is used as an example to illustrate the resolution performance of stepped-chirp burst signal. For the central ambiguity band, $p = 0$ the ambiguity function is simplified as:

$$\begin{aligned}|\chi_c(\tau,\xi)| = {} & \left|\frac{\sin \pi NT_r[(\xi - (r + \Delta f/T_r)\cdot\tau]}{\sin \pi T_r[(\xi - (r + \Delta f/T_r)\cdot\tau]}\right|, \\ & \times \left|\frac{\sin \pi(\xi - r\tau)(T - |\tau|)}{\pi(\xi - r\tau)(T - |\tau|)}\right| \cdot \left|\frac{T - |\tau|}{T}\right|, \quad |\tau| < T\end{aligned} \tag{6.44}$$

Compared with the ambiguity function of the single carrier pulse waveform in Eq. (5.14), which of the stepped-chirp burst signal in Eq. (6.44) was rotated twice. The first rotation is slope $\Delta f/T_r$ which is caused by the stepped-frequency sub-pulses, and the second rotation is slope $\Delta f/T$ which is caused by linear frequency modulation in the pulse duration. However, the rotation of ambiguity function of stepped-frequency signal only occurs for the first time, single frequency for each sub-pulse, which does not rotate second times.

If $\xi = 0$, then Eq. (6.44) can be rewritten as

$$|\chi_c(\tau,0)| = \left|\frac{\sin \pi N(rT_r + \Delta f)\tau}{\sin \pi(rT_r + \Delta f)\tau}\right| \cdot \left|\frac{\sin \pi r(T - |\tau|)\tau}{\pi r(T - |\tau|)\tau}\right| \cdot \left|\frac{T - |\tau|}{T}\right|, \quad |\tau| < T \tag{6.45}$$

Equation (6.45) shows that the ambiguity function in the central ambiguity region has the sinc shape, which decides the range resolution. The peak position of range ambiguity is $\tau = 0$, and the peak value is 1. When $|\tau| = 1/N(rT_r + \Delta f)$, the ambiguity function crosses zero for the first time. The main lobe width is $2/N\Delta f$, and the normal resolution is $\rho_\tau = 1/N\Delta f$.

If $\tau = 0$, then Eq. (6.44), the velocity ambiguity function can be rewritten as

$$|\chi_c(0,\xi)| = \left|\frac{\sin \pi NT_r\xi}{\sin \pi T_r\xi}\right| \cdot \left|\frac{\sin \pi\xi T}{\pi\xi T}\right|, \quad |\xi| < \frac{\Delta f}{T_r}T \tag{6.46}$$

The peak position of velocity ambiguity function is $\xi = 0$. Because of $T_r > T$, $1/NT_r < 1/T$ the first zero point of velocity ambiguity function is $|\xi| = 1/NT_r$, so the normal velocity resolution is $\rho_\xi = 1/NT_r$.

Parameters: B = 100 Hz, $T = 0.01$, $T_r = 0.05$, $\Delta f = 20$, N = 5. The ambiguity function, Time-frequency domain projection, range, and Doppler ambiguity at peak, profile of central ambiguity region are all shown in Figs. 6.10, 6.11, 6.12, 6.13, 6.14 and 6.15.

Comparison between Figs. 6.10 and 5.4 show that both waveforms have range ambiguity, and the ambiguity period equals to the pulse repetition period. The range ambiguity volume of stepped-chirp burst is more concentrated at the origin, and the ambiguity volume dispersed around the adjacent ambiguity region is smaller. The first ambiguity side lobe is only −26 dB. The range ambiguity of stepped-frequency is more dispersed, and the first ambiguity side lobe is as high as −8.4 dB.

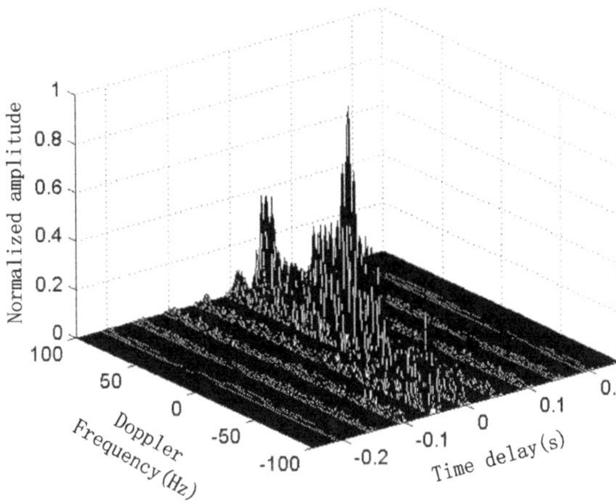

Fig. 6.10 The ambiguity function of stepped-chirp burst

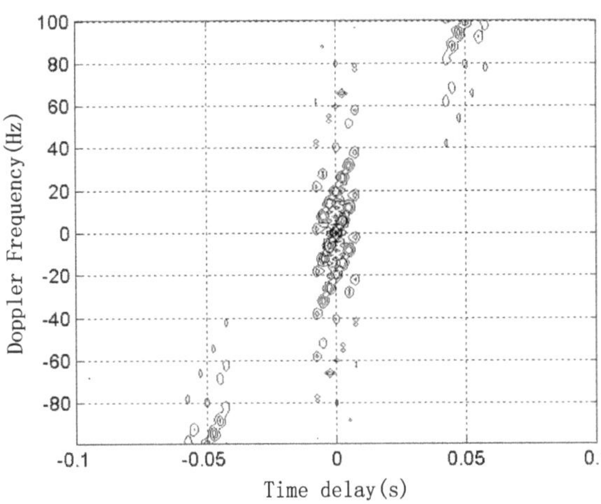

Fig. 6.11 Time-frequency domain projection of ambiguity function

Fig. 6.12 Doppler ambiguity at peak

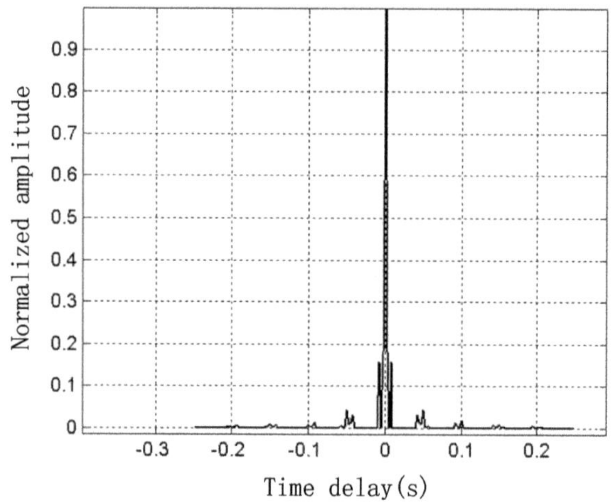

Fig. 6.13 Range ambiguity at peak

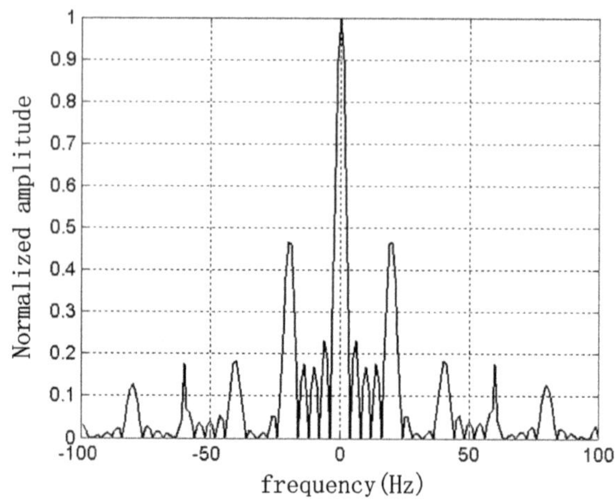

Fig. 6.14 The profile along the time delay axis in the central ambiguity region

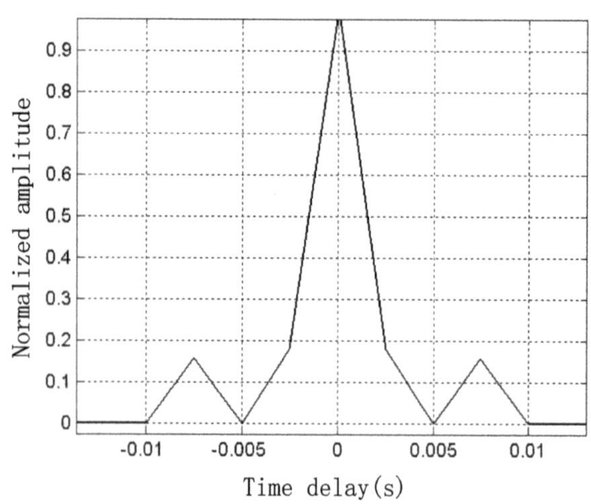

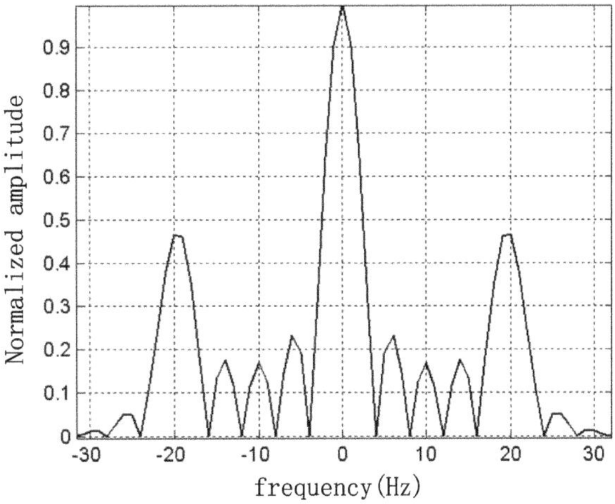

Fig. 6.15 The profile along the frequency shift axis in the central ambiguity region

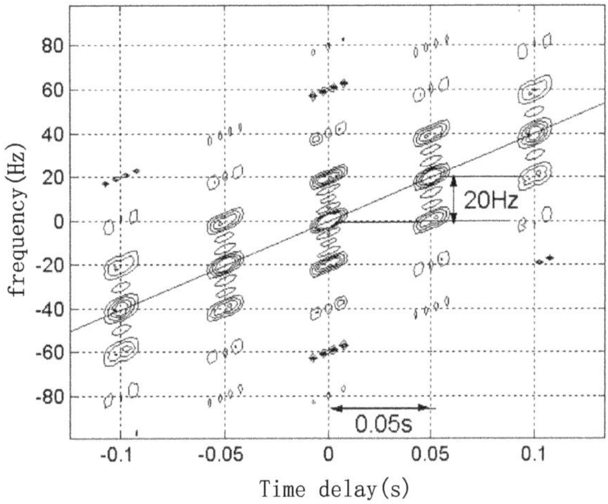

Fig. 6.16 The amplified time-frequency projection of stepped-frequency

Therefore, the main lobe of the stepped-chirp burst waveform is more sharp and narrow, and range resolution is higher. By the comparison of the velocity ambiguity function, the velocity ambiguity functions of the two waveforms are the same.

The following focuses on the inherent relationship analysis of ambiguity function between stepped-frequency and stepped-chirp. The time-frequency projection of stepped-frequency in Fig. 5.5 is amplified to be shown in Fig. 6.16, and that of stepped-chirp in Fig. 6.11 is amplified in Fig. 6.17.

Figure 6.16 shows that the counterclockwise rotation angle of ambiguity figure of stepped-frequency is $\tan^{-1}(\Delta f/T_r)$, that means that the changing of Doppler frequency $\Delta f_d = \Delta f = 20\,\text{Hz}$ is 20 Hz in the interval of 0.05 μs. Figure 6.17 shows

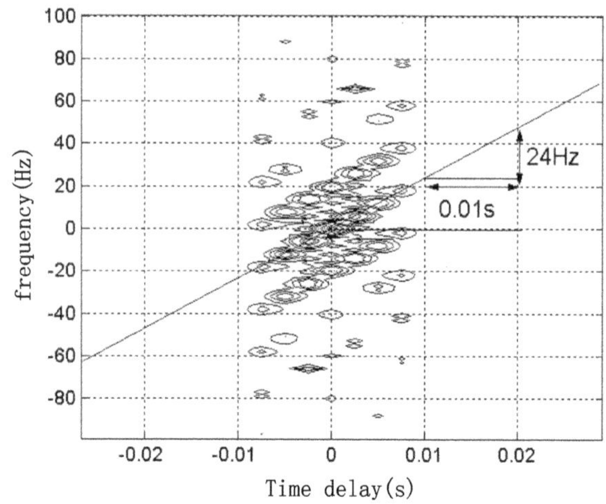

Fig. 6.17 The amplified time-frequency projection of stepped-chirp

that the counterclockwise rotation angle of ambiguity figure of stepped-chirp is $\tan^{-1}(\Delta f/T_r + \Delta f/T)$, that means that the changing of Doppler frequency is $\Delta f_d = \Delta f + \frac{T}{T_r}\Delta f = 24\,\mathrm{Hz}$ in the interval of 0.01 μs. Therefore, Fig. 6.17 can be looked as the Fig. 6.16 rotated counterclockwise by $\tan^{-1}(\Delta f/T)$.

The ambiguity figure of stepped-chirp can be looked as an anticlockwise rotation of that of the stepped-frequency waveform. It is more similar to the shear blade type, but the inside of the blade also has a nail bed type, and there is a range-Doppler coupling. Because of the periodic emission of the pulse, the ambiguity function also has the periodic ambiguity band. For the central ambiguity band, the range ambiguity volume is concentrated in one sub-pulse width, rather than directly scattered in each pulse repetition period, so the stepped-chirp has better range and velocity resolution.

6.2.3 *Conclusion*

The ambiguity function and the resolution performance of stepped-chirp are concluded as follows:

(1) The ambiguity figure of stepped-chirp can be looked as an anticlockwise rotation $\tan^{-1}(\Delta f/T)$ that of the stepped-frequency waveform.
(2) The range ambiguity main lobe at the origin of stepped-chirp is more sharp and narrow, and the ambiguity volume is more concentrated for better range resolution.

6.3 The Effect and the Compensation of Radial Velocity on Synthetic Range Profile of Stepped-Chirp Burst [8, 9]

6.3.1 *The Effect of Radial Velocity on Synthetic Range Profile of Stepped-Chirp Burst*

For stationary targets, wideband or ultra-wideband echo can be synthesized by a certain number of narrowband sub-pulses; they are completely equivalent both in time domain and in frequency domain. However, when the radial velocity between the target and the radar is not zero, the synthetic spectrum of the sub-pulses is no longer consistent *wi*th the wideband spectrum. The effect of radial velocity on the synthetic wideband spectrum is discussed in detail, and then the velocity compensation method is presented.

Assuming the initial distance between the radar and the target is R_0, the target is moving away from the radar at a radial velocity of v. The radar transmitting signal is shown ideally in Eq. (6.12). When the radial velocity between the target and the radar is not zero, the received signal is expressed as

$$r(t) = \sum_{n=0}^{N-1} r_n^v(t) = \sum_{n=0}^{N-1} \mathrm{rect}\left(\frac{t - \Delta t_n - \tau_n}{T}\right) \exp j\left[2\pi f_c(t - \tau_n) + \pi r(t - \tau_n)^2\right] \tag{6.47}$$

where

$$\tau_n = \frac{2R_0}{c} + \frac{2vt}{c} \approx \frac{2R_0}{c} + \frac{2vnT_r}{c} \tag{6.48}$$

If $\tau_0 = \frac{2R_0}{c}$, $\Delta\tau_n = \frac{2vt}{c}$ then the time delay of echo is

$$\tau_n = \tau_0 + \Delta\tau_n \tag{6.49}$$

The output of any sub-pulse echo mixed with the reference is

$$m_n^v(t) = \mathrm{rect}\left(\frac{t - \Delta t_n - \tau_n}{T}\right) \exp j\left[-2\pi f_c \tau_n + \pi r(t - \tau_n)^2\right] \tag{6.50}$$

Compared with the Eq. (6.14), the output of echo mixed with the reference of the stationary targets, the relation between the stationary target echo and the moving target echo is given by

$$m_n^v(t, \tau_n) = m_n^0(t - \Delta\tau_n, \tau_0) \exp(-j2\pi f_c \Delta\tau_n) \tag{6.51}$$

Let the phase is

$$\phi_{dn} = -j2\pi f_c \Delta\tau_n = -j2\pi \frac{2vf_c}{c} t \tag{6.52}$$

When the target is moving away from the radar at a radial velocity of v, the time delay $\Delta\tau_n$ equals to $2vt/c$, and the Doppler frequency shift f_{dn} equals to $2vf_c/c$.

From the above analysis, it can be known that the time delay within the burst duration caused by target motion is not fixed at τ_0, and the center frequency of the echo after mixing is no longer zero, but changing with the Doppler shift of the sub-pulses. If the matched filter in Eq. (6.15) is matched with the moving target echo, the matched filtering process is given as follows.

When $-T_w + \tau_n \le t \le \tau_n$, the output of sub-pulse via the matched filter is

$$\begin{aligned} y_n(t) &= \int_{-\frac{T_w}{2}+\tau_n}^{\frac{T_w}{2}+t} m_n^v(\tau) h_n(t-\tau)\mathrm{d}\tau \\ &= \int_{\Delta t_n+\tau_n-\frac{T}{2}}^{-\Delta t_n+\frac{T}{2}+t} m_n^v(\tau) h_n(t-\tau)\mathrm{d}\tau \\ &= \frac{1}{j2\pi r(t-\tau_n)}\left\{\exp\left[j2\pi r(t-\tau_n)\left(\frac{t+T-\tau_n}{2}-\Delta t_n\right)\right]\right. \\ &\quad \left.-\exp\left[-j2\pi r(t-\tau_n)\left(\frac{t+T-\tau_n}{2}-\Delta t_n\right)\right]\right\} \end{aligned} \tag{6.53}$$

The coherent superposition output of N sub-pulses compression is

$$y(t) = \sum_{n=0}^{N-1} y_n(t), \quad -T_w + \tau_0 \le t \le \tau_0$$

Obviously, since the time delay varies with sub-pulse echoes, as coherently superposed by compressed sub-pulse echoes, the peak response of the range profile could not be superimposed within a range resolution cell, however, the peak response of range dispersed into the near range cells, and the degree of dispersion depends on the target radial velocity. Finally, this will cause the resolution of synthesis range profile of target descend and the SNR loss. And the number of sub-pulse is more, the resolution of synthesis range profile is worse; the level of side lobe is more high, and loss of SNR becomes higher. The discussion about the effect of radial velocity to synthesis spectrum and the pulse compressing results is following.

Simulation:

The main parameters are same as Sect. 6.1.3. The target is moving away from the radar at the radial velocity of 200 m/s. The echo spectrum and the range profile of the target are obtained by both the wideband signal and the stepped-chirp waveform.

The first experiment:

The radar transmits a wideband signal, the center frequency is 3.2 GHz and the bandwidth is 400 MHz. The RF (Radio Frequency) echo is demodulated down to IF (Intermediate Frequency), and then sent to the low-pass filter for VF (Video Frequency) signal, eventually, the digital matched filter is used to the A/D samplings. The baseband spectrum is shown in Fig. 6.18, and the output after the matched filter is shown in Fig. 6.19.

The second experiment:

The radar transmits two sub-pulses; whose bandwidth is 200 MHz and the carrier frequency are 3.1 and 3.3 GHz, respectively. Each sub-pulse echo is mixed with the respective reference to IF and then sent to a low-pass filter. The synthetic baseband spectrum is shown in Fig. 6.20, and the synthetic range profile is shown in Fig. 6.21.

The third experiment:

The radar transmits three sub-pulses; whose bandwidth is 133.33 MHz and the carrier frequency are 3.067, 3.2, and 3.333 GHz, respectively. The synthetic baseband spectrum is shown in Fig. 6.22, and the synthetic range profile is shown in Fig. 6.23.

The fourth experiment:

The radar transmits four sub-pulses; whose bandwidth is 100 MHz and the carrier frequency are 3.05, 3.15, 3.15, and 3.35 GHz, respectively. The synthetic baseband spectrum is shown in Fig. 6.24, and the synthetic range profile is shown in Fig. 6.25.

Fig. 6.18 The spectrum of 400 MHz wideband signal

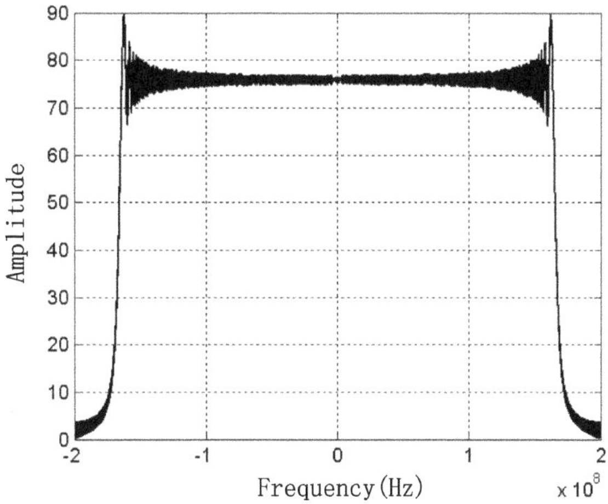

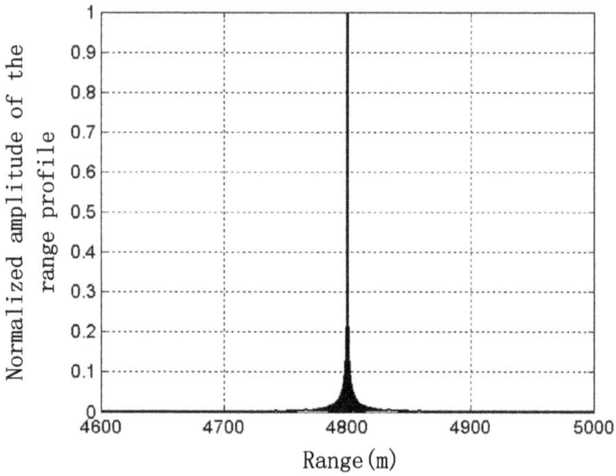

Fig. 6.19 The output of pulse compression

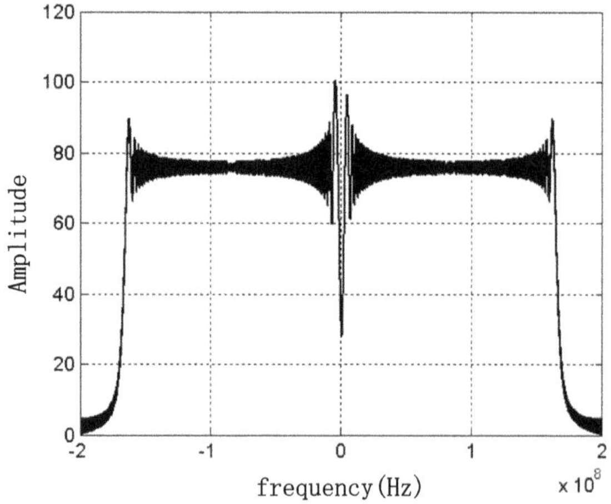

Fig. 6.20 The spectrum synthesized by two 200 MHz sub-pulses

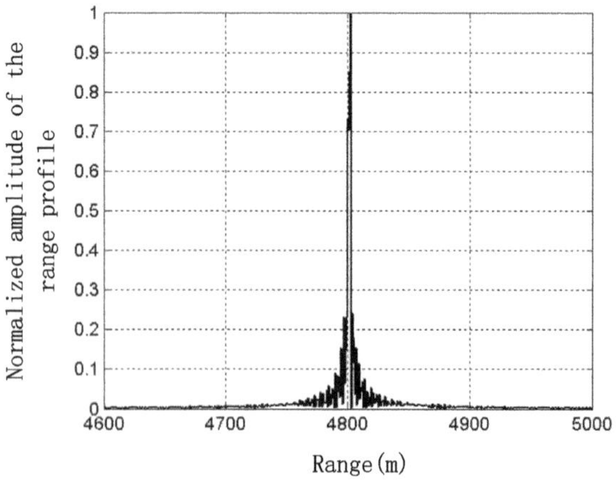

Fig. 6.21 The range profile synthesized by two 200 MHz sub-pulses

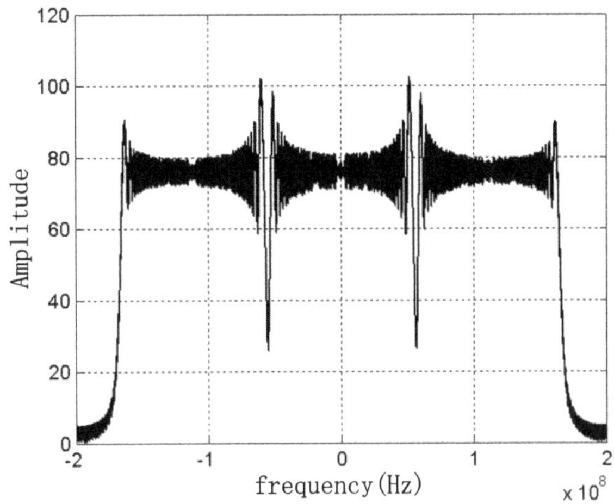

Fig. 6.22 The spectrum synthesized by three 133 MHz sub-pulses

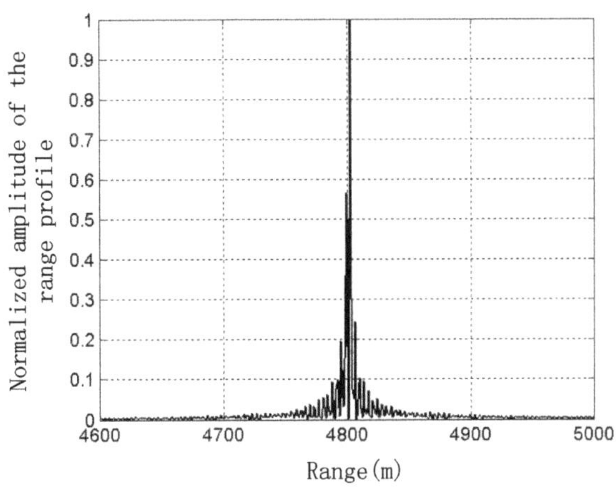

Fig. 6.23 The range profile synthesized by three 133 MHz sub-pulses

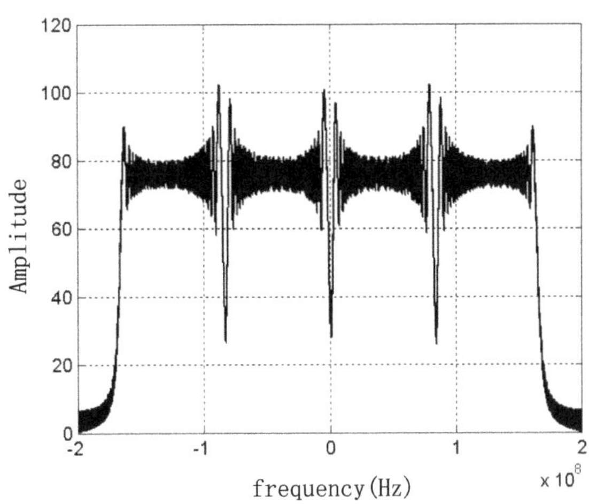

Fig. 6.24 The spectrum synthesized by four 100 MHz sub-pulses

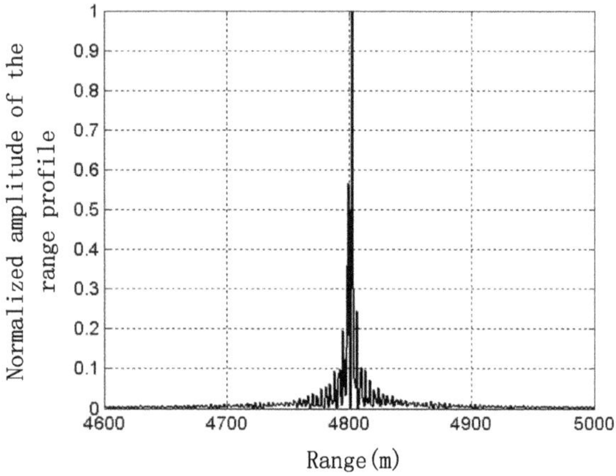

Fig. 6.25 The range profile synthesized by four 100 MHz sub-pulses

Table 6.1 The performance parameters of the range profiles synthesized via different number of sub-pulses

Number	Range resolution/m	PSLR/dB	ISLR/dB	Loss of SNR/dB
1	0.375	−13.2651	−9.8034	$-5.4289e^{-4}$
2	0.375	−3.9582	−2.7865	−1.2143
3	0.375	−4.9547	−2.2910	−1.4898
4	0.375	−4.4593	−1.5020	−2.4648

From the comparison between these four experiments can be seen that the number of sub-pulses is more, the deviation from the rectangular amplitude spectrum is greater, however, the synthetic bandwidth remains 400 MHz unchanged. The main reason is that the time–bandwidth product $TB = \frac{T_w B_w}{N^2}$ is inversely proportional to the square of the number of sub-pulses. Along with the increasing of the number of sub-pulses, the quality of the synthetic spectrum becomes worse. When the increase in the number of sub-pulses makes the time–bandwidth product less than 30, the amplitude spectrum of each sub-pulse will not be rectangular and the phase spectrum is not square law. Table 6.1 shows the performance comparison of the range profiles synthesized via different number of sub-pulses.

Table 6.1 shows that along with increasing of the number of sub-pulses, the PSLR, and the ISLR are degraded, the mismatched degree between synthetic wideband echo and the matched filter become greater, so the output of pulse compression gets worse. Therefore, when the radial velocity is not zero, it is necessary to considerate the velocity compensation, it is a common characteristic of the stepped-chirp signal.

6.3.2 The Radial Velocity Compensation of Stepped-Chirp Burst Signal

For the initial distance between the radar and the target is R_0, the target is moving away from the radar at a radial velocity of v. The output signal of sub-pulse echo mixed with reference is expressed as

$$m_{n0}^{v}(t) = \mathrm{rect}\left(\frac{t - nT_r - \tau_n}{T}\right) \exp j\left[-2\pi f_n \tau_n + \pi r(t - nT_r - \tau_n)^2\right] \tag{6.54}$$

For any echo of sub-pulse, as the time delay is τ_n, the following is obtained

$$m_{n}^{v}(t, \tau_n) = m_{n0}^{v}(t + nT_r - \Delta t_n, \tau_n) \exp(j2\pi\Delta f_n t) \exp(-j\pi r \Delta t_n^2) \tag{6.55}$$

When $\tau_n = \tau_0$, Eq. (6.55) is restored to Eq. (6.24).

From the relation between the moving target echo and the stationary target echo given by Eq. (6.51), the expression is

$$m_{n}^{0}(t, \tau_0) = m_{n}^{v}(t + \Delta\tau_n, \tau_n) \exp(j2\pi f_c \Delta\tau_n) \tag{6.56}$$

The Eq. (6.56) can be further expressed as the relation between the moving target sub-pulse echo $m_{n0}^{v}(t)$ and the stationary target sub-pulse echo $m_{n}^{0}(t)$

$$\begin{aligned} m_{n}^{0}(t, \tau_0) &= m_{n0}^{v}(t + nT_r - \Delta t_n + \Delta\tau_n, \tau_n) \\ &\quad \times \exp(j2\pi\Delta f_n t) \exp\left(j2\pi \frac{2vf_n}{c} t\right) \exp(-j\pi r \Delta t_n^2) \end{aligned} \tag{6.57}$$

Equation (6.57) indicates the relation of radial velocity compensation, and for the moving target sub-pulse echo $m_{n0}^{v}(t)$ the compensation steps are given as follows:

(1) The signal $m_{n0}^{v}(t)$ is time shifted $t + nT_r - \Delta t_n + \Delta\tau_n$ and got $m_{n0}^{v}(t + nT_r - \Delta t_n + \Delta\tau_n, \tau_n)$.
(2) Then this signal is Frequency shifted $\exp(j2\pi\Delta f_n t)$ to correct the center position of the frequency spectrum.
(3) To compensate the Doppler shift $\exp\left(j2\pi \frac{2vf_n}{c} t\right)$ caused by the radial velocity.
(4) To compensate the phase $\exp(-j\pi r \Delta t_n^2)$ and get the ideal synthetic sub-pulse echo of the stationary target.

The velocity compensation method is verified by following simulations.

The main parameters of wideband signal: the center frequency is 3.2 GHz, the bandwidth is 600 MHz, the pulse duration is 20 μs, and the PRF is 4000 Hz. The radial velocity is 200 m/s.

The main parameters of stepped-chirp burst signal: the number of the sub-pulse is 3, the bandwidth of each sub-pulse is 200 MHz, and the sub-pulse duration is 6.67 μs. The synthetic spectrum and the synthetic profile of three sub-pulses are shown in Figs. 6.26 and 6.27, and if the velocity is compensated, they are presented in Figs. 6.28 and 6.29.

It can be seen from above experiments that the synthetic spectrum of stepped-chirp burst without velocity compensation is not smooth, which deviates from the rectangle, but the velocity compensation can make it back to the ideal rectangular spectrum. Although the main lobe width of synthetic range profile without velocity compensation remains unchanged 0.25 m, the first side lobe level is up to −11.78 dB. However, the velocity compensation can make the first side lobe level lower to −13.46 dB.

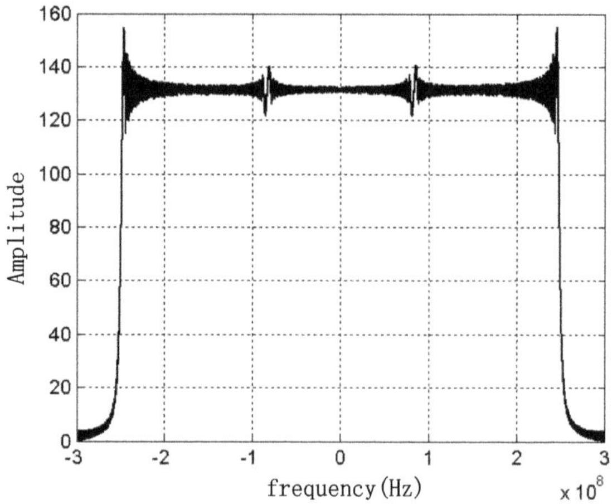

Fig. 6.26 The synthetic spectrum of stepped-chirp without velocity compensation

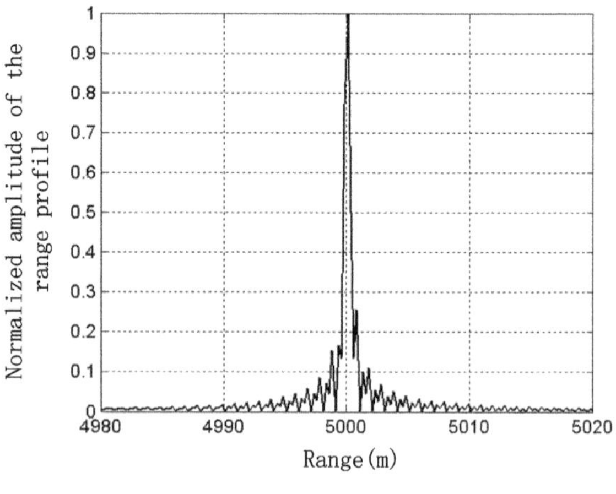

Fig. 6.27 The synthetic range profile without velocity compensation

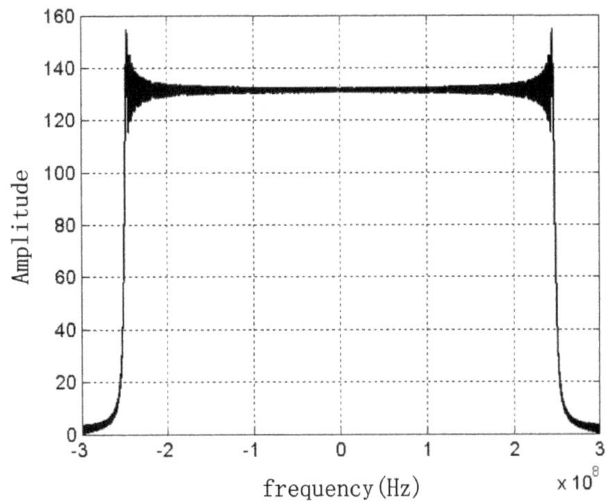

Fig. 6.28 The synthetic spectrum of stepped-chirp after velocity compensation

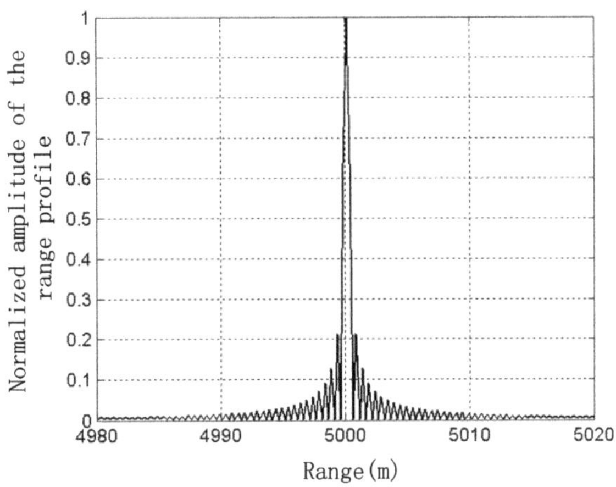

Fig. 6.29 The synthetic range profile after velocity compensation

The experimental results show that the velocity compensation method is correct and effective. On the other hand, it can be seen that the radial velocity has a certain influence on the synthetic range profile of the stepped-chirp burst, so it is necessary to compensate the radial velocity according to the actual situation and system requirements.

6.4 The Stepped-Chirp SAR System and Imaging

In all above sections, the wideband synthesis principle of narrowband stepped-chirp burst waveform is described, and the ambiguity function is expressed. In addition, the effect of the radial velocity on the matched filter is discussed. As a result, the method of stepped-chirp burst waveform design is presented in detail. Therefore, compared with stepped-frequency signal, stepped-chirp burst waveform has the advantages of design flexibility, lower side lobe, less disturbance of the radial velocity to the range profile; it is more suitable for engineering implementation. The stepped-chirp SAR system design and imaging principle will be discussed as follows, which will tell us bow to obtain the two-dimensional high-resolution SAR image without increasing the instantaneous bandwidth of the system.

6.4.1 The Stepped-Chirp SAR System

The stepped-chirp waveform can be looked as the wideband LFM signal divided by a burst of narrowband sub-pulses, and the sub-pulses are transmitted one by one, then each echo is received in turn. The RF echo is demodulated down to IF, and the output signal after IF quadrature demodulated is sent to the low-pass filter for dual I/Q (In-phase and Quadrature) video signal. Eventually, the echo of each channel is sampled and synthesized to a wideband digital signal by time shift, frequency shift, and phase compensation. The Stepped-chirp SAR system block diagram is illustrated in Fig. 6.30.

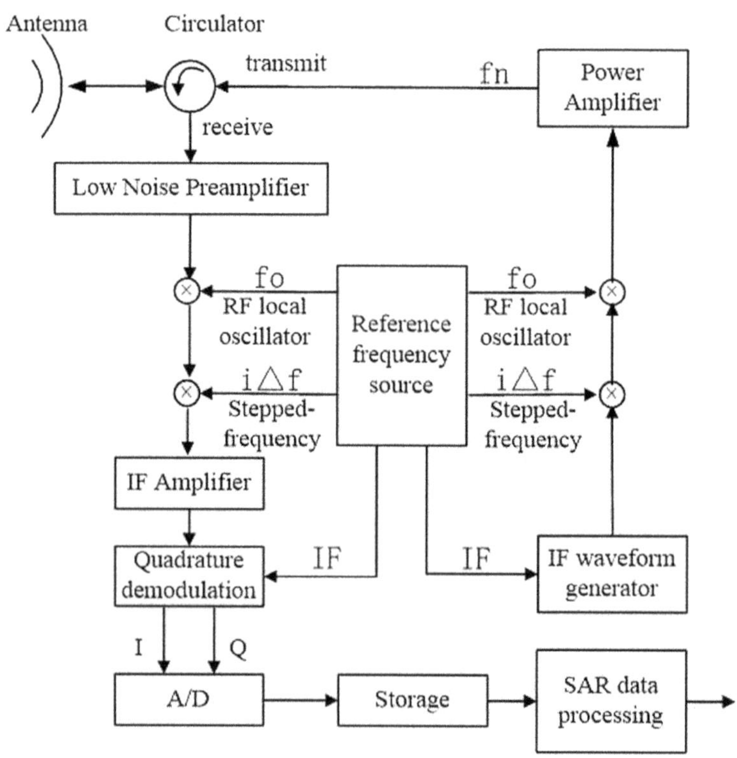

Fig. 6.30 Stepped-chirp SAR system block diagram

Reference frequency source

Taking the orthogonal side-looking SAR as an example, radar transmits M stepped-chirp bursts within a synthetic aperture time, and each stepped-chirp burst consists of N narrowband sub-pulses, the pulse repetition period of each sub-pulse is T_r, and the time duration is T. The difference from the step-frequency signal is that each sub-pulse is not a single frequency signal, but is a narrowband LF signal with step carry frequency

The transmitting waveform is expressed as

$$\begin{gathered} x_{ki} = A_{ki}\cos(2\pi f_i t + \pi r t^2 + \theta_{ki}) \\ (i + kN)T_r \leq t \leq (i + kN)T_r + T \end{gathered} \tag{6.58}$$

where $k = -M/2, \ldots, M/2$ is the index of bursts, $i = -N/2, \ldots, N/2$ is the index of sub-pulses in one burst. A_{ki} Is the initial amplitude, and θ_{ki} is the initial phase.

The received signal is

$$\begin{gathered} y_{ki} = B_{ki}\cos\left\{2\pi f_i[t - \tau(t)] + \pi r[t - \tau(t)]^2 + \theta_{ki}\right\} \\ (i + kN)T_r \leq t \leq (i + kN)T_r + T \end{gathered} \tag{6.59}$$

where

$$\tau(t) = \frac{2R(t)}{c} = \frac{2\sqrt{R_c^2 + (V_a t)^2}}{c} \tag{6.60}$$

Generally, the slant range is much larger than the synthetic aperture length, that means $|V_a t| \ll R_c$ So Fresnel approximation can be used to Eq. (6.60), the round-trip delay between radar and target can be rewritten as

$$\tau(t) = 2\left[\frac{R_c}{c} + \frac{(V_a t)^2}{2R_c c}\right] \tag{6.61}$$

The time delay in Eq. (6.61) is expressed as

$$\tau(t) = \frac{2R_c}{c} + \frac{V_a t}{R_c}\frac{V_a t}{c} \tag{6.62}$$

The antenna pattern in the slant range plane shows that

$$ctg\theta = \frac{V_a t}{R_c} \tag{6.63}$$

When $|V_a t| \ll R_c$, $ctg\theta$ approximately equals to zero, and θ approximately equals to 90°, so it can get

$$ctg\theta \approx \cos\theta$$

The round-trip delay in Eq. (6.62) can be approximately expressed as

$$\tau(t) = \frac{2R_c}{c} + \frac{V_a \cos\theta}{c} t \tag{6.64}$$

where $V_a \cos\theta$ is the component of the aircraft flight velocity in the radial direction between the radar and the target. If $R_c \gg |V_a t|$, it can be looked as unchanged within one burst.

When $\tau_0 = \frac{2R_c}{c}$, $\Delta\tau_{ki} = \frac{V_a \cos\theta}{c} t$, Eq. (6.64) can be simplified as

$$\tau(t) = \tau_0 + \Delta\tau_{ki} \tag{6.65}$$

The signal transmitted to the low-pass filter after quadrature demodulation is expressed as

$$\begin{gathered} m_{ki0}^{v} = C_{ki} \cos\left[-2\pi f_i \tau(t) + \pi r (t - \tau(t))^2\right] \\ (i + kN)T_r + \tau(t) \le t \le (i + kN)T_r + T + \tau(t) \end{gathered} \tag{6.66}$$

The phase function is

$$\phi_{ki} = -2\pi f_i \tau(t) + \pi r (t - \tau(t))^2 \tag{6.67}$$

将 (6.65) 式表示的回波延迟带入上式, 相位展开式为:

Take the time delay of Eq. (6.65) into Eq. (6.67), the phase can spread as

$$\varphi_{ki} = -2\pi f_i \tau_0 + \pi r (t - \tau_0)^2 - 2\pi f_i \Delta\tau_{ki} - 2\pi r \Delta\tau_{ki}(t - \tau_0) + \pi r \Delta\tau_{ki}^2 \tag{6.68}$$

Substituting $\Delta\tau_{ki}$ into Eq. (6.68), the following is obtained

$$\varphi_{ki} = -2\pi f_i \tau_0 + \pi r \tau_0^2 - 2\pi a(\theta) t + \pi r b(\theta) t^2 \tag{6.69}$$

where

$$a(\theta) = f_i \frac{V_a \cos\theta}{c} + r\tau_0 \left(1 + \frac{V_a \cos\theta}{c}\right) \tag{6.70}$$

$$b(\theta) = \left(1 - \frac{V_a \cos\theta}{c}\right)^2 \tag{6.71}$$

Because of $|V_a \cos\theta| \ll c$, so $\frac{|V_a \cos\theta|}{c} \ll 1$, Eqs. (6.71) and (6.72) can be approximately expressed as

$$a(\theta) \approx f_i \frac{V_a \cos\theta}{c} + r\tau_0 \tag{6.72}$$

$$b(\theta) \approx 1 \tag{6.73}$$

Using Eqs. (6.72) and (6.73) to express Eq. (6.69), the phase is given by

$$\varphi_{ki} = -2\pi f_i \tau_0 + \pi r \tau_0^2 - 2\pi f_i \frac{V_a \cos\theta}{c} t - 2\pi r \tau_0 t + \pi r t^2 \tag{6.74}$$

The carrier frequencies of sub-pulses are

$$f_i = f_c + \left(i + \frac{1-N}{2}\right) B, \quad i = 0, 1, \ldots, N-1 \tag{6.75}$$

Let $f_0 = f_c + \frac{1-N}{2} B$, then $f_i = f_0 + iB$. Equation (6.74) can be rewritten as

$$\begin{aligned}
\varphi_{ki} = & -2\pi f_c \tau_0 + \pi r \tau_0^2 - 2\pi \left(r\tau_0 + f_0 \frac{V_a \cos\theta}{c} \right) kNT_r \\
& - 2\pi \left(r\tau_0 + f_0 \frac{V_a \cos\theta}{c} + \frac{B}{T_r} \tau_0 \right) iT_r \\
& + \pi r k^2 N^2 T_r^2 + \pi r \left(rT_r - 2B \frac{V_a \cos\theta}{c} \right) i^2 T_r^2 \\
& + 2\pi \left(rT_r - B \frac{V_a \cos\theta}{c} \right) kNiT_r^2
\end{aligned} \tag{6.76}$$

In the phase expression, the first and second terms are constants; these are no meaning to pulse compression in range and the azimuth dimensions. The third term shows the relationship of the azimuth Doppler center via the radar position. Because the stepped-chirp waveform can be looked as a burst of N narrowband sub-pulses, the bandwidth is reduced to $1/N$ and PRF increases to N times, so the third term, the azimuth Doppler center frequency during the burst, can be used to synthesize the SAR azimuth Doppler bandwidth.

The fourth is required to synthesize the wideband echo, however, duo to Doppler frequency shift, the velocity compensation has to be applied to sub-pulse coherent synthesis. The velocity compensation has to be applied to sub-pulse coherent synthesis. The fifth term is the necessary phase for pulse compression in the azimuth, and the sixth term is used to pulse compression in range. The seventh is the coupling between the azimuth and range dimensions, which should be eliminated in sub-pulse echoes processing. In a word, the stepped-chirp waveform can obtain the equivalent echo as wideband LFM signal by bandwidth synthesis in range and Doppler bandwidth synthesis in the azimuth dimension; furthermore, the coupling between the azimuth and range dimensions has to be eliminated. The processing methods will be described in the next section.

6.4.2 Stepped-Chirp SAR Imaging

The signal processing block diagram of stepped-chirp SAR is shown in Fig. 6.31.

The steps of the signal processing of stepped-chirp SAR imaging are given as follows:

The first step: Compensating the coupling between the azimuth and the range dimensions.

According to the radial velocity $V_a \cos\theta$ and the pulse duration in one burst, the coupling compensation factor is defined as

$$\mu = \exp\left[-2\pi\left(rT_r - B\frac{V_a \cos\theta}{c}\right)kNiT_r^2\right] \tag{6.77}$$

The second step: the radial velocity Compensation in the range dimensions.

Since the radial velocity introduces the Doppler frequency shift to the sub-pulse echo, the center frequency of the baseband spectrum is not zero, it makes coherence synthetic frequency spectrum of sub-pulse get error. So it needs the radial velocity compensation to the sub-pulse echo for coherent synthesis. The radial velocity compensation is given by Eq. (6.57).

The third step: Coherent synthesis wideband echo in range dimensions.

The baseband sub-echoes after the radial velocity compensated, is transformed to the ideal sub-echoes by the frequency shift, time shift, and phase compensation, then these sub-echoes are coherently synthesized to equivalent wideband echo in range, and at the same time the equivalent Doppler bandwidth is coherent synthesized in the azimuth.

The fourth step: Matched filter in range and the azimuth dimensions.

Range-Doppler algorithm is used to stepped-chirp SAR imaging as conventional LFM SAR. First, the range pulse is compressed, and then the data is rearranged, and the azimuth matched filtering is carried out for different azimuth data. Finally, the two-dimensional high-resolution SAR image is obtained.

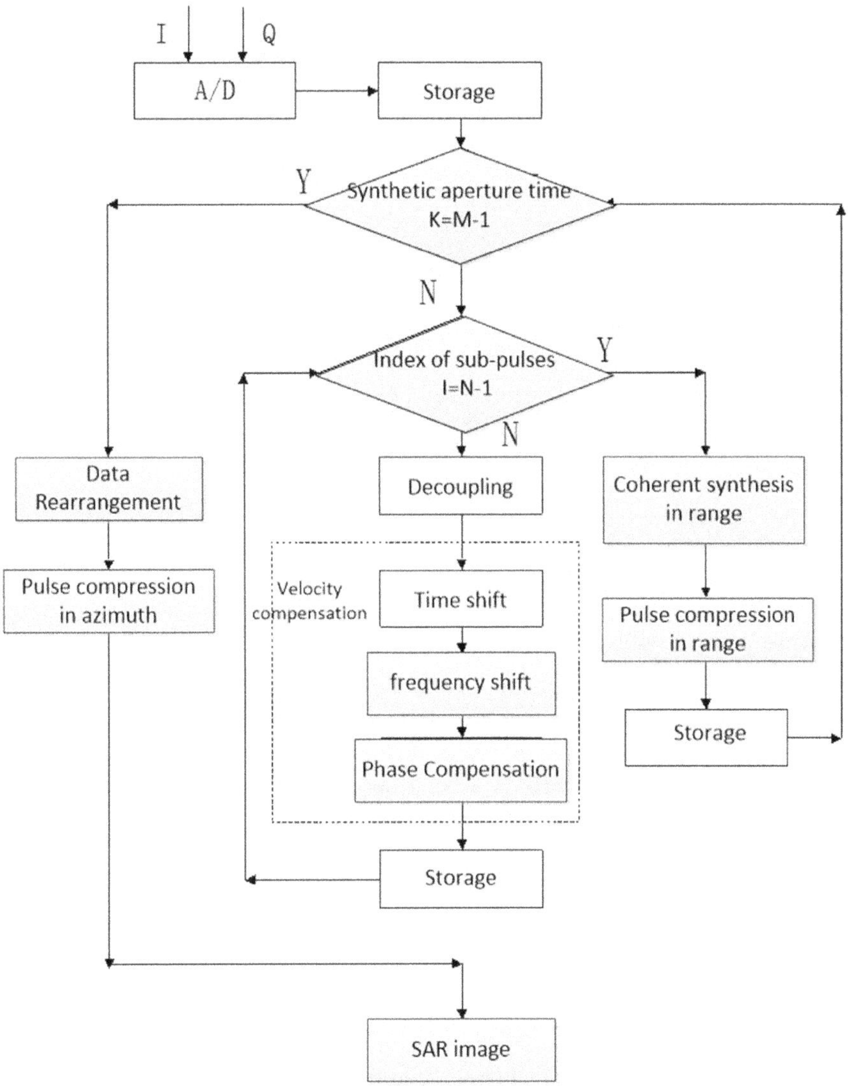

Fig. 6.31 The signal processing block diagram of stepped-chirp SAR

6.4.3 *Simulation*

The simulation is to verify the correctness of the imaging principle and method.

Suppose that the velocity of the aircraft flight V_a is 200 m/s, the vertical range from point target to airline R_c is 10 km, the center frequency f_c is 3.2 GHz, the antenna length D_a is 2 m, the swath width is 3 km, the range resolution Δr is 0.3 m. The parameters of the stepped-chirp waveform are given as follows:

(1) Pulse Repetition Frequency (PRF)

In order to avoid the ambiguity in range and the azimuth dimensions, f_{PRF} should satisfy:

$$\frac{2\beta V_a}{\lambda} \leq f_{\text{PRF}} \leq \frac{c}{2R_{\max}}$$

f_{PRF} Is set to 15 kHz based on known conditions.

(2) The synthetic aperture time T_s equals 1.17 s.
(3) The radial velocity is limited between 0 and 7.5 m/s.
(4) Signal bandwidth

$$B_w = \frac{c}{2\Delta r}$$

Signal bandwidth B_w is 500 MHz.

Considering the loss of the antenna pattern weighting and the signal processing, the actual signal bandwidth should be designed as 600 MHz.

(5) The number and the bandwidth of sub-pulses in one burst
The 200–500 MHz sampling rate is a mature technology for the existing common A/D converter, so three sub-pulses with 200 MHz bandwidth are used to synthesize the 600 MHz wideband signal. That means N equals 3, and bandwidth of sub-pulse is 200 MHz.
(6) The time duration of each sub-pulse is 5 μs.
(7) The number of the bursts

$$M = \frac{f_{\text{PRF}} * T_s}{N}$$

The number of the bursts equals to 5850.

The first experiment: The point target echo simulation without decoupling in two dimensions and compensating the radial velocity.

In order to study the effect of the coupling and the radial velocity on SAR imaging, the point target image without any compensation in the coherent synthesis is shown in Fig. 6.32, and the top view of the image is shown in Fig. 6.33. Furthermore, the profiles in range and the azimuth dimensions at the target response peak are shown in Figs. 6.34 and 6.35.

Because there is no decoupling and compensation processing, the matched filter is serious mismatched in two dimensions. The image of the point target is

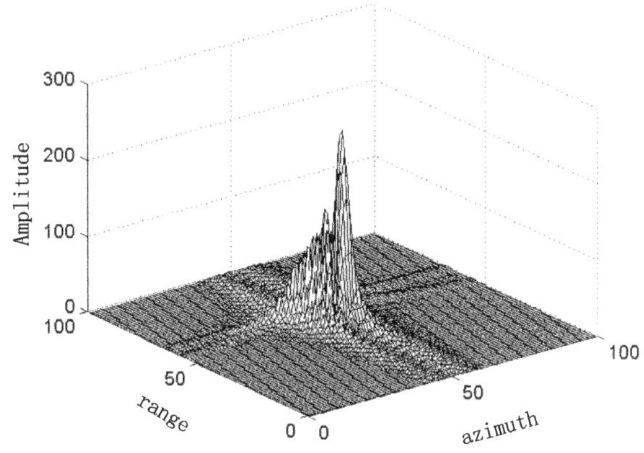

Fig. 6.32 The point target image without any compensation

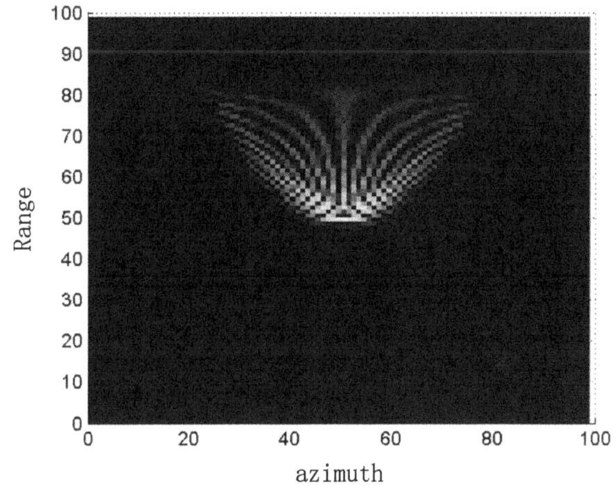

Fig. 6.33 The top view of the point target image

defocused; especially there is a large the peak side lobe level and integral side lobe level in the azimuth compressing, and the broadening of the main lobe width and the degradation of the resolution in range dimension.

The point target image decoupled in two dimensions without velocity compensation is shown in Fig. 6.36, and the top view of the image is shown in Fig. 6.37. Furthermore, the profiles in range and the azimuth dimensions at the target response peak are shown in Figs. 6.38 and 6.39.

The comparison between Figs. 6.36 and 6.32 show that the pulse responses in range and the azimuth dimensions are improved obviously by decoupling

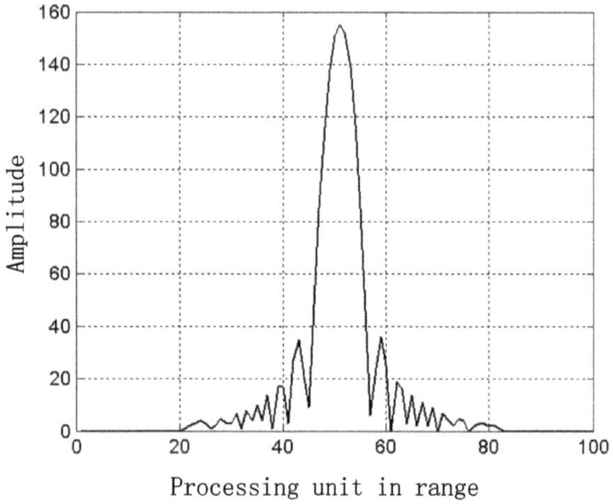

Fig. 6.34 The range profile at the pulse response peak

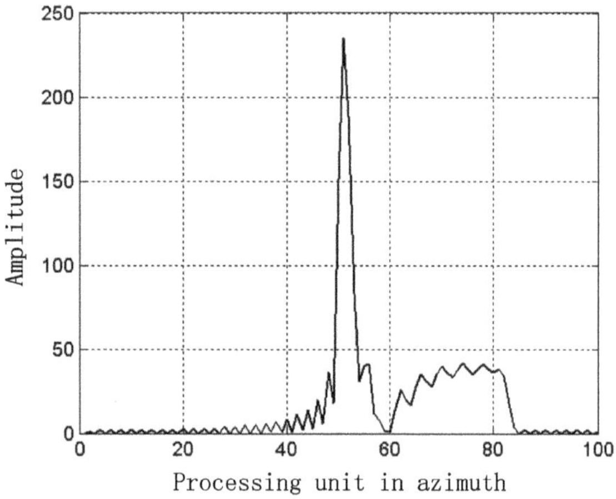

Fig. 6.35 The azimuth profile at the pulse response peak

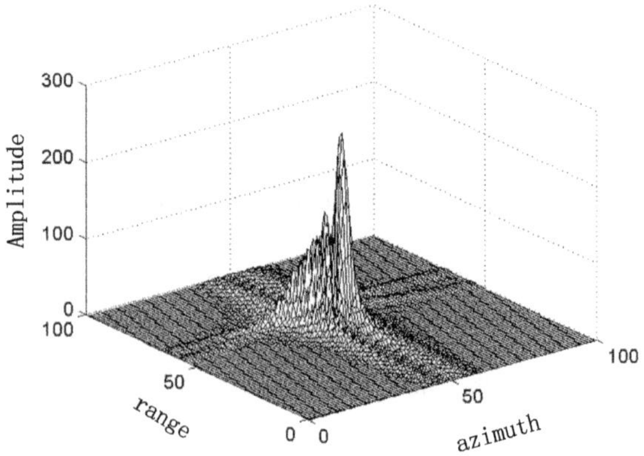

Fig. 6.36 The point target image decoupled without velocity compensation

Fig. 6.37 The top view of the point target image

Fig. 6.38 The range profile at the pulse response peak

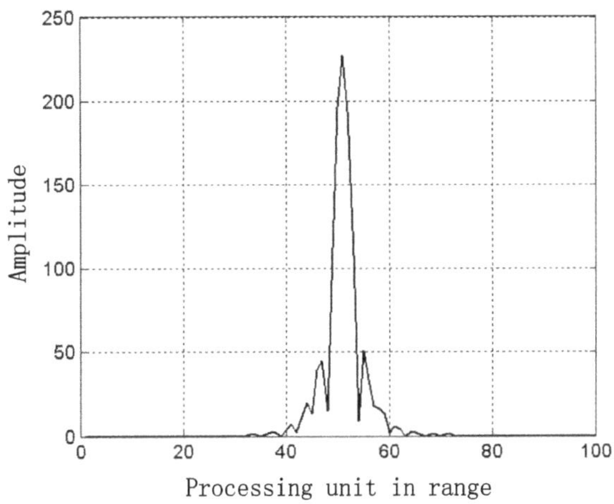

Fig. 6.39 The azimuth profile at the pulse response peak

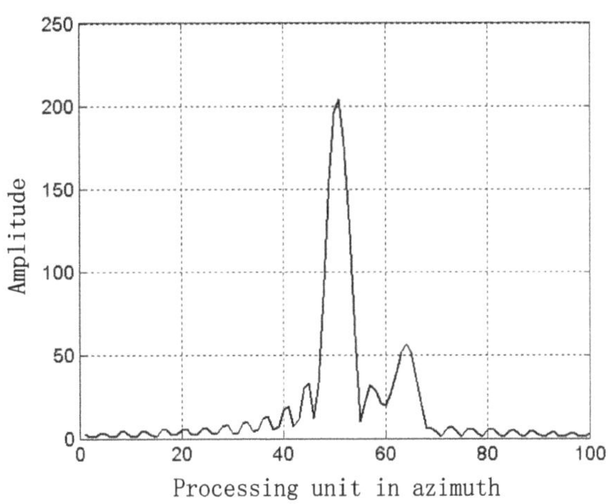

processing. The comparison of the range profile between Figs. 6.38 and 6.34 show that the main lobe width is narrowed and the peak side lobe level is raised in range, but the main lobe width is broadened and the peak side lobe level is decreased in the azimuth. It can be seen that after the decoupling processing the defocus of the SAR image is improved slightly, and the pulse response is improved obviously.

The third experiment: The point target echo simulation with decoupling in two dimensions and the radial velocity compensation

According to the processing steps described in Sect. 6.4.2, the point target image decoupled and compensated is shown in Fig. 6.40, and the top view of the image is shown in Fig. 6.41, the profiles in range and the azimuth dimensions at the target response peak are shown in Figs. 6.42 and 6.43.

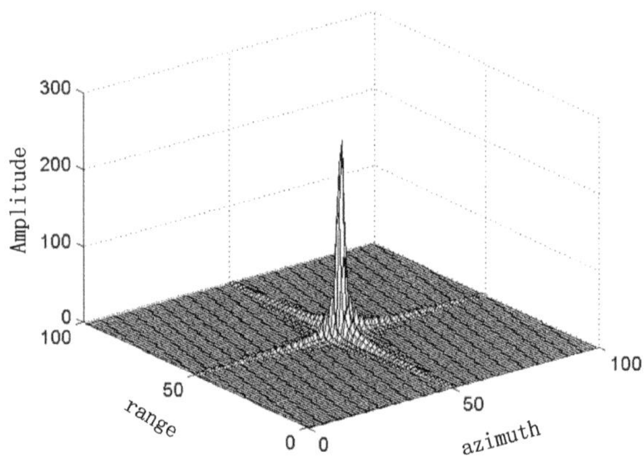

Fig. 6.40 The point target image decoupled and compensated

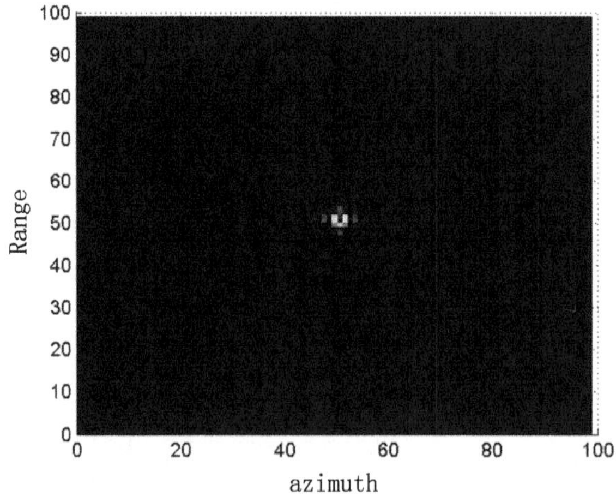

Fig. 6.41 The top view of the point target image

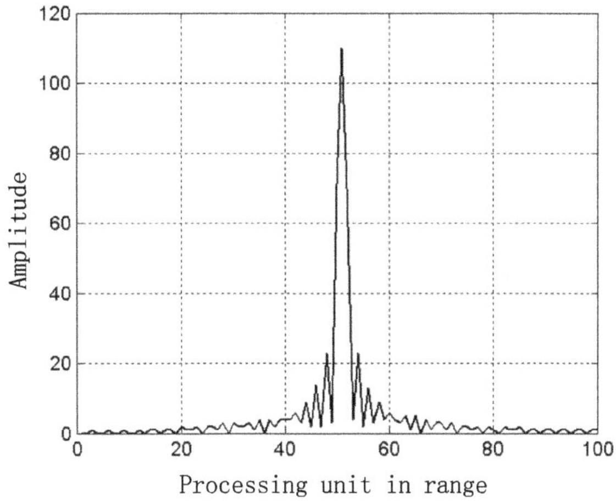

Fig. 6.42 The range profile at the pulse response peak

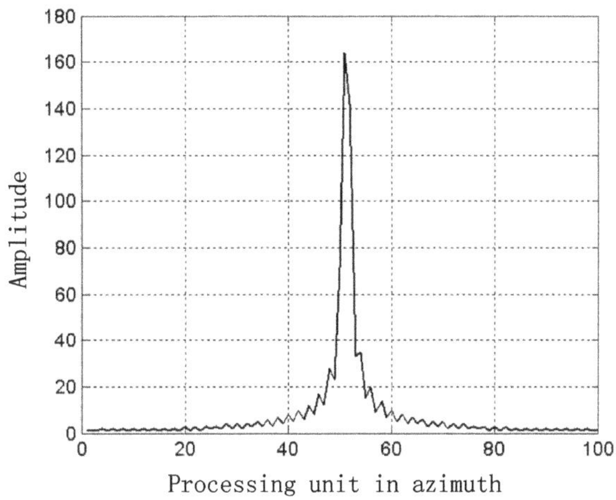

Fig. 6.43 The azimuth profile at the pulse response peak

The comparison between Figs. 6.40 and 6.36 show that the pulse responses in range and the azimuth dimensions approximately equal to sinc function. The comparisons of the range profiles between Figs. 6.42 and 6.38, Figs. 6.43 and 6.39, show that the main lobe width becomes narrower and the side lobe level becomes lower. The two-dimensional matched filtering results are ideal.

The above three groups of experiments show that the coupling between the range and the azimuth makes the SAR image defocused and the main lobe width of the matched filter response broadened. Furthermore, the radial velocity makes the synthetic spectrum away from the rectangle and the center frequency of the Doppler baseband spectrum away from zero. However, the effects of the radial velocity on range processing can be ignored generally.

In this chapter, the principle of coherent wideband signal synthesis of stepped-chirp burst is represented, and the steps of wideband signal synthesis are described first. And then the effect of the radial velocity on the synthetic spectrum and the synthetic range imaging is discussed, and the compensation method is provided. In addition, the ambiguity function of the stepped-chirp burst is deduced in detail, the relation and the performance comparison between stepped-frequency and stepped-chirp are discussed, which is the foundation of the waveform design. Finally, the block diagram, design method and processing steps of the stepped-chirp SAR system are presented and verified by the experiments.

References

1. Lord RT, Inggs MR (1996) High resolution VHF SAR processing using synthetic range profiling. In: Proceedings of IEEE geoscience remote sensing symposium, IGARSS'96, Lincoln, Nebraska, vol 1, pp 454–456
2. Gill GS (1995) Step frequency waveform design and processing for detection of moving targets in clutter. In: IEEE international radar conference, Washington D.C., pp 573–578
3. Fei Y, Su G, Mi H etc. (2002) The generating technology of wideband radar signals. National Defense Industry Press, Beijing
4. Liu Y (1999) Radar imaging technology. Harbin Institute of Technology Press, Harbin
5. Lin M, Ke Y (1984) Radar signal theory. National Defense Industry Press
6. Zhang C (1989) Synthetic aperture radar principle, system analysis and application. Science Press, Beijing
7. Li H, Yang R (2004) The wideband synthesis of the stepped-chirp burst waveform. J Electron Inf Technol 126:349–352
8. An J, Zhang R, Bai K (2000) A novel waveform applied to SAR system and its doppler phase compensation technique. J Syst Eng Electron 22(1):59–62
9. Li H (2002) The ultra-wideband signal waveform and its SAR imaging. Ph.D. thesis, Graduate University of Chinese Academy of Sciences

Chapter 7
Spotlight SAR Technology

Spotlight model is a specialties operating model of Synthetic Aperture Radar (SAR). It is an effective means of achieving high azimuth resolution. By rotation of the radar line of sight, the imaging area has always been under the radar beam irradiation, which can make Spotlight SAR obtain longer synthetic aperture length than Strip SAR, and obtain a higher azimuth resolution [1, 2].

Figure 7.1 shows the airborne Spotlight SAR operating space geometry. Throughout synthetic aperture time, the radar antenna always shines the same region by adjusting pointing continually. This can overcome azimuth resolution limit, which determined by the antenna azimuth aperture length in the Strip SAR, and get longer synthetic aperture length and higher azimuth resolution in the Spotlight SAR.

Spaceborne Spotlight SAR applies spotlight mode to the satellite platform. During the satellite flight, high azimuth resolution of radar image can be got by adjusting the radar antenna beam pointing and achieving a spotlight shines on the imaging area. Figure 7.2 shows the Spaceborne Spotlight SAR space geometry diagram. O is the center of Earth, Re is the radius of the Earth, A is the start point of synthetic aperture and C is the endpoint, B is an arbitrary point on the synthetic aperture. During Synthetic aperture, the antenna beam always irradiates an area on the surface of the Earth, and T is the center of the area, v_e is the velocity of point T.

Figure 7.3 is the azimuth geometric of Spotlight SAR and Strip SAR. It can be seen the data collection process, in Strip mode, the beam pointing is fixed, while in Spotlight mode the beam pointing is controlled real time to fixed ground scenes. This is an essential difference between Spotlight SAR and Strip SAR. Because of this difference, the echoes and Doppler frequency distribution are not the same as in Fig. 7.4, which show the azimuth frequency courses of scattering points of the two modes. Table 7.1 is the comparison table between Spotlight SAR and Strip SAR.

R. Yang et al., *High-Resolution Microwave Imaging*,
https://doi.org/10.1007/978-981-10-7138-6_7

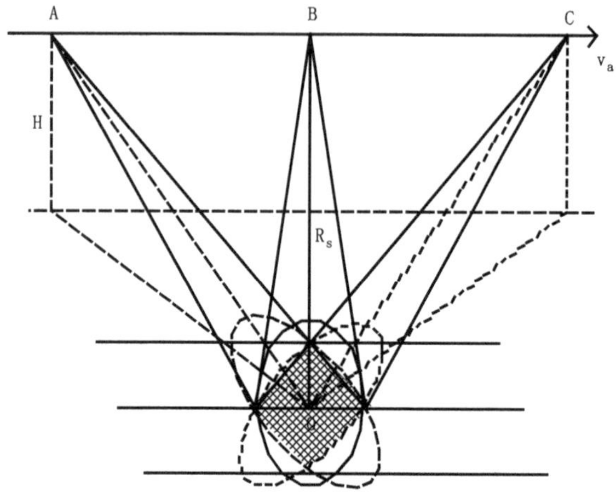

Fig. 7.1 Airborne spotlight SAR space geometry

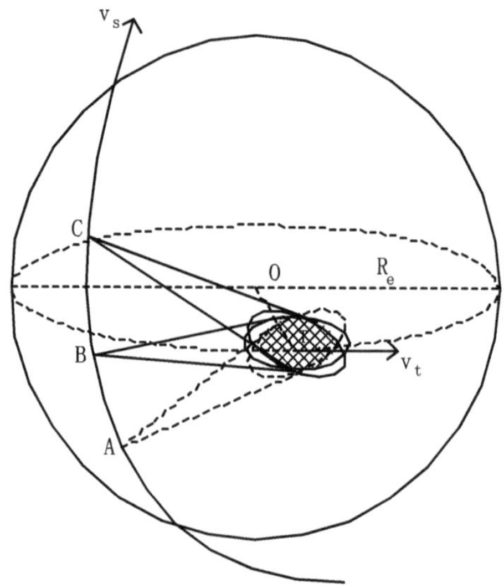

Fig. 7.2 Spaceborne spotlight SAR space geometry

7.1 Spotlight SAR Principle

Spotlight SAR can obtain longer of the synthetic aperture length by adjusting the antenna beam pointing, in principle, it can be explained by the Doppler frequency shift theory and can be carried through imaging processing by the matched filter theory.

In Spotlight SAR operating, in addition to the same translation with Strip SAR, there is a rotation moving of the antenna beam relative to the target area, it is very similar with computer-aided X-ray tomography in medicine. Therefore, the spotlight SAR imaging processing can be explained be means of the advantage of computer-aided tomography method.

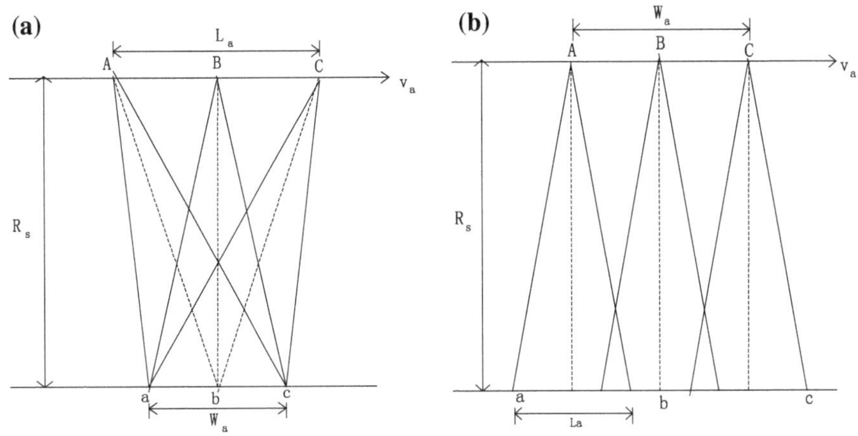

Fig. 7.3 Azimuth geometric models comparison between **a** Spotlight SAR and **b** Strip SAR

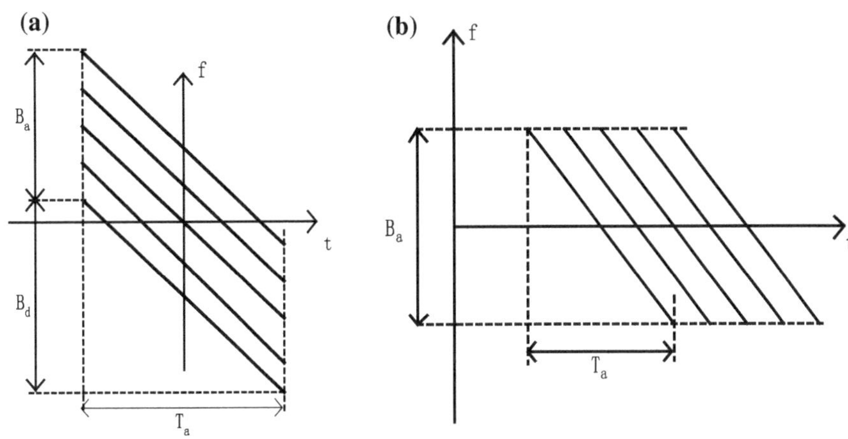

Fig. 7.4 Azimuth frequency courses of **a** Spotlight SAR and **b** Strip SAR scattering points

Both computer-aided tomography and spotlight SAR are a kind of technology of processing target area information drawn from several different angles observation to obtain high-resolution images ultimately. Computer-aided tomography is a technology to obtain two-dimensional cross-section images of three-dimensional objects by means of digital signal processing to one-dimensional projection information from a different perspective. An X-ray tube and X-ray detectors are placed at both sides of a polar radius of the imaging section, and carry through continuous circular motion around the detected objects. In a certain visual angle, X-ray tube transmitting signal is collected by the detector after the attenuation by

Table 7.1 The comparison table between spotlight SAR and strip SAR

	Spotlight SAR	Strip SAR
Antenna pointing	Real-time adjustment always irradiation with a fixed scene	Fixed keep fixed angle with platform movement direction
Synthesis aperture	All targets in scene have the same synthesis aperture	Different azimuth location target has different synthesis aperture
Azimuth resolution	Decided by data acquisition length	Decided by antenna beamwidth
Imaging scene area	Decided by antenna beamwidth	Decided by data acquisition length
Doppler center	Time-varying	Constant
Doppler bandwidth	$B_a + B_d$	B_a

detected objects, which is the "projection" information under the angle or polar radius. According to the "projection-slice principle" of the topographic image reconstruction theory, the Fourier transform of each projection constitutes a center section of the corresponding angle in Fourier transform domain of image reconstruction. The projection of different visual angle could eventually constitute the Fourier transform domain of the measured Image, thereafter by a two-dimensional Inverse FFT an unknown image can be restored.

In pulse radar, radar transmitters, and receivers are located at the same side of the targets. Radar transmitter signal will be reflected by the targets, a part of the reflection signal is received by the receiver. It is defined that the information from radar receivers as the "projection" of radar signals in the angle of view. The pulse radar projection is constituted by echo information of various scattering points, which are an equal distance away from the radar transmitter. It is intituled as "equidistant projection." If imaging target distance is far away from the radar, the projections are the equidistance lines which will perpendicular to the direction of the radar line of sight. Figure 7.5 is Spotlight SAR imaging geometry. The mathematical model of homographic reconstruction of spotlight SAR will be discussed below.

A linear frequency modulation pulse signal from radar transmitter is assumed as,

$$p_n(t) = \mathrm{rect}\left(\frac{t - n \cdot T_{\mathrm{PRT}}}{T_{\mathrm{p}}}\right) \cdot \exp\left\{j2\pi\left[f_c(t - n \cdot T_{\mathrm{PRT}}) + \frac{\gamma}{2}(t - n \cdot T_{\mathrm{PRT}})^2\right]\right\} \tag{7.1}$$

Thereinto, f_0, γ, T_{PRT}, and T_{p} are carrier frequency, linear frequency modulation ratio, and pulse repetition cycle and pulse width, respectively.

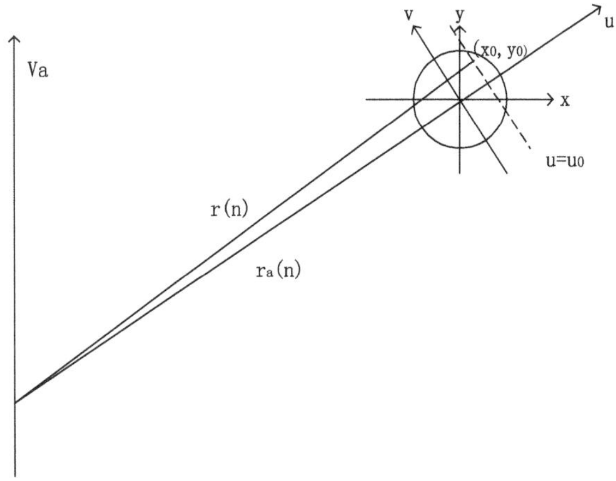

Fig. 7.5 Spotlight SAR imaging geometry

If reflection density function of the target area is $\sigma(x, y)$, the radar target echo from point target (x, y) with the distance r from radar is

$$\sigma(x_0, y_0)p_n\left(t - \frac{2r(n)}{c}\right)\mathrm{d}x_0\mathrm{d}y_0 \tag{7.2}$$

If (u, v) is the rotation of (x, y) in a certain angle of view, the echo from projection $\rho_\theta(u_0)$ with $u = u_0$ is

$$\rho_\theta(u_0) \cdot p_n\left(t - \frac{2(r_a(n) + u_0)}{c}\right)\mathrm{d}u_0 \tag{7.3}$$

Projection $\rho_\theta(u_0)$ is

$$\rho_\theta(u_0) = \int_{u=u_0} \sigma(x_0, y_0)\mathrm{d}v \tag{7.4}$$

Echoes from the entire target area can be expressed as

$$\begin{aligned} s(t;n) &= \int_D \rho_\theta(u) \cdot p_n\left(t - \frac{2(r_a(n)+u)}{c}\right)\mathrm{d}u \\ &= \int_D \rho_\theta(u) \cdot \mathrm{rect}\left(\frac{t - n \cdot T_{\mathrm{PRT}} - 2(r_a(n)+u)/c}{T_{\mathrm{p}}}\right) \\ &\quad \times \exp\left\{j2\pi\left[f_0\left(t - n \cdot T_{\mathrm{PRT}} - \frac{2(r_a(n)+u)}{c}\right) + \frac{\gamma}{2}\left(t - n \cdot T_{\mathrm{PRT}} - \frac{2(r_a(n)+u)}{c}\right)^2\right]\right\}du \end{aligned} \tag{7.5}$$

In Eq. (7.5), D is the size of the imaging area.

Select echo corresponding to the center of the target area as reference function

$$s_{\text{ref}}(t;n) = \exp\left\{j2\pi\left[f_0\left(t - n\cdot T_{\text{PRT}} - \frac{2r_a(n)}{c}\right) + \frac{\gamma}{2}\left(t - n\cdot T_{\text{PRT}} - \frac{2r_a(n)}{c}\right)^2\right]\right\} \tag{7.6}$$

Then mixing, let $\hat{t} = t - n\cdot T_{\text{PRT}} - \frac{2r_a(n)}{c}$, then

$$\begin{aligned} s_{\text{if}}(t;n) &= s(t;n)\cdot s^*_{\text{ref}}(t;n) \\ &= \int_D \rho_\theta(u)\cdot \text{rect}\left(\frac{\hat{t} - 2u/c}{T_p}\right)\cdot\exp\left\{-j\frac{4\pi\gamma}{c}\left(\frac{f_0}{\gamma} + \hat{t}\right)u\right\}\cdot\exp\left\{j\frac{4\pi\gamma}{c^2}u^2\right\}\mathrm{d}u \end{aligned} \tag{7.7}$$

Signal phase is composed by the first-order phase and the second-order phase. The first-order phase is Doppler frequency (inclusive f_0) and range frequency (inclusive $\hat{t}$), respectively. The second phase is also known as residual video, caused by a mixing operation. If the imaging area is small, the second-order phase can be negligible. There is

$$s_{\text{if}}(t;n) = \int_D \rho_\theta(u)\cdot\exp\{-j2\pi Uu\}\mathrm{d}u \tag{7.8}$$

In Eq. (7.8)

$$U = \frac{2\gamma}{c}\left(\frac{f_0}{\gamma} + \hat{t}\right) = \frac{2}{\lambda} + \frac{2\gamma\hat{t}}{c} \tag{7.9}$$

And

$$-\frac{T_{\text{p}}}{2} + \frac{D}{c} \le \hat{t} \le \frac{T_{\text{p}}}{2} - \frac{D}{c} \tag{7.10}$$

So

$$U \in \left\{\frac{2}{\lambda} - \frac{\gamma T_{\text{p}}}{c} + \frac{2\gamma D}{c^2}, \frac{2}{\lambda} + \frac{\gamma T_{\text{p}}}{c} - \frac{2\gamma D}{c^2}\right\} \tag{7.11}$$

Let $B = \gamma T_{\text{p}}$ as transmitting signal bandwidth, $U_0 = 2/\lambda$, Eq. (7.11) may be approximately expressed by

$$U \in \left\{U_0 - \frac{B}{c}, U_0 + \frac{B}{c}\right\} \tag{7.12}$$

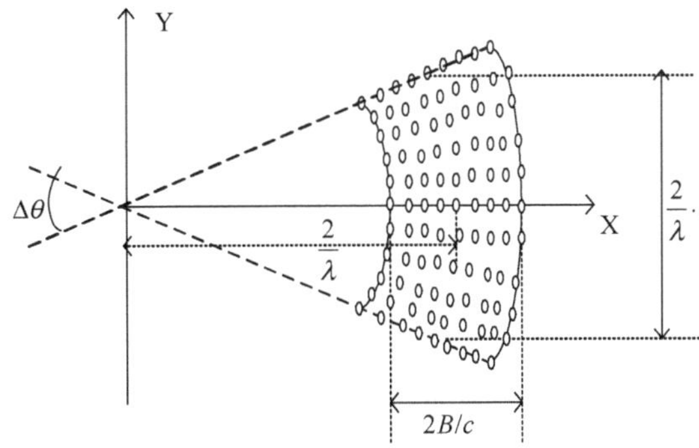

Fig. 7.6 Spotlight SAR target fourier domain

In Eq. (7.8), s_{if} (t, n) is one-dimensional Fourier transformation for projection under angle θ. According to the principle of projection-slice, in a synthetic aperture time all s_{if} (t, n) $n = 1, 2,\ldots, N-1$ received constitute a target Fourier transform domain. Figure 7.6 is Spotlight SAR target Fourier domain. The data has a polar coordinate format, and each radar pulse corresponds to a polar angle, and the sampling data of the corresponding pulse falls to the polar radius of the polar angle.

By two-dimensional interpolation processing, data in polar coordinate format can be transformed into Cartesian format, the target image can be obtained by means of two-dimensional inverses FFT after interpolation.

7.2 Spotlight SAR Parameter Analysis

7.2.1 Spotlight SAR Azimuth Resolution

(1) The azimuth resolution of airborne Spotlight SAR [3]

Airborne Spotlight SAR imaging geometry is shown in Fig. 7.7. Radar moves along the straight line. At different time t, slant range between the radar and ground targets varies.

Slant distance $r(t)$ can be expressed as a function of time t, $r(t)$ is a quadratic function of t approximately,

$$r(t) = \sqrt{r^2 + V^2t^2 - 2rVt\cos\varphi} \tag{7.13}$$

In this equation, r is the slant distance when $t = 0$, V is the equivalent radar platform speed, φ is equivalent squint angle of view. By Taylor deploying, Eq. (7.13) is expressed approximately

$$r(t) \approx r + \frac{\lambda}{2}\left(f_{\text{d}}t + \frac{1}{2}f_{\text{r}}t^2\right) \tag{7.14}$$

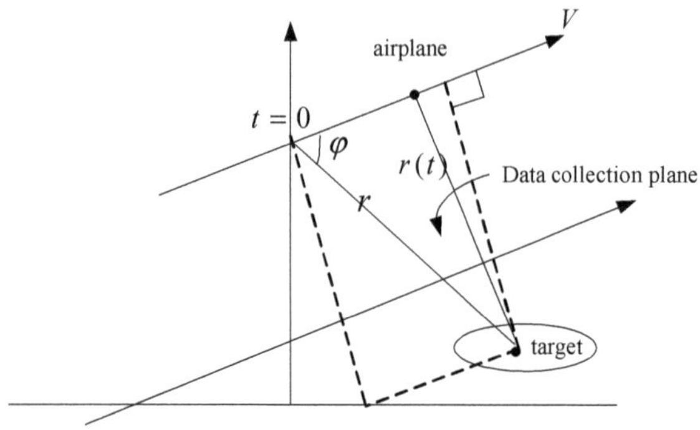

Fig. 7.7 Airborne Spotlight SAR imaging geometry

In the equation

$$f_{\mathrm{d}} = \frac{2}{\lambda}\frac{\mathrm{d}r(t)}{\mathrm{d}t}\bigg|_{t=0} = -\frac{2V}{\lambda}\cos\varphi \tag{7.15}$$

$$f_{\mathrm{r}} = \frac{2}{\lambda}\frac{\mathrm{d}^2 r(t)}{\mathrm{d}t^2}\bigg|_{t=0} = \frac{2V^2}{\lambda R}\sin^2\varphi \tag{7.16}$$

λ is radar wavelength, f_{d} is Doppler center frequency and f_{r} is Doppler frequency rate.

From Eq. (7.14), distance from point target to radar is $r(t)$, echo phase corresponding to round-trip range is

$$\phi(t) = -\frac{2\pi}{\lambda}\cdot 2r(t) \tag{7.17}$$

The derivative of phase (t) to t is an instantaneous frequency.

$$f(t) = \frac{1}{2\pi}\cdot\frac{\mathrm{d}\phi(t)}{\mathrm{d}t} = -f_{\mathrm{d}} - f_{\mathrm{r}}t \tag{7.18}$$

Figure 7.8 gives the diagram of a synthetic aperture length of airborne Spotlight SAR.

Synthetic aperture time is

$$T_{\mathrm{a}} \approx \frac{2\sin\left(\frac{\Delta\theta}{2}\right)r}{v\sin\varphi} \tag{7.19}$$

In Eq. (7.19), $\Delta\theta$ is turning angle of the radar line of sight, φ is squint angle which is between the radar line of sight (LOS) and flight direction. When

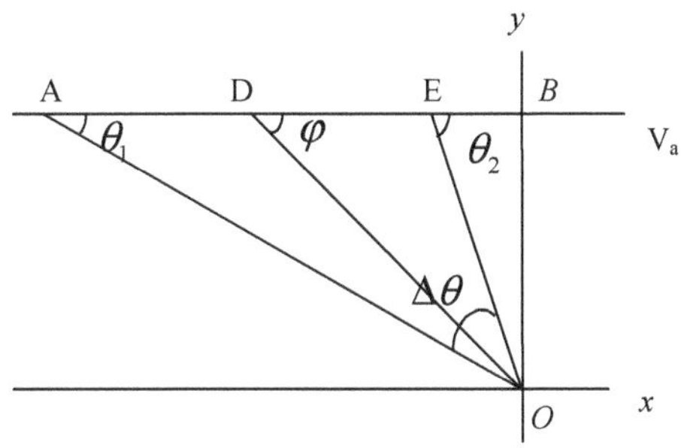

Fig. 7.8 Airborne spotlight SAR synthetic aperture length diagram

$T_a \ll \frac{r}{v \sin\phi}$, Doppler frequency shift is a linear relationship with time t. Echo signal in azimuth direction has the chirp feature. Doppler bandwidth is

$$B_d = |f_r|T = \frac{4V \sin\left(\frac{\Delta\theta}{2}\right) \sin\varphi}{\lambda} \tag{7.20}$$

Using matched filter for treating azimuth signal, and taking into account azimuth broadening factor k_a of the main lobe of antenna, the azimuth theory resolution is

$$\rho_a = \frac{V}{B_d} = \frac{k_a\lambda}{4\sin\left(\frac{\Delta\theta}{2}\right)\sin\varphi} \tag{7.21}$$

In the vertically side-looking ($\varphi = 90°$) and $\Delta\theta$ is very small, Eq. (7.21) also can be approximated as $\rho_a = \lambda/(2\Delta\theta)$. Equation (7.21) is only suitable for small synthetic aperture angle. For big synthetic aperture angle, azimuth resolution is

$$\rho_a = \frac{k_a\lambda}{4\sin(\theta_2 - \theta_1)} \cdot \frac{\frac{\sin\theta_2}{\tan\theta_1} - \frac{\sin\theta_1}{\tan\theta_2} + \sin\theta_1\sin\theta_2\left[\ln\left(tg\frac{\theta_2}{2}\right) - \ln\left(\tan\frac{\theta_1}{2}\right)\right]}{\ln\left(\tan\frac{\theta_2}{2}\right) - \ln\left(\tan\frac{\theta_1}{2}\right)} \tag{7.22}$$

Every angle in Eq. (7.22) is as shown in Fig. 7.8. Point A and point E are the beginning and end of the synthetic aperture, respectively.

(2) The azimuth resolution of spaceborne Spotlight SAR [4]

Spaceborne Spotlight SAR geometric model is shown in Fig. 7.9. Suppose the satellite orbit is a polar orbit for a constant angular velocity of circular motion. Point D locates on the equator, where orbit projected to the earth is crossed the equator, Point O is the center of the earth, Earth radius is R_e, Orbit height is H, synthetic aperture begins from the point B and ends to the point E, Point A is any point on the synthetic aperture. In order to simplify the analysis, select a point P is on the equator, and point p is the center of the target area. Squint angles at A, B, and E are

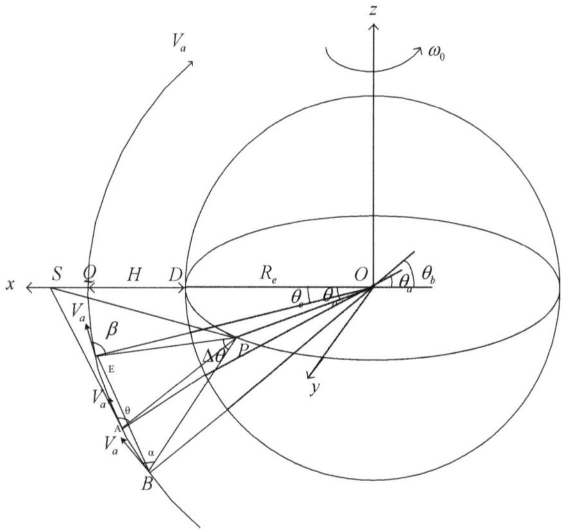

Fig. 7.9 Spaceborne spotlight SAR space geometric model

π/2−θ, π/2–α and π/2−β respectively; synthetic aperture angle is $\Delta\theta$, which is ∠*BPE*. In spherical coordinates, the point coordinates are respectively *O* (0, 0, 0),

$$A\left(R_{\mathrm{e}}+H,\theta_{\mathrm{a}}+\frac{\pi}{2},0\right),\ B\left(R_{\mathrm{e}}+H,\theta_{\mathrm{b}}+\frac{\pi}{2},0\right),\ E\left(R_{\mathrm{e}}+H,\theta_{\mathrm{e}}+\frac{\pi}{2},0\right),\ P\left(R_{\mathrm{e}},\frac{\pi}{2},\theta_{\mathrm{p}}\right).$$

It can be seen that within the whole synthetic aperture, the antenna beam moves on a curved surface. In order to facilitate the analysis, construct two triangles: Instantaneous triangle *ΔAPS* and overall triangle *ΔBPE*. In *ΔAPS*, the point $S\left((R_{\mathrm{e}}+H)/\cos\theta_{\mathrm{a}},\frac{\pi}{2},0\right)$ is the intersection of speed direction of point *A* and *OD* extensions, Rectangular coordinates are respectively:

$$\begin{cases} x_O=0 \\ y_O=0 \\ z_O=0 \end{cases} \quad \begin{cases} x_A=(R_e+H)\cos\theta_a \\ y_A=0 \\ z_A=-(R_e+H)\sin\theta_a \end{cases} \quad \begin{cases} x_B=(R_e+H)\cos\theta_b \\ y_B=0 \\ z_B=-(R_e+H)\sin\theta_b \end{cases}$$

$$\begin{cases} x_E=(R_{\mathrm{e}}+H)\cos\theta_{\mathrm{e}} \\ y_E=0 \\ z_E=-(R_{\mathrm{e}}+H)\sin\theta_{\mathrm{e}} \end{cases} \quad \begin{cases} x_P=R_{\mathrm{e}}\cos\theta_{\mathrm{p}} \\ y_P=R_{\mathrm{e}}\sin\theta_{\mathrm{p}} \\ z_P=0 \end{cases} \quad \begin{cases} x_S=(R_{\mathrm{e}}+H)/\cos\theta_{\mathrm{a}} \\ y_S=0 \\ z_S=0 \end{cases}$$

According to the geometry of Fig. 7.9, it can be analyzed that the spaceborne Spotlight SAR classical azimuth resolution formula as follows. In the overall *ΔBPE*, the various side lengths are respectively:

$$|BP| = \sqrt{H^2 + 2R_e(R_e + H)(1 - \cos\theta_p\cos\theta_b)} \tag{7.23}$$

$$|EP| = \sqrt{H^2 + 2R_e(R_e + H)(1 - \cos\theta_p\cos\theta_e)} \tag{7.24}$$

$$|BE| = (R_e + H)\sqrt{2 - 2\cos(\theta_b - \theta_e)} \tag{7.25}$$

Using the cosine theorem to get

$$\cos(\Delta\theta) = \frac{|BP|^2 + |EP|^2 - |BE|^2}{2|BP|\cdot|EP|} \tag{7.26}$$

Equations (7.23), (7.24), and (7.25) can be substituted into Eq. (7.26), and simplification is

$$\cos\Delta\theta = \frac{H^2 + R_e(R_e + H)[2 - \cos\theta_p(\cos\theta_b + \cos\theta_e)] - (R_e + H)^2[1 - \cos(\theta_b - \theta_e)]}{\sqrt{H^2 + 2R_e(R_e + H)(1 - \cos\theta_p\cos\theta_e)}\cdot\sqrt{H^2 + 2R_e(R_e + H)(1 - \cos\theta_p\cos\theta_b)}} \tag{7.27}$$

$\Delta\theta$ can be substituted into Eq. (7.21), Classical Spotlight SAR azimuth resolution is

$$\rho_a = \frac{k_a\lambda}{4\sin\left\{\frac{1}{2}\cos^{-1}\left[\frac{H^2 + R_e(R_e + H)[2 - \cos\theta_p(\cos\theta_b + \cos\theta_e)] - (R_e + H)^2[1 - \cos(\theta_b - \theta_e)]}{\sqrt{H^2 + 2R_e(R_e + H)(1 - \cos\theta_p\cos\theta_e)}\cdot\sqrt{H^2 + 2R_e(R_e + H)(1 - \cos\theta_p\cos\theta_b)}}\right]\right\}} \tag{7.28}$$

The relationship between Spaceborne Spotlight SAR azimuth resolution and Doppler bandwidth is determined, namely,

$$\rho_a = k_a\cdot\frac{v_a}{B_d} \tag{7.29}$$

In Eq. (7.29), v_a is the speed of the satellite, B_d is azimuth Doppler bandwidth, k_a is azimuth expansion factor.

In general, synthetic aperture angle $\Delta\theta$ is small. During synthetic aperture, Doppler frequency offset due to the rotation of the Earth is constant approximately. So the analysis of spaceborne Spotlight SAR azimuth resolution can regardless of the effect of Earth's rotation speed.

The analysis of azimuth Doppler frequency bandwidth is as follows. At any point, the radar flight direction is always perpendicular to the line linking between the center of the Earth and radar. In Fig. 7.9, according to the coordinates, the $\varDelta APS$ three instantaneous side lengths can be calculated as follows:

$$|AP| = \sqrt{H^2 + 2R_e(R_e + H)(1 - \cos\theta_p \cos\theta_a)} \tag{7.30}$$

$$|AS| = (R_e + H)tg\theta_a \tag{7.31}$$

$$|PS| = \sqrt{R_e^2 + \frac{(R_e + H)^2}{\cos^2\theta_a} - 2R_e(R_e + H) \cdot \frac{\cos\theta_p}{\cos\theta_a}} \tag{7.32}$$

Using the cosine theorem to get

$$\cos\theta = \frac{|AS|^2 + |AP|^2 - |PS|^2}{2|AS| \cdot |AP|} \tag{7.33}$$

Equations (7.30), (7.31), and (7.32) can be substituted into Eq. (7.33), and simplification is

$$\cos\theta = \frac{R_e \cos\theta_p \sin\theta_a}{\sqrt{H^2 + 2R_e(R_e + H)(1 - \cos\theta_p \cos\theta_a)}} \tag{7.34}$$

Instantaneous Doppler frequency is

$$f_d = \frac{2v_a \cos\theta}{\lambda} \tag{7.35}$$

Doppler frequency bandwidth for the entire synthetic aperture is

$$B_d = f_{db} - f_{de} = \frac{2v_a(\cos\alpha - \cos\beta)}{\lambda} \tag{7.36}$$

In Eq. (7.36), f_{db}, f_{de} correspond to the Doppler frequency of point B and point E respectively.

Equations (7.36) can be substituted into Eq. (7.29), the azimuth resolution is

$$\rho_a = \frac{k_a \lambda}{2(\cos\alpha - \cos\beta)} \tag{7.37}$$

In Eq. (7.37), α and β correspond to the squint angles of synthetic aperture start and endpoints, respectively.

From Eqs. (7.34) and (7.37), the ultimate expression of the azimuth resolution

$$\rho_a = \frac{k_a \lambda}{2R_e \cos\theta_p \left[\frac{\sin\theta_b}{\sqrt{H^2 + 2R_e(R_e + H)(1 - \cos\theta_p \cos\theta_b)}} - \frac{\sin\theta_e}{\sqrt{H^2 + 2R_e(R_e + H)(1 - \cos\theta_p \cos\theta_e)}} \right]} \tag{7.38}$$

If only the eccentric angle (θ_p) of the objectives, the latitudes of synthetic aperture start point θ_b and endpoint θ_e, as well as orbital altitude H are known, the spaceborne Spotlight SAR azimuth resolution can be calculated.

7.2.2 Synthetic Aperture Length of Spotlight SAR

In Spotlight SAR system design, synthetic aperture length is an important parameter affecting the resolution and imaging periods.

(1) Classical synthetic aperture length of Spotlight SAR

Traditional Spotlight SAR synthetic aperture length is calculated as [1]

$$L = \frac{R_a \cdot \Delta\theta}{\sin(\alpha_{dc})} = \frac{R_a}{\sin(\alpha_{dc})} \cdot \frac{K_a \lambda_c}{2\rho_a} \tag{7.39}$$

In Eq. (7.39), L is the synthetic aperture length of Spotlight SAR, R_a is the slant range corresponding to the aperture center, $\Delta\theta$ is synthetic aperture angle, α_{dc} is the squint angle corresponding to the aperture center, K_a is azimuth expansion factor, λ_c is radar operating wavelength, and ρ_a is azimuth resolution.

It should be pointed out that Eq. (7.39) is an approximation formula based on the small squint angle in the airborne Spotlight SAR model. There would be engendering biggish errors for large squint angle in the airborne Spotlight SAR and the spaceborne Spotlight SAR.

(2) Synthetic aperture length of airborne Spotlight SAR [5]

On-board conditions, since the airplane flying along a straight line as uniform speed and stationary ground target, the geometric relation is relatively straightforward. Figure 7.10 is the schematic diagram of inclined plane geometry in the airborne Spotlight SAR. The radar speed is v_a flying along the positive x direction from aperture beginning point A to the endpoint E. Point D is aperture center and point O is a target. According to Eq. (7.39), the calculated synthetic aperture is CF, but the true aperture is AE. CF and AE are very different, which can be seen from the graph. It means that Eq. (7.39) will create errors to calculate the synthetic aperture length, which should be amended.

Fig. 7.10 Schematic diagram of inclined plane geometry in airborne spotlight SAR

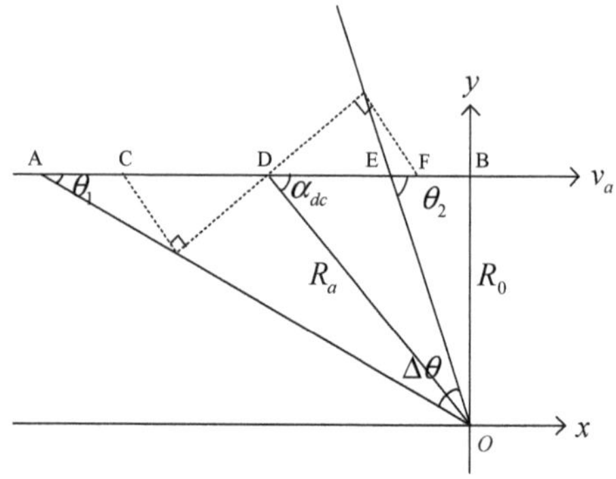

From Fig. 7.10, echo-Doppler bandwidth in a synthetic aperture is obtained [6]

$$B_{\mathrm{d}} = \frac{2v_{\mathrm{a}}\cos\theta_1}{\lambda_{\mathrm{c}}} - \frac{2v_{\mathrm{a}}\cos\theta_2}{\lambda_{\mathrm{c}}} \tag{7.40}$$

In Eq. (7.40),

$$\theta_1 = \alpha_{\mathrm{dc}} - \Delta\theta/2 \tag{7.41}$$

$$\theta_2 = \alpha_{\mathrm{dc}} + \Delta\theta/2 \tag{7.42}$$

Equations (7.41) and (7.42) can be substituted into Eq. (7.40),

$$B_{\mathrm{d}} = \frac{4v_{\mathrm{a}}\sin\alpha_{\mathrm{dc}}}{\lambda_{\mathrm{c}}} \cdot \sin\frac{\Delta\theta}{2} \tag{7.43}$$

$\Delta\theta$ is small in general, $\sin(\Delta\theta/2) \approx \Delta\theta/2$, so

$$B_{\mathrm{d}} = \frac{2v_{\mathrm{a}}\sin\alpha_{\mathrm{dc}}}{\lambda_{\mathrm{c}}} \cdot \Delta\theta \tag{7.44}$$

According to the resolution requirements, echo-Doppler bandwidth should be [7]

$$B_{\mathrm{d}} = \frac{v_{\mathrm{a}}}{\rho_{\mathrm{a}}} \tag{7.45}$$

From Eqs. (7.44) and (7.45), we can obtain

$$\Delta\theta = \frac{\lambda_{\mathrm{c}}}{2\rho_{\mathrm{a}}\sin\alpha_{\mathrm{dc}}} \tag{7.46}$$

By the transmit signal center wavelength, the required azimuth resolution and aperture center squint angle, synthetic aperture angle can be calculated. Since vertical side-looking slant range R_0, the synthetic aperture length is

$$L = \frac{R_0}{tg\theta_1} - \frac{R_0}{tg\theta_2}. \tag{7.47}$$

Equations (7.41) and (7.42) can be substituted into Eq. (7.47),

$$L = \frac{2R_0 \sin \Delta\theta}{\cos \Delta\theta - \cos 2\alpha_{dc}} \tag{7.48}$$

Using the approximate relationship of $\cos \Delta\theta \approx 1$, $\sin \Delta\theta \approx \Delta\theta$, Eq. (7.46) can be substituted in Eq. (7.48), and simplification is

$$L = \frac{\lambda_c R_0}{2\rho_a \sin^3 \alpha_{dc}} \tag{7.49}$$

Equation (7.49) is a currency calculation formula of the synthetic aperture length for airborne Spotlight SAR.

As can be seen, in nearly vertical side-looking case, the difference between Eqs. (7.39) and (7.49) is a little. So Eq. (7.39) can be used in place of Eq. (7.49) for approximate calculations. However, with the increase of squint angle, calculation errors increase gradually, so Eq. (7.49) is requested to calculate the spotlight mode synthetic aperture length accurately.

(3) Simplification model of synthetic aperture length [5] for the spaceborne Spotlight SAR

Typically, Aperture design of spaceborne Spotlight SAR is more complex than that of airborne Spotlight synthetic aperture radar. The satellite elliptical orbit, the Earth's surface curvature and rotation of the Earth will affect the performance of spaceborne Spotlight SAR. For a Spotlight synthetic aperture length, the elliptical orbit can be approximated as a line and imaging scene can be approximated as a flat surface. A polar orbit is adopted in the simplified model; Fig. 7.11 is spaceborne SAR profile in the range direction.

In Fig. 7.11, point A is the satellite, point T is the target located on the equator, point O is the center of the Earth, and v_e is the target's velocity, which can be decomposed into the slant velocity component v_t and component v_g of the vertical inclined plane. R_e is the radius of the Earth, H is the orbit height of the satellite, R_0 is the side-looking slant range, α is the angle of view which corresponding to the Earth center angle β and the incidence angle γ. The slant velocity component can be calculated as

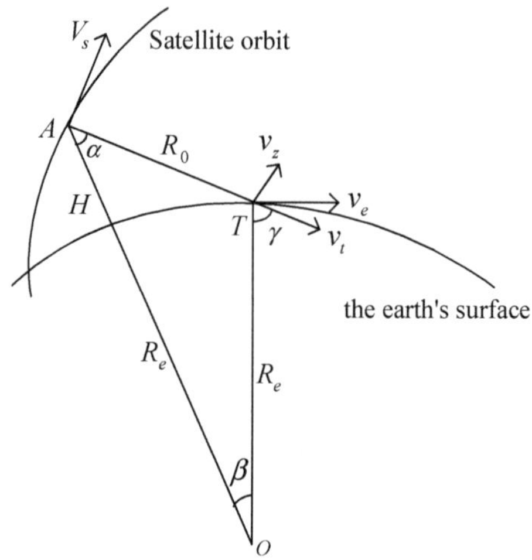

Fig. 7.11 Spaceborne SAR profile in range direction

$$v_{\mathrm{t}} = (R_{\mathrm{e}} + H)\omega_{\mathrm{e}} \sin \alpha \tag{7.50}$$

In Eq. (7.50), ω_{e} is the Earth's angular velocity.

In a synthetic aperture time, v_{t} may be considered unchanged because of the Earth just rotates only several arc seconds. Figure 7.12 is spaceborne SAR profile in the azimuth direction. Point D is the center of the aperture, L is the length of the aperture, radar velocity v_{s} is along the positive y-axis direction, target velocity v_{t} is along the positive x-axis direction. Assuming radar locates in the aperture center point D at $t = 0$, and point O is a target, R_0 is the side-looking slant range. In the synthetic aperture time, radar moves from point A to point E, and the target moves from point C to point F. the corresponding squint angles are θ_1 and θ_2 as shown in Fig. 7.12. Instantaneous Doppler frequency is [6]:

$$f_{\mathrm{d}} = \frac{2(v_{\mathrm{s}} \cos \alpha - v_{\mathrm{t}} \sin \alpha)}{\lambda_{\mathrm{c}}} \tag{7.51}$$

In Eq. (7.51), α is the instantaneous squint angle.

Doppler frequency f_{da} and f_{de} correspond to the synthetic aperture begin point and endpoint, respectively, and the Doppler signal bandwidth corresponding synthetic aperture AE is

$$B_{\mathrm{d}} = f_{\mathrm{da}} - f_{\mathrm{de}} \tag{7.52}$$

From Eqs. (7.41), (7.42), (7.51), and (7.52), we can get

$$B_{\mathrm{d}} = \frac{4\sin(\Delta\theta/2)}{\lambda_{\mathrm{c}}}(v_{\mathrm{s}} \sin \alpha_{\mathrm{dc}} + v_{\mathrm{t}} \cos \alpha_{\mathrm{dc}}) \tag{7.53}$$

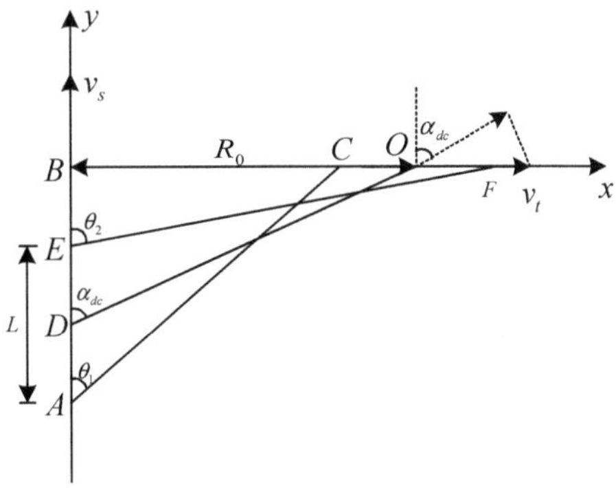

Fig. 7.12 Spaceborne SAR profile in azimuth direction

According to the resolution requirement, echo-Doppler bandwidth is [7]

$$B_\mathrm{d} = \frac{v_\mathrm{s}}{\rho_\mathrm{a}}. \tag{7.54}$$

From Eqs. (7.53) and (7.54) and approximate equation $\sin(\Delta\theta/2) \approx \Delta\theta/2$,we can get

$$\Delta\theta = \frac{v_\mathrm{s}\lambda_\mathrm{c}}{2\rho_\mathrm{a}(v_\mathrm{s}\sin\alpha_\mathrm{dc} + v_\mathrm{t}\cos\alpha_\mathrm{dc})} \tag{7.55}$$

Based synthetic aperture time of Δt, then the synthetic aperture length is

$$L = v_\mathrm{s}\Delta t = \frac{R_0 - v_\mathrm{t}\Delta t/2}{\tan\theta_1} - \frac{R_0 + v_\mathrm{t}\Delta t/2}{\tan\theta_2} \tag{7.56}$$

From Eqs. (7.41), (7.42), and (7.56), we can get

$$L = \frac{2v_\mathrm{s}R_0\sin(\Delta\theta)}{v_\mathrm{t}\sin(2\alpha_\mathrm{dc}) + v_\mathrm{s}[\cos\Delta\theta - \cos(2\alpha_\mathrm{dc})]} \tag{7.57}$$

Because of $\cos\Delta\theta \approx 1$, $\sin\Delta\theta \approx \Delta\theta$, Eq. (7.55) can be substituted into Eq. (7.57), we can get

$$L = \frac{\lambda_\mathrm{c}R_0v_\mathrm{s}^2}{2\rho_\mathrm{a}(v_\mathrm{s}\sin\alpha_\mathrm{dc} + v_\mathrm{t}\cos\alpha_\mathrm{dc})^2\sin\alpha_\mathrm{dc}} \tag{7.58}$$

This is the synthetic aperture length calculation formula of spaceborne SAR Spotlight mode with consideration of ground speed. When ground speed is zero, Eq. (7.58) reverts to Eq. (7.49).

7.3 Features of Spotlight Synthetic Aperture Radar

7.3.1 Spotlight SAR Signal Characteristics

Due to radar angle changing, Spotlight SAR Doppler center frequency changes along with azimuth time. The frequency course of the echo signals for Spotlight SAR is very different from that of Strip SAR. The azimuth echo courses of Spotlight SAR and Strip SAR are as shown in Figs. 7.13 and 7.14, respectively.

In vertical side-looking imaging geometry, for example, azimuth frequency course of the Strip SAR can be described in Fig. 7.13, where x_1, x_2, x_3 are point targets at different azimuth positions. Along with the flight of aircraft, x_1, x_2, x_3 enter and leave the irradiation area of the radar beam successively, their azimuth frequency courses have the same synthetic aperture time T_a. The time difference between frequency histories for the adjacent points is $\Delta t = \frac{\Delta x}{v}$, and the frequency

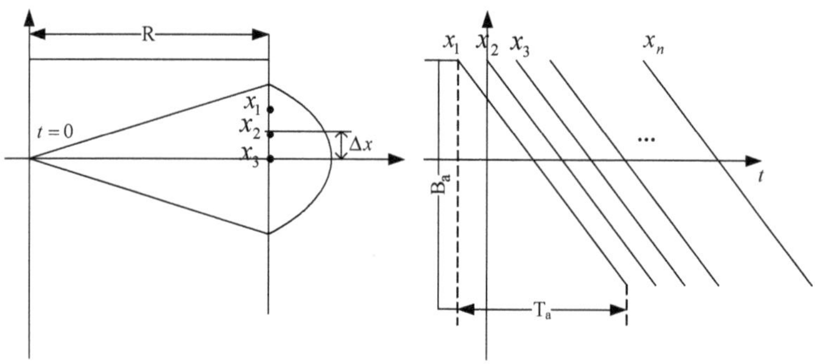

Fig. 7.13 The azimuth echo courses of strip SAR

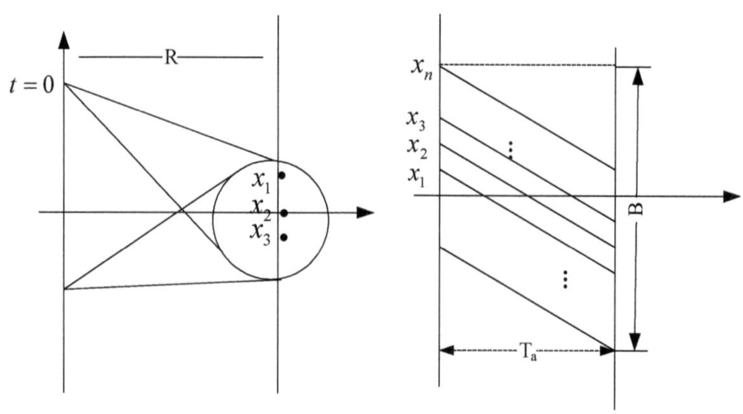

Fig. 7.14 The azimuth echo courses of spotlight SAR

difference at the same time is $\Delta f_a = \frac{2v\Delta x}{\lambda R}$. Azimuth frequency bandwidth is only associated with synthetic aperture time T_a, i.e., $B = B_a = \frac{2v^2}{\lambda R} T_a$.

In imaging processing of Spotlight mode, the point targets in the irradiation region all along remain under the radar beam irradiation, azimuth frequency courses are shown in Fig. 7.14. The duration of azimuth frequency courses for point x_1, x_2, x_3 is the synthetic aperture time T_a. It is different from Stripe mode SAR that, the frequency courses of different points have the same start and end times in the Spotlight mode. Where azimuth bandwidth does not only depend on synthetic aperture time, but also depend on azimuth image scope w_a. The bandwidth can be expressed as: $B = B_a + B_d = \frac{2vw_a}{\lambda R} + \frac{2v^2}{\lambda R} T_a$.

(1) Spotlight SAR ECHO-Doppler bandwidth

Spotlight mode SAR synthetic aperture length is relatively larger, radar echo azimuth Doppler bandwidth is wider. Figure 7.15 is the Spotlight mode echo time–frequency diagram before and after Dechirp. Spotlight SAR echoes azimuth bandwidth is composed of two parts, $B = B_a + B_d = \frac{2vw_a}{\lambda R} + \frac{2v^2}{\lambda R} T_a$, as shown in Fig. 7.15a.

In Fig. 7.15a, T_a is synthetic aperture time, B_a is instantaneous Doppler bandwidth of echo signal, and B_d is Doppler bandwidth of point target, which are as follows:

Instantaneous Doppler bandwidth of echo signal is

$$B_a = \frac{2v_a W_a}{\lambda R_0} \tag{7.59}$$

Doppler bandwidth of point target is

$$B_d = \frac{k_a v_a}{\rho_a} \tag{7.60}$$

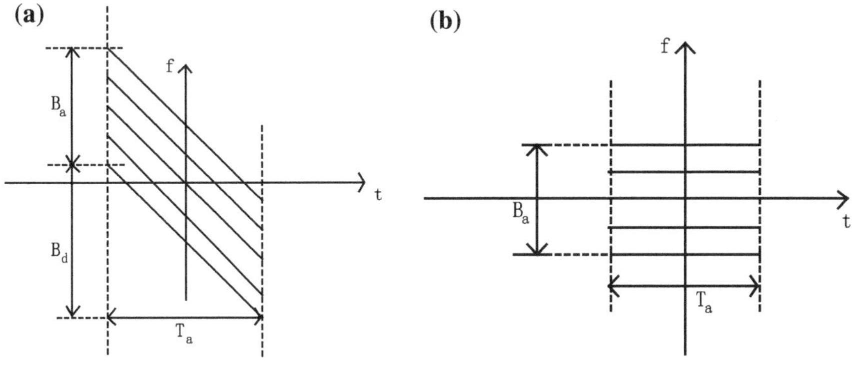

Fig. 7.15 The Spotlight mode echoes time–frequency diagram **a** Before dechirp and **b** After dechirp

In Eqs. (7.59) and (7.60), v_a is the velocity of radar platform, W_a is scene width in azimuth direction, λ is wavelength of transmitting signal, R_0 is the range between radar and imaging scene center in vertical side-looking mode, k_a is main lobe expansion factor in azimuth direction, ρ_a is image azimuth resolution.

The instantaneous Doppler bandwidth of Spotlight SAR echo signal B_a is constant and Doppler center frequency approximate to the linear frequency-modulated signal. In synthetic aperture time, the change range of Doppler center frequency B_d corresponds to Doppler bandwidth of the scene center. The frequency difference between the target point and the scene center point maintain a constant, which is proportional to the distance from the target point to the scene center in the azimuth direction.

(2) Azimuth downsampling technology

In the Spotlight SAR, wide Doppler bandwidth requires high PRF, which strictly limits the design of system parameters. If considered the instantaneous bandwidth as a lower bound to select pulse repetition frequency, pulse repetition rate does not satisfy the Nyquist sampling laws, the echo signal aliasing in the Azimuth will occurs. However, the azimuth downsampling technology can restore aliasing of azimuth signal spectrum and get the signal's spectrum before aliasing.

From the signal time-delay relationship, azimuth echo signals of point target can be expressed as

$$s_n(u) = \sigma_n \mathrm{e}^{-j\frac{4\pi}{\lambda}\left[\sqrt{R_0^2 + (u-y_n)^2}\right]} \tag{7.61}$$

In Eq. (7.61), λ is the wavelength of transmitting signal, R_0 is the range between radar and imaging scene center in side-looking mode, u is radar azimuth location variable and y_n is target azimuth coordinate.

According to the principle of stationary phase, the spectrum of $s_n(u)$ is

$$S_n(k_u) \approx \sigma_n \mathrm{e}^{\left(-j\sqrt{4k^2 - k_u^2}R_0 - j\,k_u y_n\right)} \tag{7.62}$$

In Eq. (7.62), σ_n is scattering coefficient of point target, $k = 2\pi/\lambda$ is wavenumber and $k_u \in \left[-\frac{B_a + B_d}{2}, \frac{B_a + B_d}{2}\right]$ is Doppler domain variable.

Apparently, if sampling echo signal by the frequency corresponding with instantaneous bandwidth B_a, azimuth spectrum aliasing phenomenon will occurs. Figure 7.16 is azimuth signal spectrum aliasing schematic. In Fig. 7.16a, the sampling frequency is equal to signal total bandwidth, and spectrum aliasing does not occur. In Fig. 7.16b, the sampling frequency is equal to instantaneous bandwidth, and signal spectrum aliasing occurs. The aliasing signal spectrum is

$$S_{\mathrm{ns}}(k_u) \approx \sigma_n \mathrm{e}^{-j\sqrt{4k^2 - k_u^2}R_0}\left[\mathrm{e}^{-j\,k_u y_n} + \sum_m \mathrm{e}^{-j\,(k_u \pm mB_a)y_n}\right] \tag{7.63}$$

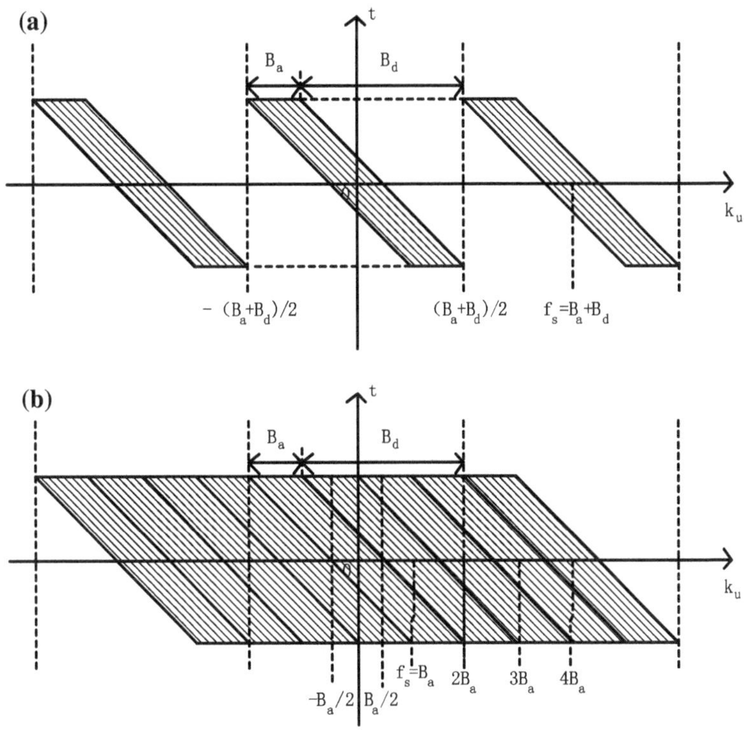

Fig. 7.16 Azimuth signal spectrum aliasing schematic, **a** $fs = B_a + B_d$ sampling non-aliasing signal frequency spread; **b** $fs = B_a$ sampling aliasing signal frequency spread

In Eq. (7.63), $m = 1, 2, 3, \ldots, M$ represents aliasing signals order coming into the main interval, the M value is determined by the ratio of B_a and B_d, the range of k_u is $\left[-\frac{B_a}{2}, \frac{B_a}{2}\right]$.

By the same principle of stationary phase, aliasing signals is

$$s_{ns}(u) = \sigma_n \left[1 + \sum_m e^{-j\,(\pm mB_a)y_n}\right] e^{-j\frac{4\pi}{\lambda}\left[\sqrt{R_0^2 + (u-y_n)^2}\right]} \tag{7.64}$$

The range of corresponding variable u is [−L/2, L/2].

From Eq. (7.61), echo signal of scene center point target, with scattering coefficient of 1, is

$$s_0(u) = e^{-j\frac{4\pi}{\lambda}\left[\sqrt{R_0^2 + u^2}\right]} \tag{7.65}$$

In Eq. (7.65), the range of u is [−L/2, L/2].

Using conjugated with $s_0(u)$ as a reference function, by means of mixing this reference function with aliasing signals $s_{ns}(u)$ carry out demodulation, and using Fresnel approximations the demodulation result can be

$$s_{nc}(u) = A(y_n)\mathrm{e}^{-j\frac{4\pi y_n}{\lambda R_0}u} \tag{7.66}$$

In Eq. (7.66)

$$A(y_n) = \sigma_n\left[1 + \sum_m \mathrm{e}^{-j(\pm mB_a)y_n}\right]\mathrm{e}^{-j\frac{4\pi}{\lambda}y_n^2} \tag{7.67}$$

The mixed signal frequency is proportional to the target azimuth position y_n. After demodulation of frequency mixing process, aliasing signals will be concentrated at a point on the spectrum, and mixed signal in different azimuth location targets will be located at different frequencies.

Up-sampling to mixed signals processing with a sampling rate of $B_a + B_d$, and then mixed-frequency modulation with a reference signal of Eq. (7.65), so that the resulting sampled signal will satisfy the Nyquist sampling laws, and signal without aliasing will return to the original appearance.

Obviously, by azimuth downsampling technology, the aliasing azimuth signal spectrum can be restored, and signal spectrum before aliasing can be got.

7.3.2 *Special Problems of Spotlight Synthetic Aperture Radar Imaging*

The specific issues of Spotlight SAR imaging include: Range Cell Migration (RCM), two-dimensional MTRC, temporal variations of Doppler center frequency, Space-Variant phase course, the coupling between range and azimuth as well as motion compensation, and so on [8].

(1) RCM

RCM phenomenon happens when the slant distance changes between radar and target preponderate over a range resolution cell in a synthetic aperture time, which makes echo signal from the same target distributes in different range units in the range direction. RCM includes range walk and range curve. Figure 7.17 is diagram of RCM, the distance changes between the aircraft and the target with time as

$$r(t) = \sqrt{r_c^2 + (vt)^2} \tag{7.68}$$

RCM brings echo signals couple in azimuth and range direction, and two-dimensional independent criteria is not met, so echo signals cannot be

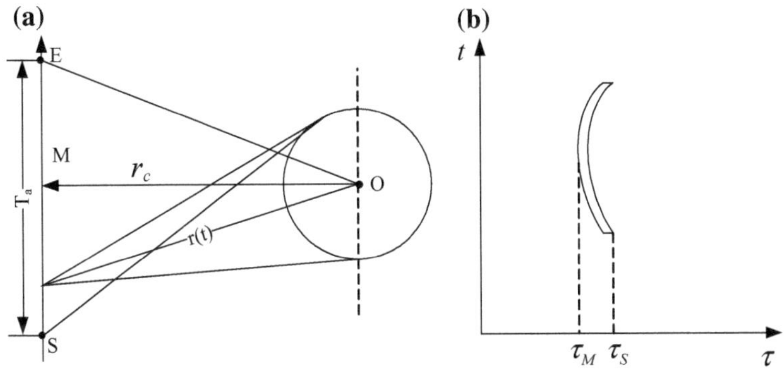

Fig. 7.17 RCM diagram, **a** Azimuth geometry of spotlight SAR; **b** Echo signal compressed result in range direction

decomposed into two one-dimensional processing directly. In Fig. 7.17 side-looking Spotlight mode, for example, point *O* locates in irradiation center. At position *S* and *E*, distances between vehicles with *O* are maximum, whereas, in synthetic aperture center *M*, the distance between vehicles and *O* is minimum. If the echo signal is compressed only in range direction, the compression results as Fig. 7.17b have shown which results in range walk and range curve.

Figure 7.18 is RCM paths comparison between Strip SAR and Spotlight SAR. Comparing with Strip mode, the RCM of Spotlight SAR has the following characteristic: Along with the resolution improvement, the volume of RCM grows gradually. The range gates number crossed by RCM paths will be much more than the number of Strip mode.

For Strip mode, in the same range cell the migration paths of targets at different azimuth position are basically the same; For Spotlight mode, because of the point target remains under beam irradiation, even if within the same range cell for point targets at different azimuth, their migration paths are not the same.

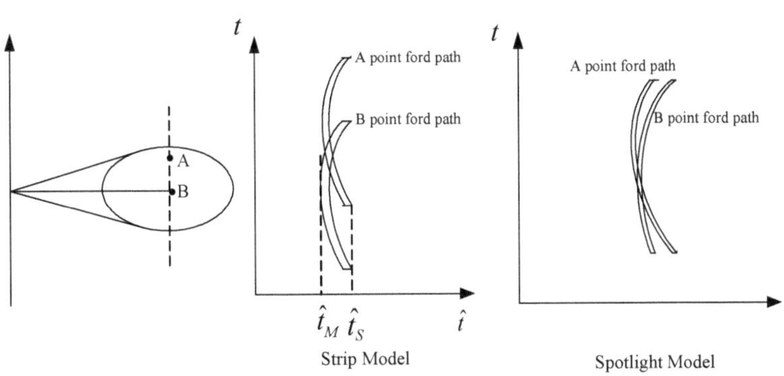

Fig. 7.18 RCM paths comparison between strip SAR and spotlight SAR

(2) MTRC

During the accumulation of synthetic aperture time, Spotlight SAR-beam gaze direction always points to target regional center, apart from the relative translation in vehicle velocity, and there is a considerable relative rotation between radar and targets. Relative rotation between the radar and targets causes the range courses of target scattering points beyond the imagination of a resolution cell at both range and azimuth direction simultaneity, that is a phenomena of two-dimensional movement through resolution cells (MTRC).

Figure 7.19 is the schematic for the two-dimensional movement through resolution cells. The movement through resolution cells generated by a rotating is the fundamentality causes of defocus of imaging.

The phenomenon of two-dimensional movement through resolution cell has the spatial variation characteristics, the rules and areas of variation are depend on scattering geometry location. This means that the imaging processing from different scattering points within the measured target area requires different reference functions. In addition, Spotlight SAR imaging with high resolution and longer synthetic aperture time, should be set up strict requirements to coherence and motion compensation of radar systems.

If the radar is far from target, the azimuth Doppler frequency is [9]

$$f_{\mathrm{d}} = \frac{2x\omega}{\lambda} \tag{7.69}$$

In Eq. (7.69), λ is the transmitting signal wavelength, ω is target's rotation angular velocity to radar, x is azimuth coordinate.

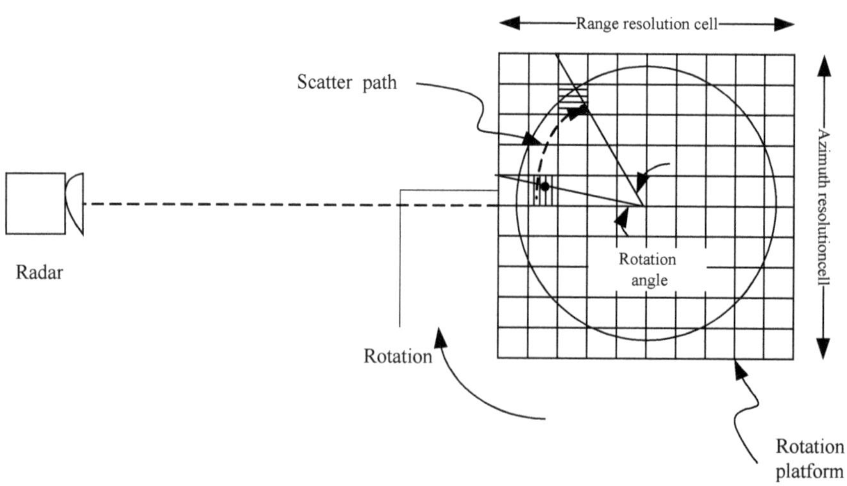

Fig. 7.19 Two-dimensional movement through resolution cells schematic

If synthetic aperture time is T, then the Doppler resolution is

$$\Delta f_{\mathrm{d}} = T^{-1} \tag{7.70}$$

The corresponding azimuth resolution is

$$\rho_{\mathrm{a}} = \Delta x = \frac{\lambda}{2\omega T} = \frac{\lambda}{2\Delta\theta} \tag{7.71}$$

In Eq. (7.71), $\Delta\theta$ is the rotation angle of target surrounding radar in a synthetic aperture time.

Assuming W_x and W_y is azimuth and range size of the target scene, respectively, then the Doppler bandwidth is

$$B_{\mathrm{d}} = 2\omega W_x/\lambda \tag{7.72}$$

To eliminate azimuth ambiguity, pulse repetition frequency should be greater than the Doppler bandwidth, that is,

$$f_{\mathrm{PRF}} > 2\omega W_x/\lambda \tag{7.73}$$

Taking into account the elimination of range ambiguity, then

$$f_{\mathrm{PRF}} < c/(2W_y) \tag{7.74}$$

W_y here needs to be transformed to the slant range direction.

The condition of avoiding azimuth and range ambiguity is

$$2\omega W_x/\lambda < \mathrm{PRF} < c/(2W_y) \tag{7.75}$$

That is

$$\omega W_x W_y < c\lambda/4 \tag{7.76}$$

Assumed range resolution of ρ_{r}, in synthetic aperture time the condition of no movement through rang resolution cells is

$$\frac{1}{2} W_x \Delta\theta < \rho_r \tag{7.77}$$

Substitute into Eq. (7.71), then

$$\rho_{\mathrm{a}}\rho_{\mathrm{r}} > \frac{1}{4}\lambda W_x \tag{7.78}$$

Similarly, the condition of no movement through azimuth resolution cells is

$$\frac{1}{2}W_y\Delta\theta < \rho_a \tag{7.79}$$

Substitute into Eq. (7.71), then [10]

$$\rho_a^2 > \frac{1}{4}\lambda W_y \tag{7.80}$$

For high-resolution Spotlight SAR, due to the distance is far away and the imaging area is wide, it can not satisfy the above restricts, so it will appear the phenomena of two-dimensional MTRC inevitably. The imaging defocusing caused by two-dimensional MTRC becomes one of the crucial problems and technical difficulties.

(3) Temporal variations of Doppler center frequency

Doppler center frequency of Spotlight SAR constantly changes along with radar line of sight and azimuth time. Spotlight SAR Doppler center frequency is:

$$f_{dc}(t) = \frac{2v_a \cos\theta_s(t)}{\lambda} \tag{7.81}$$

In Eq. (7.81), $\theta_s(t)$ is squint angle, v_a is the aircraft flight speed. By this equation, the every frequency courses of the azimuth scattering points within the target area can be calculated.

It can be seen that echoes azimuth bandwidth $B(B = B_a + B_d)$ of Spotlight SAR is far greater than the Doppler bandwidth B_a of Strip SAR. If using B_a is as a reference to design pulse repetition frequency, it will definitely cause azimuth ambiguity. If using the lower limit of B to select the pulse repetition frequency, it will debase the efficiency of imaging and produce the range ambiguity. Pulse repetition frequency should be selected to avoid azimuth ambiguity and range ambiguity simultaneity. Therefore, effective measures must be adopted to eliminate the azimuth spectrum spread infection on imaging processing since the Doppler center frequency changes.

(4) Space variety phase courses

Figure 7.20 is geometric relationships for Spotlight SAR. The relative movement between SAR and the target can be decomposed into two parts: translation and rotation. Translation indicates the translational motion between target and radar, the phase error is constant in space and as same as in Strip SAR; Rotation indicates the rotational motion of the scattering target around Imaging Center point

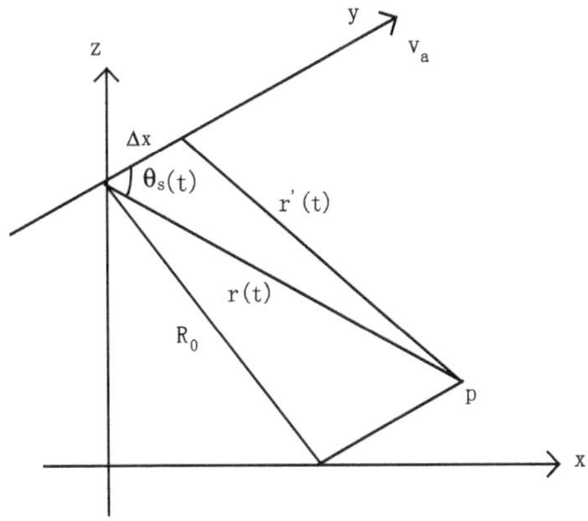

Fig. 7.20 Target range geometric relationships for spotlight SAR

(especially nonuniform rotation). Since the phase error is depended on the rotation radius of the scattering point, the rotation phase error will change with the change of scattering point's geometry location in space. Due to the relative rotation between radar and target, the phase error of Spotlight SAR has the space variety characteristics.

On an azimuth moment t, Spotlight SAR antenna beam center points to target scattering point p, beam center squint angle $\theta_s(t)$, and p can be located by ($\theta_s(t)$, $r(t)$), as shown in Fig. 7.20. If a small displacement Δx occurs on the radar flight path, instantaneous slant range of the scattering point p can be expressed as:

$$\begin{aligned} r'(t) &= \sqrt{r^2(t) + (\Delta x)^2 - 2\Delta x r(t)\cos\theta_s(t)} \\ &= r(t)\sqrt{1 + \frac{(\Delta x)^2 - 2\Delta x r(t)\cos\theta_s(t)}{r^2(t)}} \\ &\approx r(t)\left[1 + \frac{1}{2}\cdot\frac{(\Delta x)^2 - 2\Delta x r(t)\cos\theta_s(t)}{r^2(t)} - \frac{1}{8}\cdot\left(\frac{(\Delta x)^2 - 2\Delta x r(t)\cos\theta_s(t)}{r^2(t)}\right)^2 + \cdots\right] \end{aligned} \tag{7.82}$$

In Eq. (7.82)

$$r(t) = R_0/\sin\theta_s(t) \tag{7.83}$$

In Eq. (7.83), R_0 is the slant range of vertical side-looking. Then, Doppler frequency shift $f_d(t)$ is

$$\begin{aligned} f_d(t) &= \frac{2}{\lambda}\frac{dr'(t)}{dt} \\ &= \frac{2}{\lambda}\cdot\frac{1}{\sqrt{R_0^2 - 2R_0\Delta x\sin\theta_s(t)\cos\theta_s(t) + (\Delta x\sin\theta_s(t))^2}} \\ &\times \frac{-R_0^2\cos\theta_s(t) + \Delta x\sin\theta_s(t)\cos^2\theta_s(t) + \Delta x\sin^3\theta_s(t)}{\sin^2\theta_s(t)}\cdot\frac{d\theta_s(t)}{dt} \end{aligned} \tag{7.84}$$

Since $R_0 \gg \Delta x$, so

$$f_d(t) \approx -\frac{2R_0\cos\theta_s(t)}{\lambda\sin^2\theta_s(t)}\cdot\frac{d\theta_s(t)}{dt} \tag{7.85}$$

In Spotlight SAR, the line of sight of the angle $\theta_s(t)$ tracks to the Imaging Area Center continuously. The phase courses of a scattering points depend on the $\theta_s(t)$ and the change rate of $\theta_s(t)$, as well as, it depends on the locations of the scattering points, the phase courses are Space-Variant. The Space-Variant phase courses mean that the different compensation phase factors should be used to focusing processing for different scattering point, which has brought a larger difficulty to the spotlight SAR imaging.

(5) The coupling between range dimension and azimuth dimension

A case study of point target from distance $r(t)$ after I/Q channel detection, the ECHO has the following form:

$$s_r(n,\hat{t}) = \mathrm{rect}\left(\frac{t}{T_a}\right)\mathrm{rect}\left(\frac{\hat{t}-\frac{2r(t)}{c}}{T_p}\right)\exp\left(-j\frac{4\pi}{\lambda}r(t)\right)\exp\left(j\pi\gamma\left(\hat{t}-\frac{2r(t)}{c}\right)^2\right) \tag{7.86}$$

Considered just phase, that is,

$$\phi(n,\hat{t}) = -j\frac{4\pi}{\lambda}r(t) + j\pi\gamma\left(\hat{t}-\frac{2r(t)}{c}\right)^2 \tag{7.87}$$

Because of

$$r(t) \approx R_0\left[1 + \frac{(x-X)^2}{2R_0^2}\right] \tag{7.88}$$

So,

$$\begin{aligned}\phi(n,\hat{t}) &= -\frac{2\pi}{\lambda R_0}(x-X)^2 + \pi\gamma\left(\hat{t}-\frac{2R_0}{c}\right)^2 - 2\pi\gamma\left(\hat{t}-\frac{2R_0}{c}\right)\frac{(x-X)^2}{R_0 c} + \pi\gamma\left[\frac{(x-X)^2}{R_0 c}\right]^2 - \frac{4\pi R_0}{\lambda}\\ &\approx -\frac{2\pi}{\lambda R_0}(x-X)^2 + \pi\gamma\left(\hat{t}-\frac{2R_0}{c}\right)^2 - 2\pi\gamma\left(\hat{t}-\frac{2R_0}{c}\right)\frac{(x-X)^2}{R_0 c} - \frac{4\pi R_0}{\lambda}\end{aligned} \tag{7.89}$$

In formula (7.89), the first item is an azimuth Doppler signal, the second item is a linear frequency-modulated signal in range direction, and the third item is a coupled signal between azimuth dimension and range dimension, and the last item is a stationary phase.

The Fourier information of Spotlight SAR distributes along the polar RADIUS. Even if transmitting pulse bandwidth is zero, as long as the turning angle of the radar's line of sight is not zero, the frequency extension should be occurred in range dimension, to a certainty, the corresponding range dimension resolution is created. This shows that range information and azimuth information of Spotlight SAR are coupled and relevant.

(6) Motion compensation

In the airborne or spaceborne SAR, the platform motion cannot be on the ideal line, ECHO phase errors due to the flying platform motion errors will damage the imaging effect severely, never so much as envisioned. Regardless of using any imaging algorithms, the phase error due to the movement of platforms should be compensated. Along with resolution improvement, the requirements of motion compensation have become more and more strict; the motion compensation is an important link in the imaging process.

Figure 7.21 shows the main differences between Spotlight and Strip mode SAR motion compensation: In Strip mode, since the antenna pointing always is the same, the compensation for a point target is independent of the target's location in the azimuth direction. In Spotlight mode requirement, radar's line of sight should

Fig. 7.21 Motion compensation comparisons

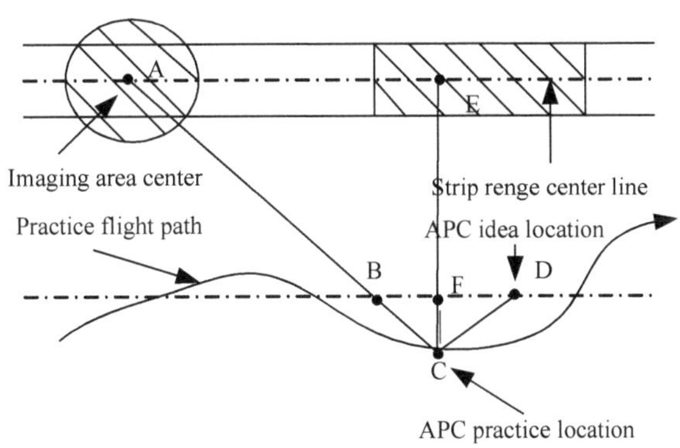

always track the Center of the irradiation area, and the radar should track distance change between the radar and the Centere of the irradiation area. In this way, after Dechirp processing of received signals, the full compensation to the center point can be achieved. However, because of the motion compensation capacity to a point target is dependent to the position of the point target in spotlight SAR; the motion compensation to other point targets in the irradiation area will create the phase errors.

In this chapter, the imaging principle and features of Spotlight SAR were discussed mainly; the Spotlight SAR imaging principle was explained using computer auxiliary tomography imaging method; the resolution, synthesis aperture length, and some other parameters of Spotlight SAR was analyzed; the echo signal features, echo-Doppler bandwidth, azimuth downsampling technology, and other features of Spotlight SAR were discussed; the special problems of Spotlight SAR Imaging were researched, such as range cell migration (RCM), two-dimensional MTRC, temporal variations of Doppler center frequency, Space-Variant phase courses, the coupling between range and azimuth as well as motion compensation, and so on.

References

1. Carrara WG, Goodman RS, Majewski RM (1995) Spotlight synthetic aperture radar: signal processing algorithms: aretch house
2. Jakowatz CV et al (1996) Spotlight-mode synthetic aperture radar: a signal processing approach. Kluwer Academic, MA
3. Cao FX, Yan JP, Zhang M (1998) Azimuth resolution study for the synthetic aperture radar spotlight model. Modern Radar 20(6):6–10
4. Gao XW, Tian XW, Yang RL (2004) The azimuth resolution study for satellite-borne synthetic aperture radar spotlight model. J Electron Inf 26(1):95–99
5. Gao XW, Lei H, Yang RL (2004) The synthetic length study for synthetic aperture radar spotlight model. J Electron Inf 26(6):935–958
6. Yan XK (2001) The echo doppler frequency calculate for satellite-borne synthetic aperture radar. Shanghai Space 18(3):1–4
7. Hong W, Mao SY (1997) The airborne spotlight synthetic aperture radar imaging technology, the science and technology report of the Chinese aviation
8. Prati C et al. SIR-C/X-SAR: the estimation of the doppler centroid ambiguity. In: IGARSS'91, pp. 1055–1058
9. Brown WM, Fredricks RJ (1969) Range-doppler imaging with motion through resolution cells. IEEE Trans Aerosp Electron Sys 5(1):98–102
10. Walker JL (1980) Range-doppler imaging of rotating objects. IEEE Trans Aerosp Electron Sys 16(1):23–52

Chapter 8
Spotlight SAR Imaging Algorithms

Imaging processing is the prominent technical characteristics and core content of synthetic aperture radar. Since 1970s, during more than 30 years' development a number of imaging algorithms for Spotlight synthetic aperture radar have emerged in succession. For example: Polar format algorithm (PFA), the range migration algorithms (RMA), Chirp Scaling algorithm (CSA), extended CSA algorithm (ECSA), frequency Scaling algorithm (FSA), the two-step algorithm (TSP) and so on. This chapter covers detailed theory, signal modeling and computational processes of various algorithms, as well as special problems of imaging algorithms.

8.1 Dechirp Technology of Spotlight SAR

Spotlight synthetic aperture radar has wider Doppler bandwidth than the simple Strip synthetic aperture radar, so it can get higher azimuth resolution. In order to avoid the mixing of azimuth spectrum, pulse repetition frequency of transmitting signal is required to be greater than the Doppler bandwidth. In order to improve range resolution, the bandwidth of the transmitted signal also needs to be increased. However, in synthetic aperture radar, the sampling frequency of range direction is determined by the signal bandwidth, usually it cannot be less than the signal bandwidth, or else it will create the alias of spectrum and affect image quality. Sampling frequency increasing will increase requirements for the A/D conversion speed, the data rate and data storage capacity. Therefore, it is urgently demanded a technology to reduce the systems' data capacity. Removing slope technology (also known as removing FM technology) meets this requirement. On Strip synthetic aperture radar, sometimes referred to as Deramp techniques. In Spotlight synthetic aperture radar, called Dechirp technology.

R. Yang et al., *High-Resolution Microwave Imaging*,
https://doi.org/10.1007/978-981-10-7138-6_8

8.1.1 Dechirp Technology of Spotlight SAR

Dechirp technology is mixing the received echo LFM signal with a LFM reference signal which has the same FM slope, turns the linear frequency-modulated signal into a single frequency signal. The reference signal is range symmetric for the center of the scene and has inverse FM slope of transmitting pulse signal. After Dechirp in range direction, the receiving signals' bandwidth B_{r2} is

$$B_{r2} = B_{r1}\frac{2w_r}{cT_p} \tag{8.1}$$

In Eq. (8.1), the w_r is scene length in slant range direction, and c is the speed of light, B_{r1} is the bandwidth of the transmitted signal. From Eq. (8.1) as can be seen, if $2w_r/c$ is less than the transmitting signal pulse width T_p, then B_{r2} is less than B_{r1}, as shown in Fig. 8.1. In addition, after Dechirp different frequency components with a range-time delay is misaligned in time, it introduces a phase in range-frequency domain, known as residual video phase (RVP), which have to be removed, or else a range phase error will occur and it will reduce the image quality.

After Dechirp in azimuth direction, the receiving signals' bandwidth B_{a2} is

$$B_{a2} = \frac{2Vw_a}{\lambda R} \tag{8.2}$$

In Eq. (8.2), w_a is the scene width in azimuth direction, and V is the speed of radar, as shown in Fig. 8.2. As can be seen, After Dechirp linear

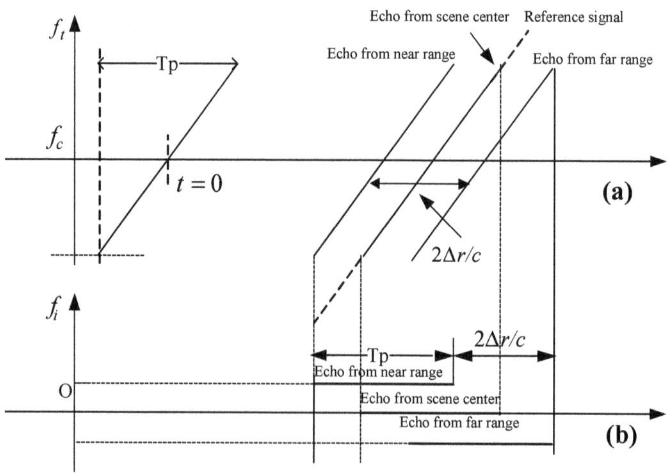

Fig. 8.1 Dechirp processing diagram in range direction. Time-frequency relationship of echoes from different distances before Dechirp. Time-frequency relationship of echoes from different distances after Dechirp

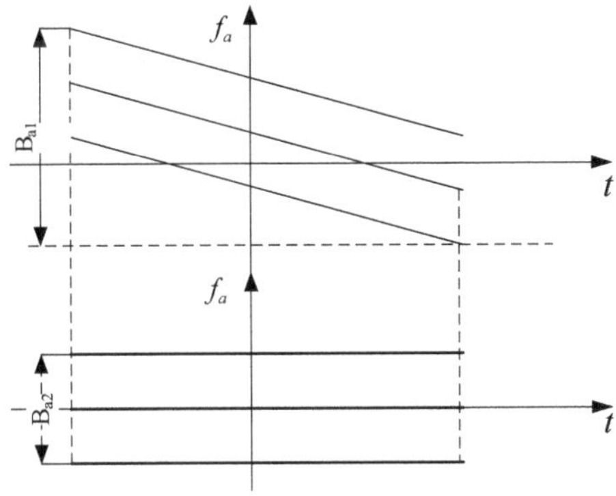

Fig. 8.2 Dechirp processing diagram in azimuth direction

frequency-modulated signal turns into a single frequency signal. The signal frequency is relative to the position in azimuth direction, that is, each frequency range on the frequency domain corresponds to the target point at a certain azimuth position.

In Spotlight SAR, there are two kinds of Dechirp ways. One way is to take echo signal from target imaging regional center points as reference demodulation signal, this way can achieve both range Dechip and azimuth Dechirp; the other way is to take echo signal from fixed distance (General select shortest distance from flight direction to irradiation Center) target as reference demodulation signal, this method only achieves range Dechirp, in azimuth it still is LFM signal. For the first way, the possibility to accurately construct reference function is from Spotlight SAR continuous tracking to the target Center.

After the I/Q channel detection (remove carrier), the echo signal received by radar can be written as

$$s_r(t,\tau) = a_t \mathrm{rect}\left(\frac{t}{T_a}\right)\mathrm{rect}\left(\frac{\tau - \frac{2r(t)}{c}}{T_\mathrm{p}}\right)\exp\left\{j2\pi f_\mathrm{c}\left[\tau - \frac{2r(t)}{c}\right] + j\pi k_\mathrm{r}\left[\tau - \frac{2r(t)}{c}\right]^2\right\} \tag{8.3}$$

In Eq. (8.3), $\mathrm{rect}(u) = \begin{cases} 1, & |u| \le \frac{1}{2} \\ 0, & |u| > \frac{1}{2} \end{cases}$, $\tau = t - n \cdot T_{\mathrm{prt}}$, n is the number of transmitted pulses, λ is the wavelength of radar signal, T_a is the time of synthetic aperture. Due to the time between radar transmitting signals and radar receiving echo is far less than pulse repetition time T_{PRT}, therefore it can be approximated that the azimuth positions of radar remains the same within this time. Generally we call τ for the fast time (range time) and t for the slow time (azimuth time).

In the first Dechirp way, the echo signal from the target center point of the imaging area is taken as a reference demodulated signal, and the reference function is

$$s_{\mathrm{ref}}(t,\tau)=\exp\left\{j2\pi f_{\mathrm{c}}\left(\tau-\frac{2r_{\mathrm{a}}(t)}{c}\right)+j\pi k_{\mathrm{r}}\left(\tau-\frac{2r_{\mathrm{a}}(t)}{c}\right)^2\right\} \tag{8.4}$$

After Dechirp, echo signal can be expressed as

$$s_{\mathrm{IF}}(t,\tau)=s_{\mathrm{r}}(t,\tau)s_{\mathrm{ref}}^{*}(t,\tau)=\mathrm{rect}\left(\frac{t}{T_a}\right)\mathrm{rect}\left(\frac{\tau-\frac{2r(t)}{c}}{T_{\mathrm{p}}}\right)\exp(j\Phi) \tag{8.5}$$

$$\Phi=-\frac{4\pi k_r}{c}\left(\frac{f_{\mathrm{c}}}{k_r}+\tau-\frac{2r_{\mathrm{a}}(t)}{c}\right)[r(t)-r_{\mathrm{a}}(t)]+\frac{4\pi k_{\mathrm{r}}}{c^2}[r(t)-r_{\mathrm{a}}(t)]^2 \tag{8.6}$$

In the second Dechirp way, the echo signal from the target, which is shortest distance from the irradiation center, is taken as a reference demodulated signal, and the reference function is

$$s_{\mathrm{ref}}(t,\tau)=\exp\left\{j2\pi f_{\mathrm{c}}\left(\tau-\frac{2r_{\mathrm{c}}}{c}\right)+j\pi k_r\left(\tau-\frac{2r_{\mathrm{c}}}{c}\right)^2\right\} \tag{8.7}$$

In Eq. (8.7), r_{c} is a fixed reference distance. Generally we select shortest distance from flight direction to irradiation center. After Dechirp, the echo signal is

$$s_{\mathrm{IF}}(t,\tau)=s_{\mathrm{r}}(t,\tau)s_{\mathrm{ref}}^{*}(t,\tau)=\mathrm{rect}\left(\frac{t}{T_a}\right)\mathrm{rect}\left(\frac{\tau-\frac{2r(t)}{c}}{T_p}\right)\exp(j\Phi) \tag{8.8}$$

$$\Phi=-\frac{4\pi k_{\mathrm{r}}}{c}\left(\frac{f_{\mathrm{c}}}{k_{\mathrm{r}}}+\tau-\frac{2r_{\mathrm{c}}}{c}\right)[r(t)-r_{\mathrm{c}}]+\frac{4\pi k_{\mathrm{r}}}{c^2}[r(t)-r_{\mathrm{c}}]^2 \tag{8.9}$$

8.1.2 Compensation to Residual Video Phase

After Dechirp processing, different frequency components of Spotlight SAR produce range-time delay which is not aligned in the time domain and introduce a residual video phase (RVP) in frequency-domain. At the first Dechirp method, residual phase video can generally be ignored, however, for the second Dechirp method, residual video phase item cannot be ignored, need to be compensated. Because residual video phase changes along with space positions and it is inconvenient to be processed in time domain. Because the delay in time domain is equal to multiply the phase shift in the frequency domain, so, as long as in the frequency domain different frequency component is multiplied by a phase related to frequency, which can compensate the phase brought by delay. Further by IFFT processing, different frequency component of echo signal can be aligned in the time domain.

Firstly, the Fourier transform of Eq. (8.8) responding to $\tau_{\mathrm{d}} = \tau - \frac{2r(t)}{c}$ is as follows:

$$s_{\mathrm{o}}(n, \tau_{\mathrm{d}}) = \mathrm{rect}\left(\frac{\tau_d - \frac{2\Delta r(t)}{c}}{T_{\mathrm{p}}}\right)\exp\left\{-j\frac{4\pi\gamma}{c}\left(\frac{f_{\mathrm{c}}}{\gamma} + \tau_{\mathrm{d}}\right)\Delta r + j\frac{4\pi\gamma}{c^2}\Delta r^2\right\} \tag{8.10}$$

In Eq. (8.10), $\Delta r = r(t) - r_c$.

By the resident phase principle, the Fourier transform of Eq. (8.10) is as follows:

$$S_{\mathrm{o}}(n, f_{\mathrm{r}}) = T_{\mathrm{p}} \sin c\left[T_{\mathrm{p}}\left(f_{\mathrm{r}} + \frac{2k_{\mathrm{r}}\Delta r}{c}\right)\right]\exp\{j\Phi'\} \tag{8.11}$$

$$\Phi' = -\frac{4\pi f_{\mathrm{c}}}{c}\Delta r - \frac{4\pi k_{\mathrm{r}}}{c^2}\Delta r^2 - \frac{4\pi f_{\mathrm{r}}}{c}\Delta r \tag{8.12}$$

In Eq. (8.11), f_{r} is range frequency.

Phase compensation item is:

$$S_{\mathrm{c}}(f_{\mathrm{r}}) = \exp(-j\pi f_{\mathrm{r}}^2/k_{\mathrm{r}}) \tag{8.13}$$

Equation (8.12) multiply with Eq. (8.13), and the IFFT processing to the result is as follows:

$$s_{\mathrm{rd}} = \mathrm{rect}\left(\frac{\tau - 2r_{\mathrm{c}}/c}{T_{\mathrm{p}}}\right)\exp\left[-j\frac{4\pi k_{\mathrm{r}}}{c}\left(\tau + \frac{f_{\mathrm{c}}}{k_{\mathrm{r}}} - \frac{2r_{\mathrm{c}}}{c}\right)\Delta r\right] \tag{8.14}$$

Comparing Eqs. (8.14) and (8.10), the third phase in Eq. (8.10), which is residual video phase, has been removed. The scattering points echo signal from different range is already aligned in the time domain. Frequency directly reflects the range positions of the various scattering points. Figure 8.3 shows echo signal before and after residual video phase elimination. Figure 8.4 shows the flow-process of residual video phase compensation.

8.2 Polar Format Algorithm

In Spotlight synthetic aperture radar, since the antenna beam always points to the fixed imaging area, radar would rotate relative to the target area. This relative rotation makes imaging of spotlight synthetic aperture radar similar to that of microwave turntable imaging. Polar format algorithm (PFA) was first put forward as an effective imaging method of rotating target, soon to be successfully used in

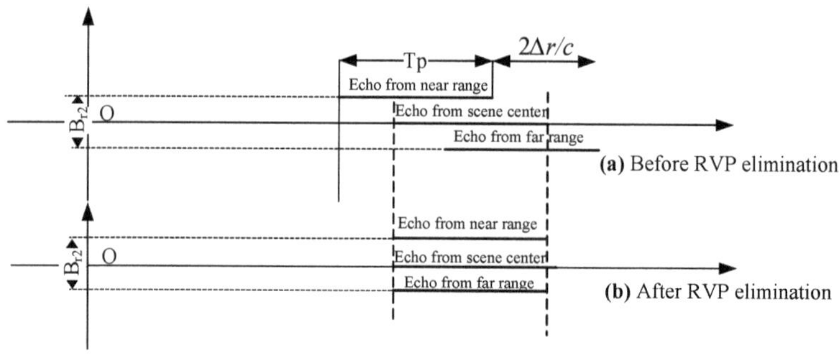

Fig. 8.3 Echo signal before and after residual video phase elimination. **a** Before removing RVP, **b** after removing RVP

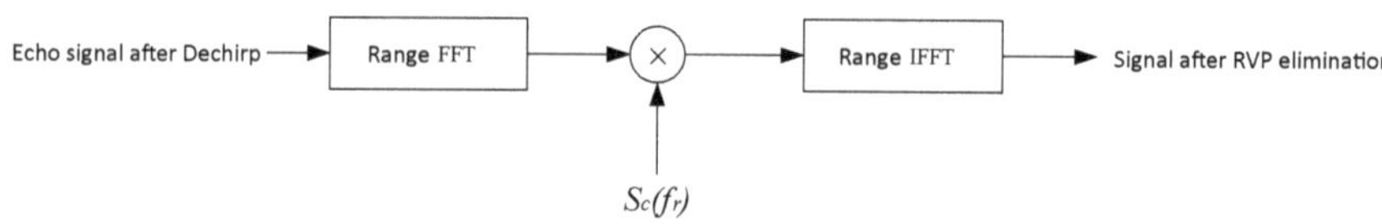

Fig. 8.4 The flow-process diagram of residual video phase compensation

the imaging of Spotlight synthetic aperture radar. This greatly improved the focus imaging range of spotlight synthetic aperture radar. This algorithm in the early Spotlight synthetic aperture radar system has been widely applied, and it is a more mature imaging algorithm in the spotlight synthetic aperture radar. Polar format algorithm process for Spotlight synthetic aperture radar is as shown in Fig. 8.5.

8.2.1 PFA Principle

Figure 8.6 is the data collection plane irradiation geometry for Spotlight synthetic aperture radar, which gives the *x*-*y* Cartesian coordinate system of the ground irradiation area in imaging process, *x* is azimuth direction, and *y* is range direction. The distance between the scatter point target *P* and irradiation regional centre *O* is a. γ is the angle between the *OP* and *x* axis. The distance between aircraft and irradiation centre *O* is r_o, and the distance from aircraft to *P* is *r*. In flight the angle between the radar beams pointing direction to the *y* axis is θ.

Suppose radar transmitting linear frequency modulation signal with pulse length T_p

$$s(t) = \mathrm{rect}\left(\frac{t}{T}\right)\exp\left(j2\pi f_c t + j\pi k_r t^2\right) \tag{8.15}$$

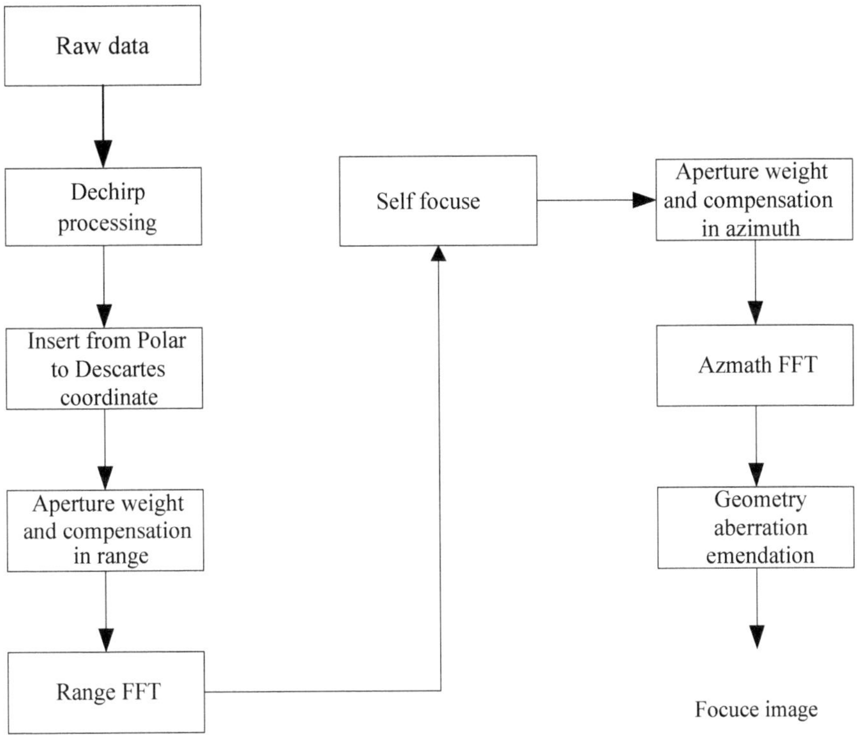

Fig. 8.5 Polar format algorithm flowchart

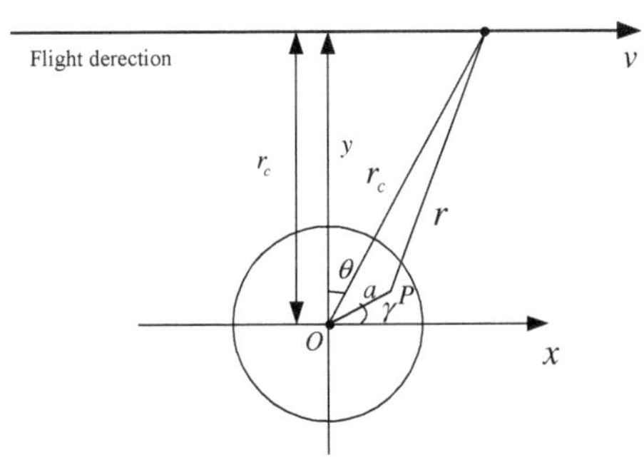

Fig. 8.6 Spotlight synthetic aperture radar data collection plane radiation geometry

Dechirp processing radar echo from point P: Take echo signal from regional centre O as a reference point and the radar echo multiplies with the conjugated reference signal. Reference signal is

$$s_{\mathrm{ref}}(t) = \exp\left\{j2\pi f_{\mathrm{c}}\left[\tau - \frac{2r_{\mathrm{o}}(t)}{c}\right] + j\pi k_{\mathrm{r}}\left[\tau - \frac{2r_{\mathrm{o}}(t)}{c}\right]^2\right\} \tag{8.16}$$

In Eq. (8.16), $r_{\mathrm{o}}\,(t)$ is the distance between the radar and the center point of the destination area. After Dechirp processing of signals

$$s_{\mathrm{IF}}(t,\tau) = s_{\mathrm{r}}(t,\tau)s_{\mathrm{ref}}^{*}(t,\tau) = \mathrm{rect}\left(\frac{t}{T_{\mathrm{a}}}\right)\mathrm{rect}\left[\frac{\tau - \frac{2r(t)}{c}}{T_{\mathrm{p}}}\right]\exp(j\Phi) \tag{8.17}$$

In Eq. (8.17), t is slow time variable, τ is the fast time variable, T_{a} is synthetic aperture time.

$$\Phi = -\frac{4\pi k_{\mathrm{r}}}{c}\left[\frac{f_{\mathrm{c}}}{k_{\mathrm{r}}} + \tau - \frac{2r_{\mathrm{a}}(t)}{c}\right][r(t) - r_0(t)] + \frac{4\pi k_r}{c^2}[r(t) - r_0(t)]^2 \tag{8.18}$$

Namely,

$$\Phi = -K[r(t) - r_{\mathrm{a}}(t)] + \frac{4\pi k_{\mathrm{r}}}{c^2}[r(t) - r_{\mathrm{a}}(t)]^2 \tag{8.19}$$

In Eq. (8.19)

$$K = \left[\frac{4\pi f_{\mathrm{c}}}{c} + \frac{4\pi k_{\mathrm{r}}}{c}\left(\tau - \frac{2r_0}{c}\right)\right] \tag{8.20}$$

From Fig. 8.6,

$$r^2 = r_0^2 + a^2 - 2ar_0\sin(\theta+\gamma) \tag{8.21}$$

(a,γ) is the polar coordinate of point P. In case of meeting $a \ll r_0$,

$$r - r_0 \approx -a\,\sin(\theta+\gamma) + \frac{a^2}{2r_0}\cos^2(\theta+\gamma) \tag{8.22}$$

Equation (8.22) can substitute Eq. (8.19),

$$\phi(\theta,K) \approx Ka\,\sin(\theta+\gamma) - \frac{Ka^2}{2r_0}\cos^2(\theta+\gamma) \tag{8.23}$$

The second item in Eq. (8.23) is error resulting from wave-front curve (or range curve) errors, which can be ignored in case of assuming plane wave-front. The first item in Eq. (8.23) can be expressed as:

$$\phi_1 = Ka(\sin\theta\cos\ + \cos\theta\sin\gamma) = (K\sin\theta)(a\cos\gamma) + (K\cos\theta)(a\sin\gamma) \tag{8.24}$$

After the above approximation, we can obtain:

$$\phi(\theta, K) \approx x_0 K_x + y_0 K_y \tag{8.25}$$

Equation (8.25) indicates that the image of targets can be obtained only by two dimensional inverse Fourier transform of the echo data. Because (x_0, y_0) is on behalf of the space points location, so (K_x, K_y) represents a two-dimensional spatial frequency domain and polar coordinate system (θ, K) is corresponding to Cartesian coordinates $K_x - K_y$. Obviously, during the process of radar echo data acquisition, different azimuth sampling time t corresponds to a different angle of view θ, and the data collected corresponds to the K direction data. Echo data in Cartesian coordinates $K_x - K_y$ is shown in Fig. 8.7 as polar distribution which corresponds to the spatial frequency domain of a fan-shaped area. At this time, the imaging results cannot be obtained directly using IFFT. Must convert it to Cartesian coordinates from polar coordinate format by interpolation, that is, rectangular data area evenly sampled in the $K_x - K_y$ coordinate system is obtained from echo data. Then by IFFT transform the imaging results can be obtained. The interpolation conversion process from polar coordinate system to Cartesian coordinates is the most critical step in the PFA. Efficiency and accuracy of interpolation affect the speed and results of imaging.

If two-dimensional IFFT processes to the sector data directly without interpolation conversion, effective focus range is very limited.

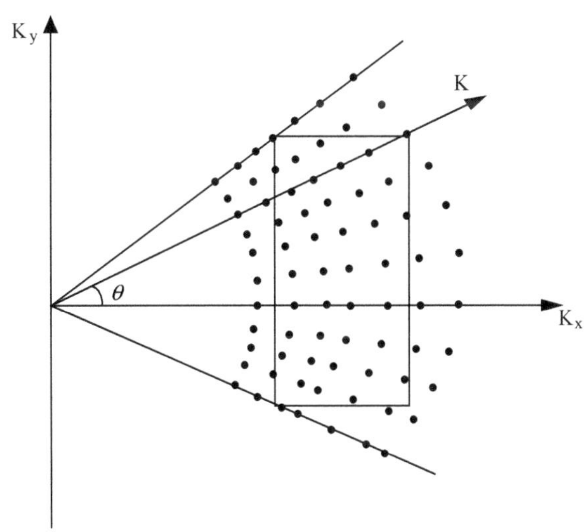

Fig. 8.7 Polar distribution of echo data

$$D_x \leq \frac{2\rho_x\rho_y}{\lambda}, \quad D_y \leq \frac{2\rho_x^2}{\lambda}$$

D_x and D_y are the maximum effective radius of imaging in spatial azimuth and range directions. ρ_x and ρ_y are spatial resolutions of azimuth and range direction respectively.

8.2.2 Data Interpolation in Polar Coordinates

Interpolation process must maintain signal integrity without signal distortion. Interpolation filter must be a linear phase filter without phase error to complex data. Filter bandwidth must be greater than the bandwidth of the signal in order to not create the signal's amplitude distortion. A full two-dimensional interpolation is the most ideal way to finish interpolation conversion from polar coordinates to Cartesian in PFA. However, two-dimensional interpolation calculation is too large, in practical applications, usually completed it by using twice one-dimensional interpolations. Figure 8.8 is for range interpolation and Fig. 8.9 is for azimuth interpolation. Firstly sector echo data is interpolated in range dimension, and then is interpolated in azimuth dimension after resampling interpolation. The required rectangular area echo data is obtained after resampling interpolation. To reduce the computational complexity, the Chirp z-transform can be applied to replace azimuth interpolation and azimuth FFT.

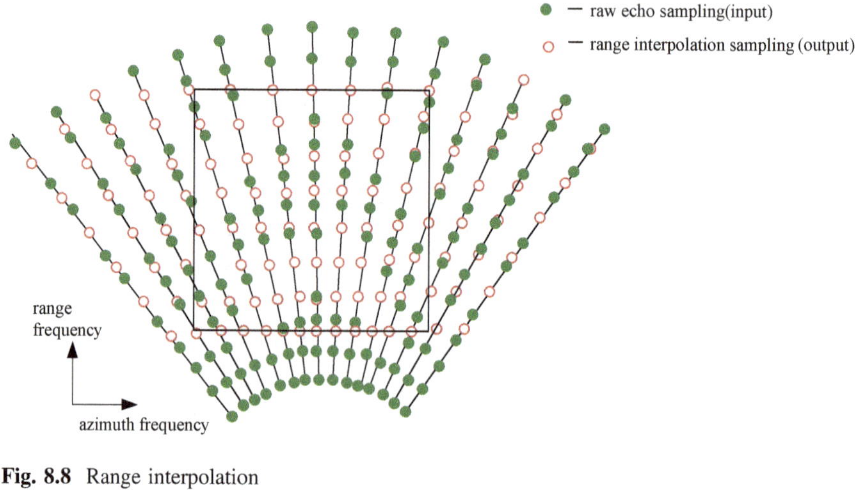

Fig. 8.8 Range interpolation

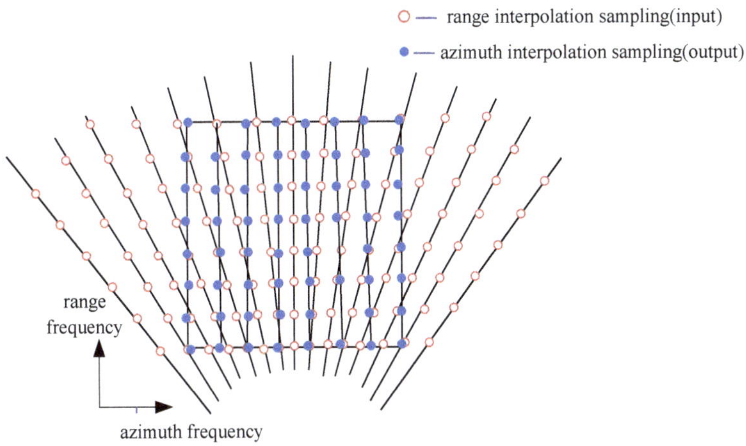

Fig. 8.9 Azimuth interpolation

8.2.3 Range Curvature Correction

When the ratio of the distance between radar and scenes to the maximum diameter of scene area is not large enough, the assumption of planar wave front approximation is no longer valid. The error arising out of the second item of Eq. (8.23) cannot be neglected; otherwise, the synthetic aperture radar images will result in significant geometric distortion.

The error that is generated by the second item in Eq. (8.23) is called range curvature error.

$$\phi_2 = -\frac{Ka^2}{2r_0}\cos^2(\theta+\gamma) \tag{8.26}$$

Suppose the coordinate of point target P is (x_0, y_0), and in the transformation range of θ, that is on the condition of accumulation angle $\Delta\theta \ll \pi/2$ and $K - K_c \ll K_c$, the Taylor expansion of Eq. (8.26) approximately equal

$$\begin{aligned}\phi_2 &\approx -\frac{Ka^2}{2r_0}\left[\cos^2\gamma - 2\theta\sin\gamma\cos\gamma - \theta^2\cos(2\gamma)\right]\\ &\approx -\frac{x_0^2}{2r_0}K_y + \frac{x_0 y_0}{r_0}K_x + \frac{x_0^2 - y_0^2}{2r_0 K_c}K_x^2\end{aligned} \tag{8.27}$$

In Eq. (8.27), the quadratic term will cause the point target defocusing in azimuth direction. When the error is not more than $\pi/4$, effective imaging focus range will provide

$$D_x \leq \rho_x \sqrt{\frac{2r_0}{\lambda}},\ D_y \leq \rho_y \sqrt{\frac{2r_0}{\lambda}} \tag{8.28}$$

In the effective focus range, ignoring more than quadratic term errors, and supposing $r_0 \approx r_c$, we can get

$$\phi_2 \approx -\frac{x_0^2}{2r_c}K_y + \frac{x_0 y_0}{r_c}K_x \tag{8.29}$$

The linear item will cause the distortion of point target. Comparing Eqs. (8.29) and (8.25), then due to range curvature errors, the imaging point target location is as follows:

$$\begin{cases} x_i = x_0 + c x_0 y_0 \\ y_i = y_0 - d x_0^2 \end{cases}, \quad \text{here } c = \frac{1}{r_c},\ d = \frac{1}{2r_c} \tag{8.30}$$

In Eq. (8.30),

$$\begin{cases} x_i = x_0 + c x_0 y_0 \\ y_i = y_0 - d x_0^2 \end{cases}, \quad \text{here } c = \frac{1}{r_c},\ d = \frac{1}{2r_c}$$

Distortion process is shown in Fig. 8.10, which is composed of two steps. Range distortion happens from Fig. 8.10a, b, and azimuth distortion only happens from Fig. 8.10b, c, which can be expressed as:

$$\begin{cases} x_m = x_0 \\ y_m = y_0 - d x_0^2 \end{cases} \qquad \begin{cases} x_i = x_m + c x_m + c d x_m^3 \\ y_i = y_m \end{cases} \tag{8.31}$$

In Eq. (8.31), the cubic term is very small and it can be ignored. Correction process is an inverse process of distortion emergence in Fig. 8.10, that is, correct (x_i, y_i) to (x_m, y_m) firstly and then to (x_0, y_0), as shown in the following:

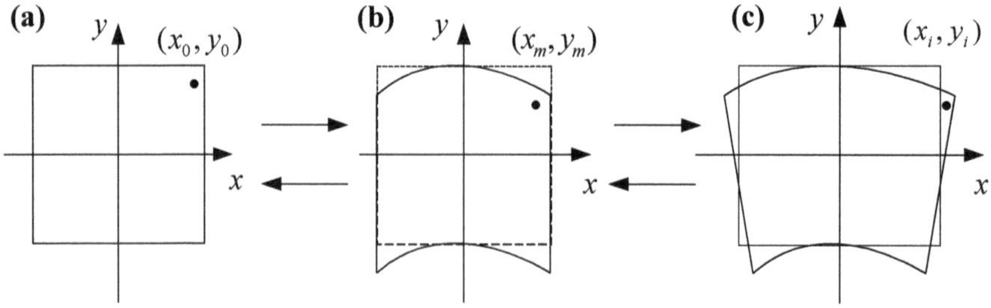

Fig. 8.10 Position distortion process caused by the range curvature. **a** Space point target, **b** range distortion, **c** range and azimuth distortion

$$\begin{cases} x_m \approx \frac{1}{1+cy_m}x_i \\ y_m = y_i \end{cases} \qquad \begin{cases} x_0 = x_m \\ y_0 = y_m + dx_0^2 \end{cases} \tag{8.32}$$

The frequency scaling characteristics and phase compensation function can be used to complete the calibration procedure above mentioned.

8.2.4 *Data Processing of Squint Spotlight SAR in Polar Coordinates*

For imaging in high squint angle, due to high squint angle, the angle between radar line of sight in different time and radar line of sight in aperture's center position is a nonlinear relationship with azimuth time. For conventional polar format algorithm it is difficult to remove the nonlinear effect.

Figure 8.11 is the data acquisition plane geometry of squint Spotlight SAR. Assume that radar track along the ideal straight line at a constant speed, and transmit radar pulse with a constant pulse repetition frequency f_{PRF}. In flight the range from A_{s} to A_{e} is synthetic aperture length, and A_{c} is the center position (synthetic aperture center). The shortest distance from imaging regional center to the flight path is r_{c}, coordinate system of the imaging area is shown in Fig. 8.11, θ_0 is the angle between radar line of sight and the y axis at the position A_{c}, and corresponding $\varphi = 90° - \theta_0$ is the squint angle of sight. The coordinate system $x'y'$ is from rotating coordinate system xy a θ_0 angle. The angle between radar line of sight in any time and radar line of sight in aperture's center position is $\hat{\theta}(n) = \theta_0 - \theta(n)$.

Figure 8.12 show the spatial frequency domain data is distributed after Dechirp processes and A/D sampling, and K_x, K_y is spatial frequency domain. According to the Fourier transforms rotational invariance,by rotating K_xK_y coordinate system a θ_0

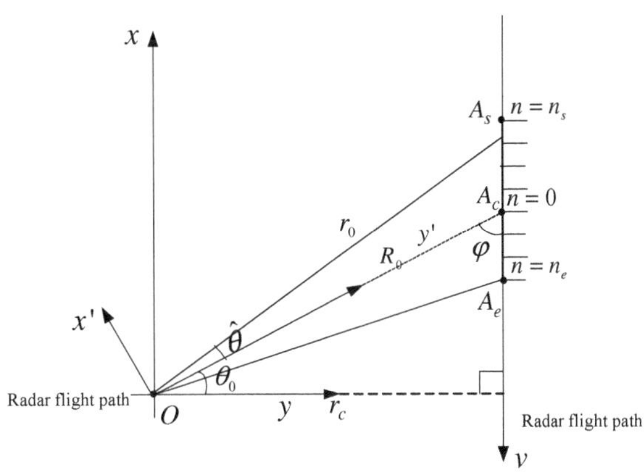

Fig. 8.11 Squint Spotlight SAR data acquisition geometry

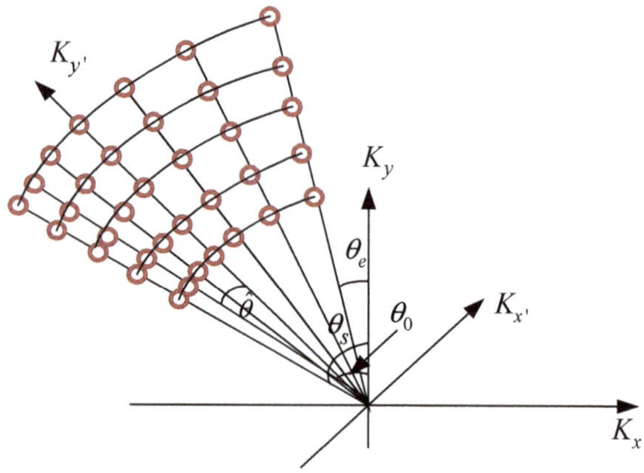

Fig. 8.12 Spatial frequency domain

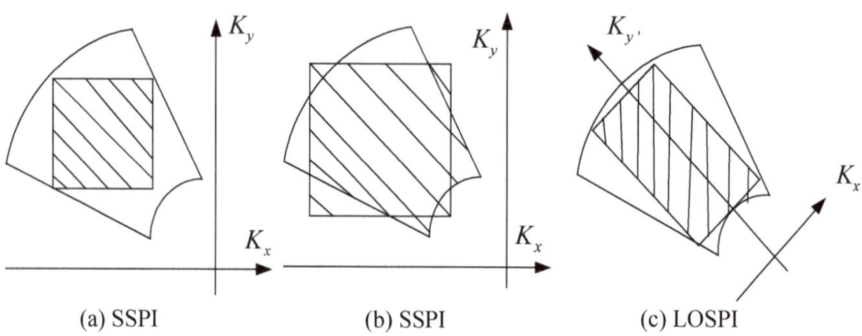

Fig. 8.13 Two kinds of polar coordinate interpolation methods. **a** and **b** SSPI; **c** LOSPI

angle we can get the frequency domain coordinate system $K_{x'}K_{y'}$ corresponding to spatial coordinate system $x'y'$.

In Fig. 8.13 [1–4] there are two kinds of interpolation methods for the squint spatial frequency domain data which is shown in Fig. 8.12, One is a stabilized scene polar interpolation (SSPI), as shown in Fig. 8.13a, b; the other is line-of-sight polar interpolation (LOSPI), as shown in Fig. 8.13c.

Polar coordinate interpolation in fixed area is a polar coordinates interpolation to squint spatial frequency domain data in a fixed area. The choice fixed area is generally a square area and the coordinate system is $K_x - K_y$; Polar coordinate interpolation of radar line of sight direction is a polar coordinates interpolation to squint spatial frequency domain data along the line of sight and the coordinate system is rotating coordinate system $K_{x'} - K_{y'}$.

8.3 Range Migration Algorithm

Range migration algorithm (RMA) is raised by Cafforio, Prati and Rocca, and so on. It was first used in the Earth's strata imaging of the seismic signal processing [5–7].

Different from PFA, range migration algorithm need not assume that wave-fronts are plane. Therefore, the Imaging results from range migration algorithm do not exist spatially variable defocusing and geometry distortion caused by ignoring wave-fronts curvature in polar format algorithm. In addition, RMA and PFA have the same amount of calculation on the whole. RMA has more advantage than PFA in larger image scene, higher resolution requirements, shorter image distance and lower center frequency of radar signal. The process of RMA is shown in Fig. 8.14.

8.3.1 *Principle of Range Migration Imaging Algorithm*

Figure 8.15 is data acquisition plane geometry. Dechirp demodulation of the echo signal by means of taking the echo signal of a fixed distance as a reference signal, and remove the RVP by residual phase compensation to obtain the time domain signal as Eq. (8.14) shown.

$$s_{\mathrm{rd}} = a_{\mathrm{p}}\mathrm{rect}\left(\frac{\tau - 2r_{\mathrm{c}}/c}{T_{\mathrm{p}}}\right)\exp\left[-j\frac{4\pi k_{\mathrm{r}}}{c}\left(\tau + \frac{f_{\mathrm{c}}}{k_{\mathrm{r}}} - \frac{2r_{\mathrm{c}}}{c}\right)\Delta r\right] \tag{8.33}$$

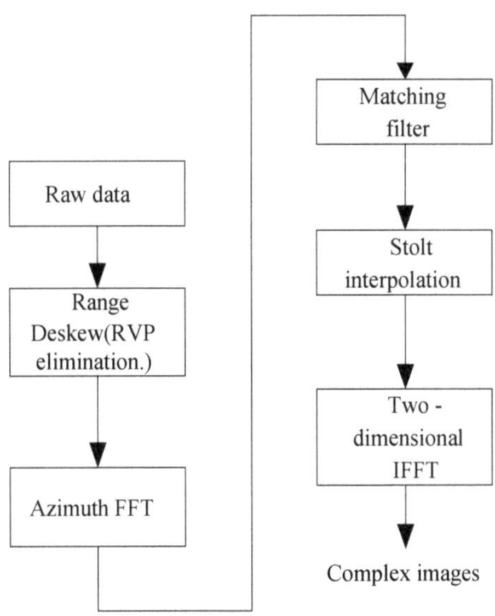

Fig. 8.14 RMA process

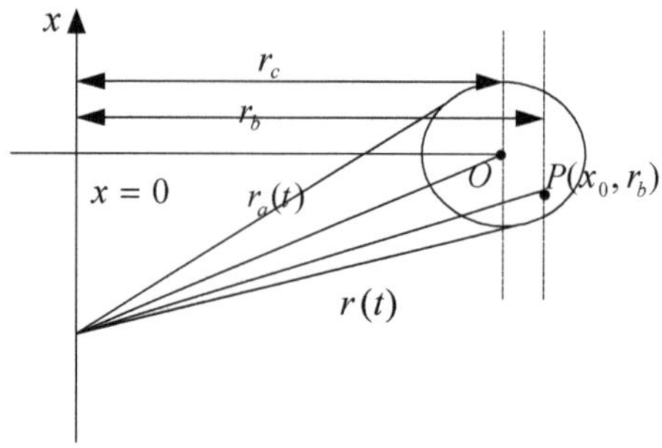

Fig. 8.15 Data acquisition plane geometry

Equation (8.33) echo data is the input data of range migration algorithm.

In Fig. 8.15, aircraft azimuth coordinates x can be used to substitute the azimuth time t ($x = Vt$, V is the flight speed), and let

$$K_R = \frac{4\pi k_r}{c}\left(\frac{f_c}{k_r} + \tau - \frac{2r_c}{c}\right) \qquad \tau' = \tau - \frac{2r_c}{c} \tag{8.34}$$

After preprocessing, two-dimensional echo signal can be represented as

$$s_{rd}(x, K_R) = a_t \mathrm{rect}\left(\frac{x - x_c}{L}\right)\mathrm{rect}\left(\frac{\tau'}{T_p}\right)\exp(-jK_R[r(x) - r_c]\} \tag{8.35}$$

From Fig. 8.15,

$$r(x) = \sqrt{(x - x_0)^2 + r_0^2} \tag{8.36}$$

According to stationary phase principle, Fourier transform of Eq. (8.35) in x direction is

$$S(K_X, K_R) = |S(K_X, K_R)|\exp\{j\Phi(K_x, K_R)\} \tag{8.37}$$

In Eq. (8.37)

$$|S(K_X, K_R)| = a_p \mathrm{rect}\left[\frac{K_R - 4\pi f_c/c}{4\pi k_r T_p/c}\right]\mathrm{rect}\left[\frac{K_X r_0}{L\sqrt{K_R^2 - K_X^2}} + \frac{x_c - x_0}{L}\right]$$

$$\Phi(K_X, K_R) = -K_X x_0 - r_0\sqrt{K_R^2 - K_x^2} + K_R r_c$$

Here, $K_X = \frac{4\pi f_c}{c}\cos\alpha$ is azimuth spatial frequency, and α is squint angle of sight.

Two-dimensional matched filtering to Eq. (8.37), the matched filter's phase is

$$\Phi_{\mathrm{mf}} = -K_{\mathrm{R}} r_{\mathrm{c}} + r_{\mathrm{c}} \sqrt{K_{\mathrm{R}}^2 - K_{\mathrm{X}}^2} \tag{8.38}$$

After filtering the signal is

$$\begin{aligned} S_2(K_{\mathrm{X}}, K_{\mathrm{R}}) &= |S(K_{\mathrm{X}}, K_{\mathrm{R}})| \exp\{j\Phi + j\Phi_{\mathrm{mf}}\} \\ &= |S(K_{\mathrm{X}}, K_{\mathrm{R}})| \exp\left\{-j\left[K_{\mathrm{X}} x_0 + (r_0 - r_{\mathrm{c}})\sqrt{K_{\mathrm{R}}^2 - K_{\mathrm{X}}^2}\right]\right\} \end{aligned} \tag{8.39}$$

In RMA, azimuth linear frequency-modulated signal is first converted to the K_{X} domain to eliminate azimuthal variability of range curvature, and get a two-dimensional signal, which is still a linear frequency-modulated signal in K_{X} domain and an approximately linear signal in K_{R} domain. RMA removes linear frequency modulation in the azimuth spatial frequency domain by Eq. (8.38), and revised all range curvature of scattered points with the nearest slant range r_{c} and there is remain the part range curvature for other scattered points which the nearest slant range is not r_{c}. In Spotlight synthetic aperture radar data, the matched filter is under compensated for slant range less than r_{c} and over compensated for slant range greater than r_{c}. Therefore, the next task is to eliminate all residual range curvature of scattered points.

Variable substitution to Eq. (8.39)

$$\sqrt{K_{\mathrm{R}}^2 - K_{\mathrm{X}}^2} -> K_{\mathrm{Y}} \tag{8.40}$$

To complete this mapping requires Stolt interpolation. Set up distance coordinates of point target relative to the center of the scene as $y_0 = r_0 - r_{\mathrm{c}}$, then the Eq. (8.39) will be

$$S(K_{\mathrm{X}}, K_{\mathrm{R}}) = |S(K_{\mathrm{X}}, K_{\mathrm{R}})| \exp\{-j[K_{\mathrm{X}} x_0 + K_{\mathrm{Y}} y_0]\} \tag{8.41}$$

Clearly, the focused images can be obtained by two-dimensional inverses Fourier transforms to the Eq. (8.41).

8.3.2 *Stolt Interpolation*

Stolt transformation is an important step of Range migration algorithm. The transformation compensates for all the range curvature of scattered points at the same time by means of the proper transformation of synthetic aperture radar data. Stolt transformation operates (K_{X}, K_{R}) domain signal data mainly.

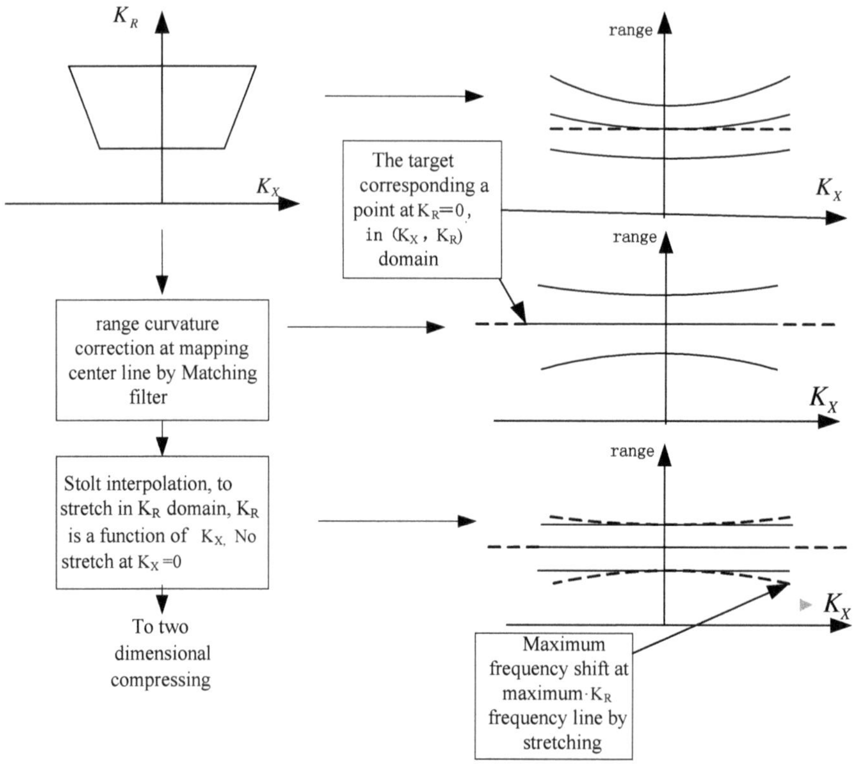

Fig. 8.16 Stolt interpolation correction range curvature process

In (K_X, K_R) domain, on K_R coordinate, the echo of a single scattering point is a sinusoidal signal whose frequency changes with K_X. Operations to stretch the one dimensional sinusoidal signal can reduce its frequency. Stolt transformation makes their sinusoidal signal frequencies constant for all K_X by appropriately reducing the range frequency of each scatter. Variable substituting by $K_Y = \sqrt{K_R^2 - K_X^2}$ and resampling to data in K_R domain, the uniformly sampling output samples can be obtained in K_Y domain. Figure 8.16 shows range curvature correction process of Stolt interpolation.

Figure 8.17 is two-dimensional signal distribution in (K_X, K_R) domain and (K_X, K_Y) domain. Clearly, the data distribution area is sector after interpolation. It is necessary to intercept a rectangular area in the region to obtain the Imaging results by using two-dimensional IFFT operation. Figure 8.18 is Stolt interpolation process. Take the center of azimuth spatial frequency K_X as reference and rectangular sample interpolated data at uniform interval in K_Y direction. Two-dimensional IFFT process to selected specific rectangular area data, and get imaging results.

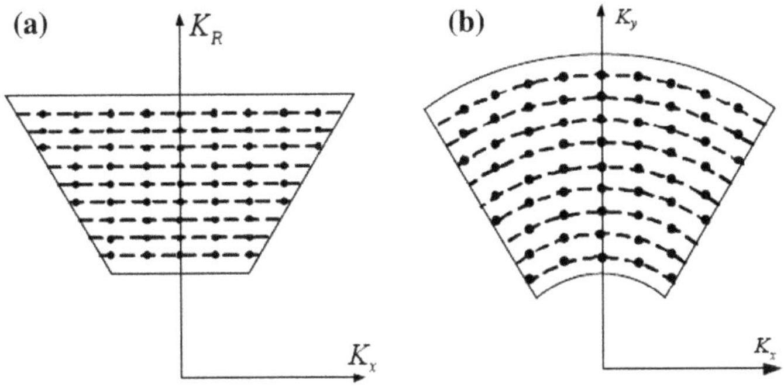

Fig. 8.17 two-dimensional signal distribution. **a** Data distribution in (KX, KR) domain before Stolt interpolation; **b** data distribution in (KX, KY) domain after Stolt interpolation

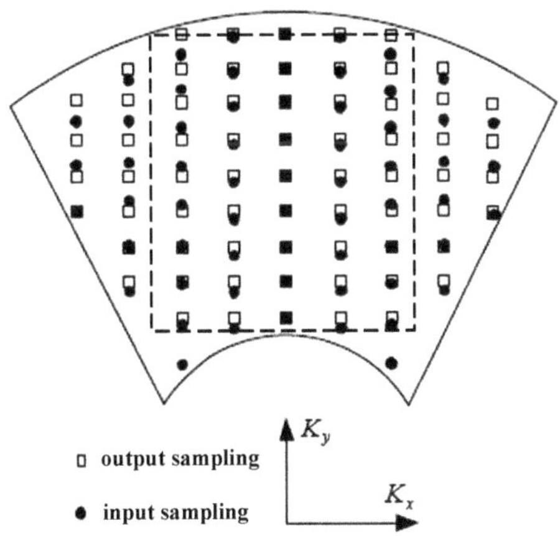

Fig. 8.18 Stolt interpolation process

8.3.3 Special Problems of Range Migration Algorithm

(1) Azimuth sampling requirements

Range migration algorithm uses a fixed distance echo signal as a demodulation reference signal and only completes range Dechirp processing. In azimuth direction, the echo signal remains in the linear frequency modulation characteristics, and the linear frequency modulated signal bandwidth as follows:

$$B_{\mathrm{d}} = \frac{2V\Delta\theta\,\sin(\varphi_{\mathrm{dc}})}{\lambda} \tag{8.42}$$

In Eq. (8.42), λ is instantaneous radar signal wavelength, V is flight speed, φ_{dc} and $\Delta\theta$ are shown in Fig. 8.19. The minimum sampling frequency in azimuth direction should be equal to B_{d} to avoid spectral aliasing.

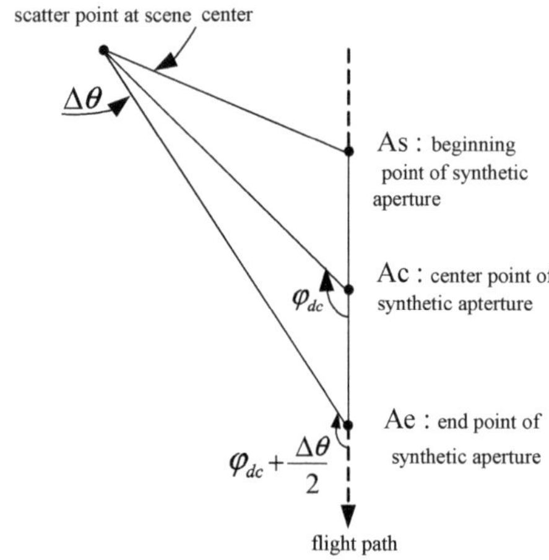

Fig. 8.19 Imaging geometry

In contrast, if we use the echo signal corresponds to the scene center as a demodulation reference signal, the range Dechirp and azimuth Dechirp processing can be completed at the same time. The bandwidth in azimuth direction is

$$B_{a2} = \frac{2Vw_a}{\lambda R_{ac}} \tag{8.43}$$

In Eq. (8.43), w_a is the azimuth mapping width, V is radar flight speed, R_{ac} is the distance from the center of synthetic aperture to the center of scene. This bandwidth is also the lowest sampling rate in azimuth direction of polar format algorithm.

Equations (8.42) and (8.43) show that the requirements to azimuth sampling rate in RMA are higher than that in PFA, which causes the azimuth sampling points and the computation capacity increasing significantly.

(2) Squint data processing of RMA

Due to existence of azimuth center frequency, squint Spotlight synthetic aperture radar data is markedly different from regular side-looking data. Figure 8.20 is squint data processing diagram of RMA. Figure 8.20a is squint data collection symmetry of SAR at X_{ac}; Fig. 8.20b represents the distance change from the antenna phase Center to the center of the scene, of which the nonlinear part introduces the azimuth chirp and linear part introduces azimuth nonzero center frequency; Fig. 8.20c is the signal distribution in wave-number domain, and there is a big difference from the side-looking signal distribution shown in Fig. 8.17; Fig. 8.20d is the signal distribution in (K_X, K_Y) after Stolt interpolation.

Before azimuth Fourier transform of RMA, the azimuth center frequency (azimuth carrier) must be removed.

In large squint angle imaging of spotlight SAR, the elimination of carrier can significantly reduce the required azimuth sampling rate and reduce the amount of calculation. Figure 8.21 is the diagram of eliminating squint operation in range

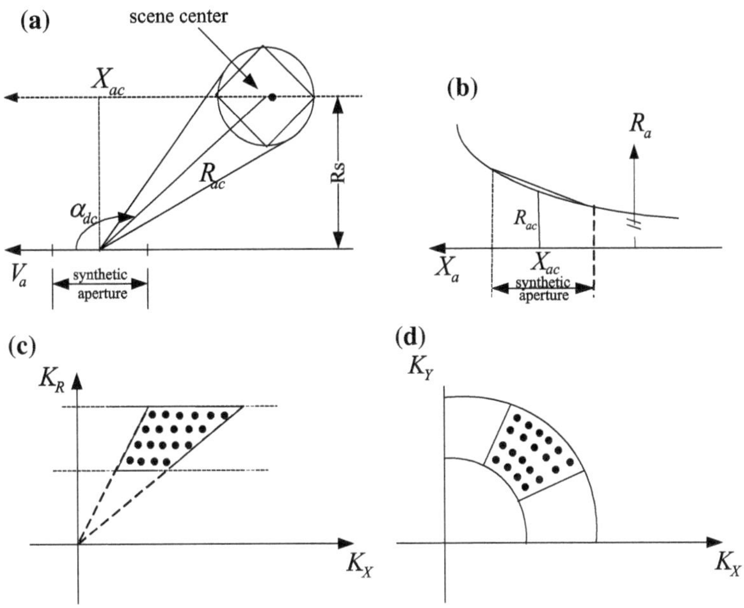

Fig. 8.20 Squint data processing diagram of RMA. **a** Squint data collection symmetry; **b** the distance change to the center of the scene; **c** the signal distribution in (K_X, K_R) domain before Stolt interpolation; **d** the signal distribution in (K_X, K_Y) after Stolt interpolation

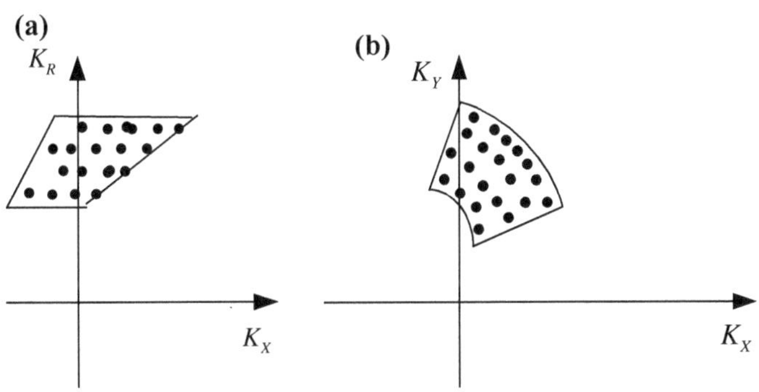

Fig. 8.21 Eliminating squint operation diagram in RMA. **a** Data distribution in (K_X, K_R) domain; **b** data distribution in (K_X, K_Y) domain

migration algorithm, which presents SAR data distribution in (K_X, K_R) domain without azimuth carrier. Comparing with Fig. 8.20c, it can be seen that azimuth carrier elimination is equal to a translation of azimuth spectrum. Figure 8.21b is the diagram of SAR data distribution in (K_X, K_Y) domain after Stolt interpolation.

(3) Space-variant impulse responses

During Spotlight synthetic aperture radar data acquisition process, the radar's angle of view changes at different scattering point, and scattering points in different locations have different range migrations which including range curvature and range walk. Range walk is the linear part of range migration, which causes azimuth spectrum frequency shift of scattering points so that the carrier is not zero. Azimuth carrier frequency is related with range walk and position of scattering point. After azimuth Fourier transform, two-dimensional signal distribution of every scattering point is not exactly the same, their spectrums have a slight twist to each other.

Figure 8.22 is simulation signal distribution of point target, this distorting effects relative to the position is magnified after matched filtering and Stolt interpolation.

The final step is a two-dimensional Fourier transform in RMA. Distorting effects make processing aperture shape changes with change of the scattering point position in the scene, which produce space-variant impulse response. However, it is visible only on very strong scattering points and it does not reduce resolution, just changing the distribution of impulse response side-lobes, so this effect is not always caused great impact on imaging results. In the data acquisition process, as long as acquiring data at larger angle of aperture than required angle by resolution will be able to eliminate this phenomenon.

(4) Stolt interpolation's approximation

When the K_X, K_R is far less than K_{RC}, the Stolt interpolation can be approximated, where K_{RC} corresponds to K_R at the range sampling center. Narrow-band small coherent angle SAR system usually meets this requirement. At this point, the following approximation can be obtained.

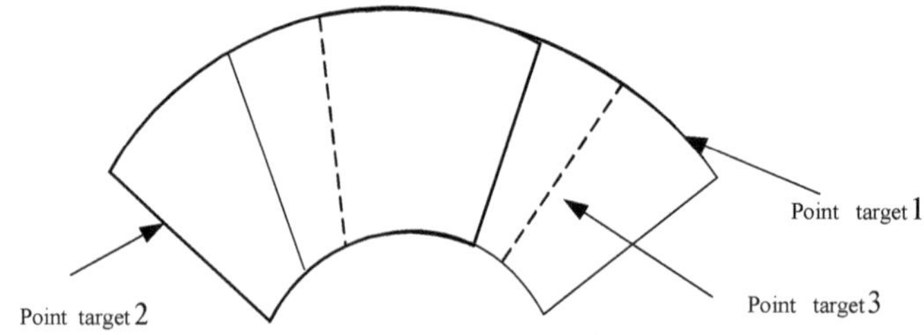

Fig. 8.22 Simulation signal distribution of point target. **a** In (K_X, K_R) domain; **b** in (K_X, K_Y) domain after Stolt interpolation

$$K_{\mathrm{Y}} \approx K_{\mathrm{R}} - \frac{K_{\mathrm{X}}^2}{2K_{\mathrm{RC}}} \tag{8.44}$$

From uniform interval distribution of K_{R}, the K_{X} location can be determined. According to Eq. (8.44), K_{Y} will be uniform interval distribution as K_{R}. There is not necessary to calculate each sample location of K_{Y} like Eq. (8.40). Only need calculate the minimum of K_{Y}, the uniform interval sequence sampling location can be obtained after interpolation, such can reduce Stolt interpolation complexity and improve calculation efficiency.

8.4 Chirp Scaling Algorithm

In 1994, based on the wave equation algorithm, R. K. Raney first proposed the Chirp Scaling algorithm (CSA). CSA is an algorithm with high computing efficiency. The algorithm does not require interpolation, only needs FFT and complex multiplication. CSA is easy to implement on the signals processor based on FFT and a fast parallel processor systems. In addition, the CSA can get very good results in the case of relatively large squint angle, and have been widely used in both airborne and space borne synthetic aperture radar [8–10].

Chirp Scaling means that the original linear frequency-modulated signal is multiplied by an LFM signal with appropriate associated frequency modulation rate (Chirp Scaling factor), which makes small changes to phase center and frequency modulation can rate of original FM signal. In SAR imaging, multiplied by Chirp Scaling factor can correct small differences of target range migration curves at different distances and compensate all of range migration curve to the same shape, and range migration correction and azimuth process can be carried out uniformly.

CSA requires echo signals to be a linear frequency-modulated signals at range dimension that is the receiving signals without Dechirp processing.

8.4.1 Receiving Signal Model Without Dechirp

Figure 8.23 is a regular side looking Spotlight synthetic aperture radar imaging geometry. According to Eq. (8.3), the echo of received signal is

$$s_{\mathrm{r}}(t,\tau) = a_t \mathrm{rect}\left(\frac{t}{T_{\mathrm{a}}}\right)\mathrm{rect}\left(\frac{\tau - \frac{2r(t)}{c}}{T_{\mathrm{p}}}\right)\exp\left\{j2\pi f_{\mathrm{c}}\left[\tau - \frac{2r(t)}{c}\right] + j\pi k_{\mathrm{r}}\left[\tau - \frac{2r(t)}{c}\right]^2\right\}$$

f_{c} is carrier, T_{p} is transmitting pulse width, k_{r} is linear frequency modulation rate, $r(t)$ is the distance from aircraft to scattering point P at the time t, r_{b} is the shortest

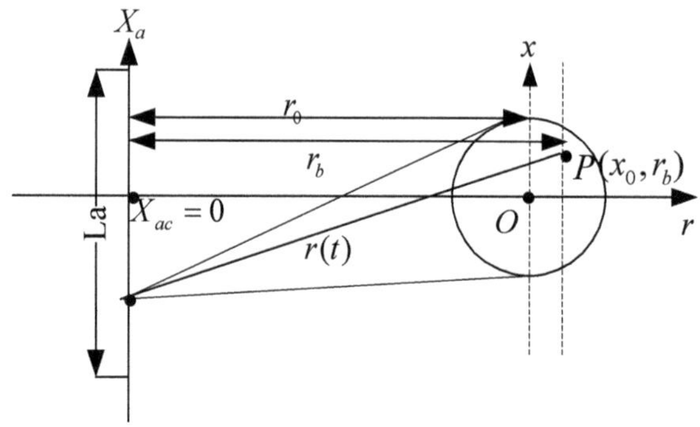

Fig. 8.23 Spotlight SAR imaging geometry

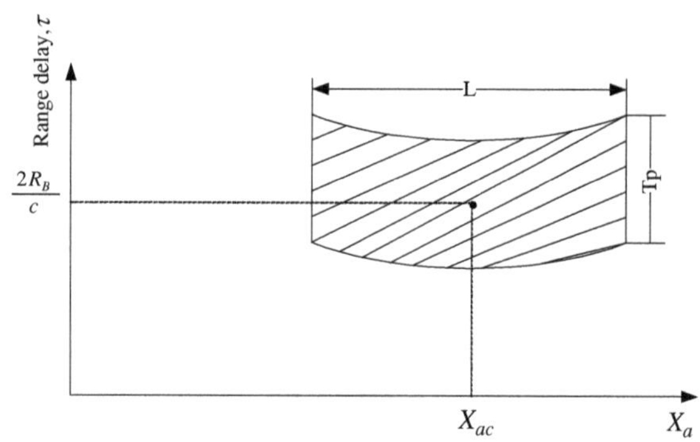

Fig. 8.24 Two dimension signal distribution with azimuth coordinate

distance from scattering point P to the flight path of aircraft, r_0 is the shortest distance from scene center O to the flight path of aircraft, X_a is the position of aircraft, X_{ac} is the midpoint of synthetic aperture.

Figure 8.24 are the two dimension signal distribution of scattering point with azimuth coordinates X_{ac}. In the case of side-looking, $X_{ac} = 0$. This distribution has a constant range delay T_p and center delay changing with aircraft location X_a.

In CSA, using reference function

$$s_{\text{ref}}(t) = \exp(j2\pi f_c t) \tag{8.45}$$

The echo signal after demodulation to raw signal is

$$\begin{aligned} s(t,\tau) &= s_r(t,\tau)s^*_{\text{ref}}(\tau) \\ &= a_t \text{rect}\left(\frac{t}{T_a}\right)\text{rect}\left(\frac{\tau - 2r(t)/c}{T_p}\right)e^{-j\frac{4\pi f_c}{c}r(t)}e^{j\pi k_r\left(\tau - \frac{2r(t)}{c}\right)^2} \end{aligned} \tag{8.46}$$

In Eq. (8.46), the first index item is associated with the distance r (t).

8.4.2 CSA Summary

CSA operates in azimuth-frequency/range-time domain and azimuth frequency/ range-frequency domain. Figure 8.25 is CSA flowchart. Firstly by azimuth Fourier transform, the echo signal is transformed to range-Doppler domain (azimuth-frequency domain), and by multiplying with Chirp Scaling factor, range migration curves of all range gates are compensated to the same shape; secondly by range Fourier transform, turn the signal to two-dimensional frequency domain, and by multiplying with range compensation factor Φ_2, complete range migration correction, secondary range-compressed and range-compressed; Thirdly by range IFFT, the signal is transformed back to range-Doppler domain, and by multiplying with azimuth compensation factor Φ_3, complete azimuth process; finally by using the azimuth inverse Fourier transform, the signal is transformed back to the time domain, and SAR image is obtained. From flowcharts can be seen, Chirp Scaling algorithm for synthetic aperture radar imaging can be accomplished just by FFT and complex multiplication.

8.4.3 CSA Calculation Procedure

(1) Azimuth Fourier transforms

The first step of CSA is a Fourier transformation to the original synthetic aperture radar signal in azimuth direction, and the signal is transformed into

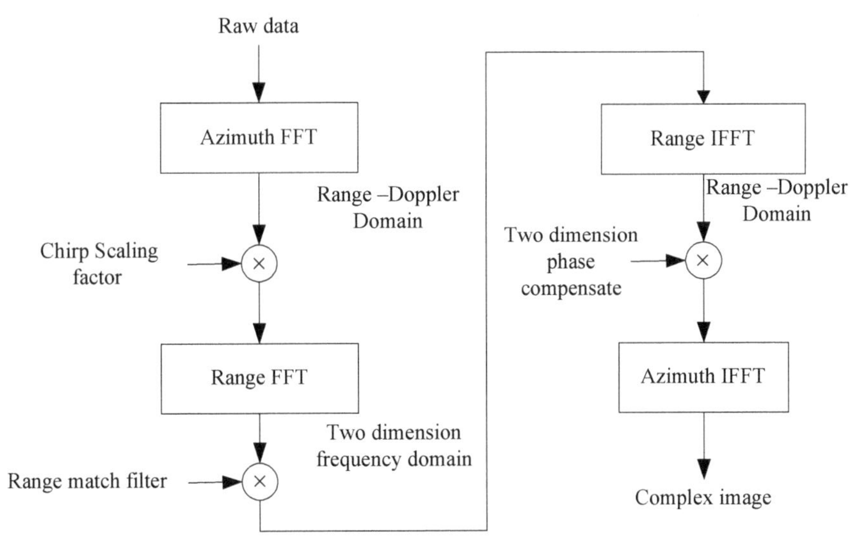

Fig. 8.25 CSA flowchart

azimuth-frequency/range-time domain. Since the exact expression of the transformation cannot be obtained, the approximations expression can be got by using stationary phase principle. Azimuth FFT calculation procedure is shown in Fig. 8.26.

By using the stationary phase principle and ignoring constant terms, a signal expression in range-Doppler domain can be obtained

$$S(\tau,f_a) = C_1 \mathrm{rect}\left[\frac{\lambda r_b}{2V^2\sqrt{1-\left(\frac{\lambda f_a}{2V}\right)^2}}\right]\mathrm{rect}\left[\tau - \frac{2}{c}R(f_a;r_b)\right]\exp\left[-j\frac{2\pi f_a x_0}{V}\right] \cdot \exp\left[j\pi k(f_a;r_b)\left(\tau - \frac{2R(f_a;r_b)}{c}\right)^2\right]\cdot\exp\left[-j\frac{4\pi}{\lambda}\cdot r_b\cdot\beta(f_a)\right] \tag{8.47}$$

In Eq. (8.47)

$$R(f_a;r_b) = \frac{r_b}{\beta(f_a)} = r_b\cdot[1+a(f_a)] \tag{8.48}$$

This is the range migration form in range-Doppler domain.

$$\frac{1}{k(f_a;r_b)} = \frac{1}{k_r} - \frac{2\lambda r_b\left(\beta^2(f_a)-1\right)}{c^2\beta^3(f_a)} \tag{8.49}$$

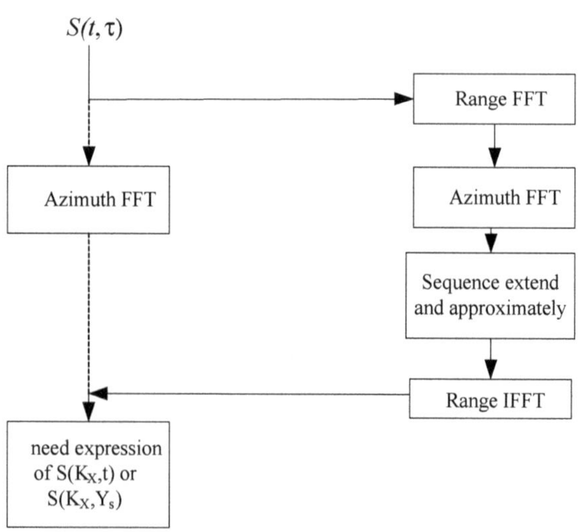

Fig. 8.26 Azimuth FFT calculation procedure

$$\beta(f_a) = \sqrt{1 - \left(\frac{f_a \cdot \lambda}{2v}\right)^2} \tag{8.50}$$

$$a(f_a) = \frac{1}{\beta(f_a)} - 1 \tag{8.51}$$

$a(f_a)$ is Chirp Scaling Factor or CS for abbreviation. If the R-D domain signal expressed by Eq. (8.47) is compressed in range direction, the peak time is

$$\tau_0 = 2R(f_a; r_b)/c = \frac{2}{c}(r_b + r_b a(f_a)) \tag{8.52}$$

(2) Chirp Scaling operation

Chirp Scaling operation is the base of CSA. For linear frequency modulation signal

$$s(\tau) = \exp\left\{j\pi k(\tau - \tau_0)^2\right\} \tag{8.53}$$

Its instantaneous frequency is $f_\tau = k(\tau - \tau_0)$. After a matched filtering, peak appears in the time τ_0. If $s(\tau)$ is multiplied by another Chirp signal,

$$s_{\text{ref}}(\tau) = \exp\left\{j\pi k_{\text{ref}}(\tau - \tau_{\text{ref}})^2\right\} \tag{8.54}$$

we can get

$$s'(\tau) = \exp\left\{j\pi\left[k(\tau - \tau_0)^2 + k_{\text{ref}}(\tau - \tau_{\text{ref}})^2\right]\right\} \tag{8.55}$$

Its instantaneous frequency is

$$f'_\tau = (k + k_{\text{ref}})\left[\tau - \left(\tau_0 + \frac{k_{\text{ref}}(\tau_{\text{ref}} - \tau_0)}{k + k_{\text{ref}}}\right)\right] \tag{8.56}$$

So the frequency modulation rate of the new linear frequency modulation signal $s'(\tau)$ is $k + k_{\text{ref}}$. After a matched filtering, peak appears in the time

$$\tau'_0 = \tau_0 + \frac{k_{\text{ref}}(\tau_{\text{ref}} - \tau_0)}{k + k_{\text{ref}}} \tag{8.57}$$

As can be seen from Eqs. (8.53)–(8.57), the original LFM signal multiply an LFM signal with an appropriate associated frequency modulation rate (Chirp

Scaling factor), which can make small changes in phase center and frequency modulation rate to the original LFM signal.

Figure 8.27 are the effects of phase and azimuth frequency of Chirp Scaling operations on three different distance scattering points. Figure 8.27a represents phase functions of three scattering echo signals for fastest time; Fig. 8.27b represents the phase of Chirp Scaling factor complex conjugate, which is also a Chirp signal with a FM rate far below that of synthetic aperture radar signals; Fig. 8.27c shows the signal phase after the Chirp Scaling operation. This operation increases the FM rate of all the signals and adding a linear phase. The linear phase increases with the distance between the center of the scene and target. It can be seen that Chirp Scaling operation is equivalent to changing the range scale factor.

The Chirp Scaling procedure is as follows: a two-dimensional complex phase-reference function is

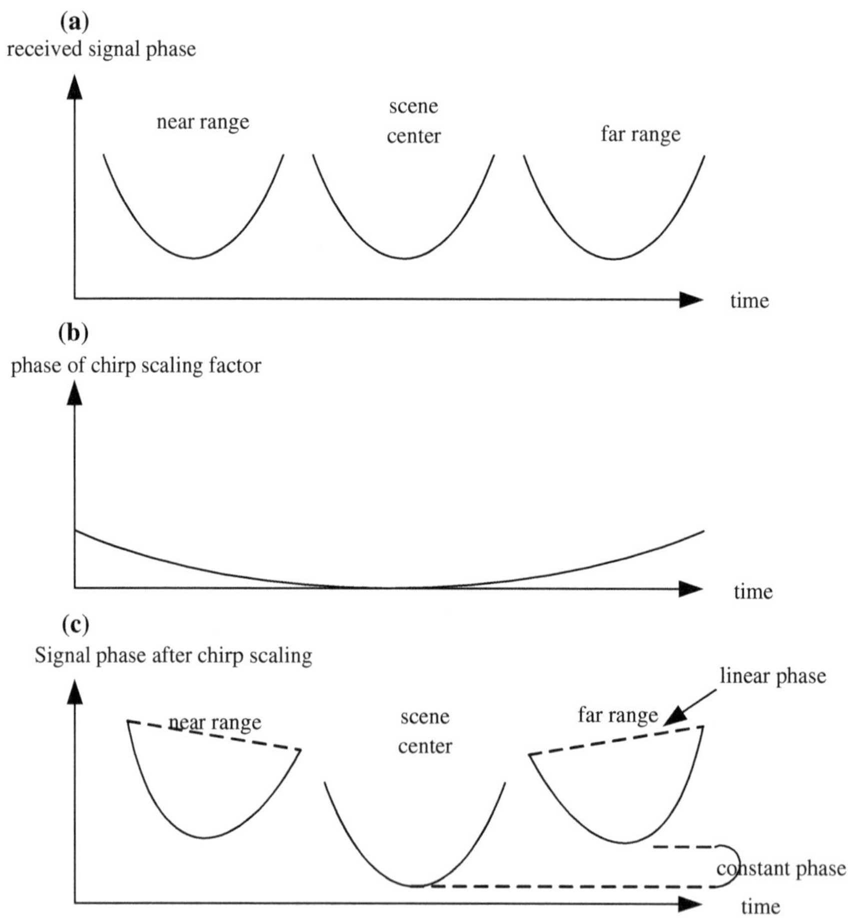

Fig. 8.27 The phase effects of Chirp Scaling operations on three different scattering points

$$H_1(f_a,\tau) = \exp\left\{ j\pi k(f_a; r_{ref}) \cdot a(f_a) \left[\tau - \frac{2R(f_a; r_{ref})}{c} \right]^2 \right\} \tag{8.58}$$

And multiplies R-D domain data after the azimuth Fourier transform, simplified as follows:

$$P \cdot H_1 = \exp\left[j\pi k(f_a; r_b)[1 + a(f_a)] \left(\tau - \frac{2r_b + 2r_{ref}a(f_a)}{c} \right)^2 \right] * \exp[-j\Theta_\Delta(f_a; r_b)] \tag{8.59}$$

In Eq. (8.59), the residual phase due to Chirp Scaling function operation is

$$\Theta_\Delta = \frac{4\pi}{c^2} k(f_a; r_b) a(f_a)[1 + a(f_a)](r_b - r_{ref})^2 \tag{8.60}$$

After range compression of Eq. (8.47), the peak time is $\tau_{ref} = 2R(f_a; r_{ref})/c$. Studying the second index item in Eq. (8.47), the LFM rate of the multiplying signal and the peak time of matching compression signal will change according to the Chirp Scaling theory. From Eqs. (8.57) and (8.48), the peak time of the product signal after range matched compression is

$$\tau_0' = \tau_0 + \frac{k_{ref}(\tau_{ref} - \tau_0)}{k + k_{ref}} = \tau_0 + \frac{a(f_a)}{1 + a(f_a)}(\tau_{ref} - \tau_0) = \frac{2}{c}(r_b + r_{ref}a(f_a)) \tag{8.61}$$

Contrast Eqs. (8.52) and (8.61), it can be seen that after Chirp Scaling operation the range migration curves of different slant range r_b and reference range r_{ref} are

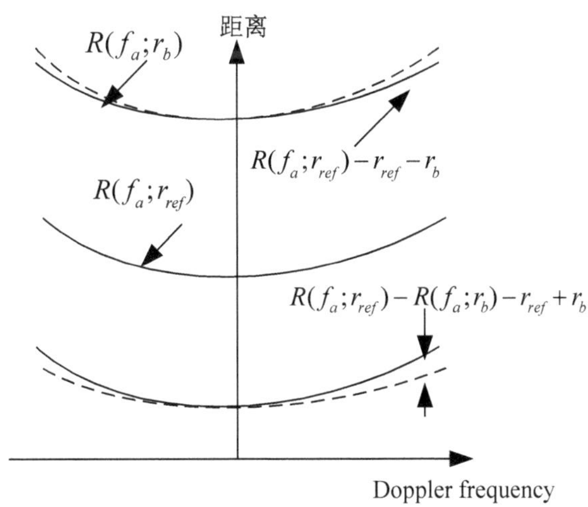

Fig. 8.28 Range migration curves in range-Doppler domain after equaling the range bend

same. Figure 8.28 is the chart of range migration curves in range-Doppler domain after equaling the range bend.

(3) Range Fourier transform

Again, using the principle of stationary phase, make range Fourier transforms to the signal after Chirp Scaling phase multiplying. The signal is transformed into a two-dimensional frequency domain, which is $f_r - f_a$ domain.

$$\begin{aligned} S(f_r, f_a) = C_2 \cdot \text{rect}\left[\frac{1}{T_a}\left(\frac{x_0}{V} - \frac{\lambda f_a r_b}{2V^2}\right)\right] \text{rect}\left\{\frac{f_r}{k \cdot [1 + a(f_a)] T_p}\right\} \cdot \exp\left[-j\frac{2\pi f_a x_0}{v}\right] \\ \cdot \exp\left\{-j\pi\frac{f_r^2}{k \cdot [1 + a(f_a)]}\right\} \cdot \exp\left\{-j\frac{4\pi}{c} f_r [r_b + r_{\text{ref}} a(f_a)]\right\} \\ \cdot \exp\left[-j\frac{4\pi}{\lambda} \cdot r_b \cdot \beta(f_a) - j\Theta_\Delta(f_a; r_b)\right] \end{aligned} \quad (8.62)$$

The second index item in Eq. (8.62) is modulation phase function in range-Doppler domain, and $r_{\text{ref}} a(f_a)$ in the third index item is the same range migration of all points after Chirp Scaling operation.

(4) Range migration correction, range compression and secondary range compression

The phase function which will be used for range compression and range migration correction is written as

$$H_2(f_r, f_a) = \exp\left\{j\pi\frac{f_r^2}{k \cdot [1 + a(f_a)]} + j\frac{4\pi}{c} f_r r_{\text{ref}} a(f_a)\right\} \quad (8.63)$$

The first item in Eq. (8.63) finishes range compression including secondary range compression, and the second item removes the range migration of all target points in different slant range.

Equation (8.63) multiplies Eq. (8.62), and range inverse Fourier Transform to the result, which can turn signal back into range-Doppler domain. At this point, the range compression and range migration correction are finished.

$$s(\tau,f_{\rm a}) = C_3 \cdot {\rm rect}\left[\frac{1}{T_a}\left(\frac{x_0}{v} - \frac{\lambda f_{\rm a} r_{\rm b}}{2v^2}\right)\right] \sin c\left[B_{\rm r}\left(\tau - \frac{2r_{\rm b}}{c}\right)\right] \cdot \exp\left[-j\frac{2\pi f_{\rm a} x_0}{v}\right]$$
$$\cdot \exp\left[-j\frac{4\pi}{\lambda}\cdot r_0 \cdot \beta(f_{\rm a}) - j\Theta_\Delta(f_{\rm a}; r_0)\right] \tag{8.64}$$

In Eq. (8.64), the computation in $\sin c(\,)$ item uses the approximate relationship $k \cdot [1 + a(f_{\rm a})]T_{\rm p} \approx k_{\rm r}T_{\rm p} = B_{\rm r}$.

(5) Azimuth compression and residual phase compensation

Azimuth compression process and compensate residual phase function due to Chirp Scaling, that is

$$\begin{aligned} H_3(\tau,f_{\rm a}) &= \exp\left[j\frac{2\pi}{\lambda}\cdot c\tau \cdot (\beta(f_{\rm a}) - 1) + j\Theta_\Delta(f_{\rm a}; r_{\rm b})\right] \\ &= \exp\left[j\frac{4\pi}{\lambda}\cdot r_{\rm b} \cdot (\beta(f_{\rm a}) - 1) + j\Theta_\Delta(f_{\rm a}; r_{\rm b})\right] \end{aligned} \tag{8.65}$$

Equation (8.65) multiplies Eq. (8.64), then azimuth inverse Fourier transform. The image result is

$$s_0(\tau, t) = C_4 \sin c\left[B_{\rm a}\left(\tau - \frac{x_0}{v}\right)\right] \sin c\left[B_r\left(\tau - \frac{2r_{\rm b}}{c}\right)\right] \cdot \exp\left[j\frac{4\pi v x_0}{\lambda r_{\rm b}}t\right] \tag{8.66}$$

In Eq. (8.66), $B_{\rm a} = 2v^2 T_{\rm a}/(\lambda r_{\rm b})$ is azimuth processing bandwidth.

(6) CSA limitations

The derivation of CSA is based on some approximate. A number of resident phase principles are used to compute approximately the integral in CSA. Furthermore, CSA uses linear interpolation instead of non-linear Stolt interpolation in range migration algorithm. When the resolution requirement is raised and scene size is increased, this approximation is somewhat insufficient and the residual phase error will cause the image defocusing.

If the CSA algorithm is used for Spotlight synthetic aperture radar, Echo cannot be performed any Dechirp operations prior to treatment. However, in many systems, in the process of data collection the data has been Dechirp processing in real time. In order to use the CSA algorithm linear frequency-modulated signal must be coupled again, which would be a heavy burden to the processor. Even without Dechirp operation, Spotlight SAR with the requirement of high azimuth resolution, the azimuth bandwidth will grow very large as the synthetic aperture length increases. None but by raising the pulse repetition frequency, azimuth signals can avoid aliasing. This is sometimes impossible to achieve under space borne conditions. Even in airborne systems, while the SAR operating on both Strip and

Spotlight mode with different pulse repetition frequencies, the complexity of systems will also increase. This is the problem in the CSA algorithm for Spotlight synthetic aperture radar imaging that must be solved.

8.5 Extended Chirp Scaling Algorithm

Extended CS algorithm (ECS) was originally proposed by Alberto Moreira, mainly used in high resolution airborne, space borne Strip mode and Scan mode SAR imaging processing. Later Josef Mittermayer applied extend CS algorithm in Spotlight synthetic aperture radar imaging [11–13].

Compared to classic CS algorithm, Extended CS algorithm has two main improvements: (1) updates the Chirp Scaling phase factors, which correct range spectrum broadening and shift from Chirp Scaling operation. Azimuth focus is realized by azimuth Scaling function. (2) Sub aperture processing method is used, which reduce the requirement for high pulse repetition frequency of the system, so that the algorithm can be adapted to a variety of applications condition.

8.5.1 Sub Aperture Processing Method

Sub aperture processing method may be used with a variety of synthetic aperture radar imaging algorithm together. Due to Spotlight SAR synthetic aperture is much larger than the corresponding Strip SAR synthetic aperture, its total Doppler bandwidth increases considerably. In order to avoid azimuth spectrum aliasing, $\mathrm{PRF} > B_{\mathrm{d,total}}$ as a basic restriction must be satisfied. High pulse repetition frequency requires the data rate is large, and can cause severe range ambiguity. In order to solve these problems, it is necessary to use sub aperture processing method. Moreover, sub aperture processing method can also effectively compensate for Doppler center frequency rapidly changing as well as reduce the range migration, which can decrease the difficulty of Imaging.

Spotlight SAR azimuth width is $B_{\mathrm{a}} = \frac{2vw_{\mathrm{a}}}{\lambda R} + \frac{2v^2}{\lambda R}T_{\mathrm{a}}$. It can be seen that this is directly proportional to its synthetic aperture time. If the synthetic aperture time is divided into a number of segments (sub aperture), the sub aperture processing can be made. Figure 8.29 is sub aperture formation diagram. Assuming that each sub aperture time is T_{sub}, the azimuth signal bandwidth for each sub aperture is

$$B_{\mathrm{a,sub}} = \frac{2vw_{\mathrm{a}}}{\lambda R} + \frac{2v^2}{\lambda R}T_{\mathrm{sub}} \tag{8.67}$$

Typically, in order to maintain the continuous phase, suppress side lobe and avoid deterioration of impulse response function, there should be some overlap

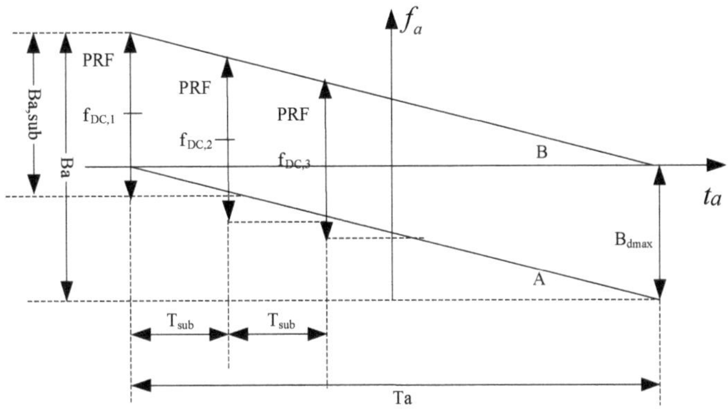

Fig. 8.29 Sub aperture formation diagram

between sub apertures and the overlapping rate is about 3–5%, which is not represented in Fig. 8.29. If a fixed aperture time T_{sub} is chosen, the bandwidth $B_{a,sub}$ is also fixed. In sub aperture processing for data, the azimuth sampling frequency f_{PRF} is only required to be greater than $B_{a,sub}$.

8.5.2 Spectral Analytical Method for Azimuth Focus

Spectral analysis method (SPECAN) is essentially a time-domain algorithm. Its calculation steps include multiplication and a Fourier transformation. Figure 8.30 represent calculation steps of spectrum analysis method. This algorithm decreases a Fourier transform comparing with fast convolution method.

Assume that the input data is sampling data at $t = m\Delta$ of linear frequency-modulated signal with frequency modulation slope K, that is,

$$s(t) = \exp\left[j\pi K(t - m\Delta)^2\right] \quad m\Delta - T/2 \le t \le m\Delta + T/2 \tag{8.68}$$

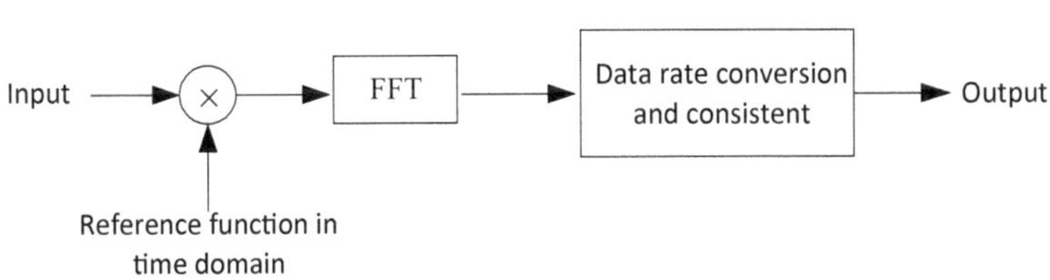

Fig. 8.30 Spectral analysis method calculation steps

The sampling data at $t = n\Delta$ of the conjugate linear frequency-modulated signal with the same frequency modulation slope K is selected as reference signal in time domain, multiply two factors

$$s(t)s^*(t) = \exp\{j\pi K[m^2\Delta^2 - n^2\Delta^2 + 2t\Delta(n-m)]\} \tag{8.69}$$

Let $t = k\Delta + n\Delta - T'/2$, Eq. (8.69) can be converted into discrete k function forms

$$\begin{aligned} s(k)s^*(k) &= \exp\{j\pi K[2k\Delta^2(n-m)]\}\exp[j\pi K(m^2\Delta^2 + T'\Delta m)] \\ &\quad \cdot \exp\{j\pi K[n^2\Delta^2 - T'\Delta n]\}\exp[j\pi K(-2mn\Delta^2)] \\ &= \exp(j\phi)\exp\{j\pi K[2k\Delta^2(n-m)]\} \end{aligned} \tag{8.70}$$

By FFT calculation to Eq. (8.70), the frequency domain expressions is

$$\begin{aligned} S(q) &= \sum_{k=0}^{T'/\Delta-1} s(k)s^*(k)\exp(-j2\pi\Delta kq/T') \\ &= \exp(j\phi)\exp\left\{j\left(\frac{T'}{\Delta}-1\right)\left[\pi\Delta\left(K\Delta(n-m)-\frac{q}{T'}\right)\right]\right\} \\ &\quad \frac{\sin\{\pi T'[K\Delta(n-m)-q/T']\}}{\sin\{\pi\Delta[K\Delta(n-m)-q/T']\}} \end{aligned} \tag{8.71}$$

After multiplying the reference function and Fourier transform, pulse compression is complete. The maximum value appears in the $q = (m-n)\Delta KT$ position. Input signal is calculated based on pulse position of the output signal and known reference function.

Spectrum analysis method overcomes the disadvantages of using longer reference function in fast convolution method, and reduces the need of sampling point's quantity in azimuth processing. However, if only use a simple spectrum analysis method, then azimuth focusing will create geometric distortion in azimuth position along with the slant range, which can be corrected by applying an azimuth Scaling method. In Extended Chirp Scaling method, azimuth focusing uses spectral analysis method combined with azimuth Scaling method.

8.5.3 *Principle and Procedure of Extended Chirp Scaling Algorithm*

(1) Imaging geometry of squint equivalent range model

Theory and practice have proven that CSA algorithm is a way to deal with large squint angle and range migration. For space-borne synthetic aperture radar, large

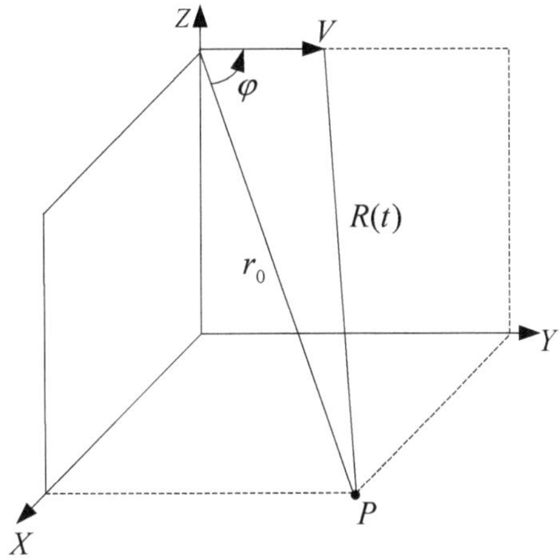

Fig. 8.31 Squint equivalent range model diagram

range walk caused by the rotation of the Earth can be equivalently processed with equivalent squint angle. For side-looking Spotlight synthetic aperture radar, most of the sub aperture is impossible to be strictly side-looking mode after dividing sub aperture. If the side-looking range model is adopted, errors will be inevitably introduced. The longer the aperture is, the larger the error is. CS algorithm using squint equivalent range model will get more accurate results.

Figure 8.31 is squint equivalent range model diagram. Assuming t is azimuth time, and the distance between radar and scattering point P at time t is $R(t)$, which can be expressed as

$$R(t;r_0)=\sqrt{r_0^2+(Vt)^2-2r_0Vt\cos\varphi} \tag{8.72}$$

In Eq. (8.72), r_0 is the slant range at synthetic aperture center point $(r=0)$, V is the equivalent speed of radar platform, φ is equivalent squint angle.

During imaging processing, equivalent speed and equivalent squint angle can be computed by accurate estimation of Doppler center frequency f_{dc} and Doppler frequency modulation slope k_{a}.

$$V=\sqrt{(\lambda f_{\mathrm{d}}/2)+\lambda r_0 k_{\mathrm{a}}/2} \tag{8.73}$$

$$\varphi=\arccos(-\lambda f_d/2V) \tag{8.74}$$

(2) Algorithm procedure

Figure 8.32 is the Extended Chirp Scaling algorithm procedure. From Eq. (8.3), the echo signal of a point target can be expressed as:

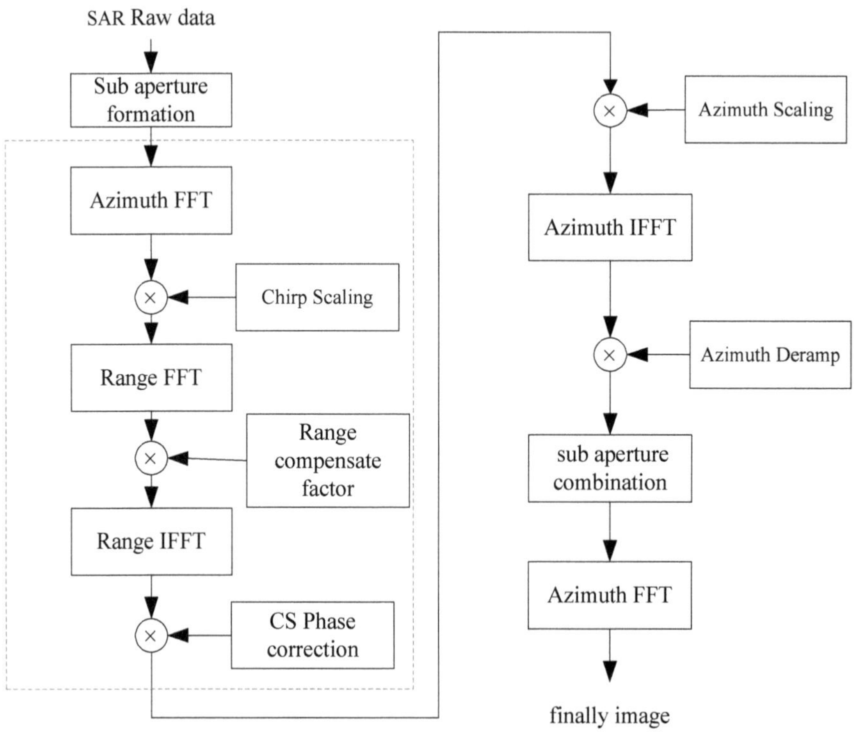

Fig. 8.32 ECS algorithm process

$$s_{\mathrm{r}}(t,\tau)=\mathrm{rect}\left(\frac{t}{T_{\mathrm{a}}}\right)\exp\left[-j\frac{4\pi}{\lambda}\cdot R(t)\right]\cdot\mathrm{rect}\left(\frac{\tau-\frac{2R(t)}{c}}{T_p}\right)\exp\left[j\pi k_{\mathrm{r}}\left(\tau-\frac{2R(t)}{c}\right)^2\right] \tag{8.75}$$

① Azimuth Fourier Transform:

After azimuth Fourier transforms, the echo signal will be converted to the range-Doppler domain

$$S(\tau,f_{\mathrm{a}})=C_1\mathrm{rect}\left[\frac{\lambda r_{\mathrm{b}}\sin\varphi}{2V^2\sqrt{1-\left(\frac{\lambda f_{\mathrm{a}}}{2V}\right)^2}}\right]\mathrm{rect}\left[\tau-\frac{2}{c}R(f_{\mathrm{a}};r_{\mathrm{b}})\right]\exp\left[-j\frac{2\pi f_{\mathrm{a}}\cos\varphi}{V}\right]$$
$$\cdot\exp\left[j\pi k(f_{\mathrm{a}};r_{\mathrm{b}})\left(\tau-\frac{2R(f_{\mathrm{a}};r_{\mathrm{b}})}{c}\right)^2\right]\cdot\exp\left[-j\frac{4\pi}{\lambda}\cdot r_{\mathrm{b}}\cdot\sin\varphi\cdot\beta(f_{\mathrm{a}})\right] \tag{8.76}$$

$$\beta(f_a) = \sqrt{1 - \left(\frac{f_a \cdot \lambda}{2V}\right)^2} \tag{8.77}$$

$$R(f_a; r_b) = \frac{r_b \sin \varphi}{\beta(f_a)} \tag{8.78}$$

$$\frac{1}{k(f_a; r_b)} = \frac{1}{k_r} - \frac{2\lambda r_b\left(\beta^2(f_a) - 1\right)\sin\varphi}{c^2\beta^3(f_a)} \tag{8.79}$$

② Chirp Scaling processing

Chirp Scaling factor is

$$H1(f_a; \tau) = \exp\left\{j\pi k(r_{ref}, f_a)\frac{\Delta\alpha(f_a)}{\alpha(f_{DC})}\left[\tau - \tau_{ref}(f)\right]^2\right\} \tag{8.80}$$

$$\tau_{ref}(f) = \frac{2}{c} r_{ref}[1 + a(f_a)] \tag{8.81}$$

$$a(f_a) = \frac{\sin \varphi_r}{\beta(f_a)} - 1 \tag{8.82}$$

$$a_e(f_a) = [1 + a(f_a)]/\alpha - 1 \tag{8.83}$$

$$k(r_{ref}, f) = \frac{1}{k_r} - \frac{2\lambda r_b\left(\beta^2(f_a) - 1\right)\sin\varphi_{ref}}{c^2\beta^3(f_a)} \tag{8.84}$$

In Eq. (8.84), r_{ref} is the distance from aperture center to reference point, φ_{ref} is equivalent squint angle of reference point at aperture center, $a_e(f_a)$ is Chirp Scaling factor in ECSA, α is stretching ratio of each sub aperture. For the convenience of sub aperture synthesis, r_{ref} of each aperture need to get the same value, and

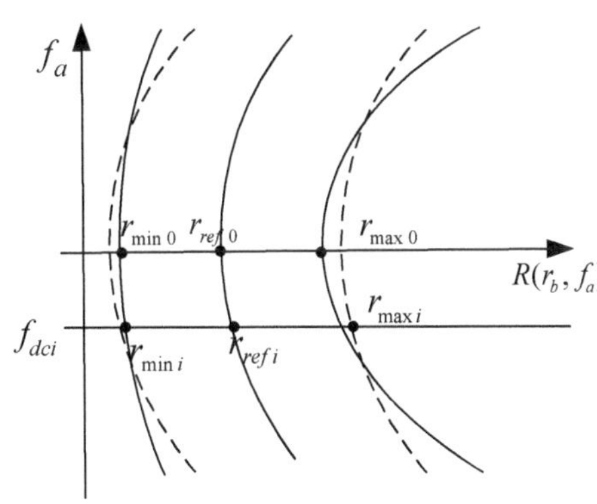

Fig. 8.33 Target migration path in range-Doppler domain

stretching ratio α of each aperture chooses sin φ_{ref}, so that each ratio of sub aperture is corrected to the scale of zero Doppler frequency. Point target migration path in range-Doppler domain is shown in Fig. 8.33. In Fig. 8.33, $f_{\text{dc}\,i}$ is Doppler center frequency of sub aperture i; $r_{\min i}$, $r_{\text{ref}\,i}$ and $r_{\max i}$ are respectively the distance of nearest point, reference point, and farthest point in swath on this Doppler Centre frequency; $r_{\min 0}$, $r_{\text{ref}\,0}$ and $r_{\max 0}$ are respectively the distance of targets on zero Doppler frequency.

If α is not introduced, after each sub aperture data multiplying CS factor, point targets will appear at the range cell of the time corresponding respective aperture center based on the respective aperture Center as standards. Point targets at the same range cell of different sub aperture will be at different range gates. After $\alpha = \sin\,\varphi_{\text{ref}}$ is introduced, range scale standards of all sub apertures will take the time corresponding to zero Doppler center frequency. In this way, just before the azimuth sub aperture synthesis, remove the shift of the reference distance of each sub aperture relative r_{ref} of zero Doppler frequency; sub aperture can be correctly synthesized.

The signal after Azimuth Fourier transform multiplies Chirp Scaling factor H_1,

$$
\begin{aligned}
S_2(\tau, f_a) = {} & \exp\left\{j\pi k(1+a_e)\left[\tau - \frac{2\alpha}{c}\left(\frac{r_b \sin\,\varphi}{\sin\,\varphi_r} + a_e r_{\text{ref}}\right)\right]^2\right\} \cdot \exp\{-j\Theta_\Delta\} \\
& \cdot \exp\left\{-j\frac{4\pi r\,\sin\,\varphi}{\lambda}\sqrt{1-(\lambda f_a/2V)^2} - j\frac{2\pi r}{v} f_a \cos\,\varphi\right\}
\end{aligned}
\tag{8.85}
$$

$$
\Theta_\Delta = \frac{4\pi k(1+a)^2}{c^2(1+a_e)}\left(\frac{r_b \sin\,\varphi}{\sin\,\varphi_r} - r_{\text{ref}}\right)^2 \tag{8.86}
$$

In Eq. (8.86), Θ_Δ is the residual phase caused by Chirp Scaling operation.

- Range compensation

Range Fourier transform to Eq. (8.85), then multiplied by the range compression factor

$$
H_2 = \exp\left\{-j\frac{\pi f_r^2}{k(f_a; r_{\text{ref}})[1+a_e(f_a)]}\right\} \cdot \exp\left[\frac{j4\pi}{c f_r r_{\text{ref}} a(f_a)}\right] \tag{8.87}
$$

Range inverse Fourier transform to the product, signal in range-Doppler domain is obtained.

- CS residual phase compensation

The compensation factor of CS residual phase is

$$H_3(\tau, f_a) = \exp(j\Theta_\Delta) \tag{8.88}$$

Θ_Δ is shown as Eq. (8.86).

Equation (8.88) is different from the compensation phase in the non-subaperture squint model CS algorithm. In subaperture method, each subaperture Doppler centric frequencies represent different time delays, which need to be preserved and will be exploited to accumulate all aperture time in order to obtain high azimuth resolution in subaperture synthesis.

⑤ Azimuth Scaling

The data after rang pulse compression multiplies azimuth scaling factor,

$$H_4(\tau, f_a) = \exp\left(j\frac{4\pi}{\lambda} r_{0,s} * \left[1 - \sqrt{1 - \left(\frac{\lambda f}{2V}\right)^2}\right] + j\frac{\pi f^2}{k_{a,s}}\right\} \tag{8.89}$$

$$r_{0,s} = r_{\text{ref}} + \left(\frac{c\tau}{2}\sin\varphi_r - r_{\text{ref}}\right)\Big/\alpha \tag{8.90}$$

In Eq. (8.89), the first index item is the hyperbolic phase history without echo, which is equivalent to match filtering to azimuth echo. For the strip synthetic aperture radar, images can be obtained by IFFT to the product of echo and this phase. For Spotlight synthetic aperture radar, if there is only this operation, after IFFT image aliasing will occur in the azimuth direction. So Azimuth Scaling combined with spectral analysis method is used in azimuth direction, multiply the second index item in Eq. (8.89) (the index item introduces a square phase-history, namely the linear frequency modulation), there will be

$$k_{a,s} = -2v^2/(\lambda r_s) \tag{8.91}$$

r_s is the reference distance of Azimuth Scaling.

In order to avoid azimuth aliasing caused by the displacement after Azimuth Scaling, also azimuth aperture time must be extended additionally. This is because the point targets in different locations have different Doppler frequencies. After introducing Azimuth scaling, although the targets have the same Doppler frequency modulation rate, the targets with distance $r \neq r_s$ will displace. In the case of squint, the point target displacement increases rapidly with the increasing degree of squint. In order to minimize such azimuth broadening due to the displacement, a new phase compensation item H_5 is needed to be multiplied.

$$H_5(\tau, f_a) = \exp(-j2\pi t_V f_a) \tag{8.92}$$

Azimuth-time displacement after azimuth data FFT can make its azimuth-time broadening be symmetry relative to the center of position. At this point

$$t_V = \frac{\lambda r_s f_{dc}}{2V^2} - \frac{\lambda r f_{dc}}{2V^2 \sin \varphi_r} \quad (8.93)$$

⑥ Azimuth Deramp

After Azimuth IFFT transformation to time domain, de-chirp operation to the obtained data need to multiply the compressed factor

$$\phi_6(t, r) = \exp\left(j\pi k_{a,s} t'^2\right) \quad (8.94)$$

$k_{a,s}$ is shown as Eq. (8.91), here $t' = t - t_V$.

⑦ Sub aperture synthesis

After the above operations, line up all sub apertures according to the time order, and remove overlap sub aperture data and take off the displacement relative to zero Doppler frequency of all sub aperture reference points. Finally obtain full aperture resolution images by azimuth FFT.

8.6 Frequency Scaling Algorithm

Frequency Scaling algorithm (FSA) is a new spotlight synthetic aperture radar imaging algorithm proposed by Josef Mittermayer and others from Germany space agency in 1998 [14–16]. FSA inherits the advantages of range migration algorithm which can directly process the Dechirp data in range direction, and at the same time overcomes the shortcomings of range migration algorithm by subaperture processing to reduce the sampling rate requirements. The maximum benefits of Frequency Scaling algorithm is to use frequency Scaling function to be strictly approximately to range migration algorithm and to make precise range migration correction, and avoids Stolt interpolation in range migration algorithm instead by FFT and complex multiplication, which greatly reduce the amount of calculation. Frequency Scaling algorithm applies to range de-chirp signal. It can eliminate effects of residual video phase and is convenient to range weighting.

8.6.1 Irradiation Geometry

Spotlight synthetic aperture radar data acquisition geometry is shown as Fig. 8.34. The vertical distance from the aircraft to the scene center O is r_c. The distance from the aircraft to point P is r, the vertical distance is r_0. The squint angles Ψ_{start} and Ψ_{end} are the start angle and end angle of the echo data, and the start time and end

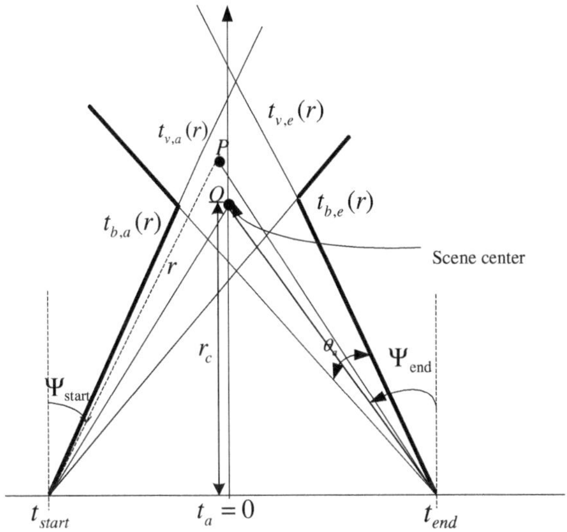

Fig. 8.34 Spotlight synthetic aperture radar data acquisition geometry

time of synthetic aperture are respectively t_{start} and t_{end}. The synthetic aperture time is T_{a}, and θ_{a} is the beam width.

$$t_{\text{a,start}} = \frac{r_{\text{c}}}{V}\tan(\Psi_{\text{start}}) \tag{8.95}$$

$$t_{\text{a,end}} = \frac{r_{\text{c}}}{V}\tan(\Psi_{\text{end}}) \tag{8.96}$$

$$T_{\text{a}} = t_{\text{start}} - t_{\text{end}} \tag{8.97}$$

In Fig. 8.34, all targets in available region are within radar ray irradiation over the entire radar synthetic aperture time. The minimum and maximum azimuth time $t_{\text{v,a}}(r_0)$ and $t_{\text{v,e}}(r_0)$ in the available target region with distance r_0 are

$$\left.\begin{aligned} t_{\text{v,a}}(r_0) &= \max\left[-\frac{r_0}{V}\cdot\tan\left(\Psi+\frac{\theta_{\text{a}}}{2}\right)+\frac{r_{\text{c}}}{V}\tan(\Psi)\right] \\ t_{\text{v,e}}(r_0) &= \min\left[-\frac{r_0}{V}\cdot\tan\left(\Psi-\frac{\theta_{\text{a}}}{2}\right)+\frac{r_{\text{c}}}{V}\tan(\Psi)\right] \end{aligned}\right\}\Psi_{\text{start}}\le\Psi\le\Psi_{\text{end}},\ t_{\text{v,a}}(r_0)\le t_{\text{v,e}}(r_0) \tag{8.98}$$

In Fig. 8.34, the thick solid line shows the irradiation theory boundary of the target area. In this area, all the targets are in whole or in part irradiated by the radar beam. The azimuth time locations of the irradiation area around the border are

$$t_{\mathrm{b,a}}(r_0) = \min\left[-\frac{r_0}{V}\cdot\tan\left(\Psi_{\mathrm{start}}+\frac{\theta_{\mathrm{a}}}{2}\right)+t_{\mathrm{a,start}};-\frac{r_0}{V}\cdot\tan\left(\Psi_{\mathrm{end}}+\frac{\theta_{\mathrm{a}}}{2}\right)+t_{\mathrm{a,end}}\right] \tag{8.99}$$

$$t_{\mathrm{b,e}}(r_0) = \max\left[-\frac{r_0}{V}\cdot\tan\left(\Psi_{\mathrm{start}}-\frac{\theta_{\mathrm{a}}}{2}\right)+t_{\mathrm{a,start}};-\frac{r_0}{V}\cdot\tan\left(\Psi_{\mathrm{end}}-\frac{\theta_{\mathrm{a}}}{2}\right)+t_{\mathrm{a,end}}\right] \tag{8.100}$$

8.6.2 Dechirp SAR Signal in Range-Doppler Domain

From Sect. 8.2, the received SAR echo signal after range Dechirp processing is

$$\begin{aligned} s(t,\tau;r_0) = {} & C\cdot\mathrm{rect}\left[\frac{\tau-2r(t;r_0)}{c}\right]\cdot\mathrm{rect}\left[\frac{t}{T_{\mathrm{a}}}\right] \\ & \cdot\exp\left[-j\frac{4\pi}{\lambda}r(t;r_0)\right]\cdot\exp\left[-j\frac{4\pi k_r}{c}(r(t;r_0)-r_c)\cdot\left(\tau-\frac{2r}{c}\right)\right] \\ & \cdot\exp\left[j\frac{4\pi k_r}{c^2}(r(t;r_0)-r_c)^2\right] \end{aligned} \tag{8.101}$$

In Eq. (8.101), τ is fast time variable, t is slow time variable, T_{a} is synthetic aperture time, λ is wavelength, k_{r} is frequency modulation rate of transmitting signal, and

$$r(t;r_0) = \sqrt{r_0^2+(Vt)^2} \tag{8.102}$$

In Eq. (8.101), the first index item is the azimuth Doppler modulation, the second index item represents the range signal, and the last one is residual video phase.

After azimuth Fourier transformation, Eq. (8.101) is transformed into range-Doppler domain

$$\begin{aligned} s(f_{\mathrm{a}},\tau;r_0) = C\cdot\Bigg\{ & \mathrm{rect}\left[\frac{\tau-\frac{2r_{\mathrm{c}}}{c}}{T_{\mathrm{p}}}\right]\cdot\exp\left[-j\frac{4\pi\cdot r_0\cdot\beta(f_{\mathrm{a}})}{\lambda}\right] \\ & \cdot\exp\left[-j\frac{4\pi k_{\mathrm{r}}}{c}\left(\frac{r_0}{\beta(f_{\mathrm{a}})}-r_c\right)\cdot\left(\tau-\frac{2\cdot r_{\mathrm{c}}}{c}\right)\right]\quad\cdot src\left(f_{\mathrm{a}},\tau-\frac{2r_{\mathrm{c}}}{c};r_0\right)\Bigg\} \\ & \cdot\exp\left[-j\pi k_{\mathrm{r}}\tau^2\right] \end{aligned} \tag{8.103}$$

In Eq. (8.103), f_a is azimuth frequency, its range is $-\frac{f_{PRF}}{2}+f_{DC}\le f_a\le\frac{f_{PRF}}{2}+f_{DC}$, $\mathrm{src}\left(f_a,\tau-\frac{2r_c}{c};r_0\right)$ is secondary range compression item, which is defined as

$$\mathrm{src}(f_a,\tau;r_0)=\exp\left[-j\cdot\frac{2\pi\cdot r_0\cdot k_r^2\cdot\lambda}{c^2}\cdot\frac{(\beta^2-1)}{\beta^3}\tau^2+j\cdot\frac{2\pi\cdot r_0\cdot k_r^3\cdot\lambda^2}{c^3}\cdot\frac{(\beta^2-1)}{\beta^5}\tau^3\right] \tag{8.104}$$

In Eq. (8.104), the first index item is the azimuth modulation, while the second index item is the sinusoidal range signal whose frequency is decided by point target distance r_0, center distance r_c, and β. β expresses the amount of range migration, which is defined as

$$\beta=\sqrt{1-\left(\frac{\lambda f_a}{2V}\right)^2} \tag{8.105}$$

The third index item in Eq. (8.104) is the key of Frequency scaling processing. It is a chirp signal, representing the residual video phase.

8.6.3 Range Cell Migration Correction by Frequency Scaling

To derive range migration correction algorithm, signals in Eq. (8.104) need to be converted to a two-dimensional wavenumber domain. The range Fourier transformation of Eq. (8.104) is

$$\begin{aligned}S(f_a,f_r;r_0)=C\cdot\Bigg\{&\exp\left[-j\cdot\frac{4\pi\cdot r_c}{c_0}\cdot f_r\right]\cdot\Bigg(\Bigg\{\exp\left[-j\cdot\frac{4\pi\cdot r_0\cdot\beta(f_a)}{\lambda}\right]\cdot T_p\\&\cdot\mathrm{sinc}\left[\pi\cdot T_p\cdot\left(f_r+\frac{2\cdot k_r}{c}\cdot\left(\frac{r_0}{\beta(f_a)}-r_c\right)\right)\right]\Bigg\}*\gamma_{SCR}(f_a,f_r;r_0)\Bigg)\Bigg\}\\&\cdot\exp\left[j\cdot\frac{\pi}{k_r}\cdot f_r^2\right]\end{aligned} \tag{8.106}$$

In Eq. (8.106), $\mathrm{sinc}(x)=\sin x/x$, $\exp\left[j\cdot\frac{\pi}{k_r}\cdot f_r^2\right]$ is the residual video phase in frequency domain. The envelope of the point target signal after range compression is the Sinc function of range frequency shift of azimuth frequency. The range cell migration is r_0/β which is concerned with the target distance r_0 and azimuth frequency.

(1) Range cell migration correction by Frequency scaling

Range cell migration correction (RCMC) is carried out in two steps. First, the range frequency is scaled to $1/\beta$ of the original frequency. The sinc function turns into

$$\mathrm{sinc}\left[\pi \cdot \frac{T_{\mathrm{p}}}{\beta}\left(f_{\mathrm{c}} + \frac{2 \cdot k_{\mathrm{r}}}{c} \cdot (r_0 - r_{\mathrm{c}} \cdot \beta)\right)\right] \tag{8.107}$$

In Eq. (8.107), the range cell migration is no longer associated with point target distance r_0, but associated with Center distance r_c. This means that all the range cell migrations are corrected to the same range cell migration of the central point. The second step is range migration correction through Block Shift. After Block Shift, the sinc function turns into

$$\mathrm{sinc}\left\{\pi \cdot \frac{T_{\mathrm{p}}}{\beta} \cdot \left[f_{\mathrm{c}} + \frac{2k_{\mathrm{r}}}{c} \cdot (r_0 - r_{\mathrm{c}})\right]\right\} \tag{8.108}$$

After Block Shift, range frequency is only associated with the distance from point target to regional center point, namely range migration correction is completed. From the above it is clear that range frequency Scaling and Block Shift are the key steps to realization of range migration correction in Frequency Scaling Algorithm.

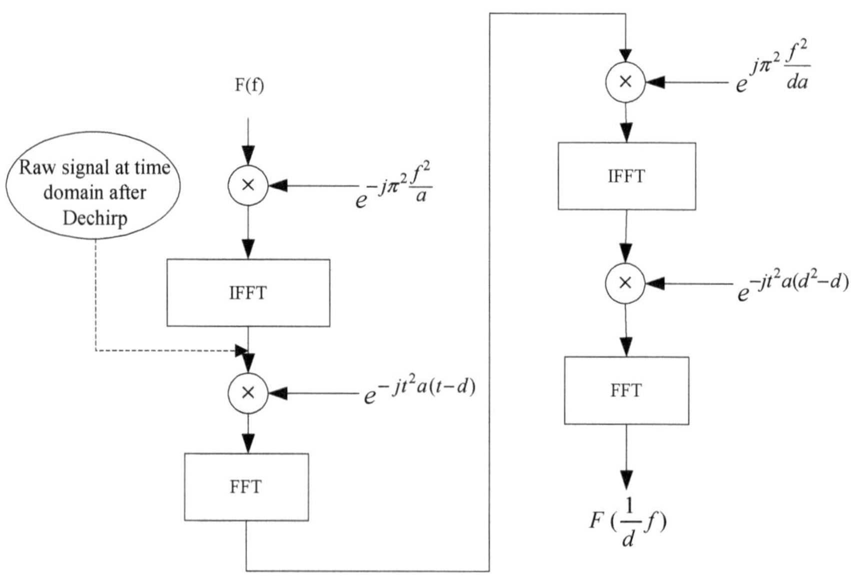

Fig. 8.35 Flow diagram of frequency Scaling

(2) Frequency Scaling steps

Figure 8.35 is the flow diagram of frequency Scaling. From Fig. 8.35, frequency Scaling is realized by four chirp signals. After a series of FFT and phase factor multiplying to $F(f)$, the frequency scaling functions $F(f/d)$ is obtained. The variable d is defined as

$$\frac{1}{\beta(f_{\mathrm{a}})} = \frac{1}{d} \tag{8.109}$$

$$\exp\left[-j \cdot \frac{\pi^2 \cdot f_{\mathrm{r}}^2}{a}\right] = \exp\left[j \cdot \frac{\pi}{k_{\mathrm{r}}} \cdot f_{\mathrm{r}}^2\right] \Rightarrow a = -\pi \cdot k_{\mathrm{c}} \tag{8.110}$$

Since residual video phase is the first multiplier in frequency Scaling, and the original data is in the range-Doppler domain, therefore, do not directly perform the first linear frequency multiplier and IFFT, in fact, effective frequency Scaling operation is started after the IFFT, specifically, begins from the part of the dashed line in Fig. 8.35. After a and d are defined, the other frequency Scaling multipliers can be calculated.

In Fig. 8.35, the second multiplier is frequency Scaling function, it functions in the range-time domain,

$$\begin{aligned} H_{FS}(f_{\mathrm{a}}, \tau) &= \exp\left[-j\tau^2 \cdot a \cdot (1-d)\right] \\ &= \exp\left[j \cdot \pi \cdot k_r \cdot \tau^2 \cdot (1-\beta)\right] \end{aligned} \tag{8.111}$$

The third multiplier is residual phase correction function; it functions in range-frequency domain.

$$H_{\mathrm{RVPC}}(f_{\mathrm{a}}, f_{\mathrm{r}}) = \exp\left[j \cdot \frac{\pi^2 \cdot f_{\mathrm{r}}^2}{a \cdot d}\right] = \exp\left[-j \cdot \frac{\pi \cdot f_{\mathrm{r}}^2}{k_{\mathrm{r}} \cdot \beta}\right] \tag{8.112}$$

H_{RVPC} as residual phase correction function corrects range-time displacement associated with range introduced by Dechirp operation. In Fig. 8.35, the last linear frequency modulation multiplier is inverse frequency Scaling function; it finishes operation in range-time domain.

$$\begin{aligned} H_{\mathrm{IFS}}(f_a, \tau) &= \exp\left[-j \cdot \tau^2 \cdot a \cdot (d^2 - d)\right] \\ &= \exp\left[j \cdot \pi \cdot k_r \cdot \tau^2 \cdot (\beta^2 - \beta)\right] \end{aligned} \tag{8.113}$$

(3) Secondary range compression and block shift

The signal of Eq. (8.106) with full frequency Scaling can be expressed as

$$S\left(f_{\mathrm{a}},\frac{f_{\mathrm{r}}}{\beta};r_0\right)=C\cdot\left\{\exp\left[-j\cdot\frac{4\pi\cdot r_{\mathrm{c}}}{c_0}\cdot\frac{f_{\mathrm{r}}}{\beta}\right]\cdot\left(\left\{\exp\left[-j\cdot\frac{4\pi\cdot r_0\cdot\beta}{\lambda}\right]\cdot T_{\mathrm{p}}\right.\right.\right.$$
$$\left.\left.\left.\cdot\,\mathrm{sinc}\left[\pi\cdot\frac{T_{\mathrm{p}}}{\beta}\cdot\left(f_{\mathrm{r}}+\frac{2\cdot k_{\mathrm{r}}}{c}\cdot(r_0-r_{\mathrm{c}}\cdot\beta)\right)\right]\right\}*\gamma_{\mathrm{SCR}}\left(\frac{f_{\mathrm{r}}}{\beta}\right)\right)\right\} \tag{8.114}$$

Range IFFT to Eq. (8.114), the signal before the final FFT in Fig. 8.35 can be got as below:

$$S(f_{\mathrm{a}},\tau,\beta;r_0)=C\cdot\mathrm{rect}\left[\frac{\tau\cdot\beta-\frac{2r_{\mathrm{c}}}{c}}{T_p}\right]\cdot\exp\left[-j\cdot\frac{4\pi\cdot k_{\mathrm{r}}}{c\cdot\beta}(r_0-r_c\cdot\beta)\right.$$
$$\left.\cdot\left(\tau\cdot\beta-2\cdot\frac{r_c}{c}\right)\right]\cdot\gamma_{\mathrm{SCR}}\left(f_{\mathrm{a}},\tau\cdot\beta-2\cdot\frac{r_{\mathrm{c}}}{c}\right)\cdot\exp\left[-j\cdot\frac{4\pi\cdot r_0}{\lambda}\cdot\beta\right] \tag{8.115}$$

Then, the secondary range compression and RCM correction can be carried out. For secondary distance compression, approximately considering $r_0\approx r_{\mathrm{c}}$, secondary range compression item is

$$\begin{aligned}H_{\mathrm{SRC}}(f_{\mathrm{a}},\tau;r_{\mathrm{c}})&=\gamma_{\mathrm{SCR}}\left(f_{\mathrm{a}},\tau\cdot\beta-\frac{2\cdot r_{\mathrm{c}}}{c};r_{\mathrm{c}}\right)^*\\&=\exp\left[j\cdot\frac{2\pi\cdot r_{\mathrm{c}}\cdot k_{\mathrm{r}}^2\cdot\lambda}{c^2}\cdot\frac{\beta^2-1}{\beta^3}\cdot\left(\tau\cdot\beta-\frac{2\cdot r_{\mathrm{c}}}{c}\right)^2\right.\\&\quad\left.-j\cdot\frac{2\pi\cdot r_{\mathrm{c}}\cdot k_{\mathrm{r}}^3\cdot\lambda^2}{c^3}\cdot\frac{\beta^2-1}{\beta^5}\cdot\left(\tau\cdot\beta-\frac{2\cdot r_{\mathrm{c}}}{c}\right)^3\right]\end{aligned} \tag{8.116}$$

Block shift phase function is

$$H_{\mathrm{BV}}(f_{\mathrm{a}},\tau)=\exp\left[j\cdot\frac{4\pi\cdot k_{\mathrm{r}}}{c}\cdot r_{\mathrm{c}}\left(\frac{1}{\beta}-1\right)\cdot\left(\tau\cdot\beta-\frac{2\cdot r_{\mathrm{c}}}{c}\right)\right] \tag{8.117}$$

In Eq. (8.116), * denotes complex conjugation. After the second range compression, block RCM shifts as well as the Range FFT, by multiplying with phase correction $H_{\mathrm{RPC}}(f_{\mathrm{a}},f_{\mathrm{r}};r_{\mathrm{c}})$, the signal processing in range direction is completed.

$$H_{\mathrm{RPC}}(f_{\mathrm{a}},f_{\mathrm{r}},\tau)=\exp\left[j\cdot\frac{4\pi\cdot r_{\mathrm{c}}}{c}\cdot\left(\frac{f_{\mathrm{r}}}{\beta}\right)\right]\cdot\exp[-2\pi\cdot\tau_0\cdot f_{\mathrm{r}}] \tag{8.118}$$

The first index item in Eq. (8.118) compensates the first index item in Eq. (8.114). The second index item compensates range-time shifting, where τ_0 is the time shifting of the first range cell. Range-Doppler signal after range processing can be represented as

$$S\left(f_{\mathrm{a}}, \frac{f_{\mathrm{r}}}{\beta}; r_0\right) = C \cdot \exp\left[-j \cdot \frac{4\pi \cdot r_0 \cdot \beta}{\lambda}\right] \cdot \operatorname{sinc}\left[\pi \cdot \frac{T_{\mathrm{p}}}{\beta} \cdot \left(f_{\mathrm{r}} + \frac{2 \cdot k_{\mathrm{r}}}{c} \cdot (r_0 - r_{\mathrm{c}})\right)\right] \tag{8.119}$$

In Eq. (8.119), compressed range signal is expressed as a Sinc function, uncompressed azimuth signal is as shown in the index.

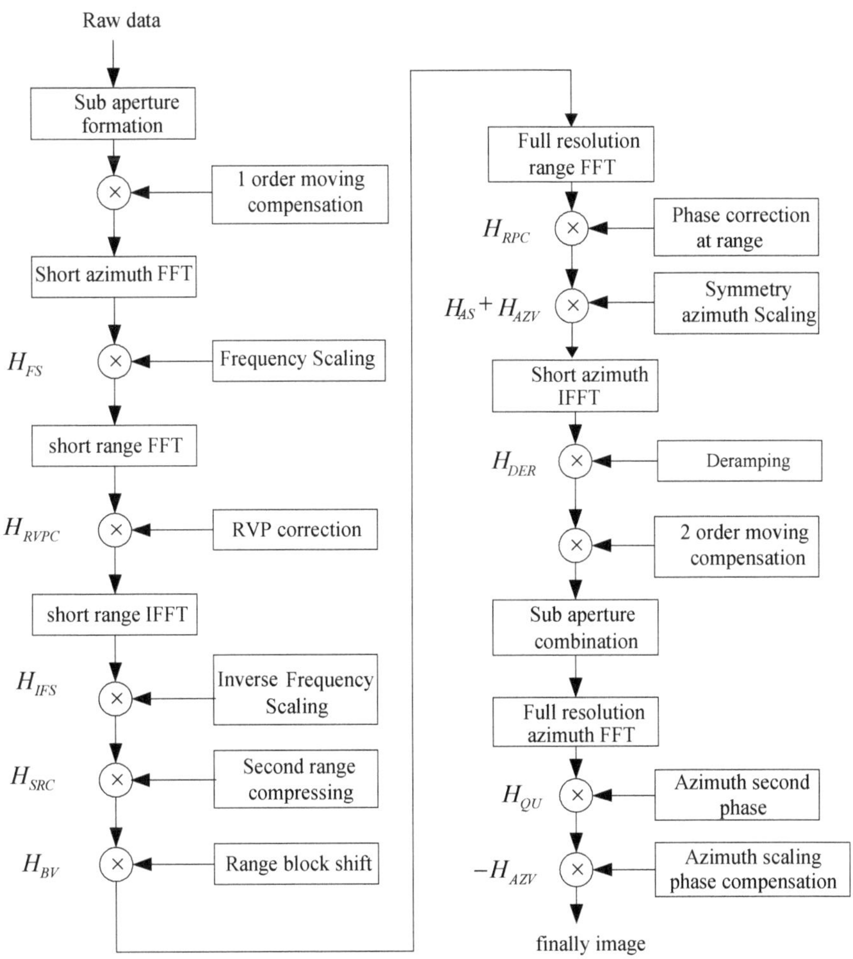

Fig. 8.36 Frequency Scaling algorithm flow diagram for Spotlight SAR data-processing

8.6.4 Frequency Scaling Algorithm

The flow diagram of Frequency Scaling algorithm for Spotlight synthetic aperture radar data-processing is as shown in the Fig. 8.36. Sub aperture forms and combinations will be discussed later. From the last section we know that after range migration correction and quadratic range compression, the processed results in range direction are obtained, namely Eq. (8.119). Azimuth processing is described as below.

In azimuth processing, the signal is first multiplied by azimuth Scaling factor in range-Doppler domain.

$$H_{\mathrm{AS}}(f_{\mathrm{a}}, r_0; \mathrm{rscl}) = \exp\left[j \cdot \frac{4\pi}{\lambda} \cdot r_0 \cdot (\beta(f_{\mathrm{a}}) - 1)\right] \cdot \exp\left[-\frac{j\pi}{k_{\mathrm{a,scl}}} f_{\mathrm{a}}^2\right] \tag{8.120}$$

Azimuth Scaling function changes the phase history depending on the target distance into a quadratic phase function independent of the target distance. $k_{\mathrm{a,scl}}$ is Doppler frequency modulation rate at the selected Scaling distance r_{scl}. Azimuth Scaling can accurately use spectrum analysis method, which avoids the extra interpolation for the geometric distortion correction in azimuth direction. However, Azimuth Scaling will result azimuth position displacements of every point targets from non-r_{scl} distance, and aperture length of each target has a certain extent of expansion. In order to avoid azimuth aliasing caused by displacements, azimuth aperture time must be extended extra a certain extent. The required azimuth aperture time extension will derive in the next section. To further reduce azimuth time extension, introduce an azimuth time displacement by linear phase multiplication, which makes required azimuth time extension symmetrical to azimuth position center. Azimuth time displacement is determined by the equation:

$$H_{\mathrm{AZV}}(f_{\mathrm{a}}, r_{\mathrm{a}}; r_{\mathrm{scl}}) = \exp[-j \cdot 2\pi \cdot t_{\mathrm{v}}(r_0; r_{\mathrm{scl}}) \cdot f_{\mathrm{a}}] \tag{8.121}$$

By azimuth inverse Fourier Transforms, the signal is transformed to azimuth-time domain. Due to the hyperbolic phase has been eliminated, azimuth linear frequency demodulation (deramp) can eliminate the second phase caused by azimuth scaling. Azimuth deramp factor is

$$H_{\mathrm{DER}}(t'_{\mathrm{a}}; r_{\mathrm{scl}}) = \exp\left[-j \cdot \pi \cdot k_{\mathrm{a,scl}} \cdot t'^2_{\mathrm{a}}\right] \tag{8.122}$$

$$t'_{\mathrm{a}} = t_{\mathrm{a}} - t_{\mathrm{v}}(r_0)$$

$$t_{\mathrm{a,start}} - \frac{T_{\mathrm{ext}}}{2} \le t_{\mathrm{a}} \le t_{\mathrm{a,end}} + \frac{T_{\mathrm{ext}}}{2} \tag{8.123}$$

8.6.5 Azimuth Scaling Time Stretching

In order to calculate the smallest symmetric azimuth time stretching T_{ext} and required time displacement $t_v(r_0)$, to take into account all the irradiation area is needed in Sect. 8.7.1, the left and right boundaries of irradiation target area in Eqs. (8.99) and (8.100) have been defined. Using irradiation geometry, hyperbolic, and quadratic azimuth-time/frequency relationship, the minimum and maximum azimuth-time location $t_{\text{a,min}}(r_0)$ and $t_{\text{a,max}}(r_0)$ of signals for all the irradiation target area without azimuth-time shift after azimuth Scaling can be calculated.

$$t'_{\text{a,max}}(r_0) = \text{MAX}\left[t_{\text{b,a}}(r_0) + \frac{r_{\text{scl}} \cdot \left(t_{\text{a,end}} - t_{\text{b,a}}(r_0)\right)}{\sqrt{r_0^2 + V^2\left(t_{\text{a,end}} - t_{\text{b,a}}(r_0)\right)^2}};\; t_{\text{b,e}}(r_0) + \frac{r_{\text{scl}} \cdot \left(t_{\text{a,end}} - t_{\text{b,e}}(r_0)\right)}{\sqrt{r_0^2 + V^2\left(t_{\text{a,end}} - t_{\text{b,e}}(r_0)\right)^2}}\right]; \tag{8.124}$$

$$t'_{a,\min}(r_0)) = \text{MIN}\left[t_{\text{b,a}}(r_0) + \frac{r_{\text{scl}} \cdot \left(t_{\text{a,start}} - t_{\text{b,a}}(r_0)\right)}{\sqrt{r_0^2 + V^2\left(t_{\text{a,end}} - t_{\text{b,a}}(r_0)\right)^2}};\; t_{\text{b,e}}(r_0) + \frac{r_{\text{scl}} \cdot \left(t_{\text{a,end}} - t_{\text{b,e}}(r_0)\right)}{\sqrt{r_0^2 + V^2\left(t_{\text{a,end}} - t_{\text{b,e}}(r_0)\right)^2}}\right]; \tag{8.125}$$

The required azimuth-time displacement $t_v(r_0)$ is the Center positions' difference of the signals before and after azimuth Scaling.

$$t_{\text{v}}(r_0) = \frac{t'_{\text{a,max}}(r_0) + t'_{\text{a,min}}(r_0)}{2} - \frac{t_{\text{a,start}}(r_0) + t_{\text{a,end}}(r_0)}{2} \tag{8.126}$$

After the azimuth-time shift, the minimum azimuth-time extension is symmetrical. Because it is block processing, the maximum azimuth stretching time of all distances is

$$T_{\text{ext}} = \text{MAX}\left[\left(t'_{\text{a,max}}(r_0), t'_{a,\min}(r_0)\right) - \left(t_{\text{a,end}} - t_{\text{a,start}}\right)\right] \tag{8.127}$$

In Eq. (8.127), T_{ext} is associated with the stretching distance r_{scl}.

8.6.6 Azimuth Sub Aperture Processing

The advantages of sub aperture processing in Sect. 8.6.1 is described. Need to be regarded that in the processing, before the last step azimuth FFT, namely full resolution FFT, it must increase sampling rate by fill up zero in azimuth-time domain, so that the sample rate after the final azimuth FFT can meet the resolution requirements.

The maximum and the minimum instantaneous azimuth frequency $f_{\mathrm{a,inst,max}}(t_{\mathrm{a}})$ and $f_{\mathrm{a,inst,min}}(t_{\mathrm{a}})$ are

$$f_{\mathrm{a,inst,max}} = -\frac{2V}{\lambda} \cdot \sin\left(a\tan\left[\frac{t_{\mathrm{a}} \cdot V}{r_{\mathrm{c}}}\right] - \frac{\theta_{\mathrm{a}}}{2}\right) \tag{8.128}$$

$$f_{\mathrm{a,inst,min}} = -\frac{2V}{\lambda} \cdot \sin\left(a\tan\left[\frac{t_{\mathrm{a}} \cdot V}{r_{\mathrm{c}}}\right] + \frac{\theta_{\mathrm{a}}}{2}\right) \tag{8.129}$$

After analysis, it is known that all target azimuth frequencies are between the minimum and the maximum instantaneous frequency.

The so-called sub aperture formation is that in two dimensional time domain the original data is divided into a number of data blocks with the same length in range direction and shorter length in azimuth direction. According to the PRF, the maximum sub aperture lasting time can be calculated by $f_{\mathrm{a,inst,max}}(t_{\mathrm{a}})$, $f_{\mathrm{a,inst,min}}(t_{\mathrm{a}})$ and their reciprocal functions $t_{\mathrm{a,inst,max}}(t_{\mathrm{a}})$,$t_{\mathrm{a,inst,min}}(t_{\mathrm{a}})$. The computation method is shown as Fig. 8.37. The starting moment of entire synthetic aperture is also the start time $t_{\mathrm{a,start,\#1}}$ of the first sub aperture (#1). From Eq. (8.129), the maximum

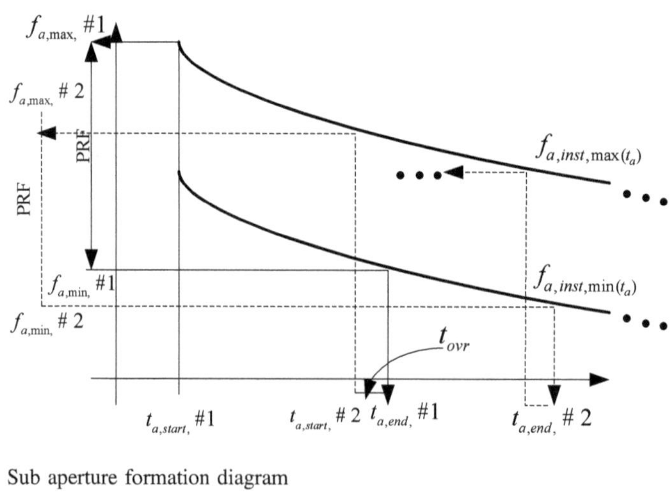

Fig. 8.37 Sub aperture formation diagram

instantaneous frequency $f_{a,max,\#1}$ of the first sub aperture can be calculated. The minimum permissible instantaneous frequency $f_{a,min,\#1}$ equals $f_{a,max,\#1}$ minus PRF. The reciprocal function of Eq. (8.129) is the maximum permissible ending time of the first sub aperture (#1). In order to ensure the smoothness of sub aperture combination, there should be a segment overlapping between sub apertures, and the overlapping rate is around 3–5%. In Fig. 8.37, overlapping time t_{ovr} is used to calculate the beginning time of the second sub aperture.

By the same way, the start and end time of the following sub apertures can be calculated. In practice, in order to achieve sub aperture method better, it would be best to make all the sub aperture length equal.

8.7 Two-Step Processing Algorithms

"Two step" algorithm (TSP) was proposed by G. Fornaro, R. Lanari and others. This method can better resolve azimuthal aliasing and image focusing of the spotlight synthetic aperture radar, and apply traditional Strip synthetic aperture radar imaging processing methods to the imaging of spotlight synthetic aperture radar successfully [17, 18].

"Two step" algorithm combines frequency spectrum analysis method and Strip mode imaging algorithms. Its first step is to use azimuth Deramp to realize data compression in azimuth direction, and make the output pixel size less than or equal to the desired azimuth resolution. The second step is to complete the residual focus by the usual stripe mode imaging algorithm in the frequency domain. This frequency domain focusing operation method is possible, because azimuthal aliasing of the original data has been completely overcome after the azimuth compression. The biggest difference between the second step method and Strip synthetic aperture radar processing method is azimuth filtering function, because it deals with the azimuth signals with the compensated quadric phase item and changed azimuth signals of azimuth pixel spacing for the input data. The advantage of "Two step" is that it does not contain the added matrix dimension operation, only need FFT and matrix multiplication.

Here the two-step algorithm which combined azimuth Deramp and Chirp Scaling algorithm in the side-looking and squint models will discuss.

8.7.1 Azimuth Deramp+Chirp Scaling Algorithm: Side-Looking Model

From the imaging geometry of Spotlight SAR introduced in Sect. 7.3.1, the distance from radar to target is

$$r(t) = \sqrt{r^2 + (Vt)^2} \approx r + \Delta r \tag{8.130}$$

In Eq. (8.130), r is the slant distance corresponding to synthetic aperture center point ($t = 0$), V is the equivalent speed of radar platform. Taylor series expansion to Eq. (8.130), ignoring higher order terms, there is

$$\Delta r \approx \frac{(Vt)^2}{2r} \tag{8.131}$$

(1) Principle analysis of azimuth Deramp

From Eq. (8.3), the echo signal is

$$\begin{aligned} s_r(t,\tau) = \sigma \cdot \operatorname{rect}\left(\frac{t}{T_{\text{spot}}}\right) \cdot \operatorname{rect}\left(\frac{\tau - \frac{2r(t)}{c}}{T_{\text{p}}}\right) \exp\left[-j\frac{4\pi}{\lambda}\cdot r\right] \\ \cdot \exp\left[-j\frac{4\pi}{\lambda}\cdot \Delta r\right] \cdot \exp\left\{j\pi k_{\text{r}}\left[\tau - \frac{2(r+\Delta r)}{c}\right]^2\right\} \end{aligned} \tag{8.132}$$

σ is the scattering coefficient of point target, c is the speed of light, T_{spot} is synthetic aperture time of Spotlight mode, T_{p} is impulse width of transmitting LFM signal, k_{r} is frequency modulation rate of transmitting signal, t is range time, τ is azimuth time.

Azimuth reference function is

$$s_{\text{ref}} = \exp\left(j\pi\frac{2V^2}{\lambda\tilde{r}}t^2\right) \qquad r_{\text{m}} \le \tilde{r} \le r_{\text{M}} \tag{8.133}$$

In Eq. (8.133), r_{m} and r_{M} are respectively the nearest and the farthest slant distance of irradiation area, $\tilde{r}$ is the distance value of a point target in the area, the reference distance value.

At first it needs azimuth Deramp processing, alone azimuth signal is

$$ss_{\text{a}}(t,r) = \operatorname{rect}\left[\frac{t}{T_{\text{spot}}}\right] \cdot \exp\left[-j\frac{4\pi}{\lambda}r\right] \cdot \exp\left[-j\frac{4\pi}{\lambda}\Delta r\right] \tag{8.134}$$

Ignore constant terms, after convolution of Eqs. (8.134) and (8.133), the obtained signal is

$$\begin{aligned}
\bar{s}s_{\mathrm{a}}(t,r) &= ss_{\mathrm{a}}(t,r)\otimes\exp\left(j\pi\frac{2V_{\mathrm{a}}^2}{\lambda\tilde{r}}t^2\right)\\
&=\left\{\mathrm{rect}\left[\frac{t}{T_{\mathrm{a}}}\right]\cdot\exp\left[-j\frac{4\pi}{\lambda}\frac{(Vt)^2}{2r}\right]\right\}\otimes\exp\left(j\pi\frac{2V^2}{\lambda\tilde{r}}t^2\right)\\
&=\int\mathrm{rect}\left[\frac{z}{T_{\mathrm{a}}}\right]\cdot\exp\left[-j\frac{4\pi}{\lambda}\frac{(Vz)^2}{2r}\right]\cdot\exp\left[j\pi\frac{2V^2}{\lambda\tilde{r}}(t-z)^2\right]\mathrm{d}z\\
&=\exp\left[j\pi\frac{2(Vt)^2}{\lambda\tilde{r}}\right]\int\mathrm{rect}\left[\frac{z}{T_{\mathrm{a}}}\right]\cdot\exp\left[-j2\pi\frac{2V^2t}{\lambda\tilde{r}}z\right]\exp\left[-j\pi\left(\frac{2V^2}{\lambda r}-\frac{2V^2}{\lambda\tilde{r}}\right)z^2\right]\mathrm{d}z
\end{aligned}\tag{8.135}$$

From Eq. (8.135) can be seen, the azimuth convolution is actually a SPECAN operations based on dechirp, including linear frequency-modulated signal multiplies with the azimuth signal, and the Fourier transform and the residual phase compensation. Comparing with SPECAN algorithm, the difference is that the frequency modulation rate of the azimuth reference function is a constant independent of range gate, and this preserve the spatially variable characteristics of the transfer function for the remaining system

For the targets at the distance $r=\tilde{r}$, ignoring complex constant, Eq. (8.135) can be simplified to

$$\bar{s}s(t;r)=\exp\left(j2\pi\frac{V^2}{\lambda\tilde{r}}t^2\right)\sin c\left[\frac{2\pi}{L}\frac{T_{\mathrm{spot}}}{T_{\mathrm{stip}}}Vt\right]\tag{8.136}$$

In Eq. (8.136), L is the antenna length in azimuth direction, T_{syn} is the synthetic aperture time of Strip mode. Obviously, azimuth precise focusing for the targets in the distance $r=\tilde{r}$ is realized.

For the point targets at the distance $r\neq\tilde{r}$, using the principle of stationary phase

$$\bar{s}s(t;r)=\mathrm{rect}\left[\frac{t}{T_{\mathrm{spot}}|r-\tilde{r}|/r}\right]\exp\left[-j2\pi\frac{V^2}{\lambda(r-\tilde{r})}t^2\right]\tag{8.137}$$

From Eq. (8.137), for the targets at the distance $r\neq\tilde{r}$, the azimuth coarse focusing is just completed, and the worst case is $|r-\tilde{r}|/r=|r_M-r_m|/r_m$. Spotlight synthetic aperture radar imaging area is generally small, it may be considered to meet the conditions $|r_M-r_m|/r_m\ll 1$, then $T_{\mathrm{spot}}|r_M-r_m|/r_m\ll T_{\mathrm{spot}}$. Even if in the worst case the azimuth compression can be realized, although azimuth direction is somewhat broadened and not like the full compression of the points at the distance $r=\tilde{r}$.

Spotlight mode SAR systems usually satisfy the following equation.

$$T_{\text{ext}} \approx \max\left[\frac{\lambda\tilde{r}}{LV} + T_{\text{spot}}\frac{|r-\tilde{r}|}{r}\right] \le \frac{\lambda r_{\text{M}}}{LV} + T_{\text{spot}}\frac{|r_{\text{M}} - r_{\text{m}}|}{r_{\text{m}}} \qquad T_{\text{stip}} \le T_{\text{ext}} \le T_{\text{spot}} \tag{8.138}$$

In Eq. (8.138), T_{ext} is synthetic aperture time after azimuth Deramp. So the above conclusion for point targets can be applied to the entire field of view of spotlight synthetic aperture radar. It ensures that the convolution operations in Eq. (8.135) can be used as the block azimuth compression, so the synthetic aperture time T_{spot} turns into T_{ext}, and the Doppler bandwidth $k_{\text{a}}T_{\text{spot}}$ will decrease to $k_{\text{a}}T_{\text{ext}}$, which overcomes the frequency spectrum aliasing of azimuth direction.

(2) Analysis in discrete domain

Suppose $\Delta t'$ and $\Delta t''$ are respectively impulse repetition time interval and output sampling time interval, then $A = T_{\text{syn}}/\Delta t''$ is the number of sampling points of echo signal in azimuth direction, $B = T/\Delta t''$ is the number of output sampling points after compression in azimuth direction. The discrete form of Eq. (8.135) is:

$$\begin{aligned}
\bar{s}s_{\text{a}}(n\Delta t'', r) &= \sum_{t=-A/2}^{A/2-1} ss_{\text{a}}(i\Delta t', r)\cdot S_{\text{ref}}(n\Delta t'' - i\Delta t', \tilde{r}) \\
&= \exp\left[j2\pi\frac{V^2}{\lambda\tilde{r}}(n\Delta t'')^2\right]\sum_{i=-A/2}^{A/2-1} ss_{\text{a}}(i\Delta t', r) \\
&\quad\cdot\exp\left[j\pi\frac{2V^2}{\lambda\tilde{r}}(i\Delta t')^2\right]\cdot\exp\left[-j2\pi\frac{V^2}{\lambda\tilde{r}}in\Delta t'\Delta t''\right] \qquad n\in(-B/2, B/2-1)
\end{aligned} \tag{8.139}$$

It can be seen that the last index item in Eq. (8.139) is a FFT transform nuclear form. In order to facilitate the use of FFT, by filling up zero operations in azimuth direction the following equation can be established:

$$\frac{\lambda\tilde{r}}{2V^2\Delta t'} = P\Delta t'' \ge T \tag{8.140}$$

Under the conditions that Eq. (8.140) is right, not only azimuth spectrum aliasing is avoided, but azimuth data loss does not occur. It is noted that in order to ensure Eq. (8.140) correctness, the sampling rate of spotlight synthetic aperture radar should be slightly higher than that meeting the Nyquist sample rate in Stripe mode. It can be proved as follows:

Substitute Eq. (8.138) into Eq. (8.140), the result is

$$\frac{\lambda\tilde{r}}{2V^2\Delta t'} = P\Delta t'' \geq T = \max\left[\frac{\lambda\tilde{r}}{LV} + T_{\mathrm{spot}}\frac{|r-\tilde{r}|}{r}\right] \tag{8.141}$$

That is

$$\frac{\lambda\tilde{r}}{2V^2\Delta t'}\cdot\frac{LV}{\lambda\tilde{r}} \geq 1 + \max\left[T_{\mathrm{spot}}\frac{|r-\tilde{r}|}{r}\cdot\frac{LV}{\lambda\tilde{r}}\right] \tag{8.142}$$

$$\frac{L/2}{V\Delta t'} \geq 1 + \max\left[\frac{T_{\mathrm{spot}}\frac{|r-\tilde{r}|}{r}}{\frac{\lambda\tilde{r}}{LV}}\right] \tag{8.143}$$

From Eq. (8.143), $L/2$ is the limits of Nyquist sampling interval in Strip mode, $V\Delta t'$ is the sampling interval of ECHO under Spotlight mode, the ratio of the two is between 1.2 and 1.3. This equation has some reference meanings to the PRF selection on Spotlight mode. Therefore, the correct selections $\tilde{r}$ can guarantee Eq. (8.140) correct. That is,

$$\max\left[\frac{|r-\tilde{r}|}{r}\right] \leq \frac{1}{T_{\mathrm{spot}}}\cdot\frac{\lambda\tilde{r}}{LV}\cdot\left(\frac{L/2}{V\Delta t'} - 1\right) \tag{8.144}$$

So, if $\frac{r_{\mathrm{M}}-r_{\mathrm{m}}}{r_{\mathrm{m}}} \leq \frac{T_{\mathrm{strip}}}{T_{\mathrm{spot}}}\cdot\left(\frac{L/2}{V\Delta t'} - 1\right)$ is satisfied, then any $\tilde{r}$ between r_{m} and r_{M} can satisfy Eq. (8.140). Thus, Eq. (8.139) can be written as

$$\begin{aligned}\bar{s}s_{\mathrm{a}}(n\Delta t'', r) &= \sum_{t=-P/2}^{P/2-1} ss_{\mathrm{a}}(i\Delta t', r)\cdot S_{\mathrm{ref}}(n\Delta t'' - i\Delta t', \tilde{r}) \\ &= \exp\left[j2\pi\frac{V^2}{\lambda\tilde{r}}(n\Delta t'')^2\right]\sum_{i=-P/2}^{P/2-1} ss_{\mathrm{a}}(i\Delta t', r) \\ &\quad\cdot\exp\left[j\pi\frac{2V^2}{\lambda\tilde{r}}(i\Delta t')^2\right]\cdot\exp\left[-j2\pi\frac{V^2}{\lambda\tilde{r}}\frac{in}{P}\right] \\ &= \exp\left[j2\pi\frac{V^2}{\lambda\tilde{r}}(n\Delta t'')^2\right]\cdot\mathrm{DFT}\left\{ss_{\mathrm{a}}(i\Delta t', r)\exp\left[j2\pi\frac{V^2}{\lambda\tilde{r}}(i\Delta t')^2\right]\right\}\end{aligned} \tag{8.145}$$

In General, p is an integral power of 2. As can be seen from Eq. (8.145), azimuth Deramp processing is accomplished through two complex multiplications in azimuth direction and an FFT.

(3) Chirp Scaling focusing processing

Processed by azimuth pre-filter, point targets at reference distance are totally focused in azimuth direction, and other point targets at other distance will have a certain amount of stretching, just the coarse focus is completed. Using Chirp Scaling principle, the remaining range focus and azimuth precise focus can be completed. The algorithm is derived as following. It ignores complex constants which will not affect the entire algorithm.

Azimuth Fourier transformation to the Deramp data is

$$\begin{aligned} \bar{s}s(\tau, f_{\rm a}; r) &= FT[ss(\tau, t; r) \otimes s_{\rm ref}(t; \tilde{r})] \\ &= sS(\tau, f_{\rm a}; r) \cdot \exp\left(-j\pi \frac{\lambda \tilde{r}}{2V^2} f_{\rm a}^2\right) \end{aligned} \tag{8.146}$$

In Eq. (8.146), FT[] is Fourier transform, $sS(\tau, f_{\rm a}; r)$ is Fourier transform of $ss(\tau, t; r)$, $f_{\rm a}$ is azimuth frequency. The echo of point target is

$$s_{\rm r}(t, \tau) = \sigma \cdot \text{rect}\left(\frac{t}{T_{\rm spot}}\right) \cdot \text{rect}\left(\frac{\tau - \frac{2r(t)}{c}}{T_{\rm p}}\right) \exp\left[-j\frac{4\pi}{\lambda} \cdot r\right] \exp\left\{j\pi k_{\rm r}\left[\tau - \frac{2r(t)}{c}\right]^2\right\} \tag{8.147}$$

According to the principle of stationary phase, azimuth Fourier transform of the echo signal, $sS(\tau, f_{\rm a}; r)$ is

$$\begin{aligned} sS(\tau, f_{\rm a}; r) = \sigma\, \text{rect}\left[\frac{1}{T_{\rm spot}} \frac{\lambda f_{\rm a} r}{2V^2\sqrt{1 - \left(\frac{\lambda f_{\rm a}}{2V}\right)^2}}\right] \text{rect}\left[\tau - \frac{2}{c} r_{\rm f}(f_{\rm a}; r)\right] \\ \cdot \exp\left[j\pi k_{\rm s}(f_{\rm a}; r)\left(\tau - \frac{2 r_{\rm f}(f_{\rm a}; r)}{c}\right)^2\right] \cdot \exp\left[-j\frac{4\pi}{\lambda}\sqrt{1 - \left(\frac{\lambda f_{\rm a}}{2V}\right)^2}\right] \end{aligned} \tag{8.148}$$

In Eq. (8.148),

$$r_{\rm f}(f_{\rm a}; r) = \frac{r}{\sqrt{1 - \left(\frac{f_{\rm a} \cdot \lambda}{2v}\right)^2}} \tag{8.149}$$

$$\frac{1}{k_{\rm s}(f_{\rm a}; r)} = \frac{1}{k_{\rm r}} - \frac{2\lambda r_{\rm b}\left(\beta^2(f_{\rm a}) - 1\right)}{c^2 \beta^3(f_{\rm a})} \tag{8.150}$$

$$\beta(f_a) = \sqrt{1 - \left(\frac{f_a \cdot \lambda}{2v}\right)^2} \tag{8.151}$$

Multiplied by the Chirp Scaling factor in the range-Doppler domain

$$\Phi_1(\tau, f; r_{ref}) = \exp\left\{j\pi k_s(f_a; r_{ref}) \cdot a(f_a)[\tau - \tau_{ref}(f)]^2\right\} \tag{8.152}$$

$$a(f_a) = \frac{1}{\beta(f_a)} - 1 \tag{8.153}$$

$$\tau_{ref} = \frac{2}{c} r_f(f_a; r_{ref}) \tag{8.154}$$

After CS factor multiplying, range migration curves of different distance lines has the same form with that of the reference distance. Doppler parameters variation along with the slant range has less influence to range focusing, so the same reference slant range can be used for the entire swath focusing. After multiplying the CS factor, the signal has been transformed into the wave-number domain by means of range Fourier transform, and multiplying range compensation factor $\Phi_2(f_r, f_a; r_{ref})$ the range migration correction and range focusing has completed

$$\Phi_2(f_r, f_a) = \exp\left\{j\pi \frac{f_r^2}{k_s(f_a; r_{ref}) \cdot [1 + a(f_a)]} + j\frac{4\pi}{c} f_r r_{ref} a(f_a)\right\} \tag{8.155}$$

In Eq. (8.155), f_r is range frequency, the first item completes quadric range compression and range focusing, and the second item completes range migration correction. Range inverse Fourier transform to Eq. (8.155), and multiplying azimuth compensation factor $\Phi_3(\tau, f_a)$ in range-Doppler domain.

$$\Phi_3(\tau, f_a) = \exp\left[j\frac{2\pi}{\lambda} \cdot c\tau \cdot (\beta(f_a) - 1) + j\frac{\pi}{k_{ref}} f_a^2\right] \cdot \exp[j\Theta_\Delta(f_a; r)] \tag{8.156}$$

$$\Theta_\Delta = \frac{4\pi}{c^2} k_s(f_a; r) a(f_a)[1 + a(f_a)](r - r_{ref})^2 \tag{8.157}$$

$k_{ref} = \frac{2V^2}{\lambda \tilde{r}}$ is Doppler frequency modulation rate at dechirp reference distance $\tilde{r}$. Due to range migration is completely corrected, so $r = c\tau/2$, the first index item of $\Phi_3(\tau, f_a)$ compensates the residual azimuth errors after azimuth pre-filter and completes azimuth precise focusing; the second index item compensates the residual phase which is failed to compensate in CS phase multiplication. The azimuth inverse Fourier transform completes the azimuth compression, and obtains synthetic aperture radar image. It should be noted that it always assume that squint

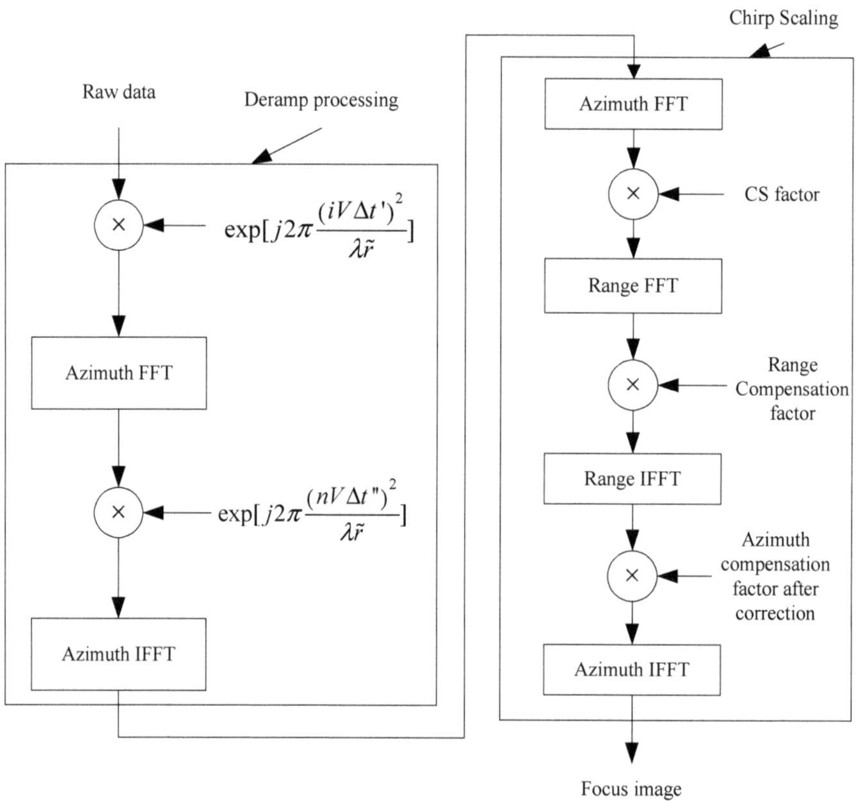

Fig. 8.38 Azimuth Deramp+Chirp Scaling (side-looking model) algorithm flow chart

angle has no effect in the above derivation process, so the algorithm above is only for cases of the side-looking or small squint angle.

(4) Algorithm flow diagram

Based on the analysis above, we get the azimuth Deramp+Chirp Scaling algorithm flow as shown in Fig. 8.38.

8.7.2 *Azimuth Deramp+Chirp Scaling Algorithm: Squint Model*

(1) Principle analysis of azimuth Deramp

Reference the squint equivalent model described in Sect. 7.2.1,

$$R(t;r_0) = \sqrt{r_0^2 + (Vt)^2 - 2r_0Vt\cos\varphi} = r_0 + \Delta r \tag{8.158}$$

Taylor series expansion to Eq. (8.158), and ignore three and more higher order terms,

$$R(t;r_0) \approx r + \frac{\lambda}{2}f_{\mathrm{dc}}t + \frac{\lambda}{4}k_{\mathrm{a}}t^2 = r + \Delta r \tag{8.159}$$

In Eq. (8.159), f_{dc} and k_{a} are respectively Doppler center frequency and Doppler frequency modulation rate.

$$f_{\mathrm{dc}} = -\frac{2V}{\lambda}\cos\varphi, \quad k_{\mathrm{a}} = \frac{2V^2}{\lambda r}\sin^2\varphi$$

The echo signal is

$$\begin{aligned} s_r(t,\tau) &= \sigma\cdot\mathrm{rect}\left(\frac{t}{T_{\mathrm{spot}}}\right)\cdot\mathrm{rect}\left[\frac{\tau-\frac{2R(t;r_0)}{c}}{T_{\mathrm{p}}}\right]\exp\left[-j\frac{4\pi}{\lambda}\cdot R(t;r_0)\right]\cdot\exp\left\{j\pi k_r\left[\tau-\frac{2R(t;r_0)}{c}\right]^2\right\} \\ &= \sigma\cdot\mathrm{rect}\left(\frac{t}{T_{\mathrm{spot}}}\right)\cdot\mathrm{rect}\left[\frac{\tau-\frac{2R(t;r_0)}{c}}{T_{\mathrm{p}}}\right]\cdot\exp\left[-j\frac{4\pi}{\lambda}\cdot r_0\right]\cdot\exp\left[-j\frac{4\pi}{\lambda}\cdot\Delta r\right] \\ &\quad\cdot\exp\left\{j\pi k_r\left[\tau-\frac{2(r_0+\Delta r)}{c}\right]^2\right\} \end{aligned} \tag{8.160}$$

Azimuth reference function is

$$s_{\mathrm{ref}} = \exp\left(j\pi k_{\mathrm{a},\tilde{r}}t^2\right) \quad r_{\mathrm{m}} \le \tilde{r} \le r_{\mathrm{M}} \tag{8.161}$$

$k_{\mathrm{a},\tilde{r}}$ is Doppler frequency modulation rate at reference distance $\tilde{r}$. Studying on azimuth signal alone:

$$ss_{\mathrm{a}}(t,r_0) = \mathrm{rect}\left[\frac{t}{T_{\mathrm{spot}}}\right]\cdot\exp\left[-j\frac{4\pi}{\lambda}r_0\right]\cdot\exp\left[-j\frac{4\pi}{\lambda}\Delta r\right] \tag{8.162}$$

For squint imaging, in order to avoid azimuth aliasing caused by Deramp convolution operation, a linear phase item need to be introduced into the echo according to the Doppler center frequency $f_{\mathrm{dc},\tilde{r}}$ at reference distance $\tilde{r}$ of synthetic aperture center, Eq. (8.162) after amendment is as follows:

$$ss_{\rm a}(t,r_0)=\text{rect}\left[\frac{t}{T_{\rm spot}}\right]\cdot\exp\left[-j\frac{4\pi}{\lambda}r_0\right]\cdot\exp\left[-j\frac{4\pi}{\lambda}\Delta r\right]\cdot\exp\left[j2\pi f_{{\rm dc},\tilde{r}}t\right]\quad(8.163)$$

Convolution of Eqs. (8.163) and (8.161), ignoring constant terms, it is

$$\begin{aligned}\bar{s}s(t,r_0)&=ss_{\rm a}(t,r_0)\otimes\exp\left(j\pi k_{{\rm a},\tilde{r}}t^2\right)\\&=\left\{\text{rect}\left[\frac{t}{T_{\rm spot}}\right]\cdot\exp\left[-j\frac{4\pi}{\lambda}\left(\frac{\lambda}{2}f_{\rm dc}t+\frac{\lambda}{4}k_{\rm a}t^2\right)\right]\right\}\otimes\exp\left(j\pi k_{{\rm a},\tilde{r}}t^2\right)\\&=\left\{\text{rect}\left[\frac{t}{T_{\rm spot}}\right]\cdot\exp\left[-j2\pi\left(f_{\rm dc}t+\frac{1}{2}k_{\rm a}t^2\right)\right]\right\}\otimes\exp\left(j\pi k_{{\rm a},\tilde{r}}t^2\right)\\&=\int\text{rect}\left[\frac{z}{T_{\rm spot}}\right]\cdot\exp\left[-j2\pi\left(f_{\rm dc}z+\frac{1}{2}k_{\rm a}z^2\right)\right]\cdot\exp\left[j\pi k_{{\rm a},\tilde{r}}(t-z)^2\right]{\rm d}z\\&=j\pi k_{{\rm a},\tilde{r}}t^2\int\text{rect}\left[\frac{z}{T_{\rm spot}}\right]\cdot\exp\left[-j2\pi\left(f_{\rm dc}-f_{{\rm dc},\tilde{r}}\right)z-j\pi\left(k_{\rm a}-k_{{\rm a},\tilde{r}}\right)z^2\right]\\&\quad\cdot\exp\left(-j2\pi k_{E,\tilde{r}}tz\right){\rm d}z\end{aligned}\quad(8.164)$$

For the targets at the distance $r=\tilde{r}$, ignoring complex constant, Eq. (8.164) can be simplified into

$$\bar{s}s(t;r)=\exp\left(j\pi k_{{\rm a},\tilde{r}}t^2\right)\sin c\left[\frac{2\pi}{L}\frac{T_{\rm spot}}{T_{\rm stip}}V\sin\varphi_{\tilde{r}}t\right]\quad(8.165)$$

In Eq. (8.165), L is the antenna length in azimuth direction, $T_{\rm syn}$ is the synthetic aperture time of Strip mode. It can be seen from above equation, azimuth precise focusing for the targets at the distance $r=\tilde{r}$ is realized.

For the point targets at the distance $r\neq\tilde{r}$, using the principle of stationary phase

$$\begin{aligned}\bar{s}s(t;r)=&\,\text{rect}\left[\frac{t-\frac{\tilde{r}}{V\sin^2\varphi_{\tilde{r}}}}{T_{\rm spot}|r_0-\tilde{r}|/r}\right]\exp\left[-j2\pi\frac{\left(k_{\rm a}-k_{a,\tilde{r}}\right)}{\lambda\left(r_0-\tilde{r}\frac{\sin^2\varphi}{\sin^2\varphi_{\tilde{r}}}\right)/r_0}t^2\right]\\&\cdot\exp\left[-j\pi\frac{k_{\rm a}}{\left(r_0-\tilde{r}\frac{\sin\varphi}{\sin\varphi_{\tilde{r}}}\right)/r_0}t^2\right]\end{aligned}\quad(8.166)$$

$$\begin{aligned}&T_{\rm ext}\approx\max\left[\frac{\lambda\tilde{r}}{LV\sin\varphi_{\tilde{r}}}+T_{\rm spot}\frac{\left|r-\tilde{r}\frac{\sin^2\varphi}{\sin^2\varphi_{\tilde{r}}}\right|}{r}\right]\leq\frac{\lambda r_{\rm M}}{LV\sin\varphi_{\rm M}}+T_{\rm spot}\frac{\left|r_M\frac{\sin^2\varphi_{\rm m}}{\sin^2\varphi_{\rm M}}-r_{\rm m}\right|}{r_{\rm m}}\\&T_{\rm stip}\leq T_{\rm ext}\leq T_{\rm spot}\end{aligned}\quad(8.167)$$

In Eq. (8.167), $T_{\rm ext}$ is synthetic aperture time after azimuth Deramp. So we can apply the above conclusion for point targets to the entire field of view of spotlight synthetic aperture radar. It ensures the convolution operations in Eq. (8.164) can be looked as block azimuth compression, so the synthetic aperture time T_{spot} is turned into T_{ext}, and the Doppler bandwidth $k_{\text{a}}T_{\text{spot}}$ will decrease to $k_{\text{a}}T_{\text{ext}}$, which overcomes the frequency spectrum aliasing of azimuth direction.

(2) Analysis in discrete domain

Suppose $\Delta t'$ and $\Delta t''$ are respectively impulse repetition time and output sampling time interval, then $A = T_{\text{syn}}/\Delta t''$ is the number of sampling points of echo signal in azimuth direction, $B = T/\Delta t''$ is the number of output sampling points after compression in azimuth direction. The discrete form of Eq. (8.166) is:

$$\begin{aligned}\bar{s}s_{\text{a}}(n\Delta t'', r_0) &= \sum_{t=-A/2}^{A/2-1} ss_{\text{a}}(i\Delta t', r_0)\cdot S_{\text{ref}}(n\Delta t'' - i\Delta t', \tilde{r}) \\ &= \exp\left[j\pi k_{\text{a},\tilde{r}}(n\Delta t'')^2\right]\sum_{i=-A/2}^{A/2-1} ss_{\text{a}}(i\Delta t', r_0) \\ &\quad\cdot \exp\left[j\pi f_{\text{dc},\tilde{r}}(i\Delta t')^2 + j\pi k_{\text{a},\tilde{r}}(i\Delta t')^2\right]\cdot\exp\left[-jk_{\text{a},\tilde{r}}in\Delta t'\Delta t''\right] \\ &\quad n\in(-B/2, B/2-1)\end{aligned} \tag{8.168}$$

It can be seen that the last index item in Eq. (8.168) is a FFT transform nuclear form. In order to facilitate the use of FFT, by filling up zero operations in azimuth direction the following equation can be established:

$$\frac{1}{k_{\text{a},\tilde{r}}\Delta t'} = P\Delta t'' \geq T \tag{8.169}$$

So Eq. (8.168) can be turned into

$$\begin{aligned}\bar{s}s_{\text{a}}(n\Delta t'', r_0) &= \sum_{t=-A/2}^{A/2-1} ss_{\text{a}}(i\Delta t', r_0)\cdot S_{\text{ref}}(n\Delta t'' - i\Delta t', \tilde{r}) \\ &= \exp\left[j\pi k_{\text{a},\tilde{r}}(n\Delta t'')^2\right]\sum_{i=-P/2}^{P/2-1} ss_{\text{a}}(i\Delta t', r_0)\cdot\exp\left[-j\pi k_{\text{a},\tilde{r}}\frac{in}{P}\right] \\ &\quad\cdot\exp\left[j2\pi f_{\text{dc},\tilde{r}}(i\Delta t')^2 + j\pi k_{\text{a},\tilde{r}}(i\Delta t')^2\right] \\ &= \exp\left[j\pi k_{\text{a},\tilde{r}}(n\Delta t'')^2\right]\cdot\text{DFT}\left\{\sum_{t=-A/2}^{A/2-1} ss_{\text{a}}(i\Delta t', r_0)\cdot\exp\left[j2\pi k_{\text{a},\tilde{r}}(i\Delta t') + j\pi k_{\text{a},\tilde{r}}(i\Delta t')^2\right]\right\}\end{aligned} \tag{8.170}$$

In General, p is an integral power of 2. As can be seen from Eq. (8.170), azimuth Deramp processing is accomplished through two complex multiplications in azimuth direction and an FFT.

(3) Chirp Scaling focusing processing

Processed by azimuth pre-filter, point targets at reference distance are totally focused in azimuth direction, and other point targets at other distance will have a certain amount of stretching, just the coarse focus is completed. Using Chirp Scaling principle, the remaining range focus and azimuth precise focus can be completed. The algorithm is derived as following. It ignores complex constants which will not affect the entire algorithm.

Azimuth Fourier transformation to the Deramp data is

$$\begin{aligned}\bar{s}s(\tau,f_{\mathrm{a}};r_0) &= FT\left\{\left[ss(\tau,t;r_0)\cdot\exp(j2\pi f_{\mathrm{dc},\tilde{r}}t)\right]\otimes s_{\mathrm{ref}}(t;\tilde{r})\right\}\\ &= FT\left\{\left[ss(\tau,t;r_0)\cdot\exp(j2\pi f_{\mathrm{dc},\tilde{r}}t)\right]\cdot\exp\left[-j\pi\left(f_{\mathrm{a}}^2/k_{\mathrm{a},\tilde{r}}\right)\right]\right.\\ &= sS(\tau,f_{\mathrm{a}};r_0)\cdot\exp\left[-j\pi f_{\mathrm{a}}^2/k_{\mathrm{a},\tilde{r}}\right)]\end{aligned} \tag{8.171}$$

In Eq. (8.171), FT[] is Fourier transform, $sS(\tau,f_{\mathrm{a}};r)$ is Fourier transform of $ss(\tau,t;r)$, f_{a} is azimuth frequency.

According to the principle of stationary phase, Fourier transform of $ss(\tau,t;r_0)\cdot\exp(j2\pi f_{\mathrm{dc},\tilde{r}}t)$ is

$$\begin{aligned}sS(\tau,f_{\mathrm{a}};r) &= \sigma\,\mathrm{rect}\left[\frac{1}{T_{\mathrm{spot}}}\frac{\lambda f_{\mathrm{a}} r\sin\varphi}{2V^2\sqrt{1-\left(\frac{\lambda f_{\mathrm{a}}}{2V}\right)^2}}\right]\mathrm{rect}\left[\tau-\frac{2}{c}r_{\mathrm{f}}(f_{\mathrm{a}};r)\right]\\ &\quad\cdot\exp\left[j\pi k_{\mathrm{s}}(f_{\mathrm{a}};r_0)\left(\tau-\frac{2r_f(f_{\mathrm{a}};r)}{c}\right)^2\right]\\ &\quad\cdot\exp\left\{-j\frac{4\pi r_0\sin\varphi}{\lambda}\sqrt{1-\left(\frac{\lambda f_{\mathrm{a}}}{2V}\right)^2}\right\}\cdot\exp\left(-j\frac{2\pi r_0 f}{V}\cos\varphi\right)\end{aligned} \tag{8.172}$$

In Eq. (8.172),

$$r_{\mathrm{f}}(f_{\mathrm{a}};r) = \frac{r\sin\varphi}{\sqrt{1-\left(\frac{f_{\mathrm{a}}\cdot\lambda}{2v}\right)^2}} \tag{8.173}$$

$$\frac{1}{k_s(f_a;r)} = \frac{1}{k_r} - \frac{2\lambda r_b \sin\varphi\left(\beta^2(f_a)-1\right)}{c^2\beta^3(f_a)} \tag{8.174}$$

$$\beta(f_a) = \sqrt{1-\left(\frac{f_a\cdot\lambda}{2v}\right)^2} \tag{8.175}$$

Multiplied by the Chirp Scaling factor in the range-Doppler domain

$$\Phi_1(\tau,f;r_{\text{ref}}) = \exp\left\{j\pi k_s(f_a;r_{\text{ref}})\cdot a(f_a)[\tau-\tau_{\text{ref}}(f)]^2\right\} \tag{8.176}$$

$$a(f_a) = \frac{\sin\varphi_{\text{ref}}}{\beta(f_a)} - 1 \tag{8.177}$$

$$\tau_{\text{ref}} = \frac{2}{c}\tilde{r}[1+a(f_a)] \tag{8.178}$$

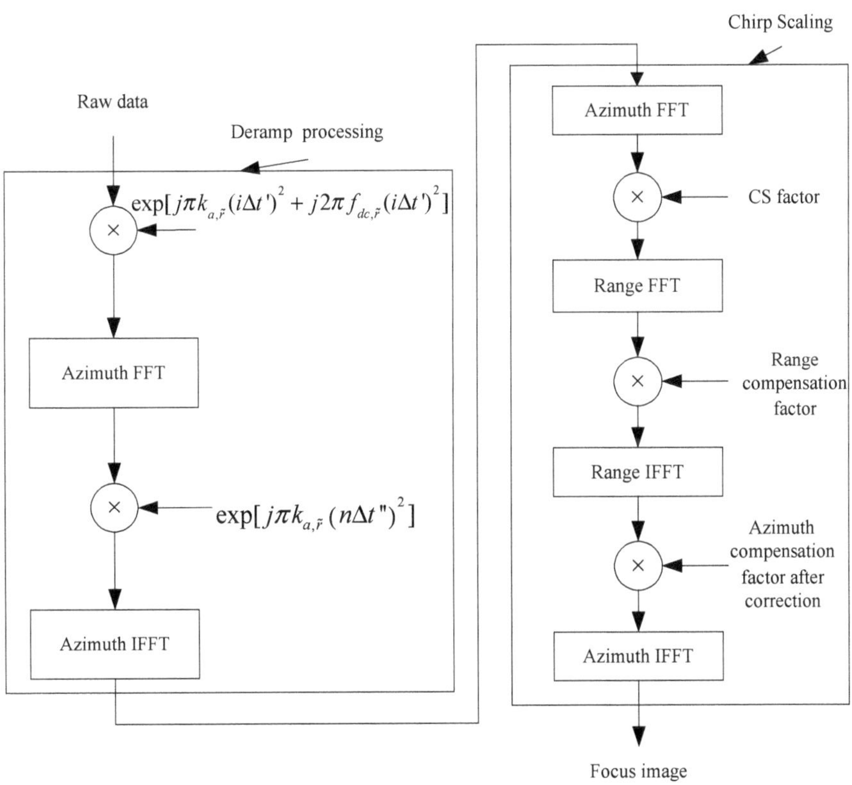

Fig. 8.39 Azimuth Deramp+Chirp Scaling (squint mode) algorithm flow chart

After CS factor multiplying, range migration curves of different distance lines has the same form with that of the reference distance. Doppler parameters variation along with the slant range has less influence to range focusing, so it can be used the same reference slant range for the entire swath to focus. According to squint equivalent model to modify CS factors, the aliasing of the range spectrum is avoided.

After multiplying CS factor, the signal is turned into wave-number domain by range Fourier transforms, then multiplies range compensation factor $\Phi_2(f_r, f_a; \tilde{r})$ to complete range migration correction and range focusing.

$$\Phi_2(f_r, f_a) = \exp\left\{j\pi \frac{f_r^2}{k_s(f_a; \tilde{r}) \cdot [1 + a(f_a)]} + j\frac{4\pi}{c} f_r \tilde{r} a(f_a)\right\} \tag{8.179}$$

In Eq. (8.179), f_r is range frequency, the first item completes quadric range compression and range focusing, and the second item completes range migration correction. Range inverse Fourier transform to Eq. (8.179), and multiplying azimuth compensation factor $\Phi_3(\tau, f_a)$ in range-Doppler domain.

$$\Phi_3(\tau, f_a) = \exp\left[j\frac{2\pi}{\lambda} \cdot c\tau \cdot (\sin\varphi_{\tilde{r}}\beta(f_a) - 1) + j\frac{\pi}{k_{a,\tilde{r}}} f_a^2\right] \cdot \exp[j\theta_1(f_a) + j\theta_2(f_a; \tilde{r})] \tag{8.180}$$

In Eq. (8.180)

$$\theta_1(f_a; r_0) = \frac{4\pi}{c^2} k_s(f_a; r_0) a(f_a)[1 + a(f_a)] \left(r\frac{\sin\varphi}{\sin\varphi_{\tilde{r}}} - \tilde{r}\right)^2 \tag{8.181}$$

$$\theta_2(f_a; r_0) = \frac{2\pi r_0 f_a}{V} \cos\varphi \tag{8.182}$$

Due to range migration is completely corrected, so $r = \frac{c\tau}{2}\frac{\sin\varphi_{\tilde{r}}}{\sin\varphi}$, the first index item of $\Phi_3(\tau, f_a)$ compensates the residual azimuth errors after azimuth pre-filter and completes azimuth precise focusing; the second index item compensates the residual phase which is failed to compensate in CS phase multiplication. The azimuth inverse Fourier transform completes the azimuth compression, and obtains synthetic aperture radar image.

(4) Algorithm flow diagram

Based on the analysis above, the azimuth Deramp+Chirp Scaling (squint mode) algorithm flow as shown in Fig. 8.39.

In this chapter, the several commonly used imaging algorithm theories, algorithmic processes and their respective advantages and disadvantages of the spotlight synthetic aperture radar were introduced. Among these algorithms, the PFA and the RMA need to be interpolated, and interpolation can be avoided in CSA, ECSA,

FSA, as well as two step processing algorithm. These algorithms take advantage of spotlight synthetic aperture radar signal characteristics adequately, and can achieve high resolution imaging. In practical application, it should compromise performance and operation capacity, and select the appropriate algorithm.

References

1. Carrara WG, Goodman RS, Majewski RM (1995) Spotlight synthetic aperture radar signal processing algorithms. Aretch House, Norwood
2. Jakowatz CV et al (1996) Spotlight-mode synthetic aperture radar: a signal processing approach. Kluwer Academic Publishers, Boston
3. Wen H (1997) Focus processing study of airborne spotlight synthetic aperture radar imaging. Doctor Thesis, Beijing University of Aviation and Astrospace
4. Sun JP (2001) Imaging algorithms study of airborne spotlight synthetic aperture radar. Doctor thesis, Beijing University of Aviation and Astrospace
5. Rocca F, Cafforio C, Prati C (1989) Synthetic aperture radar: A new application for wave equation techniques. Geophys Prospect 7(37):809–830
6. Cafforio C, Prati C, Rocca F (1991) SAR data focusing using seismic migration techniques. IEEE Trans Aerosp Electron Syst 27(2):194–206
7. Prati C, Guarnieri AM, Rocca F (1991) Spot mode SAR focusing with the ω-K technique. In: Proceedings of the 1991 IEEE international geoscience and remote sensing symposium, Espoo, Finland
8. Moreira A, Huang Y (1994) Airborne SAR processing of highly squinted data using a chirp scaling approach with integrated motion compensation. IEEE Trans Geosci Remote Sens 32 (5):1029–1040
9. Raney RK et al (1994) Precision SAR processing using Chirp Scaling. IEEE Trans Geosci Remote Sens 32(4):786–799
10. Runge H, Bamler R (1992) A novel high precision SAR focusing algorithm based on chirp scaling. In: Proceedings of the 1992 IEEE international geoscience and remote sensing symposium
11. Moreira A, Mittermayer J, Scheiber R (1996) Extended chirp scaling algorithm for air- and space borne SAR data processing in strip map and scan SAR imaging modes. IEEE Trans Geosci Remote Sens 34(5):1123–1136
12. Mittermayer J, Moreira A (1997) Spotlight SAR processing using the extended chirp scaling algorithm. In: Proceedings of IGARSS '97
13. Moreira A, Mittermayer J, Scheiber R (2000) Extended chirp scaling SAR data processing in Stripmap, ScanSAR and Spotlight imaging mode. In: EUSAR 2000, Munich, Germany
14. Mittermayer J, Moreira A, Loffeld O (1999) Spotlight SAR data processing using the frequency scaling algorithm. IEEE Trans Geosci Remote Sens 37(5):2198–2214
15. Mittermayer J, Moreira A (1998) Spotlight processing of wide-beam strip map data using the frequency scaling algorithm. In: Proceedings of IGARSS '98
16. Mittermayer J, Moreira A, Loffeld O (1998) The frequency scaling algorithm for spotlight processing. In: IRS '98
17. Fornaro G et al (2000) A two-step spotlight SAR data focusing approach. In: Proceedings of IGARSS '00
18. Lanari R et al (2000) Spotlight SAR data focusing based on a two-step processing approach. IEEE Trans Geosci Remote Sens 39(9):1993–2004

Chapter 9
Sliding Spotlight Synthetic Aperture Radar

Sliding Spotlight synthetic aperture radar is an operating mode between Strip synthetic aperture radar and spotlight synthetic aperture radar.

In Strip SAR, radar antenna beam is perpendicular to the flight path of the platform, pointing to a fixed direction. As the platform movement, antenna footprint makes the strip ground imaging. In theory, there is no limit extending along the azimuth direction, but the antenna length limits the azimuth resolution.

In order to improve azimuth resolution, Spotlight SAR antenna beam is controlled to the same point on the ground throughout the data collection period to increase the effective azimuth length. However, this resolution improvement is at the cost of losing the length of azimuth irradiation.

The main feature of Sliding Spotlight mode arising recently is to control the beam pointing a point away from the scene center. This has the potential to azimuth images wider than Spotlight mode, and azimuth resolution better than Strip mode. If Sliding Spotlight beam pointing is controlled to the ground center, this is pure Spotlight. If the beam pointing is controlled at infinity, then it is the Strip mode. Strip mode and pure Spotlight can be thought of as two special cases of Sliding Spotlight.

Sliding Spotlight mode has been successfully used in Germany synthetic aperture radar satellite TerraSAR-X. This chapter includes imaging geometry model, target irradiation time and effective irradiated area, ECHO signal characteristics, resolution, ECHO Doppler process, and imaging algorithms of Sliding Spotlight SAR.

9.1 Sliding Spotlight Mode [1–4]

The diagram of Sliding Spotlight SAR imaging is shown in Fig. 9.1, and the diagram of Spotlight SAR imaging and Strip SAR imaging is shown in Fig. 9.2. The antenna beam of Spotlight SAR points to the center of the imaging scene. The

R. Yang et al., *High-Resolution Microwave Imaging*,
https://doi.org/10.1007/978-981-10-7138-6_9

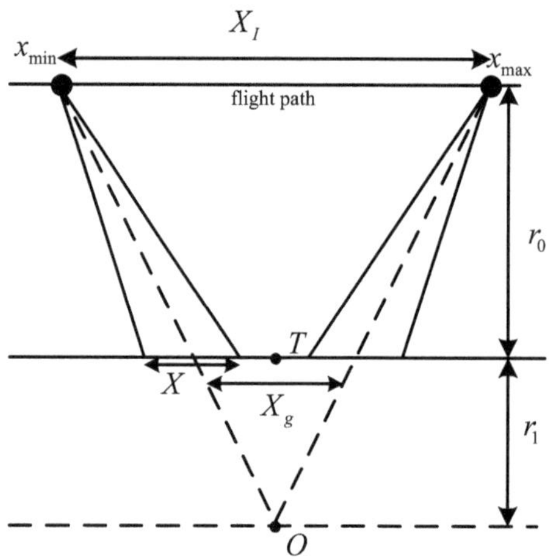

Fig. 9.1 Sliding spotlight SAR imaging diagram

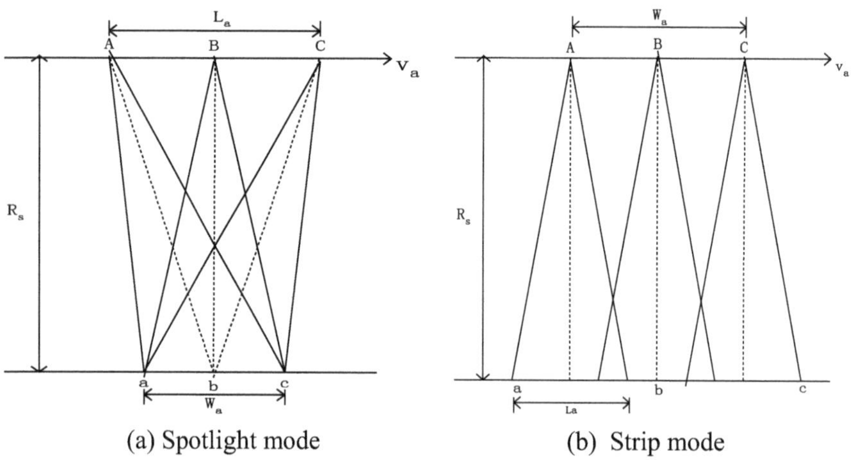

Fig. 9.2 Spotlight SAR imaging and strip SAR imaging diagram

antenna beam of Sliding Spotlight SAR points to the focus beyond the center of the imaging scene. Strip mode SAR beam points to the point at infinity.

The schematic diagram of Sliding Spotlight SAR imaging is shown in Fig. 9.1, where X_{I} is the track of spacecraft, X is the length of antenna irradiation, and X_{g} is the length of ground irradiation swath.

In Sliding Spotlight mode, Sliding factor A is defined as

$$A = \frac{r_1}{r_1 + r_0} = \frac{v_{\mathrm{f}}}{v_{\mathrm{s}}} \tag{9.1}$$

In Eq. (9.1), v_s is the speed of platform, v_f is antenna beam movement speed on the ground, r_1 is the shortest distance between the scene and focus point, and $r_0 + r_1$ is the shortest distance between the radar and the focal point.

Sliding Spotlight factor determines the characteristics of sliding Spotlight SAR:

When A = 0, it is pure Spotlight mode. In this case, $r_1 = 0$, antenna beam points to the center of the surface imaging scene, and its movement speed on the ground is 0. Azimuth signal bandwidth is the maximum and optimal azimuth resolution can be obtained, but the azimuth image width is strictly limited.

When 0 < A < 1, it is Sliding Spotlight mode. The focus of antenna beam center is far to the imaging center of the scene. The antenna beam movement speed v_f on the ground is less than platform speed v_s, and the direction of the movement speed of antenna beam on the ground and the direction of Platform movement are the same Azimuth imaging width is greater than pure Spotlight mode and less than Strip mode. Azimuth resolution is inferior to pure Spotlight mode and is better than Strip mode.

When A = 1, it is Strip mode. In the Focused imaging process, the directivity of the antenna beam center focuses at infinity. The movement speed of antenna beam on the ground is the speed of the platform, and azimuth resolution is half of the antenna length in azimuth direction.

Sliding Spotlight must make a trade-off between the azimuth resolution and the azimuth image width.

9.2 Sliding Spotlight SAR Azimuth Resolution and Imaging Width

Sliding Spotlight is an operating mode between Strip mode and Spotlight mode. It has better azimuth resolution than Strip mode and longer imaging width in azimuth than Spotlight mode. In this section, the irradiation time of point target, effective irradiation area, azimuth width of surveying and mapping, and the azimuth resolution are analyzed.

Figure 9.3 shows beam direction control schematic diagram of sliding Spotlight SAR.

In Fig. 9.3, θ_a is the antenna 3 dB beamwidth in azimuth direction, $\Psi_{s,o}$ is the antenna beam squint angle of aperture start point for a point target, $\Psi_{e,o}$ is the antenna beam squint angle of aperture endpoint for a point target. $T_{a,s,o}$ is the time of aperture start point for a point target, $T_{a,e,o}$ is the time of aperture endpoint for a point target. v_s is the speed of platform, v_f is antenna beam movement speed on the ground. The range from antenna to target is r_0, and the range from antenna to the spotlight beam center focus point is r_{rot}. Beam irradiation time of target is related to its azimuth and range location ($t_{a,o}$, r_o) and the azimuth and range location ($t_a = 0$, r_{rot}) of rotation point [5, 6].

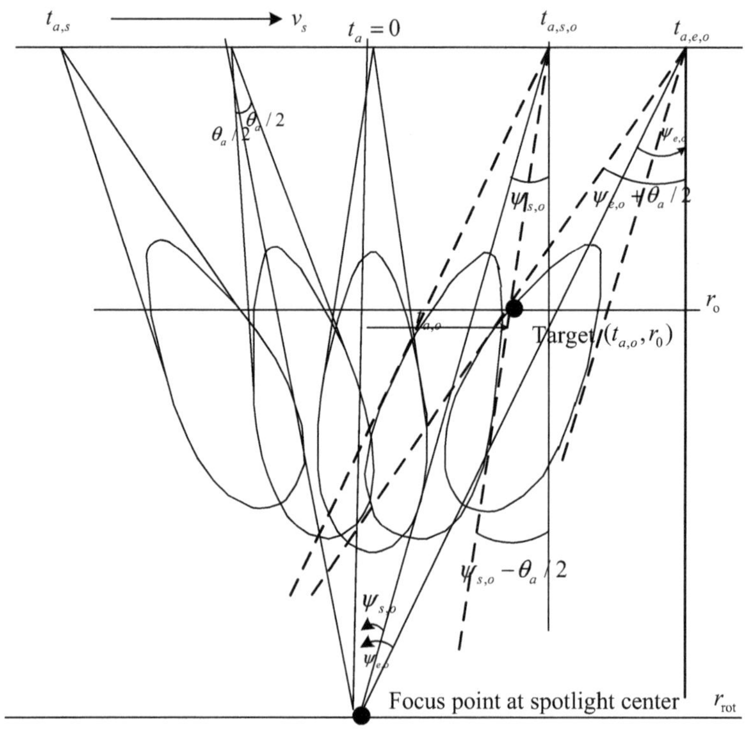

Fig. 9.3 Beam direction control schematic diagram of sliding spotlight SAR

9.2.1 Irradiation Time of Point Target

Beam irradiation time of target is related to its azimuth and range location $(t_{a,o}, r_o)$. From Fig. 9.3,

$$\begin{aligned} \frac{v_s \cdot t_{a,e,o}}{r_{rot}} &= \tan(\Psi_{e,o}) \\ \frac{v_s \cdot (t_{a,e,o} - t_{a,o})}{r_o} &= \tan\left(\Psi_{e,o} + \frac{\theta_a}{2}\right) \end{aligned} \tag{9.2}$$

$$\begin{aligned} \frac{v_s \cdot t_{a,s,o}}{r_{rot}} &= \tan(\psi_{s,o}) \\ \frac{v_s \cdot (t_{a,s,o} - t_{a,o})}{r_o} &= \tan\left(\psi_{s,o} - \frac{\theta_a}{2}\right) \end{aligned} \tag{9.3}$$

Derivation is

$$\tan(\psi_{e,o}) = \frac{v_s \cdot t_{a,o}}{2r_{rot}} + \frac{1 - r_o/r_{rot}}{2\tan(\theta_a/2)} (+)/- \sqrt{\left(\frac{v_s \cdot t_{a,o}}{2r_{rot}} + \frac{1 - r_o/r_{rot}}{2\tan(\theta_a/2)}\right)^2 - \frac{r_o}{r_{rot}} - \frac{v_s \cdot t_{a,o}}{r_{rot} \cdot \tan(\theta_a/2)}} \tag{9.4}$$

$$\tan(\psi_{s,o}) = \frac{v_s \cdot t_{a,o}}{2r_{rot}} + \frac{r_o/r_{rot} - 1}{2\tan(\theta_a/2)} (+)/- \sqrt{\left(\frac{v_s \cdot t_{a,o}}{2r_{rot}} + \frac{r_o/r_{rot} - 1}{2\tan(\theta_a/2)}\right)^2 - \frac{r_o}{r_{rot}} - \frac{v_s \cdot t_{a,o}}{r_{rot} \cdot \tan(\theta_a/2)}} \tag{9.5}$$

In Sliding Spotlight mode, the target irradiated time is

$$T_s(t_{a,o}; r_o) = t_{a,e,o} - t_{a,s,o} = \frac{r_{rot}}{v_s}\left(\tan(\psi_{e,o}) - \tan(\psi_{s,o})\right) \tag{9.6}$$

9.2.2 *Effective Irradiation Area*

Effective target area and irradiation area are shown in Fig. 9.4. According to characteristics of beam irradiation, as can be seen in the dark shaded area shown in Fig. 9.4 is the effective target area, light shaded areas for irradiation target area.

At slant distance r_0, the minimum and maximum effective target azimuth locations $t_{a,o,valid,min}$ and $t_{a,o,valid,max}$ are

$$t_{a,o,valid,min} = \frac{r_{rot}}{v_s}\tan(\psi_s) - \frac{r_o}{v_s}\tan(\psi_s - \theta_a/2) \tag{9.7}$$

$$t_{a,o,valid,max} = \frac{r_{rot}}{v_s}\tan(\psi_e) - \frac{r_o}{v_s}\tan(\psi_e + \theta_a/2) \tag{9.8}$$

From $t_{a,o,valid,min}$ to $t_{a,o,valid,max}$, at the slant range r_o, Sliding Spotlight scene size X_{scene} is

$$X_{scene} = r_{rot}[\tan(\psi_e) - \tan(\psi_s)] - r_o \cdot [\tan(\psi_e + \theta_a/2) - \tan(\psi_s - \theta_a/2)] \tag{9.9}$$

Compared with traditional Spotlight mode, X_{scene} of far distance is smaller and X_{scene} of near distance is bigger under the Sliding Spotlight mode.

At the slant distance r_o, the minimum and maximum azimuth irradiation location $t_{a,o,illu,max}$ and $t_{a,o,illu,min}$ are

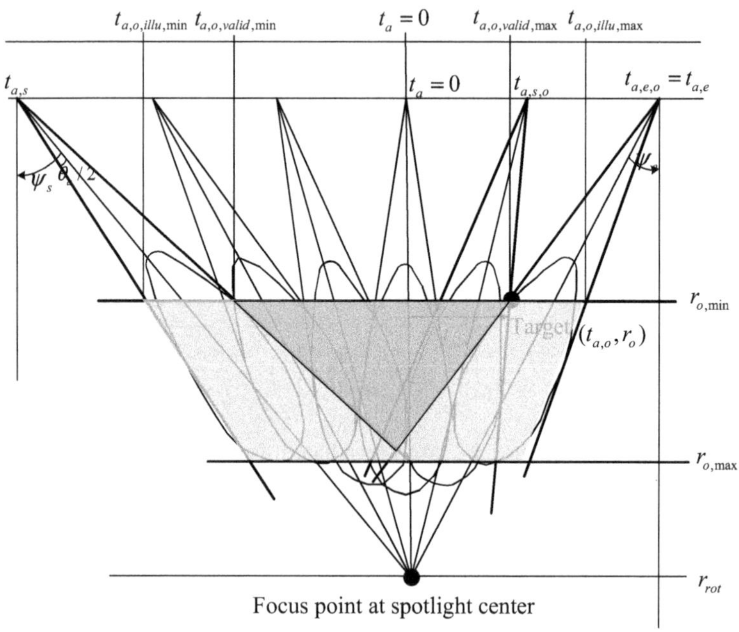

Fig. 9.4 Effective target area and irradiation area

$$t_{\text{a,o,illu,min}} = \frac{r_{\text{rot}}}{v_{\text{s}}}\tan\left(\psi_{\text{s}}\right) - \frac{r_{\text{o}}}{v_{\text{s}}}\tan\left(\psi_{\text{s}} + \theta_{\text{a}}/2\right) \tag{9.10}$$

$$t_{\text{a,o,illu,max}} = \frac{r_{\text{rot}}}{v_{\text{s}}}\tan\left(\psi_{\text{e}}\right) - \frac{r_{\text{o}}}{v_{\text{s}}}\tan\left(\psi_{\text{e}} - \theta_{\text{a}}/2\right) \tag{9.11}$$

9.2.3 Azimuth Resolution

Azimuth resolution can be decided by a single target irradiation length, that are $t_{\text{a,s,o}}$ and $t_{\text{a,e,o}}$. For the target at the location $(t_{\text{a,o}}; r_{\text{o}})$, its start and end Doppler frequency $f_{\text{a,s,o}}$ and $f_{\text{a,e,o}}$ are

$$\begin{aligned} f_{\text{a,s,o}} &= -\frac{2v_{\text{s}}}{\lambda}\cdot\sin(\psi_{\text{s}} - \theta_{\text{a}}/2) = -\frac{2v_{\text{s}}}{\lambda}\cdot\frac{\tan\left(\psi_{\text{s}} - \theta_{\text{a}}/2\right)}{\sqrt{1+\tan^2(\psi_{\text{s}} - \theta_{\text{a}}/2)}} \\ &= -\frac{2v_{\text{s}}^2}{\lambda}\cdot\frac{t_{\text{a,s,o}} - t_{\text{a,o}}}{\sqrt{r_{\text{o}}^2 + v_{\text{s}}^2\left(t_{\text{a,s,o}} - t_{\text{a,o}}\right)^2}} \end{aligned} \tag{9.12}$$

$$f_{\mathrm{a,s,o}} = -\frac{2v_{\mathrm{s}}^2}{\lambda} \cdot \frac{t_{\mathrm{a,e,o}} - t_{\mathrm{a,o}}}{\sqrt{r_{\mathrm{o}}^2 + v_{\mathrm{s}}^2 (t_{\mathrm{a,e,o}} - t_{\mathrm{a,o}})^2}} \tag{9.13}$$

From Eqs. (9.12) and (9.13), azimuth resolution can be deduced as below.

$$\rho_{\mathrm{az}}(t_{\mathrm{a,o}}; r_{\mathrm{o}}) = \frac{\lambda}{2v_{\mathrm{s}}} \left(\frac{t_{\mathrm{a,e,o}} - t_{\mathrm{a,o}}}{\sqrt{r_{\mathrm{o}}^2 + v_{\mathrm{s}}^2 (t_{\mathrm{a,e,o}} - t_{\mathrm{a,o}})^2}} - \frac{t_{\mathrm{a,s,o}} - t_{\mathrm{a,o}}}{\sqrt{r_{\mathrm{o}}^2 + v_{\mathrm{s}}^2 (t_{\mathrm{a,s,o}} - t_{\mathrm{a,o}})^2}} \right)^{-1} \tag{9.14}$$

From Eq. (9.14), azimuth resolution is related to the target position, and azimuth resolution gets worse with increase of the slant range.

9.3 The Model of Sliding Spotlight SAR Echo Signal

9.3.1 The Echo Signal of Sliding Spotlight SAR

Sliding Spotlight SAR imaging geometry is shown in Fig. 9.1.

In the analysis, the slant range plane imaging in cylindrical coordinates is only considered as shown in Fig. 9.5.

Suppose the frequency modulation rate of transmitting LFM is K_{r}, the wavelength of the carrier is λ, the pulse duration is T_{p}, azimuth time is t_{a}, range time is τ, the speed of light is c, the speed of platform is v_{s}. The echo of point target (t_0, r_{c}) is

$$\begin{aligned} S(t_{\mathrm{a}}, \tau; t_0, r_{\mathrm{c}}) = & \exp\left[jK_{\mathrm{r}}\pi \left(\tau - \frac{2R(t_{\mathrm{a}}; r_{\mathrm{c}})}{c} \right)^2 \right] \exp\left(-j4\pi \frac{R(t_{\mathrm{a}}; r_{\mathrm{c}})}{\lambda} \right) \\ & \cdot \operatorname{rect}\left[\frac{\tau - \frac{2R(t_{\mathrm{a}}; r_{\mathrm{c}})}{c}}{T_{\mathrm{p}}} \right] \operatorname{rect}\left[\frac{t_{\mathrm{a}} - t_0/A}{L_{\mathrm{a}}/v_{\mathrm{s}}} \right] \end{aligned} \tag{9.15}$$

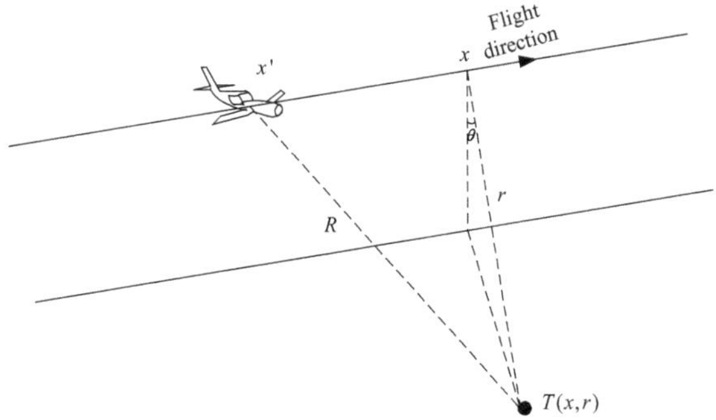

Fig. 9.5 Imaging geometry in cylindrical coordinates

In Eq. (9.15), $R(t_a; t_0, r_c) = \sqrt{r_c^2 + v_s^2(t_a - t_0)^2} \approx r_c + \Delta R$, L_a is beamwidth at the distance of R. The geometric relationships show that the effective synthetic aperture length L_s at the scene center point T of Sliding Spotlight mode is related to beamwidth L_a, and the relationship is

$$L_s = L_a \cdot \frac{v_s}{v_f} = \frac{L_a}{A} \tag{9.16}$$

Since $0 < A < 1$ in Sliding Spotlight SAR, so $L_a < L$.

By the principle of stationary phase, the two-dimensional spectrum of Sliding Spotlight synthetic aperture radar is as

$$\begin{aligned} S(f_a, f_r; t_0, r_c) = &\exp\left[-j\frac{4\pi r_c f_0}{c}\sqrt{\left(\frac{f_r}{f_c}+1\right)^2 - \left(\frac{\lambda f_a}{2v_s}\right)^2} - j\frac{\pi f_r^2}{K_r} - j2\pi t_0 f_a\right] \\ &\cdot \mathrm{rect}\left[\frac{f_r}{B_r}\right] \cdot \mathrm{rect}\left[\frac{f_a - t_0 k_r(1/A - 1)}{L_a k_a / v_s}\right] \end{aligned} \tag{9.17}$$

In Eq. (9.17), f_a is azimuth frequency, f_r is range frequency, $k_a = -\frac{2v_s^2}{\lambda r_c}$ is azimuth frequency modulation rate, B_r is transmitting signal bandwidth, f_c is carrier frequency.

9.3.2 *The Doppler Frequency Course of Sliding Spotlight SAR Echo*

Since the beam pointing is continuously variable during Sliding Spotlight SAR imaging, the instantaneous Doppler centroid of Sliding Spotlight SAR echo is a function of azimuth time t_a.

$$f_{dc} = \frac{2v_s}{\lambda}\sin\theta_s = \frac{2v_s}{\lambda}\frac{v_s t_a}{\sqrt{v_s^2 t_a^2 + r_{rot}^2}} \tag{9.18}$$

In Eq. (9.18), θ_s is squint angle.

Echo-Doppler bandwidth of SAR consists of beam bandwidth and Doppler center frequency offset. Echo-Doppler frequency course of Sliding Spotlight SAR is shown in Fig. 9.6.

$$B = B_a + B_{shift} = T_a k_{a,r} + T k_{rot} \tag{9.19}$$

In Eq. (9.19), T_a is synthetic aperture time, $k_{a,r} = -\frac{2v_a^2}{\lambda r}$ is Doppler frequency at the slant range vector r, T represents the total time of imaging, $k_{rot} = -\frac{2v_a^2}{\lambda r_{rot}}$

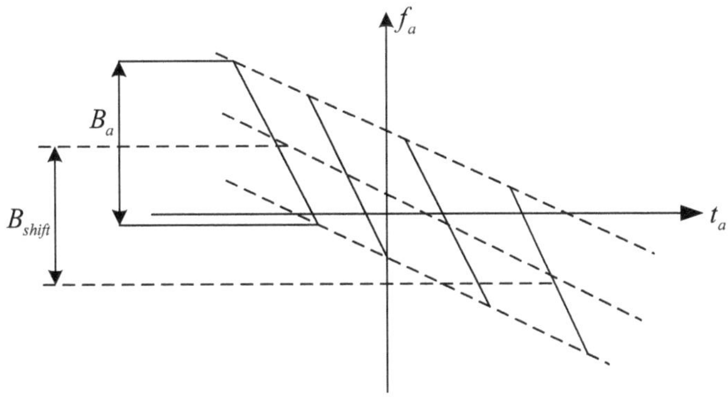

Fig. 9.6 Echo Doppler frequency course of sliding spotlight SAR

represents the change rate of Doppler centroid, which is related to the slant range of rotation center r_{rot}, v_{a} is the speed of platform.

From Eq. (9.19) can see that the linear change of Doppler centroid will cause the azimuth frequency spectrum expanded, sometimes it is far greater than the pulse repetition frequency. Therefore, azimuth frequency spectrum aliasing needs to be removed in image processing.

9.4 The PRF Selection of Sliding Spotlight SAR

According to the sampling theorem it is known that, to restore signal without distortion, the sampling frequency must be greater than the signal bandwidth. In order to avoid azimuth spectral aliasing pulse repetition frequency is required greater than echo-Doppler bandwidth. That is,

$$f_{\mathrm{PRF}} \geq B_{\mathrm{a}} \tag{9.20}$$

Spaceborne Sliding Spotlight SAR with large Doppler bandwidth requires a high pulse repetition frequency, which brings severe restricts to the other parameters design of the system. Signal instantaneous azimuth Doppler bandwidth of Sliding Spotlight SAR B_{a} is constant. Doppler center frequency is an approximate linear frequency-modulated signal. In synthetic aperture, Doppler center frequency variation range B_{d} corresponds to the Doppler bandwidth of the scene Center point target. The frequency difference between target point on the distribution of azimuth direction and center point of the scene maintains a constant and the frequency difference is directly proportional to the distance from this point to the center of the scene.

According to the Doppler frequency relationship of Sliding Spotlight SAR echo signal, it might be considered using instantaneous bandwidth as the lower limit of a pulse repetition frequency. This pulse repetition frequency doesn't satisfy the law of Nyquist sampling, echo azimuth frequency aliasing occurs. However, using azimuth downsampling processing, the aliasing azimuth signal spectrum can be recovered to the spectrum before aliasing, and Sliding Spotlight SAR echo signal can be recovered without distortion from the undersampling signal.

9.5 Antenna Beam Control of Sliding Spotlight SAR

In Strip mode, antenna pointing is constant. In Spotlight SAR, antenna scans with fixed angular velocity. The antenna scanning speed of Spotlight SAR is related to the aircraft flight speed and the distance to the swath center. Sliding Spotlight SAR controls the moving speed of antenna radiation zone on the ground by controlling the speed of antenna scanning, so the scanning speed of Sliding Spotlight SAR is not only associated with aircraft flight speed and the distance to swath Center, but also with moving speed of antenna radiation area on the ground. Figure 9.7 is the antenna scanning schematic. In Sliding Spotlight SAR, assume that the vertical distance from antenna to the center of radiation of the antenna is R, the antenna scanning speed is

$$\omega = \Delta\theta/\Delta t \approx (v_s - v_f)/R \tag{9.21}$$

The deduction of Eq. (9.21) ignores the change of R and L_s during antenna scanning.

Set scanning range of the antenna is $(-\theta_{max}, \theta_{max})$, Sliding Spotlight SAR azimuth swath width is

$$W_a = \frac{2\theta_{max}}{|\omega|} v_f + L_s \tag{9.22}$$

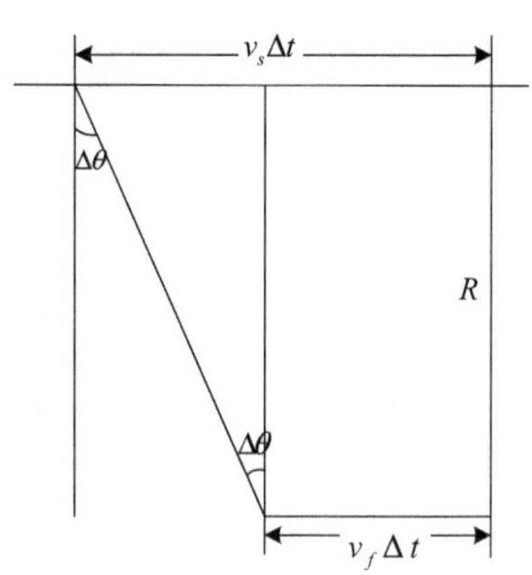

Fig. 9.7 Antenna scanning schematic

Equation (9.21) is inserted in Eq. (9.22), then

$$W_a = \frac{2\theta_{\max} \cdot R}{|v_s - v_f|} v_f + L_s \tag{9.23}$$

In a practical system, due to the digit discrete characteristic of phase shifter for the phased-array antenna and adoptive virtual technology, continuous change beam direction cannot be achieved, so the antenna scans take a step-and-scan mode that every beam pointing is revised at regular intervals of time, and beam remains unchanged during the same period.

In Sliding Spotlight SAR the step-and-scan mode of beam points has certain effects on ECHO amplitude. The beam points adjust once at intervals, and periodic saw tooth wave-like fluctuations are produced on ECHO amplitude; and the irradiation strength of point target on the ground appears changed related to parameter A like antenna pattern. This kind of azimuth echo signal amplitude modulation will affect imaging quality, and amplitude compensation for antenna pattern is necessary.

Scanning step length is the minimum value of azimuth beam angle change, and the smaller the value is, the smaller impact it has on image quality, but it is more difficult to system implementing. Usually resident pulse, which is the transmitting pulse number in the time of beam pointing remaining fixedness, is used to describe system scan step. The expression of the resident pulse

$$N \approx \Delta\theta \frac{r_0 \cdot f_{\text{PRF}}}{v} \tag{9.24}$$

In Eq. (9.24), $\Delta\theta$ expresses scan step.

9.5.1 Antenna Beam Control of Spaceborne Spotlight SAR

In Spotlight mode SAR, in order to ensure the continuous observation of target in designated areas, it is needed to implement the beam pointing real-time control in azimuth direction and to know the beam pointing angle at every moment. Using simplified models, get the geometry model shown in Fig. 9.8.

In Fig. 9.8, establish the time coordinate system. When t = 0, the satellite is located at point P, and the positive axis of time is along satellite speed v_s direction. At time t_0, satellite moves to point S, and instantaneous beam pointing angle is θ. Beam pointing control is to calculate real-time angle θ. Then, beam control system adjusts the direction of antenna beam center and makes the angle between beam direction and radar motion speed direction equal to θ.

Instantaneous slant range calculation formula is:

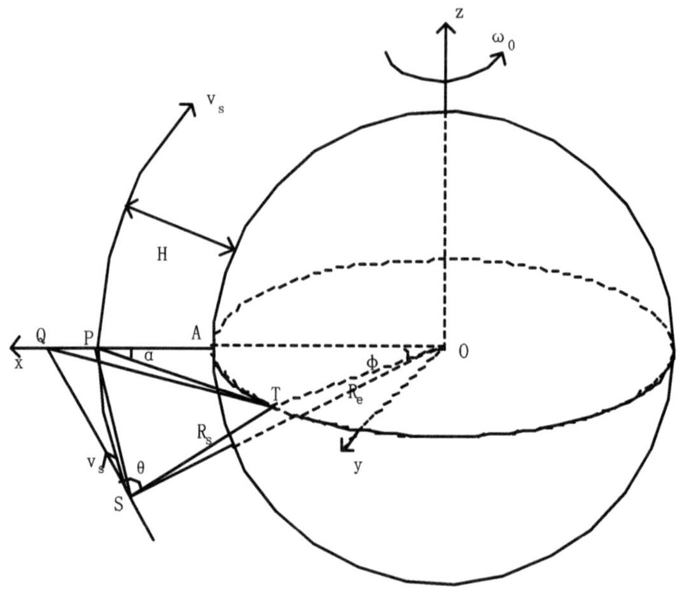

Fig. 9.8 Spaceborne spotlight simplified geometrical model

$$R_{\mathrm{ST}}(t) = \sqrt{R_e^2 + (R_e + H)^2 - 2R_e(R_e + H)\cos(\Omega t)\cos(\omega_0 t + \phi)} \tag{9.25}$$

Within a synthetic aperture time, Earth's rotation angle is too small, this will not be considered so that Eq. (9.25) is approximately

$$R_{\mathrm{ST}}(t) = \sqrt{R_e^2 + (R_e + H)^2 - 2R_e(R_e + H)\cos(\phi)\cos(\Omega t)} \tag{9.26}$$

Based on triangle geometry in Fig. 9.8, we can get

$$|QT|^2 = |QS|^2 + |ST|^2 - 2|QS||ST|\cos\theta \tag{9.27}$$

$$|QT|^2 = |QO|^2 + |OT|^2 - 2|QO||OT|\cos\phi \tag{9.28}$$

In Eqs. (9.27) and (9.28),

$$|QS| = |SO|\tan(\Omega t_0) = (R_e + H)\tan(\Omega t_0) \tag{9.29}$$

$$|ST| = \sqrt{R_e^2 + (R_e + H)^2 - 2R_e(R_e + H)\cos(\phi)\cos(\Omega t_0)} \tag{9.30}$$

$$|QO| = \frac{|SO|}{\cos(\Omega t_0)} = \frac{(R_e + H)}{\cos(\Omega t_0)} \tag{9.31}$$

$$|OT| = R_e \tag{9.32}$$

Here, Ω is the angular velocity of the satellites movement surrounding the Earth. By from Eq. (9.27) to Eq. (9.32), Eq. (9.33) can be obtained.

$$\cos\theta = \frac{R_e \cos\phi \, \sin(\Omega t_0)}{\sqrt{R_e^2 + (R_e + H)^2 - 2R_e(R_e + H)\cos(\phi)\cos(\Omega t_0)}} \tag{9.33}$$

Provided that the orbit parameters, angle of view, and aperture location time are known, the antenna beam pointing angle can be calculated. Figure 9.9 shows antenna beam pointing angles corresponding to different orbit altitudes and different aperture times when side-looking angle is π/4.

9.5.2 *Antenna Beam Control of Sliding Spotlight SAR*

Figure 9.10 is Sliding Spotlight SAR beam pointing control schematic. Beam Center no longer points to the center T-point of the scene, but to the far-off W-point on the extension line of PT.

Assuming Spotlight factor of Sliding Spotlight SAR is A, then

$$|TW| = |PT|\left(\frac{1}{A} - 1\right) \tag{9.34}$$

In Eq. (9.34),

$$|PT| = \sqrt{R_e^2 + (R_e + H)^2 - 2R_e(R_e + H)\cos(\phi)} \tag{9.35}$$

Fig. 9.9 Beam pointing angle of different orbit altitude and different times

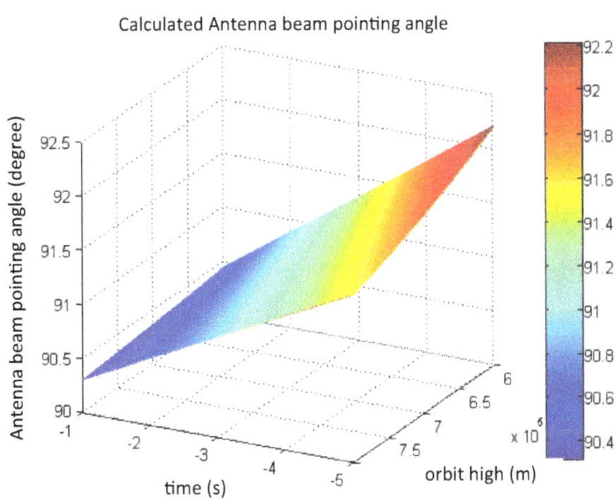

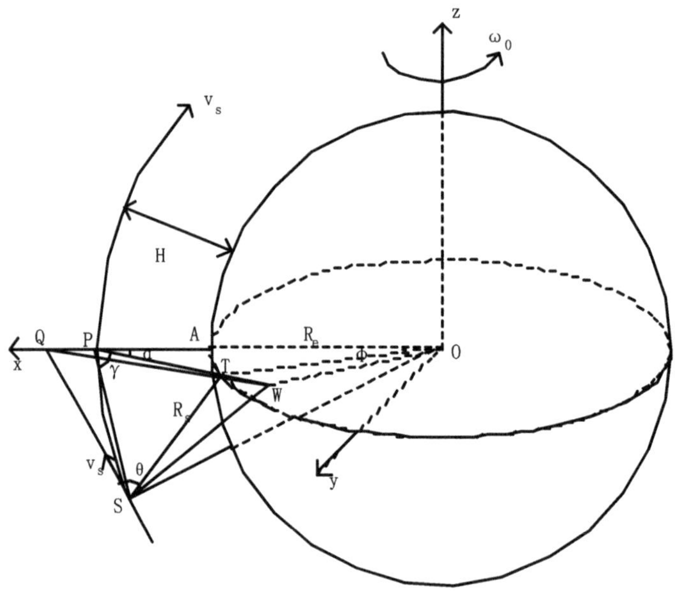

Fig. 9.10 Schematic diagram of sliding spotlight SAR beam pointing

Then,

$$|PW| = \frac{|PT|}{A} = \frac{\sqrt{R_e^2 + (R_e + H)^2 - 2R_e(R_e + H)\cos\varphi}}{A} \tag{9.36}$$

By geometric relationship in the diagram can be drawn

$$|QW|^2 = |QS|^2 + |SW|^2 - 2|QS||SW|\cos\theta \tag{9.37}$$

$$|QW|^2 = |QP|^2 + |PW|^2 + 2|QP||PW|\cos\alpha \tag{9.38}$$

$$|SW|^2 = |PS|^2 + |PW|^2 - 2|PS||PW|\cos\gamma \tag{9.39}$$

In Eq. (9.37), Eq. (9.38) and Eq. (9.39),

$$|QS| = |SO|\tan(\Omega t_0) = (R_e + H)\tan(\Omega t_0) \tag{9.40}$$

$$|PS| = 2(R_e + H)\sin\left(\frac{\Omega t_0}{2}\right) \tag{9.41}$$

$$|QP| = (R_e + H)\left[\frac{1}{\cos(\Omega t_0)} - 1\right] \tag{9.42}$$

By triangular relationships,

$$\cos\gamma = \cos\alpha \cdot \cos\beta = \cos\alpha \cdot \sin\left(\frac{\Omega t_0}{2}\right) \tag{9.43}$$

By from Eq. (9.37) to Eq. (9.43), the antenna beam pointing angle of sliding Spotlight SAR can be calculated.

Similarly, as long as we know the orbit parameters, the angle of view, aperture location time, as well as Spotlight factor, the antenna beam pointing angle can be calculated. Figure 9.11 shows antenna beam pointing angles corresponding to different orbit altitudes and different aperture times when the side-looking angle is $\alpha = \pi/4$, and Spotlight factor is $\eta = 0.5$. Figure 9.12 shows antenna beam pointing angles corresponding to different Spotlight factors and different aperture times when the orbital altitude is $H = 600$ km, and the side-looking angle is $\alpha = \pi/4$.

As can be seen, along with the Spotlight factor A changing from 1 to 0, that is, from the Spotlight mode changing to Strip mode passing Sliding Spotlight mode, the change range of antenna beam pointing gradually decreases in synthetic aperture time.

9.6 The Imaging Algorithm of Sliding Spotlight SAR

The imaging geometry model and echo-Doppler frequency characteristics of Sliding Spotlight SAR have been analyzed as mentioned earlier, azimuth beam of Sliding Spotlight SAR continuously changes during the imaging process, and the Doppler centroid also changes during the imaging process. The azimuth bandwidth of Sliding Spotlight SAR is no longer equal to beam bandwidth. It increases with the imaging time, which will cause severe azimuth spectral aliasing and

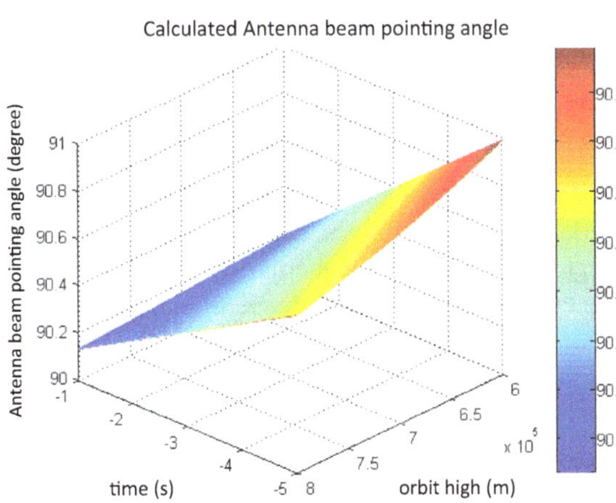

Fig. 9.11 Different orbit spotlight SAR antenna beam pointing angle ($\eta = 0.5$)

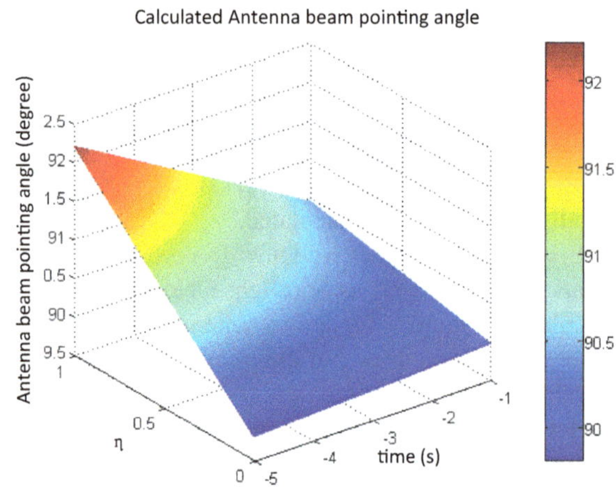

Fig. 9.12 Sliding spotlight SAR antenna beam pointing angle

range-azimuth two-dimensional coupling. The main difficulty of Sliding Spotlight SAR imaging processing is to resolve azimuth spectral aliasing and range-azimuth two-dimensional coupling, and so on.

9.6.1 Extended Chirp Scaling (ECS) Algorithm

In 2003, Josef Mittermayer and others applied Extended Chirp Scaling algorithm, which was initially applied to Strip mode and Scan mode SAR. To Sliding Spotlight SAR imaging processing, this algorithm reduces the demand for pulse repetition frequency by sub-aperture partition, and uses Chirp Scaling principle for accurate range migration correction of different slant range units, and uses azimuth scale transformation spectrum analysis (SPECAN) method to eliminate slant range pertinence of frequency sampling intervals, and to implement azimuth precise focusing. Figure 9.13 shows the ECS algorithm flowchart [7, 8].

(1) Sub-aperture partition

Sub-aperture partition divides azimuth data into some blocks. To a target, it is equivalent to divide the full synthetic aperture into sub-apertures, and ensure sub-aperture bandwidth B_{sub} is below pulse repetition frequency. A length of overlap between adjacent sub-apertures is required in order to ensure smooth integration between sub-apertures. Overlap rate depends on pulse repetition frequency, and greater the pulse repetition frequency is, higher the overlap rate should be.

Sliding Spotlight SAR echo instantaneous Doppler bandwidth boundaries are as follows:

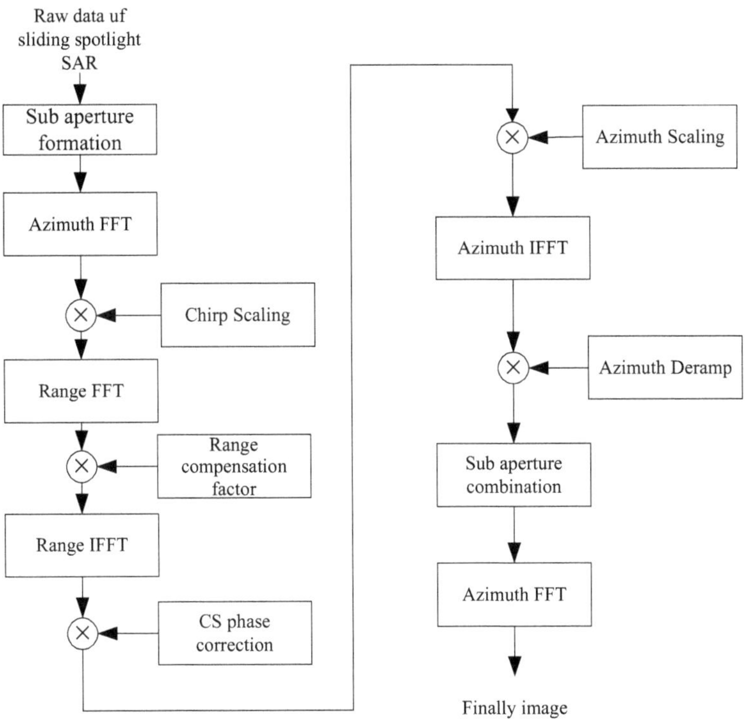

Fig. 9.13 ECS algorithm flow chart

$$f_1(t_a) = -\frac{2v}{\lambda}\cos\left(\text{arc cot}\left(-\frac{t_a \cdot v_a}{r_{rot}}\right) - \frac{\theta_a}{2}\right) \tag{9.44}$$

$$f_2(t_a) = -\frac{2v}{\lambda}\cos\left(\text{arc cot}\left(-\frac{t_a \cdot v_a}{r_{rot}}\right) + \frac{\theta_a}{2}\right) \tag{9.45}$$

In Eq. (9.44) and Eq. (9.45), θ_a is azimuth beam width, t_a is azimuth time, v_a is the flight speed of platform.

If azimuth data is divided into N sub-blocks, data repetition rate between sub-blocks is ε. The Doppler bandwidth of sub-block data can be approximated as

$$B_{sub} = \frac{1}{N(1-\varepsilon)+\varepsilon} k_{rot} T + B_a \tag{9.46}$$

In Eq. (9.46), $B_a = |f_1(t_a) - f_2(t_a)|$ is Echo instantaneous bandwidth.

(2) Sub-aperture range migration correction

The synthetic aperture radar echo signal is transformed to the range-Doppler domain by azimuth FFT operation, then complete the Chirp Scaling operations. After CS process, range migration curves of different slant range units in the range-Doppler domain have the same shape.

For echo data with a certain squint angle, after processing it does not necessarily correct contour line to the nearest range r_c, so range migration coefficient is introduced in order to reduce the range spectrum expansion.

(3) Azimuth Scaling

After range migration correction and CS phase compensation to Sub-aperture data, the azimuth phase course can be expressed as

$$S(f_a; r_c) = \exp\left[-j\frac{4\pi r_c}{\lambda}\beta(f_a)\right] \quad (9.47)$$

In Eq. (9.47), $\beta(f_a) = \sqrt{1 - \left(\frac{\lambda f_a}{2v}\right)^2}$, f_a is azimuth frequency.

Since signal phase course is related to slant range, if linear frequency-modulated signal is transformed into a sinusoidal signal whose frequency is proportional to the azimuth location by the direct de-chirp operation, azimuth sampling interval of different range cells will be different, which needs interpolation process to resampling frequency domain image. ECS introduces azimuth scale transformation in the algorithm, with uniform quadratic phase course instead of the original phase process. Azimuth scale function is as follows:

$$H_{az}(f_a; r_{scl}) = \exp\left[j\frac{4\pi r_c}{\lambda}\beta(f_a) - j\frac{\pi}{k_{scl}}f_a^2\right] \quad (9.48)$$

In Eq. (9.48), k_{scl} is Doppler frequency modulation rate at the selected scale transformation slant range r_{scl}.

(4) Azimuth Deramp

After azimuth scaling, the azimuth signal results in two variations:

① Signal time Center offset

In sub-block data azimuth time $[t_s, t_e]$, the ECHO time center variation of target (t_0, r_c) before and after azimuth scaling is

$$\Delta t_a = t_c(r_{scl}) - t_c(r_c) = \left(\frac{1}{k_{scl}} - \frac{1}{k_r}\right) f_{dc\_p} \quad (9.49)$$

Equation (9.49) shows that time center offset Δt_a is related to f_{dc_p}, so after azimuth scaling transformation, signals are likely to offset out of the time interval of data blocks $[t_s, t_e]$. In order to prevent this happening, the azimuth time $[t_s, t_e]$ needs to be extended. Because the echo signal start and end times of different azimuth position target are different, target Doppler center frequencies of Sliding Spotlight SAR vary greatly, therefore, the time interval is extended to $[t_s - \Delta t_{max}, t_e + \Delta t_{max}]$, to ensure signal still in the processing time range after azimuth scaling transformation.

$$\Delta t_{max}\left(\frac{1}{k_{scl}} - \frac{1}{k_r}\right)f_{dc\_max} \tag{9.50}$$

$$f_{dc\_max} = \max\left(f_{dc}\right) \tag{9.51}$$

② Signal time length variation

After scaling transformation, time length is k_r/k_{scl} times the original signal. In order to reduce range extension of the time-domain signal, r_{scl} can select nearest slant range of scenes.

$$k_{scl} = \frac{2v^2}{\lambda r_{min}} \tag{9.52}$$

After scaling transformation of the azimuth signal, inverse Fourier transform and multiply azimuth Deramp function.

$$H_{deramp} = \exp\left[j\pi k_{scl} t_a^2\right] \tag{9.53}$$

After Deramp, the signals are transformed into sinusoidal signals whose frequencies are related to target azimuth position. Full aperture data can be obtained by putting sub-block data together, and frequency-domain image can be obtained through the unified Fourier transform.

9.6.2 Wavenumber Domain Algorithm

Wavenumber domain algorithm processes data in two-dimensional frequency domains, and completes target imaging by Stolt interpolation in frequency domain. In this algorithm, there is no approximation of the system transfer function in the two-dimensional frequency domain, so it is an accurate reconstruction imaging algorithm. SAR Echo signal processing effect of Wavenumber domain algorithm is only concerned with Stolt interpolation precision, not aggravated by the increased as the squint angle and beam angle, and it is an ideal imaging algorithm. Wavenumber domain algorithm specific imaging procedure is shown in Fig. 9.14.

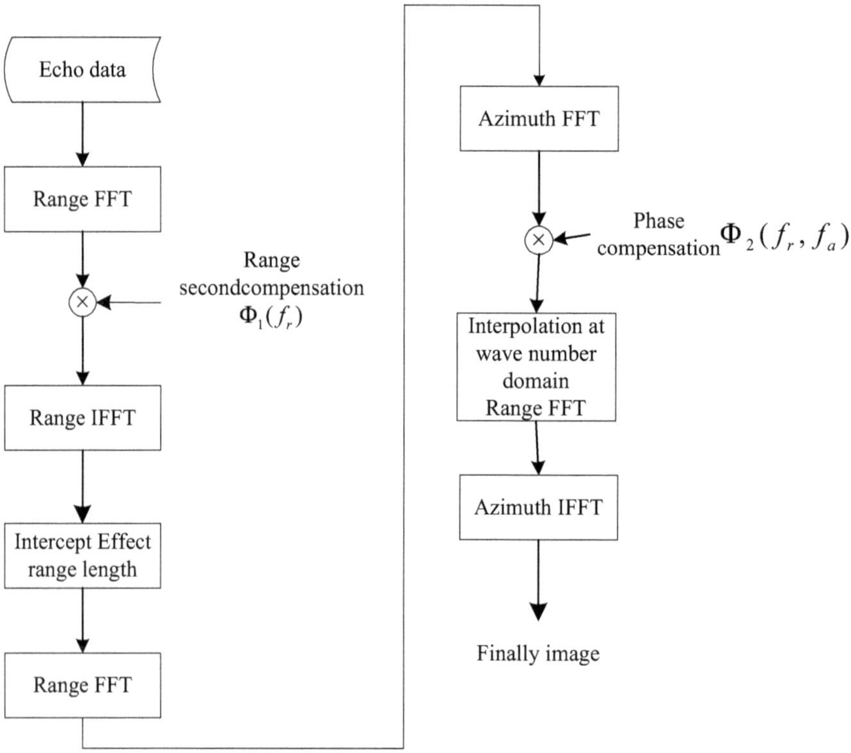

Fig. 9.14 Wavenumber domain algorithm imaging procedure

After azimuth Fourier transform, two-dimensional frequency-domain expression is

$$S_{\mathrm{r}}(f_{\mathrm{r}},f_{\mathrm{a}}) = C \cdot \exp\left(-j2\pi f_{\mathrm{a}}\frac{x_0}{v_{\mathrm{s}}}\right) \cdot \exp\left[-j\frac{4\pi r_0}{c}\sqrt{(f_{\mathrm{c}}+f_{\mathrm{r}})^2 - \frac{c^2 f_{\mathrm{a}}^2}{4v_{\mathrm{s}}^2}}\right] \tag{9.54}$$

Two-dimensional phase compensation to Eq. (9.54), the phase compensation factor is

$$\Phi_2(f_{\mathrm{r}},f_{\mathrm{a}}) = \exp\left[j\frac{4\pi R_{\mathrm{ref}}}{c}\sqrt{(f_{\mathrm{c}}+f_{\mathrm{r}})^2 - \frac{c^2 f_{\mathrm{a}}^2}{4v_{\mathrm{s}}^2}}\right] \tag{9.55}$$

Compensated signal expression is

$$S_{\mathrm{r}}(f_{\mathrm{r}},f_{\mathrm{a}}) = C \cdot \exp\left(-j2\pi f_{\mathrm{a}}\frac{x_0}{v_{\mathrm{s}}}\right) \cdot \exp\left[-j\frac{4\pi (r_0 - R_{\mathrm{ref}})}{c}\sqrt{(f_{\mathrm{c}}+f_{\mathrm{r}})^2 - \frac{c^2 f_{\mathrm{a}}^2}{4v_{\mathrm{s}}^2}}\right] \tag{9.56}$$

Interpolation to Eq. (9.56) that is the transformation of $\sqrt{(f_c + f_r)^2 - \frac{c^2 f_a^2}{4v_s^2}} \sim f_c + \hat{f}_r$. The transformed expression is

$$S_r(f_r, f_a) = C \cdot \exp\left(-j2\pi f_a \frac{x_0}{v_s}\right) \cdot \exp\left[-j\frac{4\pi(r_0 - R_{ref})}{c}(f_c + \hat{f}_r)\right] \tag{9.57}$$

Two-dimensional IFFT to Eq. (9.57), the complex image expression of the target can be obtained.

$$S_r(\tau, t) = C \cdot \text{sinc}\left[T_p \cdot K_r\left(\tau - \frac{2r_0}{c}\right)\right] \cdot \text{sinc}\left[B_a\left(t_a - \frac{x_0}{v_s}\right)\right] \tag{9.58}$$

In Eq. (9.58), B_a is azimuth Doppler bandwidth.

9.6.3 Base-Band Azimuth Scaling Algorithms

Base-band azimuth Scaling (BAS) algorithm is a more effective Sliding Spotlight SAR imaging algorithm presented in recent years. This algorithm overcomes the shortcoming of low azimuth processing efficiency in Extended CS algorithm, and is a new azimuth Scaling algorithm. This algorithm flow path is shown in Fig. 9.15 [9].

The first several steps of Base-band azimuth Scaling algorithm are: sub-aperture partition, range migration correction, range compression, and other factors are exactly the same as those of ECS algorithm described in Sect. 9.6.1, not repeat them here.

After range compression, quadratic phase course is needed to replace the original hyperbolic phase course, correction factor is

$$\begin{aligned} H_4(f_a; r) = &\exp\left[j\frac{4\pi}{\lambda} \cdot r \cdot (\beta(f_a, r) - 1)\right] \\ &\exp[-j2\pi f_a t_v(r)] \cdot \exp\left[-j\frac{\pi}{k_{scl}(r)} f_a^2\right] \end{aligned} \tag{9.59}$$

In Eq. (9.59), f_a is azimuth frequency, r is the nearest range vector.

$$\beta(f_a, r) = \sqrt{1 - \left(\frac{\lambda f_a}{2v_{eff}(r)}\right)} \tag{9.60}$$

In Eq. (9.60), $v_{eff}(r)$ is the effective speed related to range. The quadratic phase course in Eq. (9.59) can be expressed by $k_{scl}(r)$.

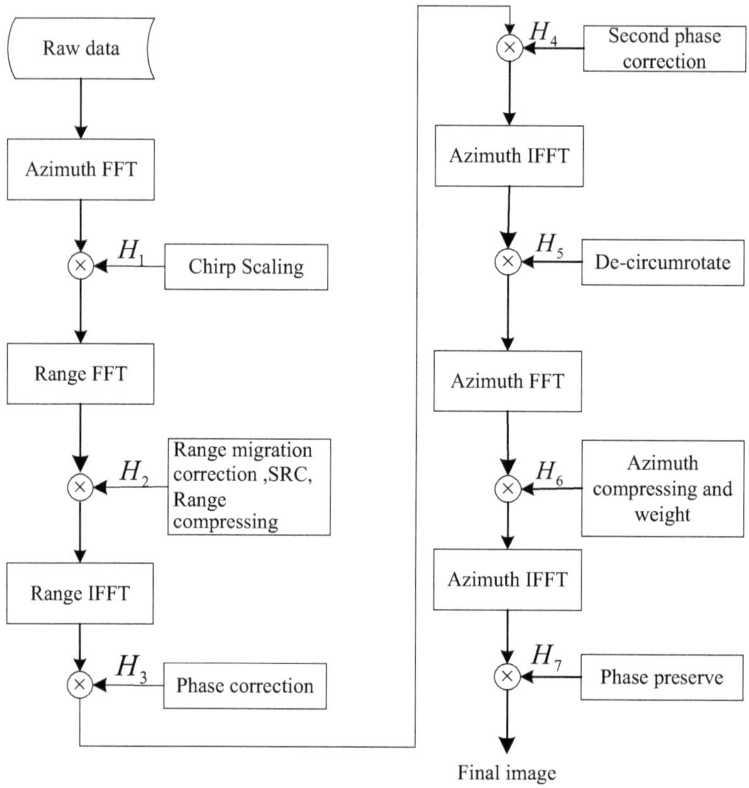

Fig. 9.15 Base-band azimuth scaling (BAS) algorithm flow chart

$$k_{\mathrm{scl}}(r) = -\frac{2v_{\mathrm{eff}}^2(r_{\mathrm{mid}})}{\lambda r_{\mathrm{scl}}(r)} \tag{9.61}$$

In Eq. (9.61), $v_{\mathrm{eff}}(r_{\mathrm{mid}})$ is the effective speed at middle range, r_{scl} is a variant related to range.

Due to H_4 causes shifting of azimuth signal not at the distance of r_{scl}, azimuth-time domain signals need to do minor extensions, which is similar to ECS algorithm. Subsequently, the signals are transformed back to the azimuth-time domain. However, all the sub-aperture azimuth bandwidth is still larger than pulse repetition frequency, a de-rotation demodulation operation is required, de-rotation factor is

$$H_5(t_{\mathrm{a}}; r) = \exp\left[-j\pi K_{\mathrm{rot}}(r) \cdot (t_{\mathrm{a}} - t_{\mathrm{mid}})^2\right] \tag{9.62}$$

In Eq. (9.62), t_{a} is azimuth time, t_{mid} is the time corresponding to scene center, and $K_{\mathrm{rot}}(r)$ is a variant related to range, it is

$$K_{\text{rot}}(r) = -\frac{2v_{\text{eff}}^2(r_{\text{mid}})}{\lambda r_{\text{rot}}(r)} \tag{9.63}$$

At this point, the effective frequency modulation slope of the signal is

$$K_{\text{eff}}(r) = K_{\text{scl}}(r) - K_{\text{rot}}(r) \tag{9.64}$$

The matched filter is

$$H_6(f_{\text{a}}, r) = W(f_{\text{a}}) \cdot \exp\left[j\frac{\pi}{K_{\text{eff}}(r)}f_{\text{a}}^2\right], -f_{\text{PRF}} + f_{\text{dc}} < f_{\text{a}} < f_{\text{PRF}} + f_{\text{dc}} \tag{9.65}$$

In Eq. (9.65), f_{dc} is the average Doppler center of obtained data. However, in order to keep the phase, the data must multiply phase as follows:

$$H_7(t_{\text{a}}, r) = \exp\left[j\pi K_{\text{t}}(r)\left(1 - \frac{r_{\text{scl0}}}{r_{\text{rot0}}}\right)^2 (t_a - t_{\text{mid}})^2\right] \tag{9.66}$$

In Eq. (9.66)

$$K_{\text{t}}(r) = -\frac{2v_{\text{eff}}^2(r_{\text{mid}})}{\lambda \cdot (r_{\text{rot}}(r) - r_{\text{scl}}(r))} \tag{9.67}$$

Through these operations, without interpolation, the phase preserving compression results can be obtained. However, in the algorithm above, the definition of r_{rot}, r_{scl} is not given.

In order to get the definition of r_{rot}, r_{scl}, need to take into account the following two factors [10]:

① after the base-band azimuth Scaling algorithm, azimuth image sampling interval turns into

$$\Delta x_{\text{final}} = \Delta x_{\text{orig}} \cdot \left(1 - \frac{r_{\text{scl}}}{r_{\text{rot}}}\right) \tag{9.68}$$

In Eq. (9.68), r_{scl} is scale transform range used in H_4, r_{rot} is rotation range used in H_5, Δx_{orig} is the original image interval, it is v_{g}/PRF, v_{g} is ground speed. In addition, scale transform factor H_4 amends the rotation bound.

$$r_{\text{rot}}(r) = r_{\text{rot0}} - (r - r_{\text{scl}}) \tag{9.69}$$

In equation, r_{rot0} is the vector range to the rotation center.

② When using a constant scale, and a constant rotation bound must be selected, otherwise, the azimuth sampling will be a variable related to range. However, due to valid rotation bound is changed, such as Eq. (9.69) shows, in this case, not all the targets are in the baseband, this will cause aliasing. In order to maintain a constant azimuth image sampling on all range gates, by combining Eq. (9.68) and Eq. (9.69) can be got,

$$r_{\text{scl}}(r) = \frac{r_{\text{scl0}}}{r_{\text{rot0}}} r_{\text{rot}}(r) \tag{9.70}$$

$$r_{\text{rot}}(r) = \frac{r_{\text{rot0}} - r}{1 - r_{\text{scl0}}/r_{\text{rot0}}} \tag{9.71}$$

In Eq. (9.70) and Eq. (9.71), r_{rot0} is the selected Scale range with consideration of the required azimuth image sampling, and it must be in each sub-band to make required extending sub-aperture minimum. At this time, azimuth image sampling is

$$\Delta x_{\text{final}} = \Delta x_{\text{orig}} \cdot \left(1 - \frac{r_{\text{scl0}}}{r_{\text{rot0}}}\right) \tag{9.72}$$

This chapter, the principle of Siding Spotlight SAR, imaging geometry model, target irradiation time, effective irradiation area, ECHO signal characteristics, resolution, ECHO Doppler history, and reconstruction algorithms have been discussed. Through controlling antenna radiation zones movement speed on ground, Sliding Spotlight SAR increases azimuth coherent integration time, so as to get higher azimuth resolution than Strip SAR with the same size antenna, and get bigger imaging area than Spotlight SAR. With the development of wide scan angle technology of the antenna, the application of Sliding Spotlight SAR will be further expanded.

References

1. Gao XW (2004) Spotlight model studying for space borne synthetic aperture radar system. Doctor Thesis Institute of Electronics Chinese Academy of Sciences
2. Belcher DP, Baker CJ (1994) Hybrid stripmap/spotlight SAR. IEEE Colloquium on Radar and Micorwave Imaging
3. Belcher DP, Baker CJ (1996) High resolution processing of hybrid stripmap/spotlight mode SAR. IEE Proc. Radar, Sonar Navigation 143(6):366–374
4. Henrion S, Savy L, Plaubs J-G Properties of hybrid strip-map/spotlight space borne SAR processing. Proc. of IGARSS 99, 530–534
5. Tang Y, Wang YF, Zhang BC (2007) Imaging model studying for slide spotlight synthetic aperture radar. Electron Inf Trans 29(1):26–29
6. Jia XX (2010) Imaging technology studying for multiple model synthetic aperture radar. Doctor thesis Institute of Electronics Chinese Academy of Sciences

7. Moreira Alberto, Mittermayer Josef, Scheiber Rolf (1996) Extended chirp scaling algorithm for air- and space borne SAR data processing in strip map and scanSAR imaging modes. IEEE Trans Geosci Remote Sens 34(5):1123–1136
8. Mittermayer J, Lord R, Borner E (2003) Processing for TerraSAR-X using a new formulation of the extended chirp scaling algorithm. Proc. of IGARSS p 1462–1464
9. Prats P, Scheiber R, Mittermayer J et al. (2010) Processing of Sliding Spotlight and TOPSSAR Data Using Baseband Azimuth Scaling. IEEE Trans Geosci Remote Sens 48 (2):770–780
10. Ossowska A, Speck R (2009) Processing of sliding spotlight mode data with consideration of orbit geometry. Proc. of SPIE, p 7502

Chapter 10
Single Phase Center Multi-beam SAR

One of the fundamental limitations in the design of space borne SAR systems is the contradiction between the swath and azimuth resolution of the survey. This contradiction arises from their differing demands on the system pulse repetition frequency (PRF). Increasing the swath requires increasing the pulse repetition period, thus reducing the PRF. To increase the azimuth resolution, the azimuth sampling frequency needs to be increased due to the increase of the echo Doppler bandwidth, thereby increasing the PRF. So in the design of space borne synthetic aperture radar systems, there is a compromise between the swath and azimuth resolution [1].

In principle, the techniques of high resolution wide swath SAR can be single phase center multiple azimuth beam, multiple phase center multiple azimuth beam, multiple beams in elevation, and multiple-input multiple-output (MIMO) technique. In this chapter, the single phase center multiple azimuth beam (SPCMAB) techniques will discussed

In the SPCMAB technique, multiple receiving beams are formed in the azimuth direction of the radar antenna, Adjacent regions were covered respectively, and the Doppler spectra of the echo signals of the respective beams are also adjacent to each other. The azimuthal echo signal with wide Doppler bandwidth can be synthesized by frequency domain processing, and the azimuth resolution obtained by image processing is higher than that of single beam azimuth resolution.

10.1 Single-Phase Center Multi-beam Technique

10.1.1 Principle of SPCMAB

In order to obtain a relatively high azimuth resolution, the synthetic aperture length covered by the radar azimuth beam should be long enough, echo Doppler bandwidth to be wide enough; on the other hand, in order to obtain a wide swath, the

R. Yang et al., *High-Resolution Microwave Imaging*,
https://doi.org/10.1007/978-981-10-7138-6_10

system PRF cannot be too high. To satisfy the need for azimuthal sampling at lower PRF, multiple receive beams can be utilized in azimuth, the Doppler bandwidth of each beam is only a fraction of the total required Doppler bandwidth, the synthesis of each beam data can produce the total Doppler bandwidth, which is the basic idea for SPCMAB technique.

In the SPCMAB technique, multiple adjacent received sub-beams (N_B denote the number of sub-beams) with a common phase center are generated in the azimuth direction of the radar antenna, and these sub-beams have different oblique angles, covering only a portion of the total synthetic aperture length. Each receive beam is contiguous to each other and covers the entire synthetic aperture length. Since each receive beam echo only occupies a part of the total Doppler bandwidth, The system pulse repetition frequency can simply satisfy the Nyquist sampling rate of the partial Doppler bandwidth corresponding to each sub-beam.

Using N_B independent receive channels to receive echoes, for each point on the ground target, N_B group echo can be obtained. Because of the different oblique angle of each beam, N_B group echo has different Doppler center frequency. Although the echo signal spectrum of each sub-beam after being sampled in azimuth direction will be folded to near the zero-Doppler frequency, however since the sub-beam relative oblique angle is known in advance, the sub-beam real Doppler center frequency can be calculated accordingly, so then the echo data can be restored back to their actual spectral position during data processing. After frequency correction, each sub-beam echo data can be synthesized to obtain a Doppler bandwidth that is N_B times larger than the bandwidth of each sub-beam. After azimuth compression, the azimuth resolution can be improved by N_B times. Therefore, by using of multiple azimuth beams, without reducing the system operating PRF, the azimuth resolution can be fined by approximately equal to the multiple of the beam number [5–6].

There are two ways to achieve SPCMAB, the first one is the "wide-transmit narrow-receive" (WTNR) approach, a wide transmit beam in the azimuth is used to cover the entire synthetic aperture length, while multiple narrow beams are used for reception. The second one is "narrow-transmit narrow-receive" (NTNR) mode. When transmitting and receiving, the azimuth direction adopts multiple narrow beams, which are adjacent to each other and have a common phase center. Multiple beams are simultaneously transmitted and received at the same time.

Figure 10.1 shows a schematic diagram of a SPCMAB SAR in two ways, where Fig. 10.1a is the "wide-transmit narrow receive" approach and Fig. 10.1b is the "narrow-transmit narrow-receive" approach. Two approaches have the same principle, while their round trip antenna gain, ambiguity characteristics are different, and their signal processing methods are basically the same. Therefore, in the discussion that follows this chapter, there is no distinction between the two implementations unless otherwise stated.

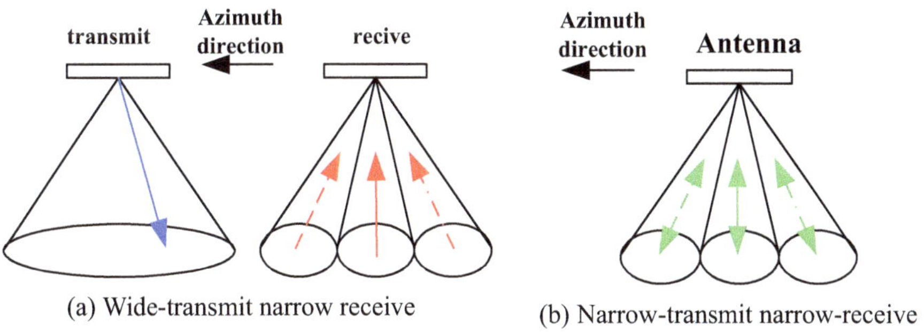

Fig. 10.1 Schematic diagram of SPCMAB SAR

10.1.2 Azimuthal Signal Characteristics of SPCMAB SAR

In order to study the range history of SPCMAB SAR, and then study the every beam echo signal characteristics, Fig. 10.2 shows the range model of SPCMAB SAR. In the case of a three-beam case, the middle beam is an orthogonal side-looking beam, the left beam is a forward-looking beam, the right beam is a backward-looking beam, and the oblique angle is θ [7, 8]. Each beam adjoins each other that mean the half-power point of the adjacent beam is in the same direction. For an antenna with a length of L_a, the beam width of the middle beam is $\frac{\lambda}{L_a}$, the beam width of the beam on both sides is $\frac{\lambda}{L_a \cos\theta}$.

Taking the left beam as an example, the range history of SPCMAB SAR is explained. In Fig. 10.2, there is a point P in the left beam illumination area. When $t = 0$, the target is located at the center of the beam, and the slope range of the point target to the radar can be expressed as [9]:

$$R(t) = \sqrt{R_c^2 + v_s^2 t^2 - 2R_c v_s t \sin\theta} \approx R_c - v_s t \sin\theta + \frac{v_s^2 t^2 \cos^2\theta}{2R_c} \tag{10.1}$$

Fig. 10.2 Range model of SPCMAB SAR

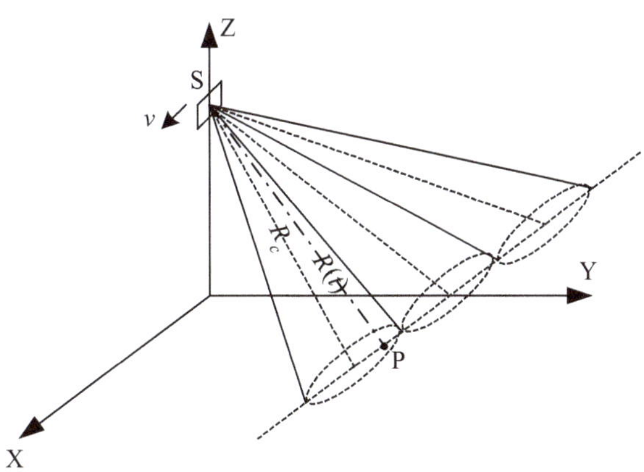

where R_c is the slant range of the point target to the radar when $t = 0$, v_s is the velocity of satellite relative to the ground point target. The instantaneous Doppler frequency of the echo signal is:

$$f_d(t) = \frac{1}{2\pi}\frac{d}{dt}\left[-\frac{4\pi}{\lambda}R(t)\right] = \frac{2}{\lambda}\left(v_s \sin\theta - \frac{v_s^2\cos^2\theta}{R_c}t\right) \tag{10.2}$$

where λ is the radar wavelength. The Doppler centroid and Doppler slopes of the left beam echo signals are expressed respectively as:

$$f_{dcL} = \frac{2}{\lambda}v_s \sin\theta \qquad f_{drL} = -\frac{2}{\lambda R_c}v_s^2\cos^2\theta \tag{10.3}$$

Accordingly the left-beam echo Doppler bandwidth is

$$B_{aL} = \frac{2v_s\cos\theta}{L_a} \tag{10.4}$$

By changing the value of the oblique angle θ, the echo signal characteristics of the middle beam and the right beam can be obtained respectively.

The middle beam:

$$f_{dcM} = 0, \quad f_{drM} = -\frac{2v_s^2}{\lambda R_c}, \quad B_{aM} = \frac{2v_s}{L_a} \tag{10.5}$$

The right beam:

$$f_{dcR} = -\frac{2}{\lambda}v_s\sin\theta, \quad f_{drR} = -\frac{2}{\lambda R_c}v_s^2\cos^2\theta, \quad B_{aR} = \frac{2v_s\cos\theta}{L_a} \tag{10.6}$$

As the oblique angle is generally very small, so $f_{drL} \approx f_{drM} \approx f_{drR}$, $B_{aL} \approx B_{aM} \approx B_{aR}$, and $f_{dcL} - f_{dcM} = f_{dcM} - f_{dcR} \approx B_{aM}$, i.e. the difference of Doppler center frequencies between adjacent beams are about the Doppler bandwidth of each sub-beam. The instantaneous Doppler frequency and signal spectrum of each beam echo signal are shown in Fig. 10.3.

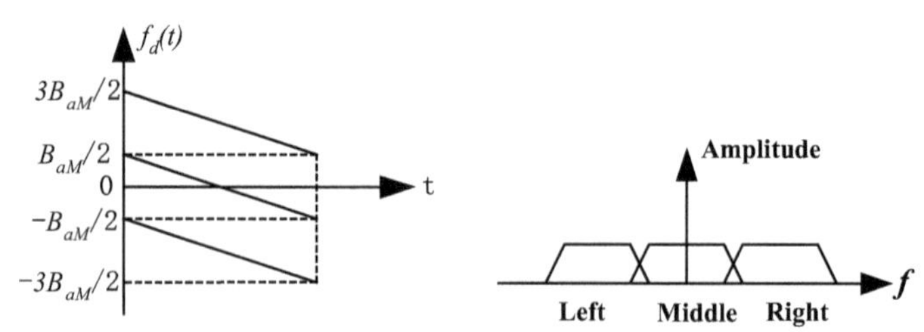

Fig. 10.3 Instantaneous doppler frequency and spectral distribution of SPCMAB SAR

10.2 Azimuthal Signal Processing of SPCMAB SAR

The SPCMAB SAR requires no special signal processing before imaging, that is the same as that of a conventional single-beam system, but the sub-beam echo data are needed to be synthesized in azimuth direction before the imaging process to form the Doppler bandwidth which is expanded N_B (beam number) times. While the bandwidth is widened, the azimuth sampling rate should be increased by N_B times to ensure the integrity of azimuthal sampling.

10.2.1 Interpolation Theory of Complex Signal

Signal interpolation theory is one of the commonly used tools of the multi-sampling rate of digital signal processing, which is an important branch of digital signal processing. Many literatures [11–11] introduced the real signal interpolation theory and commonly used algorithms. SAR echo signal can be transformed to I, Q two parts after quadrature demodulation, which represent the real part and the imaginary part of the echo signal respectively, so SAR echo signals processed in the SAR signal processor is complex signals. In order to illustrate the azimuthal echo signal processing of SPCMAB SAR, the complex signal interpolation theory is introduced.

In [12–14], it is possible to define a complex signal in which the real part is Hilbert transform of the imaginary part and vice versa, which is called complex analytic signal. The spectrum of complex analytic signal is single-side spectrum, including only positive frequency component, or only negative frequency component, which brings a lot of convenience for signal processing. For example, when the band-pass sampling is performed, the sampling frequency can be selected as long as it is larger than its bandwidth, which greatly facilitates the selection of sampling frequency. In [15], for a narrow-band signals with a relative bandwidth much less than 50%, the complex exponential signal can be used to approximate its complex analytic signal, and the approximation error is negligible. Usually the radar signal can satisfy this narrowband signal condition. In fact, the purpose of quadrature demodulation in radar receivers is to generate complex exponential signals for approximating complex analytic signals.

Figure 10.4 shows the spectral transformation during complex analytic signal interpolation [16]. Let a continuous band-pass complex analysis signal is $x(t)$, and the center frequency is f_0, bandwidth is Δf, the spectrum is shown in Fig. 10.4a. The sampling sequence $x(n)$ is generated according to the sampling theorem, and the spectrum $X(e^{j\omega})$ arising from periodical extension is shown in Fig. 10.4b, where the shaded portion is the actual spectral position corresponding to the analog spectrum and the other is the result of the periodic extension caused by the sampling. The horizontal axis in the figure is the normalized frequency ω of the digital domain, the

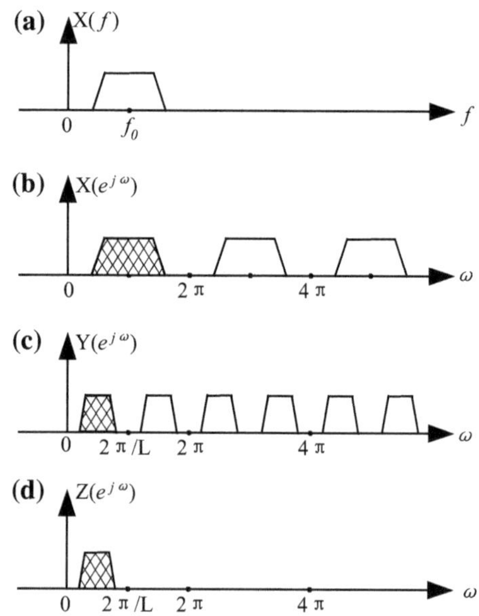

Fig. 10.4 Spectral transformations in interpolation of complex analytic signals

corresponding center frequency is $\omega_0 + 2\pi k$, the bandwidth is $\Delta\omega$, and the sampled signal has the following relationship with its spectrum:

$$X(\mathrm{e}^{j\omega}) = \sum_{n=-\infty}^{+\infty} x(n)\mathrm{e}^{-j\omega n} \tag{10.7a}$$

$$x(n) = \frac{1}{2\pi}\int_{-\pi}^{\pi} X(\mathrm{e}^{j\omega})\mathrm{e}^{j\omega n}\mathrm{d}\omega \tag{10.7b}$$

In order to increase the sampling frequency of signal $x(n)$, we added $L - 1$ zeros between adjacent two samples of $x(n)$, and L times interpolation. The interpolated signal is denoted as $y(n)$, then

$$y(n) = \begin{cases} x\left(n/L\right) & n = 0, \pm L, \pm 2L, \ldots \\ 0 & \text{其他} \end{cases} \tag{10.8}$$

The signal spectrum after interpolation is

$$Y(\mathrm{e}^{j\omega}) = \sum_{n=-\infty}^{+\infty} y(n)\mathrm{e}^{-j\omega n} = \sum_{n=-\infty}^{+\infty} x(n)\mathrm{e}^{-j\omega Ln} = X(\mathrm{e}^{j\omega L}) \tag{10.9}$$

That is, the signal spectrum after interpolation is compressed L times on the frequency axis of the normalized digital domain. Figure 10.4b shows the interpolated signal spectrum, $Y(\mathrm{e}^{j\omega})$ becomes L cycles, the center frequency becomes

$\frac{\omega_0+2\pi k}{L}$, and the bandwidth becomes $\frac{\Delta\omega}{L}$. In order to remove the extra spectral image, the spectrum must be filtered to truncate $Y(e^{j\omega})$ of a cycle. The characteristics of the filter used are as below:

$$H(e^{j\omega})=\begin{cases} L & \left|\omega-\omega_0/L\right|\leq\Delta\omega/L \\ 0 & \text{其它} \end{cases} \tag{10.10}$$

The output signal after filtering is $z(n)$, and its spectrum is

$$Z(e^{j\omega}) = H(e^{j\omega})Y(e^{j\omega}) = LX(e^{j\omega L}) \tag{10.11}$$

The filtered spectrum is shown in Fig. 10.4d.

It should be noted that, as shown in Fig. 10.4, the signal spectrum in the normalized digital frequency domain is compressed L times after interpolation, in fact, the corresponding analog frequency does not change, the spectral compression is due to increased sampling rate after interpolation.

10.2.2 Frequency Domain Analysis of Azimuth Signal Processing

With SPCMAB technique, the Doppler center frequencies of the receiving beams are different due to the different oblique angles of the receiving beams. For the space borne SAR system, the Doppler center frequency is not zero even for the middle orthogonal side-looking beam due to the effect of the Earth's rotation. From the analysis of Sect. 10.1, it can be seen that the difference of the Doppler center frequency of adjacent beams is about the Doppler bandwidth of each sub-beam in the case of small oblique angle, so that the Doppler center frequency of middle orthogonal side-looking beam can be calculated according to the parameters of platform attitude, or estimated from the echo data of the middle orthogonal side-looking beam [17–23], and then the Doppler center frequency of other sub-beams is calculated according to the above relation.

The SAR systems operate in pulsed mode, so the echo signals received by each beam have been sampled in azimuth direction, and the sampling frequency is PRF. In order to synthesize the beam echoes, it is necessary to increase the sampling frequency by a factor of N_B. For this reason, the azimuthal echo signal needs to be interpolated by N_B times. The center frequency of each sub-beam filter is designed to correspond to Doppler center frequency, and the filter bandwidth is designed to match the Doppler bandwidth of each sub-beam. The spectrum of azimuthal signals after interpolation filtering is adjacent to each other, and their spectrum combination can form a complete Doppler spectrum with N_B times of bandwidth extension, and the azimuth signal is synthesized. The azimuth resolution can be improved by processing the azimuth signal after the synthesis.

According to the above analysis, the theoretical model of SPCMAB azimuth signal processing is shown in Fig. 10.5 (in the case of three-beam case). If the $l(n)$, $m(n)$, $r(n)$ represent left, middle and right beam azimuth echo signal respectively, which the Doppler center frequency are f_{dL}, f_{dM}, f_{dR}, and Doppler bandwidth is B_{aM}, and the corresponding normalized digital domain frequencies are ω_{dL}, ω_{dM}, ω_{dR} and $\Delta\omega_M$ respectively, the azimuthal echo signals of the left, middle and right beams after interpolation are expressed as $l_i(n)$, $m_i(n)$, $r_i(n)$, The relationship between azimuth spectra before and after N_B times interpolation is thereby expressed as follows:

$$LI(e^{j\omega}) = L(e^{j\omega N_B}) \tag{10.12a}$$

$$MI(e^{j\omega}) = M(e^{j\omega N_B}) \tag{10.12b}$$

$$RI(e^{j\omega}) = R(e^{j\omega N_B}) \tag{10.12c}$$

The azimuth sampling frequency after N_B times interpolation is the system equivalent pulse repetition frequency PRF_e, and the relationship between PRF_e and the system operation pulse repetition frequency PRF is $\text{PRF}_e = N_B \cdot \text{PRF}$. Then the Doppler center frequency of the beam in normalized digital domain frequency becomes $\frac{\omega_{dL}}{N_B}, \frac{\omega_{dL}}{N_B}, \frac{\omega_{dR}}{N_B}$, and the Doppler bandwidth becomes $\frac{\Delta\omega_M}{N_B}$. The relationship between the azimuth spectrum of each beam after filtering and the spectrum before interpolation is:

$$LF(e^{j\omega}) = N_B L(e^{j\omega N_B}) \tag{10.13a}$$

$$MF(e^{j\omega}) = N_B M(e^{j\omega N_B}) \tag{10.13b}$$

$$RF(e^{j\omega}) = N_B R(e^{j\omega N_B}) \tag{10.13c}$$

where $LF()$, $MF()$, $RF()$ are left, middle and right beam echo filtered spectrum respectively.

Figure 10.6 shows a schematic representation of the azimuth signal processing for a single-phase center three-beam system, where Fig. 10.6a shows the azimuthal

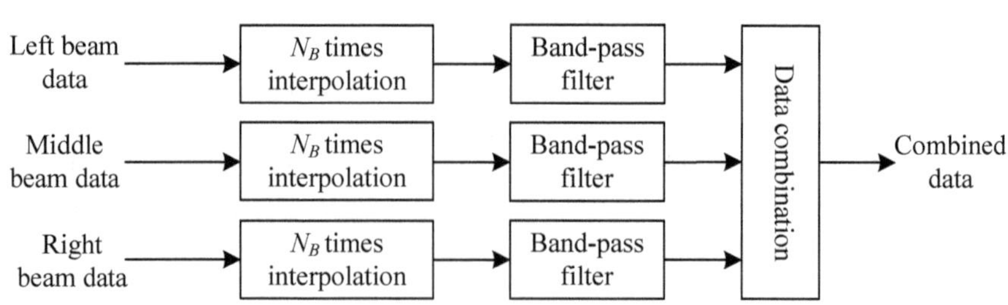

Fig. 10.5 The theoretical model of azimuthal signal processing

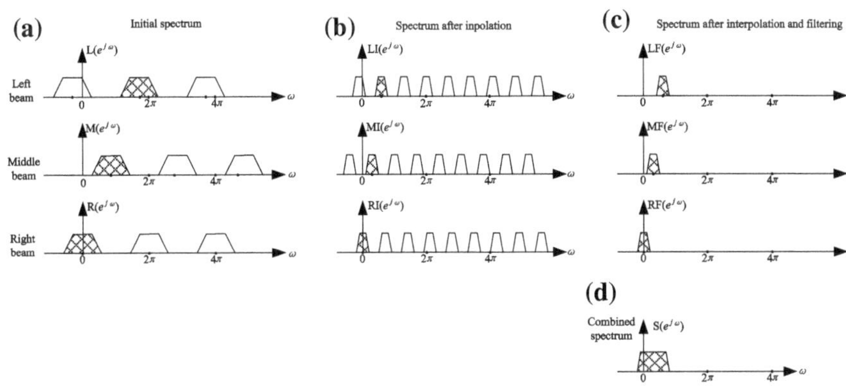

Fig. 10.6 Spectrum changes in azimuthal signal processing of SPCMAB SAR

Doppler spectra of the three beam echoes, Fig. 10.6b shows the Doppler spectra after interpolation for the three beams, and Fig. 10.6c shows the output azimuth spectra after bandpass filtering, and Fig. 10.6d shows azimuth spectra after combination of three beams data. The azimuth Doppler spectra are broadened by a factor of N_B and the resultant signals are compressed to obtain high azimuth resolution.

10.2.3 FFT Implementation of Azimuth Signal Processing

The azimuth signal of SPCMAB SAR can be synthesized by interpolation and filtering. In fact, due to SAR range ambiguity restrictions, the PRF is generally not much higher than the azimuthal Doppler bandwidth. After being sampled by PRF, each beam echo is extended and the periods of the spectrum will be close to each other. This requires that the filter transition band after interpolation be narrow enough that the attenuation in the transition band is fast enough. The design of this kind of filter is more difficult.

FFT is a fast computation tool for discrete Fourier transform. According to the nature of discrete Fourier transform, adding zeroes in the output spectrum of Fourier transform is equivalent to the time domain interpolation. Resulting from this feature, FFT can be used to complete the interpolation, filtering and other operations in the multi-beam echo signal azimuth signal processing to avoid the difficulties of filter design.

The azimuthal spectrum of each sub-beam is periodically extended after sampled by the PRF so that they partially overlap each other on the frequency axis. In order to obtain a finer azimuth resolution, the sub-beam spectrum must be spliced before azimuth compression to form a wide azimuth spectrum. The Doppler center frequency of each beam should be first calculated to make the beam spectrum splicing is correct.

With the satellite flight, each sub-beam of the SPCMAB SAR system receives echoes at the same time, and the azimuth echo signals are on different positions in the spectrum due to the different oblique angle of each sub-beam. The azimuth Doppler centroid of the echo is not zero even for a space borne SAR with side look geometry due to the effects of the Earth's rotation [24–27]. When the beam angle is not too large, the adjacent beam Doppler centroid difference is about the Doppler bandwidth of each sub-beam, so the Doppler centroid of all beams can be known so long as the Doppler centroid of middle beam is calculated. The Doppler centroid of the middle beam can be calculated using the following formula [26, 27] with the satellite position, the ground point target position, and the satellite flight velocity, etc.

$$f_{dc} = (-2/\lambda R_c)[\mathbf{V} \cdot (\mathbf{R_s} - \mathbf{R_t}) + \boldsymbol{\omega}_\mathbf{e} \cdot (\mathbf{R_s} \times \mathbf{R_t})] \quad (10.14)$$

where R_c is the closest slant range of the radar to the ground point target, $\mathbf{V}$ is the satellite velocity vector, $\mathbf{R_s}$ is the satellite position vector in the inertial coordinate system, $\mathbf{R_t}$ is the ground point target position vector, and $\boldsymbol{\omega}_\mathbf{e}$ is the angular velocity vector of the Earth's rotation.

The azimuth sampling frequency of each beam echo is the pulse repetition frequency PRF of the system operation. The azimuth sampling frequency of the combined signal is the system equivalent pulse repetition frequency $N_B \cdot f_{\mathrm{PRF}}$. The idea of azimuthal signal processing with FFT is to calculate by FFT the spectrum between $0 \sim f_{\mathrm{PRF}}$ of each beam echo data, and take account of the actual Doppler center frequency of each beam to generate the spectrum between $0 \sim N_B \cdot f_{\mathrm{PRF}}$ of the combined signal. The specific steps of azimuthal signals processing are as follows.

STEP I FFT Operation: The azimuth spectrum of each beam is calculated by FFT, and the center of the output spectrum is obtained by means of modulo the PRF of the actual Doppler centroid frequency of each beam.

STEP II Spectrum Broadening: Adding zeroes to the FFT output spectrum, the spectrum points are increased to N_B times the original to complete the spectrum broadening of each beam.

STEP III Frequency Shift: The spectral center of each beam after spectrum broadening is the modulus of operation of the actual Doppler center frequency of each beam on $N_B \cdot f_{\mathrm{PRF}}$. The frequency shift of each beam spectrum can be determined accordingly, and after frequency shift, the frequency spectrum of each beam echo sampled by the equivalent PRF is provided.

STEP IV Spectral Combination: The beam spectrum is superimposed to form the azimuth spectrum of the final combined signal.

STEP V IFFT Operation: The IFFT operation is performed on the combined spectrum to obtain the combined azimuth echo signal.

Repeat the above steps for the echo data of each range gate, and finally complete the combination of each beam echo data.

The combined data is compressed to generate radar image. In the imaging process, the PRF used to form azimuth reference function is the equivalent $f_{\mathrm{PRF_e}}$. $f_{\mathrm{PRF_e}} = N_B \cdot f_{\mathrm{PRF}}$.

10.3 Ambiguity Analysis of SPCMAB SAR

In the SPCMAB SAR, the range ambiguity characteristics are the same as that of the conventional single-beam SAR, and the azimuth ambiguity characteristics are much more complicated. As mentioned above, the implementation of SPCMAB SAR has two ways named by "wide-transmit narrow-receive" and "narrow-transmit narrow-receive" respectively. Since the azimuth antenna patterns of the two implementations are different, their azimuth ambiguity sources are different, and the formula for calculating azimuth ambiguities is different. The source of the SPCMAB SAR ambiguities in the two implementations will study, and the formula of the azimuth ambiguity to signal ratio in the two modes will propose. In addition, the range ambiguity of SPCMAB SAR is also simulated and compared with conventional single-beam SAR system.

10.3.1 Different Operation Modes of SPCMAB SAR

As mentioned above, there are two main ways to achieve single-phase center multi-beam: "wide-transmit narrow-receive" and "narrow-transmit narrow-receive" approach.

Figure 10.1a shows the transmit and receive beam patterns of the "wide-transmit narrow-receive", and Fig. 10.7 shows the transmitted and the received beam pattern for the three beam conditions in this operation mode, where the solid line portion is the transmitted beam pattern and the dotted line portion is the left, middle and right receive beam patterns (plotted as a function of the azimuthal antenna pattern as a function of the Doppler frequency). The transmitted beam is widened by N_B times (N_B is the number of beams). The shape of the pattern is basically the same as the

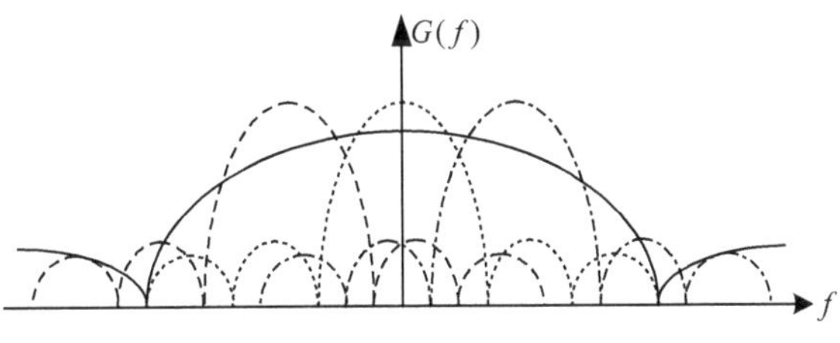

Fig. 10.7 Antenna pattern for wide-transmit narrow-receive beam

antenna pattern with the azimuth length equal to L_a/N_B and uniform weighted, and the error that this approximation brings to the ambiguity to signal ratio calculation is tolerable. Under this condition, the one-way normalized azimuthal power pattern can be approximated as:

$$G_t(\theta) = \text{sinc}^2\left(\frac{\pi L_a}{N_B \lambda}\sin\theta\right) \tag{10.15}$$

where θ is the azimuthal off-axis angle and L_a is the antenna azimuth length.

The solid line is transmitting pattern, and the dashed lines are receive pattern.

Each of the received beams is assumed to be uniformly weighted, and then the normalized one-way power pattern can be expressed as

$$G_{ri}(\theta) = \text{sinc}^2\left(\frac{\pi L_a \cos(i\theta_s)}{\lambda}\sin(\theta - i\theta_s)\right), \quad i = 0, \pm 1, \pm 2, \ldots \tag{10.16}$$

where θ_s is the amount of squint angle change between adjacent beams, i is the beam number, $i < 0$ represents the left beam, $i = 0$ represents the middle beam, and $i > 0$ indicates the right beam.

For the three-beam case, since θ_s are generally small, there are $L_a \cos\theta_s \approx L_a$, thus

$$\theta_s = \theta_H = K_A \frac{\lambda}{L_a} \tag{10.17}$$

where θ_H is the 3 dB width of the received beam, K_A is the constant associated with the antenna weight, and for uniform weighting $K_A = 0.886$. In this three-beam case, the one-way power pattern of the left, middle and right receive beams can be approximated

$$\text{Left beam}: G_{rL}(\theta) = \text{sinc}^2\left(\frac{\pi L_a}{\lambda}\sin(\theta + \theta_H)\right) \tag{10.18a}$$

$$\text{Middle beam}: G_{rM}(\theta) = \text{sinc}^2\left(\frac{\pi L_a}{\lambda}\sin\theta\right) \tag{10.18b}$$

$$\text{Right beam}: G_{rR}(\theta) = \text{sinc}^2\left(\frac{\pi L_a}{\lambda}\sin(\theta - \theta_H)\right) \tag{10.18c}$$

As shown in Fig. 10.1b, In the "narrow-transmit narrow-receive" mode the transmit and receive beams are the same, and the same signal is transmitted simultaneously with the three beams in accordance with the uniform weighting, while receiving the echo simultaneously. Figure 10.8 shows the shape of each beam pattern in this mode of operation. The solid line in the figure is the middle beam

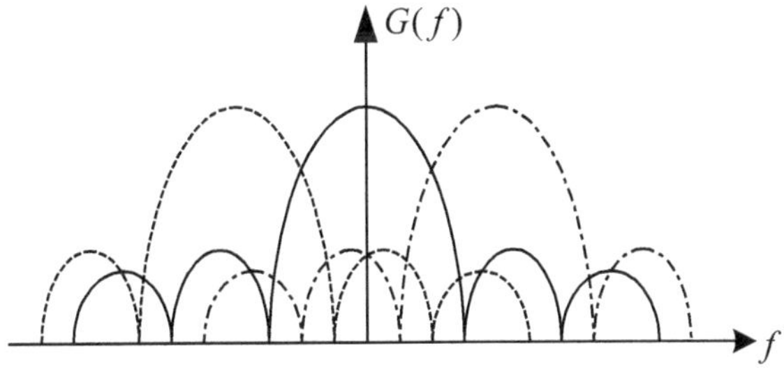

Fig. 10.8 The antenna pattern of "narrow-transmit narrow-receive" mode

pattern, the dashed line is the right beam pattern, and the dotted line is the left beam pattern. The normalized one-way power pattern of each beam can be expressed as

$$G_i(\theta) = \mathrm{sinc}^2\left(\frac{\pi L_{\mathrm{a}} \cos(i\theta_{\mathrm{s}})}{\lambda} \sin(\theta - i\theta_{\mathrm{s}})\right) \quad i = 0, \pm 1, \pm 2, \ldots\ldots \tag{10.19}$$

where θ_{s} is the amount of squint angle change between adjacent beams, i is the beam number, $i < 0$ represents the left beam, $i = 0$ represents the middle beam, and $i > 0$ represents the right beam. Referring to (10.16), the normalized one-way power pattern of the left, middle and right beams in the case of three beams can be approximated as

$$\text{Left beam}: G_{\mathrm{L}}(\theta) = \mathrm{sinc}^2\left(\frac{\pi L_{\mathrm{a}}}{\lambda} \sin(\theta + \theta_{\mathrm{H}})\right) \tag{10.20a}$$

$$\text{Middle beam}: G_{\mathrm{M}}(\theta) = \mathrm{sinc}^2\left(\frac{\pi L_{\mathrm{a}}}{\lambda} \sin\theta\right) \tag{10.20b}$$

$$\text{Right beam}: G_{\mathrm{R}}(\theta) = \mathrm{sinc}^2\left(\frac{\pi L_{\mathrm{a}}}{\lambda} \sin(\theta - \theta_{\mathrm{H}})\right) \tag{10.20c}$$

10.3.2 Azimuthal Ambiguity Characteristics of "Wide-Transmit Narrow-Receive" Mode

In order to calculate the azimuth ambiguity to signal ratio of each beam, at first the azimuth signal processing bandwidth should be confirmed, usually selected as azimuth signal Doppler bandwidth, and secondly the beam Doppler center frequency should be determined. There is a one-to-one correspondence between the azimuth position and the Doppler frequency of the radar target. The Doppler spectrum of the azimuth signal is weighted by the antenna azimuthal power pattern.

For echo from a large area uniform ground, the Doppler spectral shape is exactly the same as the antenna power pattern, so the azimuthal Doppler bandwidth corresponds to the 3 dB bandwidth of the antenna pattern. The 3-dB beam width of the antenna pattern is generally defined for the one-way power pattern. In the "wide-transmit narrow-receive" mode, taking into account the difference between transmission and reception patterns, the equivalent one-way power pattern is the product of transmission and reception voltage patterns. Then the equivalent one-way power pattern of the i-th received beam is

$$G_{\rm i}(\theta) = {\rm sinc}\left(\frac{\pi L_{\rm a}}{N_B \lambda}\sin\theta\right)\cdot{\rm sinc}\left(\frac{\pi L_{\rm a}\cos(i\theta_{\rm s})}{\lambda}\sin(\theta - i\theta_{\rm s})\right) \tag{10.21}$$

The 3-dB beam width is derived as

$$\theta_{\rm H} = K_{\rm A}\frac{\lambda}{L_{\rm a}} = 1.159\frac{\lambda}{L_{\rm a}} \tag{10.22}$$

That is about 30% wider than the beam width of the uniform weight $0.886\frac{\lambda}{L_{\rm a}}$, and the corresponding azimuth signal processing bandwidth is

$$B_{\rm p} = K_{\rm A}\frac{2v_{\rm s}}{L_{\rm a}} = 1.159\frac{2v_{\rm s}}{L_{\rm a}} \tag{10.23}$$

In general, the squint angle of each beam is not large, so the beam Doppler center frequency can be approximated as

$$f_{dci} = iB_{\rm p}, \quad i = 0, \pm 1, \pm 2, \ldots \tag{10.24}$$

For the "wide-transmit narrow-receive" mode, the formula to compute the i-th receive beam azimuth ambiguity to signal ratio is as follows

$${\rm AASR}_i = \frac{\sum_{\substack{m=-\infty \\ m\neq 0}}^{+\infty}\int_{(i-0.5)B_{\rm p}}^{(i+0.5)B_{\rm p}} G_i^2(f + m\cdot f_{\rm PRF}){\rm d}f}{\int_{(i-0.5)B_{\rm p}}^{(i+0.5)B_{\rm p}} G_i^2(f){\rm d}f} \tag{10.25}$$

According to the literature [12], the relationship between the azimuthal off-bore sight angle θ and the Doppler frequency f can be approximated as

$$\theta \cong \frac{\lambda}{2v_{\rm s}}\cdot f \tag{10.26}$$

Applying the relationship above into Eq. (10.15), (10.16), and taking into account Eq. (10.25), the azimuthal ambiguity to signal ratio of "wide-transmit narrow-receive" mode can be calculated as [28]

$$\mathrm{AASR}_i = \frac{\sum_{\substack{m=-\infty \\ m \neq 0}}^{+\infty} \int_{(i-0.5)B_\mathrm{p}}^{(i+0.5)B_\mathrm{p}} \sin c^2\left[\frac{\pi L_\mathrm{a}}{N_B \cdot 2v_\mathrm{s}}(f + m \cdot f_\mathrm{PRF})\right] \sin c^2\left[\pi L_\mathrm{a} \cos(i\theta_\mathrm{s})\left(\frac{1}{2v_\mathrm{s}}(f + m \cdot f_\mathrm{PRF}) - i \cdot \frac{K_\mathrm{A}}{L_\mathrm{a}}\right)\right] \mathrm{d}f}{\int_{(i-0.5)B_\mathrm{p}}^{(i+0.5)B_\mathrm{p}} \sin c^2\left[\frac{\pi L_\mathrm{a}}{N_B \cdot 2v_\mathrm{s}} f\right] \sin c^2\left[\pi L_\mathrm{a} \cos(i\theta_\mathrm{s})\left(\frac{1}{2v_\mathrm{s}} f - i \cdot \frac{K_\mathrm{A}}{L_\mathrm{a}}\right)\right] \mathrm{d}f} \tag{10.27}$$

10.3.3 *Azimuthal Ambiguity Characteristics of "Narrow-Transmit Narrow-Receive" Mode*

If the antenna of each beam is uniformly weighted in the "narrow-transmit narrow-receive" mode, the 3 dB width of each beam is

$$\theta_\mathrm{H} = K_\mathrm{A} \frac{\lambda}{L_\mathrm{a}} = 0.886 \frac{\lambda}{L_\mathrm{a}} \tag{10.28}$$

The corresponding azimuthal Doppler bandwidth is

$$B_\mathrm{p} = K_\mathrm{A} \frac{2v_\mathrm{s}}{L_\mathrm{a}} = 0.886 \frac{2v_\mathrm{s}}{L_\mathrm{a}} \tag{10.29}$$

The sources of desired signals and azimuth ambiguities interfere in "narrow-transmit narrow-receive" mode are complicated, both the components introduced by the beam and the other beams. For the desired signal, the main source can be divided into two parts: the first part is the echo components emitted by the main beam, and reflected by the ground and then received by the beam, which is the same as that in the conventional single-beam synthetic aperture radar. The second part is the echo components emitted by other beams outside the main lobe 3 dB part or side lobe part, and received by the main beam.

There are also two types of ambiguities in "narrow-transmit narrow-receive" mode: the first is introduced by the beam itself, which is the same as the conventional single-beam synthetic aperture radar, that is, the echo signal from the coverage area outside the 3 dB of the beam itself is folded into the azimuth signal processing bandwidth due to PRF sampling. The second type is the ambiguity components introduced by other beams, that is, the echo components which are emitted by other beams, and reflected by the outer part of the main lobe, and folded into the azimuth signal processing bandwidth after PRF sampling.

According to the above analysis, combined with the antenna pattern shape of "narrow-transmit narrow-receive" mode, the formula to compute azimuth ambiguity to signal ratio of the i-th receive beam is as follows [28]

$$\mathrm{AASR}_i = \frac{\sum_{\substack{m=-\infty \\ m \neq 0}}^{+\infty} \left[\int_{(i-0.5)B_p}^{(i+0.5)B_p} G_i^2(f + m \cdot f_{\mathrm{PRF}}) \mathrm{d}f + \sum_{j \neq i} \int_{(i-0.5)B_p}^{(i+0.5)B_p} G_i(f + m \cdot f_{\mathrm{PRF}}) G_j(f + m \cdot f_{\mathrm{PRF}}) \mathrm{d}f \right]}{\int_{(i-0.5)B_p}^{(i+0.5)B_p} G_i^2(f) \mathrm{d}f + \sum_{j \neq i} \int_{(i-0.5)B_p}^{(i+0.5)B_p} G_i(f) G_j(f) \mathrm{d}f} \tag{10.30}$$

Let Eq. (10.26) into (10.19), then the approximate formula to compute azimuth pattern of each beam can be obtained

$$G_i(f) = \sin c^2 \left[\pi L_\mathrm{a} \cos(i\theta_\mathrm{s}) \left(\frac{f}{2v_\mathrm{s}} - i \cdot \frac{K_\mathrm{A}}{L_\mathrm{a}} \right) \right] \tag{10.31}$$

Substituting Eq. (10.31) into Eq. (10.30), the azimuth ambiguity to signal ratio of "narrow-transmit narrow-receive" mode can be calculated.

"In discussion of the principle of SPCMAB technique, we it has been proposed that the SPCMAB technique can reduce the N_B pulse repetition frequency (PRF) while obtaining the given azimuth resolution. The premise of this argument is that the deterioration of azimuthal ambiguity while adopting SPCMAB technique is not taken into account. In fact, just as the previous analysis, the azimuthal ambiguity characteristics of SPCMAB technique are more complex, and the system pulse repetition frequency cannot be reduced N_B times if considering the ambiguity requirements. Figure 10.9a, b show the azimuthal ambiguity distributions of the "wide-transmit narrow-receive" and "narrow-transmit narrow-receive" modes respectively, where Fig. 10.9a shows only the azimuthal ambiguity distribution of the middle receiving beam. For the "wide-transmit narrow-receive" approach, there is an azimuthal ambiguity within the range of the main lobe of the transmitting beam. As can be seen from the figure, this part of the ambiguity echo has a higher transmit gain than the conventional single-beam system, making the azimuth ambiguity to signal ratio worse. Therefore, if the SPCMAB system is realized by "wide-transmit narrow-receive" approach, the system pulse repetition frequency cannot be reduced by N_B times in order to achieve the same azimuth resolution under the same azimuth ambiguity performance, and the PRF should be selected appropriately higher. For the "narrow-transmit narrow-receive" approach, the middle beam as an example, it can be seen from the figure, part of the azimuth ambiguities is transmitted by the adjacent beam is received by the middle beam. Some of these ambiguity signals are located in the main lobe of the adjacent beam, the transmission antenna gain is higher, making the azimuth ambiguity stronger and azimuth ambiguity to signal ratio worse. Therefore, if the SPCMAB system is realized by "narrow-transmit narrow-receive" approach, the system pulse repetition frequency cannot be reduced by N_B times in order to achieve the same azimuth resolution under the same azimuth ambiguity performance, and the PRF should be selected appropriately higher.

Fig. 10.9 The distribution diagram of azimuthal ambiguities in two modes

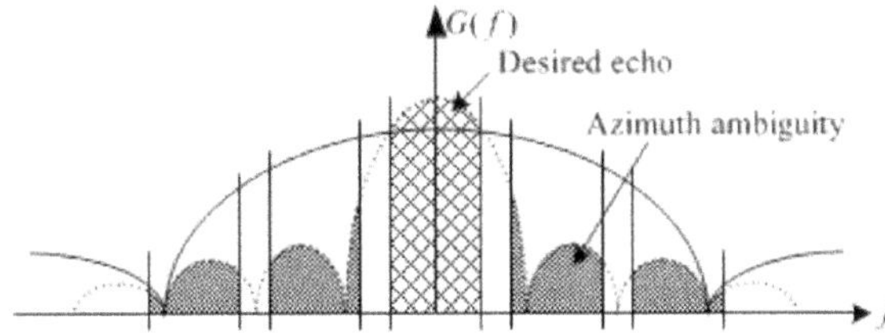

(a) "Wide-transmit narrow-receive" mode

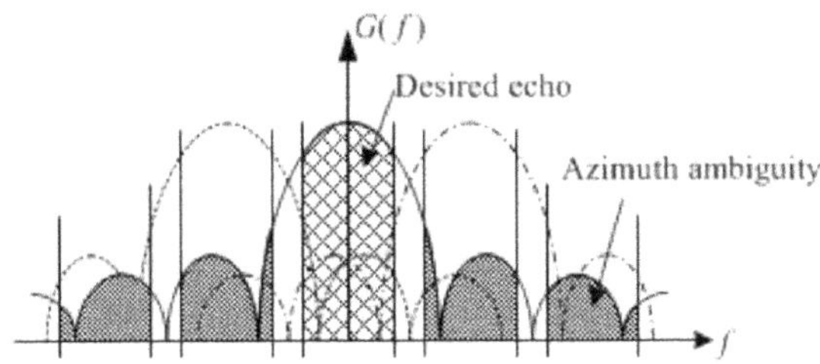

(b) "Narrow-transmit narrow-receive" mode

10.3.4 Simulation of Azimuth Ambiguity

Based on the azimuth ambiguity to signal ratio formula given above, the azimuth ambiguity characteristics of SPCMAB SAR systems are simulated. The main parameters of the system in simulation are as below: The satellite orbit height is 490 km. The elliptical orbit model is used, and the orbital eccentricity is 0.001146, the orbital inclination is 98.55°, the argument of perigee is 90°, the red path of ascending node is 89.37°, the radar operates in the S-band and the wavelength is 0.09375 m. The antenna length in azimuth direction is 12 m, the system has three beams, and azimuth resolution is 2 m.

When operating in the "Wide-transmit Narrow-receive" mode, the azimuthal ambiguity of the middle beam under various PRFs is shown in Fig. 10.10a, and the azimuthal ambiguity of the beam on both sides is shown in Fig. 10.10b (symmetry on the left and right sides, and has the same ambiguity characteristics). When operating in the "Narrow-transmit Narrow-receive" mode, the azimuthal ambiguity of the middle beam under various PRFs is shown in Fig. 10.11a, and the azimuthal ambiguity of the beam on both sides is shown in Fig. 10.11b. For convenience of comparison, the azimuthal ambiguity characteristics of conventional single-beam SAR with the same azimuth resolution are also given as shown in Fig. 10.12.

The following conclusions can be drawn from the above simulation results:

(1) When using SPCMAB technique, the PRFs required for both implementations are lower than that in conventional single-beam situations to achieve the same ambiguity performance at the same azimuth resolution. This is just the result of SPCMAB technique, which facilitates the broadening of the swath.

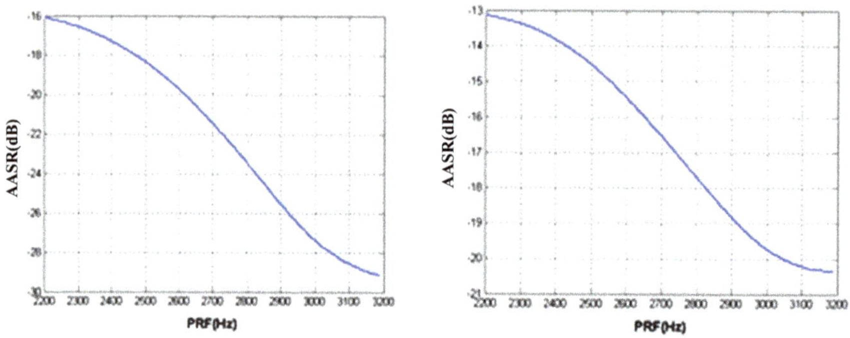

(a) The middle beam (b) The side beams

Fig. 10.10 Azimuthal ambiguity Characteristics of "Wide-transmit narrow-receive" mode

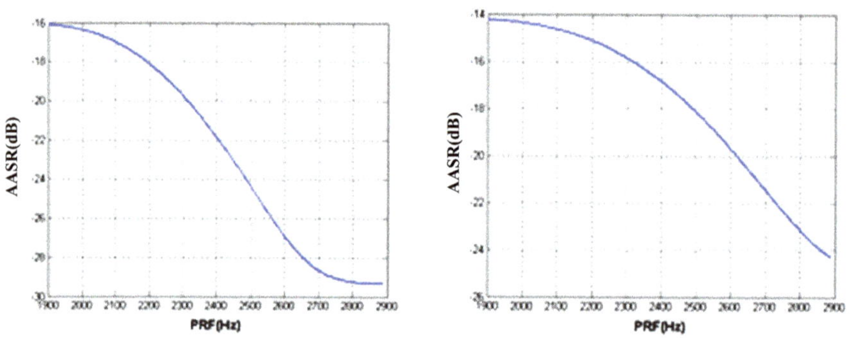

(a) The middle beam (b) The side beams

Fig. 10.11 Azimuthal ambiguity Characteristics of "Narrow-transmit narrow-receive" mode

Fig. 10.12 Azimuthal ambiguity characteristics of conventional single-beam SAR

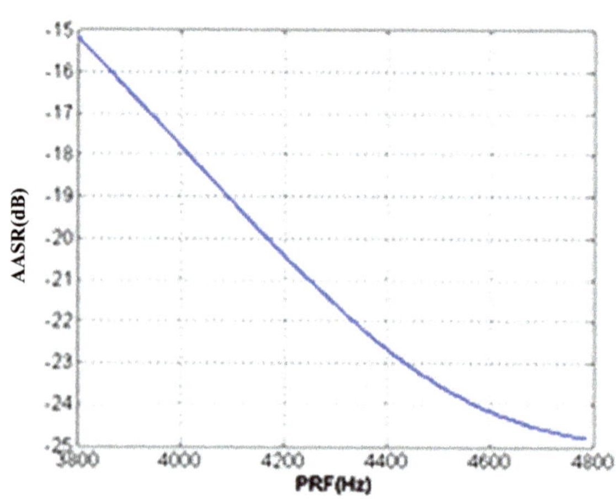

(2) The azimuthal ambiguity of "Wide-transmit narrow-receive" mode is more serious at the same PRF value. This is because that the transmitting beam uses a broadening lobe in this way, and the more areas outside the main lobe of the receiving beam fall on the main lobe of the transmitting beam, and the beam gain in these areas decreases slowly, making the azimuth ambiguity stronger. Therefore during the system design, it can prefer to use "narrow-transmit narrow-receive" approach to achieve single-phase center multi-beam to get a lower azimuth ambiguity to signal ratio.

(3) The azimuthal ambiguity of both side beams is more serious than that of the middle beam under the two modes. This is because in the "wide-transmit narrow-receive" mode, the main lobe of both side receive-beams correspond to the rolling down part of antenna transmission pattern main lobe, and the echoes from the main lobe peak part (corresponding to the middle beam main lobe) of the antenna transmission pattern are ambiguity components of both side receive-beams, which makes the ambiguity on both side receive-beams more serious. In the "narrow-transmit narrow-receive" mode, although the azimuthal ambiguity components on side beams are almost the same as that on the middle beam, but the components forming the desired signal have been reduced. In addition to the desired signal components introduced by side beam itself, only the adjacent middle beam contributes to the desired signal components of the side beams, unlike the middle beam where the two beams on both sides of the beam contribute the desired signal components.

10.3.5 Analysis of Range Ambiguity

From the principle and system structure of SPCMAB technique, it can be seen that the purpose of using SPCMAB technique is to reduce the system operating PRF to obtain a certain azimuth resolution by means of changing the azimuthal beam structure and echo signal characteristics. Therefore, since the formation mechanism and the source of SPCMAB system range ambiguity are the same with the conventional single-beam system, so the range ambiguity to signal ratio (RASR) calculation method is the same with the conventional single-beam system. The only difference is that the system PRF is reduced while obtaining a certain azimuth resolution, making the range ambiguity to signal ratio improved, or allowing the system to broaden the swath while maintaining the same range ambiguity to signal ratio.

The range ambiguity to signal ratio is calculated using the system parameters in Sect. 10.3.4 simulation, and compared with conventional single-beam systems.

First the range ambiguity to signal ratio of the two implementations models under the same mapping bandwidth are calculated. Assuming that the antenna is uniformly weighted in range direction, the antenna width is chosen to be 1.42 m, and the corresponding ground swath width is 45.3 km at incident angle of 35°.

When the "wide-transmit narrow-receive" mode is adopted to implement the SPCMAB system, the pulse repetition frequency is selected as 2826 Hz, and the range ambiguity to signal ratio is calculated as shown in Fig. 10.13a with the incident angle changing from 20° to 50°. When the "narrow-transmit narrow-receive" mode is used, the pulse repetition frequency is selected as 2492 Hz, and the range ambiguity to signal ratio is as shown in Fig. 10.13b with the incident angle in the range of 20° ~ 50°. When the conventional single-beam system is used, the pulse repetition frequency is selected as 4016 Hz, and the range ambiguity to signal ratio is shown in Fig. 10.13c when the incident angle is in the range of 20° ~ 50°. With the central incident angle of 35° as a comparison reference, it can be seen that the conventional single-beam system has a RASR of −23.48 dB, and the SPCMAB system with "wide-transmit narrow-receive" mode has a RASR of −26.02 dB, the SPCMAB system implemented by "narrow-transmit narrow-receive" mode has a RASR of −28.65 dB. Therefore, when the SPCMAB technique is adopted, the RASR is improved obviously under the same swath width.

The following describes the swath width of the SPCMAB system in the two modes under the same RASR. When the "wide-transmit narrow-receive" mode is

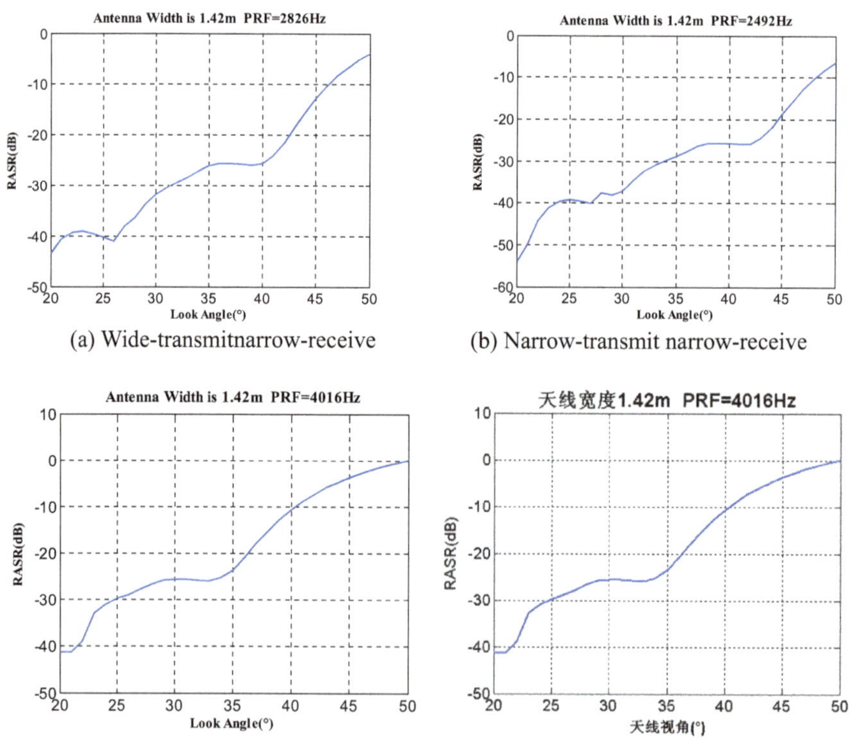

(a) Wide-transmitnarrow-receive (b) Narrow-transmit narrow-receive

(c) Conventional single-beam

Fig. 10.13 RASR under the same swath width

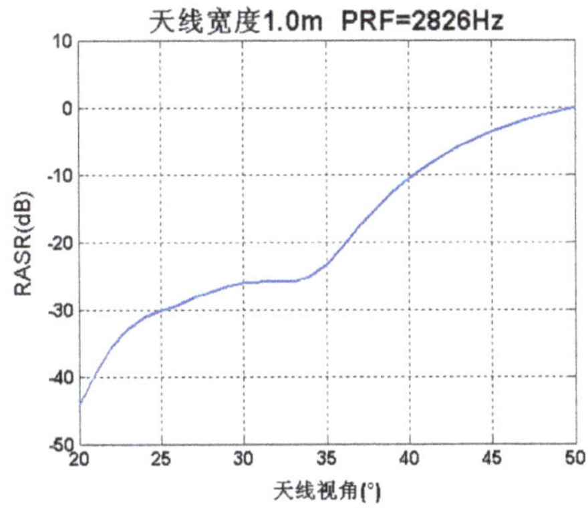

(a) Wide-transmit narrow-receive (b) Narrow-transmit narrow-receive

Fig. 10.14 RASR after changing the antenna width

adopted, the pulse repetition frequency is selected as 2826 Hz, and the antenna width is reduced to 1.0 m, the RASR is shown in Fig. 10.14a when the incident angle is in the range of 20° ~ 50°. When the "narrow-transmit narrow-receive" mode is used, the pulse repetition frequency is maintained at 2492 Hz and the antenna width is reduced to 0.89 m. The RASR is shown in Fig. 10.14b with the incident angle in the range of 20° ~ 50°. Still with the central incident angle of 35° as a comparison reference, the RASR of "wide-transmit narrow-receive" mode is −23.22 dB, the corresponding ground swath width is 64.42 km, and the RASR of "narrow-transmit narrow-receive" mode is −23.60 dB, the corresponding ground swath width is 72.38 km. Compared with the conventional single-beam system, it can be seen that when the SPCMAB technique is adopted, the swath width of the system can be significantly expanded under the same RASR.

10.4 Simulation Verification of SPCMAB SAR

In this section, the space borne point target simulation is used to verify the feasibility of SPCMAB technique to implement high-resolution wide-swath SAR.

Considering the three-beam situation, the basic radar parameters are as follows: the orbit altitude is 450 km, the argument of perigee is 90°, the inclination of the orbit is 98.55°, the right ascension of ascending node is 89.37° [24–26], the operating frequency is S band, the wavelength $\lambda = 0.09375$ m, the antenna azimuth length $L_{aSPC} = 12$ m, the antenna uses uniform weight with incident angle of 35°. The pulse repetition frequency of the system is PRF = 1500 Hz, and the corresponding equivalent pulse repetition frequency is $f_{\mathrm{PRF_e}} = 4500\,\mathrm{Hz}$. The Doppler centroid frequency of the middle beam is calculated using Eq. (10.14) as $f_{dc} = f_{dcM} = -5966.7\,\mathrm{Hz}$.

According to the above parameters, the three-beam situation of the computer simulation has been carry out, and the azimuth spectrums of left, middle and right beam are as shown in Fig. 10.15a, b, c. The data of the three beams are synthesized by the signal processing method proposed above. The azimuth spectrum after synthesis is shown in Fig. 10.15d. The azimuth response of the point target after azimuth compression is shown in Fig. 10.16a, b shows the single beam azimuth response for the same antenna length for comparison.

The analysis shows that the azimuth response of the 4 dB main lobe after the three-beam synthesis corresponds to the azimuth resolution of 2 m, and the single-beam case is about 6 m, so the azimuth resolution of SPCMAB SAR has been improved times equal to the number of beams than that of the conventional SAR. It can be seen that the azimuth resolution can be improved using the SPCMAB technique compared with the conventional single-beam system with the same antenna length in the case of the same PRF. On the other hand, if the conventional single-beam system is used to obtain the azimuth resolution of 2 m, the system PRF should be larger than 3800 Hz because the Doppler bandwidth is about 3800 Hz. Simulation using SPCMAB technique, the system PRF is selected as 1500 Hz that can get 2 m azimuth resolution. Therefore, compared with the conventional single-beam system with the same azimuth resolution, the SPCMAB technique can reduce the system PRF and obtain the wider swath width.

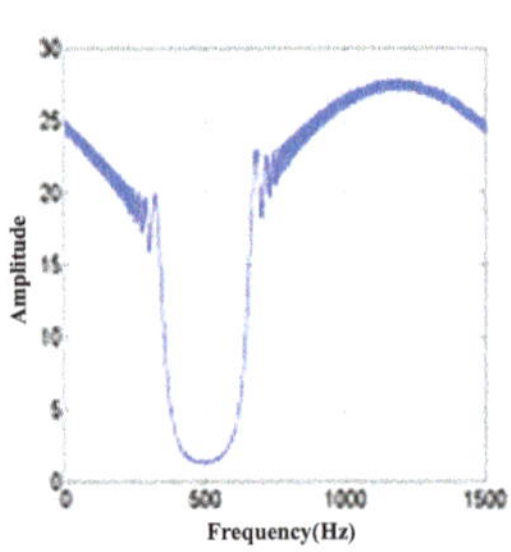

(a) Spectrum of left beam

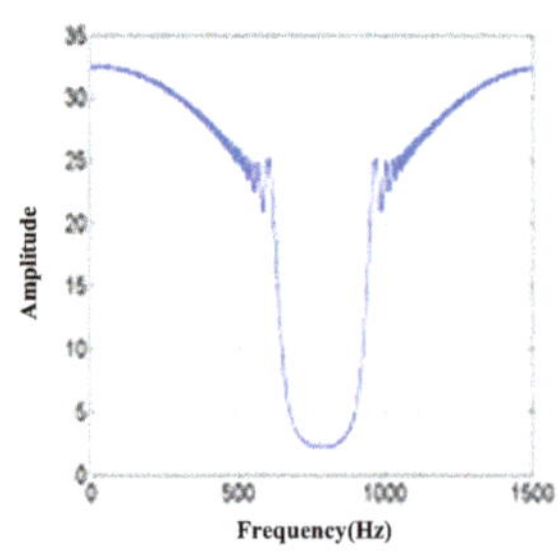

(b) Spectrum of middle beam

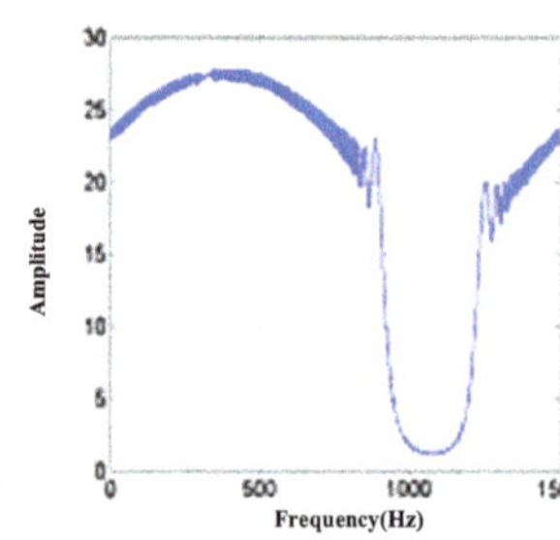

(c) Spectrum of right beam

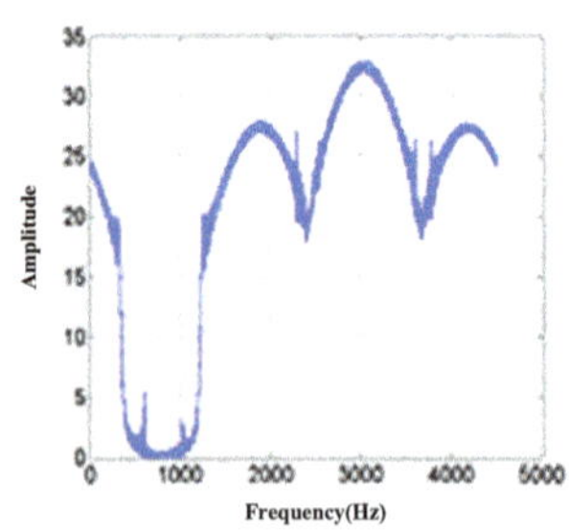

(d) Spectrum of synthesized beam

Fig. 10.15 Spectrum of each sub-beam and synthesized beam of SPCMAB SAR

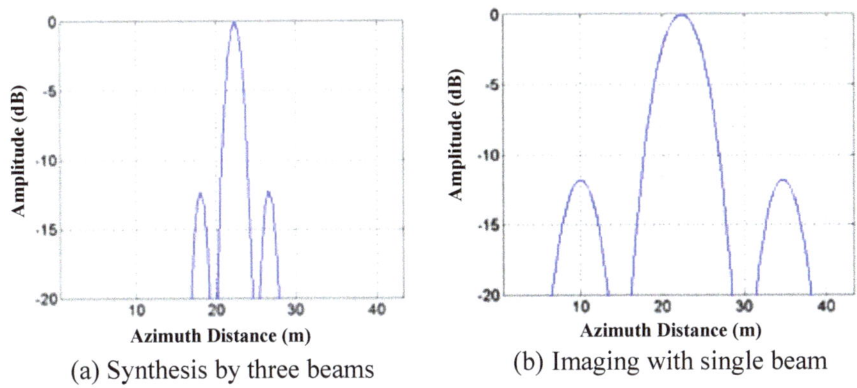

Fig. 10.16 Azimuthal response of point target of single beam and synthesized beam by three beams

10.5 System Model of SPCMAB SAR

In the previous sections, the principle and signal processing method of SPCMAB technique have been introduced. In this section, the system model of SPCMAB SAR will be provided. Section 10.1 points out that there are two ways to achieve SPCMAB, one is "wide-transmit narrow-receive" mode, and the other is "narrow-transmit narrow-receive" mode. The system models of the two modes are shown in Fig. 10.17a, b respectively, where the three-beam case is taken as an example.

For the "wide-transmit narrow-receive" approach, there is only one transmission path, the chirp signal is sent into the circulator corresponding to the middle beam after amplified by the power amplifier, and then is transmitted to the space through the middle beam. The signal is transmitted with a weighted broadened beam, and the broadening times is equal to the number of beams. When the echo is received, the beams are uniformly weighted, and there is a receive path corresponding to each receiving beam. The echo signals received by each beam are azimuthal interpolated and filtered after quadrature demodulation and A/D sampling, the output data from each sub-path will be synthesized. Whole echo data will be formed through synthesis, and the radar images will be generated by means of two-dimensional compression.

For the "narrow-transmit narrow-receive" approach, there are multiple transmission paths. The linear frequency modulation signal is amplified and then sent into the power divider into multiple output, each output is sent into a transmission beam. Each beam is transmitted simultaneously, and the antenna is uniformly weighted in both transmit and receive direction. The reception of the signal and the subsequent processing flow are the same as the previous approach.

For the space borne SAR system, the signal processing and subsequent post-processing in Fig. 10.17 is generally accomplished on the ground.

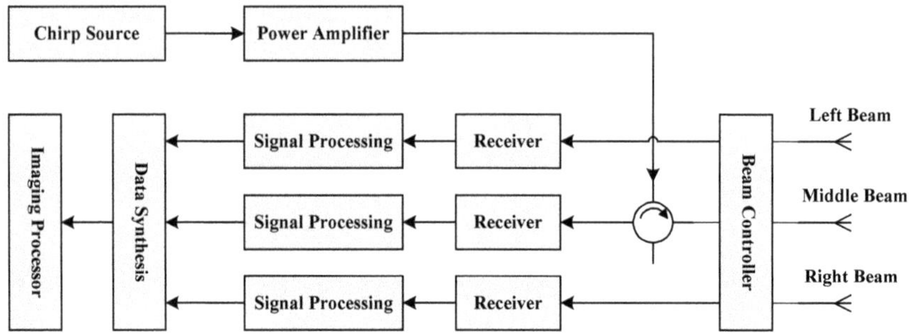

(a) System model of "wide-transmit narrow-receive" approach

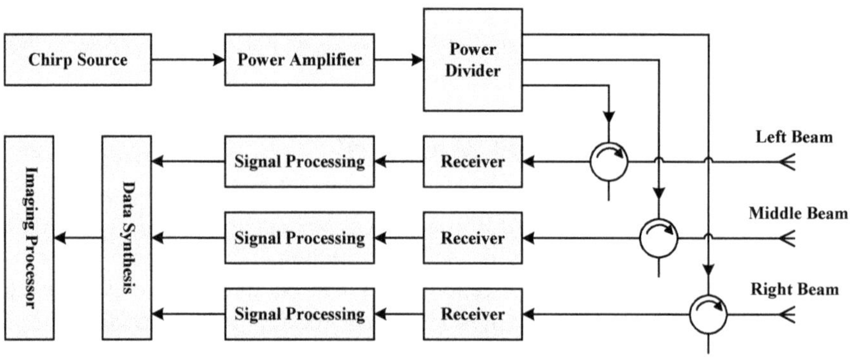

(b) System model of "narrow-transmit narrow-receive" approach

Fig. 10.17 System model of SPCMAB SAR

By adopting the SPCMAB technique, where multiple sub-beams are used in azimuth direction, the azimuthal Doppler bandwidth can be increased without increasing the system operating PRF, and the high resolution can be obtained in azimuth direction. It is a new idea to implement high-resolution wide-swath space borne SAR. In this chapter, the principle of SPCMAB technique has been introduced, the signal characteristics of SPCMAB system have been studied, and a SPCMAB SAR azimuth signal processing method has been proposed, and the method of azimuth signal processing has been analyzed theoretically. The ambiguity characteristic analysis method of SPCMAB SAR has been provided. The feasibility of the SPCMAB technique to realize the high-resolution wide-swath SAR system and the validity of the azimuth signal processing method proposed in this chapter have been proved by space borne point target simulation. A high-resolution imaging algorithm based on conventional single-beam SAR [12, 27–41] can be used to process the synthesized azimuth echo signals after azimuth signal processing with wide Doppler bandwidth to obtain a high-resolution wide-swath radar image.

References

1. Li S (2004) Study on High-resolution wide-swath synthetic aperture radar systems. Ph.D. Thesis of Institute of Electronics, Chinese Academy of Sciences, Beijing
2. Currie A (1989) Wide-swath SAR imaging with multiple azimuth beams. In: IEE colloquium on synthetic aperture radar, 29 Nov 1989
3. Currie A, Hall CD (1990) A synthetic aperture radar technique for the simultaneous provision of high-resolution wide-swath coverage. In: (Conference) Proceedings Military Microwaves'90, Microwave Exhibitions & Publishers Ltd., July 1990
4. Currie A, Brown MA (1992) Wide-swath SAR. IEE Proceedings-F 139(2):122–135
5. [日]藤坂贵彦. (2000) Multi-beam synthetic aperture radar to realize high azimuthal resolution and wide swath. Translated by Wang S, Airborne Radar (4):36–44
6. Davidson GW, Cumming I (1997) Signal properties of space borne squint-mode SAR. IEEE Trans Geosci Remote Sen 35(3):611–617
7. Liu G (2002) Imaging and algorithm study on slant-looking and forward-looking synthetic aperture radar systems. PhD thesis of University of Electronic Science and Technology of China
8. Yuan X (2003) Introduction to space borne synthetic aperture radar. National Defense Industry Press, Beijing, 18 Sept 2003
9. Hu G (2002) Digital signal processing: theory: algorithm and implements. Tsinghua University Press, Beijing
10. Orfanidis SJ (1999) Introduction to signal processing. Tsinghua University Press, Beijing
11. Yang X, Lou C, Xu J (2001) Principles and applications of software radio. Publishing House of Electronics industry, Beijing
12. Zhang C (1989) Synthetic aperture radar: principles, system analysis and applications. Science Press, Beijing
13. Lin M, Ke Y (1984) Radar signal theory. National Defense Industry Press, Beijing
14. Cohen L (1998) Translated by Bai J. Time-frequency analysis: theory and applications. Xi'an Jiaotong University Press, Xi'an
15. Wehner DR (1995) High-resolution radar, 2 Revised ed. Artech House Boston, London, 1995
16. Li S, Yang R (2004) Theory analysis of azimuthal signal processing of single phase center multiple azimuth beams synthetic aperture radar. J Electron Inform Technol Beijing 26 Supplement
17. Li FK, Held Daniel N, Curlander JC, Chialin Wu (1985) Doppler parameter estimation for space borne synthetic-aperture radars. IEEE Trans Geosci Remote Sens GE-23(1):47–55
18. Madsen SN (1989) Estimating the Doppler centroid of SAR data. IEEE Trans Aerosp Electron Syst AES-25(2):134–140
19. Wahl DE, Jakowatz CV, Thompson PA (1994) New approach to strip-map SAR autofocus. In: 1994 sixth IEEE digital signal processing workshop, pp 53–56
20. Chang CY, Curlander JC (1992) Application of the multiple PRF technique to resolve Doppler centroid estimation ambiguity for space borne SAR. IEEE Trans Geosci Remote Sens 30(2):941–949
21. Peng D, Xiang J (1999) Performance analysis of typical auto-focusing algorithm of synthetic aperture radar. J Air Force Radar Acad 13(4):24–27
22. Raney RK (1986) Doppler properties of radars in circular orbits. Int J Remote Sens 7 (9):1153–1162
23. de Vries F P Ph (1991) Estimation of the parameters used in the azimuth compression. AD-A245679
24. Chen F, Jia N (1992) Handbook of satellite telemetry and telecontrol. Science Press, Beijing
25. Zhang S (1999) Fundamentals of external ballistic measurement and satellite orbit measurement. National Defense Industry Press, Beijing
26. Wei Z (2001) Synthetic aperture radar satellite. Science Press, Beijing
27. Curlander JC (1991) Synthetic aperture radar systems and signal processing. Wiley

28. Li S, Yang R (2005) Azimuthal ambiguity analysis of single phase center multiple azimuth beams synthetic aperture radar. J Electron Inform Technology Beijing 27(10):1569–1572
29. Liu Y (1999) Radar imaging technology. Harbin Institute of Technology Press, Harbin
30. Franceschetti G (1999) Synthetic aperture radar processing. CRC Press, New York
31. Zhang J, Xu R, Liu Y (1999) Extended range-doppler algorithm suited to space borne SAR systems with large slant-looking angle. J Syst Eng Electron 21(8):9–11
32. Cafforio C, Prati C, Rocca F (1991) SAR data focusing using seismic migration techniques. IEEE Trans Aero Elec Syst 27(2):194–207
33. Yuan X (1994) Imaging data processing of synthetic aperture radar. Aerosp Shanghai 3:2–8
34. Huang Y, Li C et al (2000) Extended chirp scaling imaging algorithm of high-resolution space borne SAR. Chin J Electron 28(3):35–38
35. Jin MY, Wu C (1984) A SAR correlation algorithm which accommodates large-range migration. IEEE Trans Geosci Remote Sens GE-22(6):592–597
36. Raney RK, Runge H et al (1994) Precision SAR processing using chirp scaling. IEEE Trans Geosci Remote Sens 32(4):786–799
37. Bamler Richard (1992) A comparison of range-Doppler and wavenumber domain SAR focusing algorithm. IEEE Trans Geosci Remote Sens 30(4):706–713
38. Moreira A, Mittermayer J, Scheiber R (1996) Extended chirp scaling algorithm for air- and space borne SAR data processing in strip map and ScanSAR imaging modes. IEEE Trans Geosci Remote Sens 34(5):1123–1136
39. Li C, Zhou Y. (2000) Imaging processing algorithm of high-resolution space borne synthetic aperture radar. J Telemetry Track Command 21(2):15–18,25
40. Eldhuset K (1998) A new fourth-order processing algorithm for space borne SAR. IEEE Trans Aerosp Electron Syst 34(3):824–835
41. Wu C, Liu KY, Jin M (1982) Modeling and a correlation algorithm for space borne SAR signals. IEEE Trans Aerosp Electron Syst AES-18(5):563–574

Chapter 11
Multi-phase Centers Multi-beam SAR

As mentioned in the previous chapter, in order to obtain a wide swath and high azimuth resolution at the same time, it is possible to reduce the system pulse repetition frequency (PRF) and use multiple beams in the azimuth dimension to receive echoes. Using signal processing to increase Doppler bandwidth and make the azimuth resolution improved.

The single phase center multi-beam technique introduced in the previous chapter utilizes multiple azimuth narrow beams to receive echoes so that the Doppler bandwidth of each beam is small, allowing to reduce the system PRF and obtain a wider swath. In the process of azimuth signal processing, the azimuth Doppler spectrum of each beam is spliced to form a wide spectrum. So a higher azimuth resolution is obtained. That is, the single phase center multi-beam technique achieves wide swath and high resolution at the same time through signal processing in frequency domain.

The multi-phase centers multi-beam technology (MPCMB) discussed in this chapter implements high resolution and wide swath of synthetic aperture radar (SAR) through signal processing in time domain. Radar transmits a wide beam, covering a wide Doppler bandwidth. In order to make the azimuth sampling to satisfy the requirements of the sampling theorem, a higher pulse repetition frequency should be used. When the multi-phase centers multi-beam technology is adopted, on the one hand, the PRF is reduced in order to obtain the wide swath. On the other hand, multiple sub-antennas are arranged at a certain interval along the azimuth dimension. The echoes are received at the same time so as to compensate the azimuth insufficient sampling caused by PRF decrease. As long as the distance between the sub-antennas is appropriate, the different sub-antenna can be used to receive the echo instead of the conventional antenna system. A complete azimuth Doppler bandwidth can be got by the signal processing, and the azimuth echo sampling satisfies the requirements of the sampling theorem, allowing the system to obtain higher azimuth resolution with lower PRF [31].

In this chapter, the principle of multi-phase centers multi-beam technology is introduced, and the azimuth signal model is given. The errors introduced by

R. Yang et al., *High-Resolution Microwave Imaging*,
https://doi.org/10.1007/978-981-10-7138-6_11

multi-phase centers multi-beam technology, the influence on imaging and ambiguity characteristics are studies. At last, the azimuth phase correction and signal processing methods are proposed. These methods are verified by simulation experiment.

11.1 Principles of Multi-phase Centers Multi-beam Technique

The basic idea of the multi-phase centers multi-beam technique is to reduce the time sampling rate by increasing space samplings. The SAR pulse mode is equivalent to using PRF to sample echo signal, the process can be seen as a radar echo sampling in the time dimension.

The radar is always moving with the platform, so different sampling times correspond to different radar azimuth positions. This sampling process can also be seen as a spatial dimension sampling along the azimuth direction. When the multi-phase centers multi-beam technology is adopted, the radar arranges multiple sub-antennas along the azimuth direction. Each sub-antenna has the same beam width, covering the same region. The mid-sub-antenna transmit the signal (or the weighted full-aperture beam), and each sub-antennas receive echo signals from the same area.

Figure 11.1 shows the operation principle of multi-phase centers multi-beam technology. For each transmitting pulse, the N_B (beam number) group sampled value of echo signal can be obtained simultaneously along the azimuth direction. As long as the distance of each sub-antenna d is set appropriate, each group of echo samples is independent of each other. So the PRF that is reduced N_B times can ensure the correct sampling of echo signal. In this way, through reducing the time sampling by the increasing the space sampling in the azimuth direction, it is possible to reduce the system PRF to ensure the azimuth resolution, and to increase the swath. Or under the certain PRF (corresponding to a certain swath) condition, the wider azimuth Doppler bandwidth is allowed, so that the azimuth resolution can be improved.

For the convenience of the following description, several concepts are introduced. One is the pulse repetition frequency PRF, that is, the actual pulse repetition

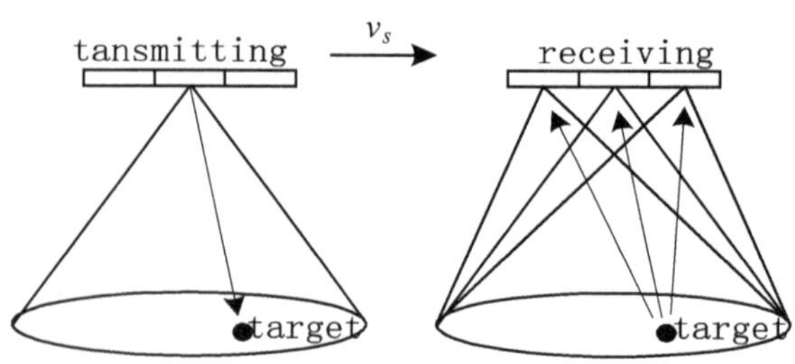

Fig. 11.1 Multi-phase centers multi-beam operation principle

frequency of the system. The equivalent PRFe, that is, after the multi-phase centers multi-beam technique, the increased equivalent sampling rate in the azimuth direction. It corresponds to the N_B times of PRF. The reciprocal is called the equivalent pulse repetition period. And another is an equivalent single-beam system, which refers to the conventional single-beam system with the same orbital parameters, operating frequency and spatial resolution as the multi-phase centers multi-beam system described in this chapter.

11.1.1 The Multi-phase Centers Multi-beam Operation Principle [1–4]

Figure 11.1 is the multi-phase centers multi-beam operation principle diagram. Take three-beams as an example to illustrate the principle of multi-phase centers multi-beam technology. The range of mid-beam phase center to any position on the ground is into the Taylor series:

$$\begin{aligned} R_1(t) &= \sqrt{R_0^2 + v_s^2(t-t_0)^2 - 2R_0 v_s (t-t_0)\sin\theta} \\ &\approx R_0 + \dot{R}_1(t_0)(t-t_0) + \tfrac{1}{2}\ddot{R}_1(t_0)(t-t_0)^2 \\ &= R_0 + \alpha(t-t_0) + \beta(t-t_0)^2 \end{aligned} \qquad t \in \left[-\frac{T_S}{2}, \frac{T_S}{2}\right] \tag{11.1}$$

Where R_0 is the slope distance at the reference time t_0, θ is the angle between R_0 and the normal direction of the antenna,

v_s is the satellite speed, and

$$\alpha = -v_s \sin\theta \quad \beta = \frac{v_s^2 \cos^2\theta}{2R_0} \tag{11.2}$$

The transmitting and receiving round-trip phase history of the mid-beam is:

$$\phi_1(t) = \frac{4\pi}{\lambda} R_1(t) \tag{11.3}$$

The received signal path of the left receiving beam is the mid-beam to the target point to the left beam phase center. If the center distance of the adjacent sub-antenna is d, the distance from the target point to phase center of the left beam is:

$$\begin{aligned} R_2(t) &= R_1\left(t - \frac{d}{v_s}\right) \approx R_0 + \alpha\left(t - t_0 - \frac{d}{v_s}\right) + \beta\left(t - t_0 - \frac{d}{v_s}\right)^2 \\ &= R_1(t) - \left[\frac{\alpha d}{v_s} + \frac{2\beta d(t-t_0)}{v_s} - \frac{\beta d^2}{v_s^2}\right] \end{aligned} \tag{11.4}$$

The corresponding round trip phase history is:

$$\phi_2(t) = \frac{2\pi}{\lambda}R_1(t) + \frac{2\pi}{\lambda}R_1(t - \frac{d}{v_s}) = \phi_1(t) - \frac{4\pi}{\lambda}\left[\frac{\alpha d}{2v_s} + \frac{\beta d(t-t_0)}{v_s} - \frac{\beta d^2}{2v_s^2}\right] \quad (11.5)$$

Similarly, the round trip phase history of the right beam is:

$$\phi_3(t) = \frac{2\pi}{\lambda}R_1(t) + \frac{2\pi}{\lambda}R_1(t + \frac{d}{v_s}) = \phi_1(t) + \frac{4\pi}{\lambda}\left[\frac{\alpha d}{2v_s} + \frac{\beta d(t-t_0)}{v_s} + \frac{\beta d^2}{2v_s^2}\right] \quad (11.6)$$

Now the phase history of received signal at the adjacent three receiving/transmitting positions of the equivalent single-beam system is considered. The operating pulse repetition frequency of the system is PRF^c. Assuming that the flight distance of the antenna is X within a pulse repetition time PRT^c, and the phase of received signal in the mid receiving/transmitting position is:

$$\phi_1^c(t) = \frac{4\pi}{\lambda}R_1(t) \quad (11.7)$$

The path of the received echo at the previous position on its left side is the round trip distance from the center of the antenna phase to the target point, and the one-way distance is:

$$R_2(t) = R_1(t - \frac{X}{v_s}) = R_1(t) - \left[\frac{\alpha X}{v_s} + \frac{2\beta X}{v_s}(t-t_0) - \frac{\beta X^2}{v_s^2}\right] \quad (11.8)$$

The corresponding phase history is:

$$\phi_2^c(t) = \frac{4\pi}{\lambda}R_2(t) = \phi_1^c(t) - \frac{4\pi}{\lambda}\left[\frac{\alpha X}{v_s} + \frac{2\beta X}{v_s}(t-t_0) - \frac{\beta X^2}{v_s^2}\right] \quad (11.9)$$

Similarly, the phase history of the right side of the next position of the received signal is:

$$\phi_3^c(t) = \frac{4\pi}{\lambda}R_3(t) = \frac{4\pi}{\lambda}R_1(t + \frac{X}{v_s}) = \phi_1^c(t) + 2k\left[\frac{\alpha X}{v_s} + \frac{2\beta X}{v_s}(t-t_0) + \frac{\beta X^2}{v_s^2}\right] \quad (11.10)$$

Through comparison of Eqs. (11.5) and (11.9) and (11.6) and (11.10), it can be seen, as long as the distance of beam phase center is set reasonable, $d = 2X$, the echo on both sides of the receiving beam can be used to instead of equivalent single-beam system echo. This process introduces a constant phase error:

$$\Delta\phi = \frac{4\pi}{\lambda} \cdot \frac{\beta X^2}{v_s^2} \tag{11.11}$$

As long as the echo data on both sides of the receiving beam is corrected appropriately, the three-beam echo data can be synthesized to obtain the complete azimuth echo phase history.

Figure 11.2 shows a phase center position diagram of the multi-phase centers multi-beam system operating antenna. For easily understanding, the system is operating in the Nyquist sampling rate in the figure. For a conventional single-beam system operating at the Nyquist sampling rate, the PRF is exactly equal to the azimuth Doppler bandwidth. The azimuth sampling interval is exactly equal to half the antenna length. If the azimuth antenna length is *L*a, the flight distance of antenna is $L_a/2$ in one PRT. In the multi-phase centers multi-beam system, the phase distance of each sub-antenna is equal to the length of the sub-antenna $d = L_a$, and the total length of the azimuth antenna is $L_{\mathrm{aMPC}} = N_B{\cdot}L_a$. Because its operating PRF is lower than the conventional single-beam system by N_B times, the flight distance of antenna in a PRT is $L_{\mathrm{aMPC}}/2$. The position relationship between the multi-phase centers multi-beam system and the equivalent single-beam antenna phase center system is shown in Fig. 11.2. It can be seen that "vacancy" in the echo phase history caused by the low PRF can just be supplement by echo data received by multi-beam on both sides of antenna. It is multi-phase centers multi-beam technology operating mechanism.

In the actual system, in order to restrain the azimuth ambiguity, it is common to set the system PRR greater than the azimuth Doppler bandwidth so that the azimuth sampling has a certain oversampling rate. Each sub-antenna also must overlap with each other so that phase center distance of each beam is no longer the length of the sub-antenna, and the total length of the antenna is no longer N_B times of sub-antenna.

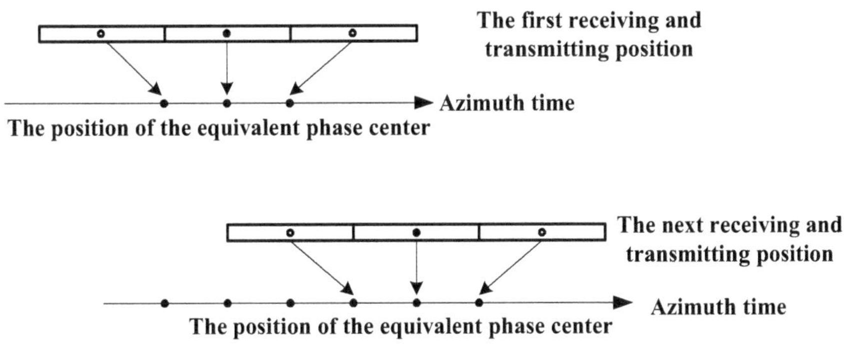

Fig. 11.2 Multi-phase centers multi-beam system antenna location diagram

11.1.2 The Selection of Antenna Phase Center Distance and Total Antenna Length

When the multi-phase centers multi-beam technique is used, the azimuth sampling vacancy due to the decrease of the pulse repetition frequency is filled with the echo signal of the receiving beam only. Figure 11.3 shows the geometric relationship between the beams, which shows the schematic diagram of the receiving beam and the sampling vacancy position needed to fill the multi-phase centers multi-beam. The figure shows only the transmitting beam A and the adjacent receiving beam B on the left side. The echo propagation path of the receiving B is from the A point transmitting through the AP to reach the target P point. The reflecting echo is through PB to the point B, and the range is $(R_1 + R_2)$. Its corresponding "vacancy" position is the midpoint C between A and B. The signal propagation path is the round trip propagating along the CP. That is, the echo of the only the receiving beam B can be approximated by the echo transmitted and received at point C with the conventional single-beam system. If the approximated error of the azimuth phase is small enough, the multi-phase centers multi-beam technique can be used to realize the high resolution wide swath SAR system.

If the flight speed of platform is v_s, the beam center distance d should satisfy the following equation:

$$d = \frac{2v_s}{N_B \cdot f_{\mathrm{PRF}}} \tag{11.12}$$

In where, the PRF is the pulse repetition frequency for the actual operation of the system.

In the actual system, in order to restrain the azimuth ambiguity, the azimuth sampling generally have a certain oversampling rate μ_a:

$$N_B \cdot f_{\mathrm{PRF}} = \mu_a \cdot B_a \tag{11.13}$$

In where, B_a is the azimuth Doppler bandwidth. In this case, the adjacent sub-antennas are partially overlapped. Setting the overlapped length is w, and the

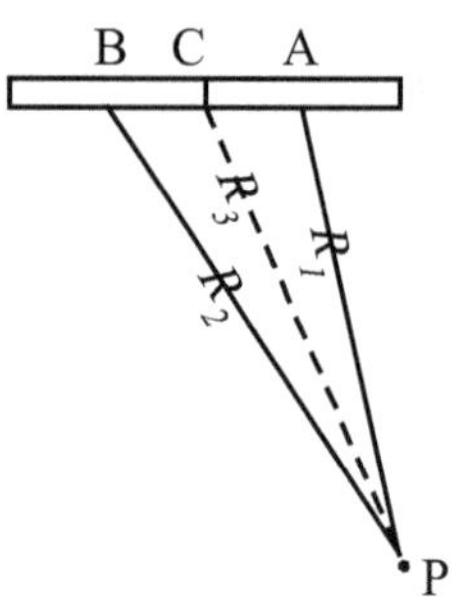

Fig. 11.3 Geometry diagrams of the beams

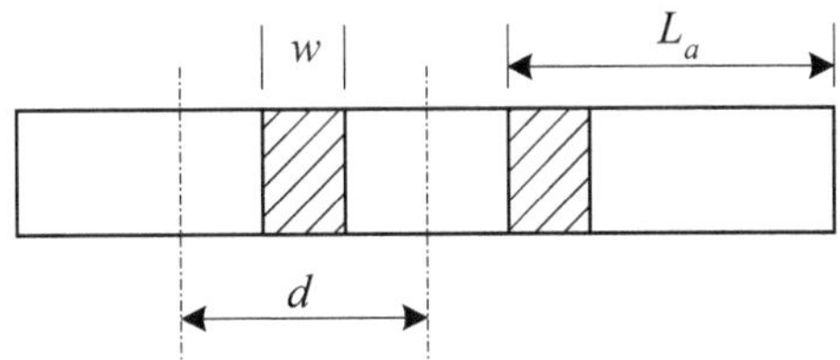

Fig. 11.4 The azimuth antenna size for multi-phase centers multi-beam SAR

actual azimuth antenna size is shown in Fig. 11.4. The antenna phase center interval can be expressed as:

$$d = L_a - w \tag{11.14}$$

From the previous analysis, it can be seen that d is equivalent to the azimuth interval sampling by 2 times of single-beam system, so it is:

$$d = 2 \cdot \frac{v_s}{N_B \cdot f_{\text{PRF}}} = 2 \cdot \frac{v_s}{\mu_a \cdot B_a} = \frac{L_a}{\mu_a} \tag{11.15}$$

From the relationship between the relationship between the Eq. (10.14) and the Eq. (10.15) the size of overlapping part can be obtained:

$$w = \frac{\mu_a - 1}{\mu_a} L_a \tag{11.16}$$

The total length of the antenna is:

$$L_{\text{aMPC}} = N_B L_a - (N_B - 1)w = \frac{N_B + (\mu_a - 1)}{\mu_a} L_a \tag{11.17}$$

11.2 The Echo Signal Mathematical Model of Multi-phase Centers Multi-beam SAR

The mathematical model of echo signal is a useful tool to understand the SAR system from the mathematical point of view. Literature [5] proposed a mathematical model of echo signal applied to spaceborne SAR. In this section, a mathematical model of echo signal for multi-phase centers multi-beam synthetic aperture radar (SAR) is derived, taking a three-beam as an example.

Assumed that the radar transmitting pulse is:

$$p(t) = \text{rect}(\frac{t}{T_p}) \cos(\omega t - \varphi(t)) \tag{11.18}$$

Where $\varphi(t) = -\pi k_r t^2$ is the linear frequency modulation term, k_r is the modulation frequency slope, T_p is the duration of transmitting pulse, rect(·) is the rectangular window function:

$$\mathrm{rect}(\frac{t}{T_p}) = \begin{cases} 1 & |t| \le \frac{T_p}{2} \\ 0 & |t| > \frac{T_p}{2} \end{cases} \tag{11.19}$$

The transmitting pulse can be expressed as:

$$f(t) = \sum_n p(t - n \cdot T_{\mathrm{PRT}}) \tag{11.20}$$

If the distance from the target to the phase center of the mid-beam is R (t), the echo signal after quadrature demodulation can be expressed as:

$$\begin{aligned} s(t) &= s_{-1}(t) + s_0(t) + s_{+1}(t) \\ &= \sigma W_a(t - \frac{R(t-\frac{d}{v_s})}{c}) f(t - \frac{R(t)}{c} - \frac{R(t-d/v_s)}{c}) + \sigma W_a(t - \frac{R(t)}{c}) f(t - \frac{2R(t)}{c}) \\ &\quad + \sigma W_a(t - \frac{R(t+d/v_s)}{c}) f(t - \frac{R(t)}{c} - \frac{R(t+d/v_s)}{c}) \\ &= \sum_n \sigma W_a(t - \frac{R(t-d/v_s)}{c}) \,\mathrm{rect}\left[\frac{t - n \cdot T_{\mathrm{PRT}} - \frac{R(t)}{c} - \frac{R(t-d/v_s)}{c}}{T_p}\right] \cdot \exp\left[-j\frac{2\pi R(t)}{\lambda} \right. \\ &\quad \left. -j\frac{2\pi R(t-d/v_s)}{\lambda} - j\phi(t - n \cdot T_{\mathrm{PRT}} - \frac{R(t)}{c} - \frac{R(t-d/v_s)}{c})\right] \\ &\quad + \sum_n \sigma W_a(t - \frac{R(t)}{c}) \,\mathrm{rect}\left[\frac{t - n \cdot T_{\mathrm{PRT}} - \frac{2R(t)}{c}}{T_p}\right] \cdot \exp\left[-j\frac{4\pi R(t)}{\lambda} \right. \\ &\quad \left. - j\phi(t - n \cdot T_{\mathrm{PRT}} - \frac{2R(t)}{c})\right] \\ &\quad + \sum_n \sigma W_a(t - \frac{R(t+d/v_s)}{c}) \,\mathrm{rect}\left[\frac{t - n \cdot T_{\mathrm{PRT}} - \frac{R(t)}{c} - \frac{R(t+d/v_s)}{c}}{T_p}\right] \cdot \exp\left[-j\frac{2\pi R(t)}{\lambda} \right. \\ &\quad \left. -j\frac{2\pi R(t+d/v_s)}{\lambda} - j\phi(t - n \cdot T_{\mathrm{PRT}} - \frac{R(t)}{c} - \frac{R(t+d/v_s)}{c})\right] \end{aligned} \tag{11.21}$$

Where d is the center distance between the adjacent beams, v_s is the satellite flight speed, σ is the target backscatter coefficient, W_a (t) is the antenna pattern function in the azimuth direction.

R (t) and W_a (t) in the above equation can be approximated as $R(n \cdot PRT)$ and $W_a(n \cdot PRT)$, since R (t) and W_a (t) are slow-changing functions of time t relative to the transmitting signal during satellite flight. So the above equation becomes:

$$
\begin{aligned}
s(t) = & \sum_n \sigma W_a(n \cdot T_{\mathrm{PRT}}) \operatorname{rect}\left[\frac{t - n \cdot T_{\mathrm{PRT}} - \frac{R(n \cdot T_{\mathrm{PRT}})}{c} - \frac{R(n \cdot T_{\mathrm{PRT}} - d/v_s)}{c}}{T_p}\right] \cdot \exp\left[-j\frac{2\pi R(n \cdot T_{\mathrm{PRT}})}{\lambda}\right. \\
& \left. - j\frac{2\pi R(n \cdot T_{\mathrm{PRT}} - d/v_s)}{\lambda} - j\phi(t - n \cdot T_{\mathrm{PRT}} - \frac{R(n \cdot T_{\mathrm{PRT}})}{c} - \frac{R(n \cdot T_{\mathrm{PRT}} - d/v_s)}{c})\right] \\
& + \sum_n \sigma W_a(n \cdot T_{\mathrm{PRT}}) \operatorname{rect}\left[\frac{t - n \cdot T_{\mathrm{PRT}} - \frac{2R(n \cdot T_{\mathrm{PRT}})}{c}}{T_p}\right] \cdot \exp\left[-j\frac{4\pi R(n \cdot T_{\mathrm{PRT}})}{\lambda}\right. \\
& \left. - j\phi(t - n \cdot T_{\mathrm{PRT}} - \frac{2R(n \cdot T_{\mathrm{PRT}})}{c})\right] \\
& + \sum_n \sigma W_a(n \cdot T_{\mathrm{PRT}}) \operatorname{rect}\left[\frac{t - n \cdot T_{\mathrm{PRT}} - \frac{R(n \cdot T_{\mathrm{PRT}})}{c} - \frac{R(n \cdot T_{\mathrm{PRT}} + d/v_s)}{c}}{T_p}\right] \cdot \exp\left[-j\frac{2\pi R(n \cdot T_{\mathrm{PRT}})}{\lambda}\right. \\
& \left. - j\frac{2\pi R(n \cdot T_{\mathrm{PRT}} + d/v_s)}{\lambda} - j\phi(t - n \cdot T_{\mathrm{PRT}} - \frac{R(n \cdot T_{\mathrm{PRT}})}{c} - \frac{R(n \cdot T_{\mathrm{PRT}} + d/v_s)}{c})\right]
\end{aligned}
\tag{11.22}
$$

Let $t = K \cdot T_{\mathrm{PRT}} + \tau$, $t_d \le \tau \le t_d + T_{\mathrm{PRT}}$. t_d is the minimum delay time for the echo. At this point the echo signal can be expressed as:

$$
\begin{aligned}
s(K, \tau) = & \sum_n \sigma W_a(n \cdot T_{\mathrm{PRT}}) \operatorname{rect}\left[\frac{K \cdot T_{\mathrm{PRT}} + \tau - n \cdot T_{\mathrm{PRT}} - \frac{R(n \cdot T_{\mathrm{PRT}})}{c} - \frac{R(n \cdot T_{\mathrm{PRT}} - d/v_s)}{c}}{T_p}\right] \\
& \cdot \exp\left[-j\frac{2\pi R(n \cdot T_{\mathrm{PRT}})}{\lambda} - j\frac{2\pi R(n \cdot T_{\mathrm{PRT}} - d/v_s)}{\lambda}\right. \\
& \left. - j\phi(K \cdot T_{\mathrm{PRT}} + \tau - n \cdot T_{\mathrm{PRT}} - \frac{R(n \cdot T_{\mathrm{PRT}})}{c} - \frac{R(n \cdot T_{\mathrm{PRT}} - d/v_s)}{c})\right] \\
& + \sum_n \sigma W_a(n \cdot T_{\mathrm{PRT}}) \operatorname{rect}\left[\frac{K \cdot T_{\mathrm{PRT}} + \tau - n \cdot T_{\mathrm{PRT}} - \frac{2R(n \cdot T_{\mathrm{PRT}})}{c}}{T_p}\right] \\
& \cdot \exp\left[-j\frac{4\pi R(n \cdot T_{\mathrm{PRT}})}{\lambda} - j\phi(K \cdot T_{\mathrm{PRT}} + \tau - n \cdot T_{\mathrm{PRT}} - \frac{2R(n \cdot T_{\mathrm{PRT}})}{c})\right] \\
& + \sum_n \sigma W_a(n \cdot T_{\mathrm{PRT}}) \operatorname{rect}\left[\frac{K \cdot T_{\mathrm{PRT}} + \tau - n \cdot T_{\mathrm{PRT}} - \frac{R(n \cdot T_{\mathrm{PRT}})}{c} - \frac{R(n \cdot T_{\mathrm{PRT}} + d/v_s)}{c}}{T_p}\right] \\
& \cdot \exp\left[-j\frac{2\pi R(n \cdot T_{\mathrm{PRT}})}{\lambda} - j\frac{2\pi R(n \cdot T_{\mathrm{PRT}} + d/v_s)}{\lambda}\right. \\
& \left. - j\phi(K \cdot T_{\mathrm{PRT}} + \tau - n \cdot T_{\mathrm{PRT}} - \frac{R(n \cdot T_{\mathrm{PRT}})}{c} - \frac{R(n \cdot T_{\mathrm{PRT}} + d/v_s)}{c})\right]
\end{aligned}
\tag{11.23}
$$

The echo delay time of the three receive beams satisfy:

$$t_d \le \frac{R(n \cdot T_{\mathrm{PRT}})}{c} + \frac{R(n \cdot T_{\mathrm{PRT}} - d/v_s)}{c} < t_d + T_{\mathrm{PRT}} - T_p \tag{11.24a}$$

$$t_d \le \frac{2R(n \cdot T_{\mathrm{PRT}})}{c} < t_d + T_{\mathrm{PRT}} - T_p \tag{11.24b}$$

$$t_d \le \frac{R(n \cdot T_{\mathrm{PRT}})}{c} + \frac{R(n \cdot T_{\mathrm{PRT}} + d/v_s)}{c} < t_d + T_{\mathrm{PRT}} - T_p \tag{11.24c}$$

There are:

$$0 \le \tau - \frac{R(n \cdot T_{\mathrm{PRT}})}{c} - \frac{R(n \cdot T_{\mathrm{PRT}} - \frac{d}{v_s})}{c} \le T_p \tag{11.25a}$$

$$0 \le \tau - \frac{2R(n \cdot T_{\mathrm{PRT}})}{c} \le T_p \tag{11.25b}$$

$$0 \le \tau - \frac{R(n \cdot T_{\mathrm{PRT}})}{c} - \frac{R(n \cdot T_{\mathrm{PRT}} + \frac{d}{v_s})}{c} \le T_p \tag{11.25c}$$

Thus, only $K \cdot T_{\mathrm{PRT}} = n \cdot T_{\mathrm{PRT}}$, the rect [] are not zero and the echo signal can be written as:

$$\begin{aligned} s(K,\tau) = {} & \sum_n \sigma W_a(n \cdot T_{\mathrm{PRT}}) \operatorname{rect}\left[\frac{\tau - \frac{R(n \cdot T_{\mathrm{PRT}})}{c} - \frac{R(n \cdot T_{\mathrm{PRT}} - d/v_s)}{c}}{T_p}\right] \delta(K \cdot T_{\mathrm{PRT}} - n \cdot T_{\mathrm{PRT}}) \\ & \cdot \exp\left[-j\frac{2\pi R(n \cdot T_{\mathrm{PRT}})}{\lambda} - j\frac{2\pi R(n \cdot T_{\mathrm{PRT}} - d/v_s)}{\lambda}\right] \cdot \exp\left[-j\phi\left(\tau - \frac{R(n \cdot T_{\mathrm{PRT}})}{c}\right.\right. \\ & \left.\left. - \frac{R(n \cdot T_{\mathrm{PRT}} - d/v_s)}{c}\right)\right] + \sum_n \sigma W_a(n \cdot T_{\mathrm{PRT}}) \operatorname{rect}\left[\frac{\tau - \frac{2R(n \cdot T_{\mathrm{PRT}})}{c}}{T_p}\right] \cdot \delta(K \cdot T_{\mathrm{PRT}} - n \cdot T_{\mathrm{PRT}}) \\ & \exp\left[-j\frac{4\pi R(n \cdot T_{\mathrm{PRT}})}{\lambda}\right] \cdot \exp\left[-j\phi(\tau - \frac{2R(n \cdot T_{\mathrm{PRT}})}{c})\right] \\ & + \sum_n \sigma W_a(n \cdot T_{\mathrm{PRT}}) \operatorname{rect}\left[\frac{\tau - \frac{R(n \cdot T_{\mathrm{PRT}})}{c} - \frac{R(n \cdot T_{\mathrm{PRT}} + d/v_s)}{c}}{T_p}\right] \\ & \cdot \delta(K \cdot T_{\mathrm{PRT}} - n \cdot T_{\mathrm{PRT}}) \cdot \exp\left[-j\frac{2\pi R(n \cdot T_{\mathrm{PRT}})}{\lambda} - j\frac{2\pi R(nT_{\mathrm{PRT}} + d/v_s)}{\lambda}\right] \\ & \cdot \exp\left[-j\phi(\tau - \frac{R(n \cdot T_{\mathrm{PRT}})}{c} - \frac{R(n \cdot T_{\mathrm{PRT}} + d/v_s)}{c})\right] \end{aligned} \tag{11.26}$$

Setting $x = v_s \cdot n \cdot \mathrm{PRT}$, the echo signal can be written as:

$$
\begin{aligned}
s(x,\tau) = {} & \sigma W_a(x)\exp\left[-j\frac{2\pi R(x)}{\lambda}-j\frac{2\pi R(x-d)}{\lambda}\right]\otimes\left\{\frac{1}{v_s}\delta(x)\,\mathrm{rect}\left[\frac{\tau-\frac{R(x)}{c}-\frac{R(x-d)}{c}}{T_p}\right]\right.\\
& \left.\cdot\exp\left[-j\phi(\tau-\frac{R(x)}{c}-\frac{R(x-d)}{c})\right]\right\}\\
& +\sigma W_a(x)\exp\left[-j\frac{4\pi R(x)}{\lambda}\right]\otimes\left\{\frac{1}{v_s}\delta(x)\,\mathrm{rect}\left[\frac{\tau-\frac{2R(x)}{c}}{T_p}\right]\cdot\exp\left[-j\phi(\tau-\frac{2R(x)}{c})\right]\right\}\\
& +\sigma W_a(x)\exp\left[-j\frac{2\pi R(x)}{\lambda}-j\frac{2\pi R(x+d)}{\lambda}\right]\otimes\left\{\frac{1}{v_s}\delta(x)\,\mathrm{rect}\left[\frac{\tau-\frac{R(x)}{c}-\frac{R(x+d)}{c}}{T_p}\right]\right.\\
& \left.\cdot\exp\left[-j\phi(\tau-\frac{R(x)}{c}-\frac{R(x+d)}{c})\right]\right\}\\
= {} & \sigma W_a(x)\exp\left[-j\frac{2\pi R(x)}{\lambda}-j\frac{2\pi R(x-d)}{\lambda}\right]\otimes\left\{\frac{1}{v_s}\delta(x)\delta\left(\tau-\frac{R(x)}{c}-\frac{R(x-d)}{c}\right)\right\}\\
& \otimes\left\{rect\left(\frac{\tau}{T_p}\right)\cdot\exp[-j\phi(\tau)]\right\}\\
& +\sigma W_a(x)\exp\left[-j\frac{4\pi R(x)}{\lambda}\right]\otimes\left\{\frac{1}{v_s}\delta(x)\delta\left(\tau-\frac{2R(x)}{c}\right)\right\}\otimes\left\{\mathrm{rect}\left(\frac{\tau}{T_p}\right)\cdot\exp[-j\phi(\tau)]\right\}\\
& +\sigma W_a(x)\exp\left[-j\frac{2\pi R(x)}{\lambda}-j\frac{2\pi R(x+d)}{\lambda}\right]\otimes\left\{\frac{1}{v_s}\delta(x)\delta\left(\tau-\frac{R(x)}{c}-\frac{R(x+d)}{c}\right)\right\}\\
& \otimes\left\{\mathrm{rect}\left(\frac{\tau}{T_p}\right)\cdot\exp[-j\phi(\tau)]\right\}
\end{aligned}
\tag{11.27}
$$

Setting $r=\frac{c\tau}{2}$, the echo signal can be expressed as:

$$
\begin{aligned}
s(x,r) = {} & \sigma W_a(x)\exp\left[-j\frac{2\pi R(x)}{\lambda}-j\frac{2\pi R(x-d)}{\lambda}\right]\otimes\left\{\frac{2}{cv_s}\delta(x)\delta[r-R(x)-R(x-d)]\right\}\\
& \otimes\{\mathrm{rect}(r)\cdot\exp[-j\phi(r)]\}\\
& +\sigma W_a(x)\exp\left[-j\frac{4\pi R(x)}{\lambda}\right]\otimes\left\{\frac{2}{cv_s}\delta(x)\delta[r-2R(x)]\right\}\otimes\{\mathrm{rect}(r)\cdot\exp[-j\phi(r)]\}\\
& +\sigma W_a(x)\exp\left[-j\frac{2\pi R(x)}{\lambda}-j\frac{2\pi R(x+d)}{\lambda}\right]\otimes\left\{\frac{2}{cv_s}\delta(x)\delta[r-R(x)-R(x+d)]\right\}\\
& \otimes\{\mathrm{rect}(r)\cdot\exp[-j\phi(r)]\}
\end{aligned}
\tag{11.28}
$$

As mentioned in the previous section, the condition for using the multi-phase centers multi-beam technique is that the following approximation (that is the approximated error is sufficiently small compared to the wavelength):

$$
\left.\begin{aligned}
R(x)+R(x-d)\approx 2R(x-\tfrac{d}{2})\\
R(x)+R(x+d)\approx 2R(x+\tfrac{d}{2})
\end{aligned}\right\}
\tag{11.29}
$$

Thus, the echo signal expression can be approximated as:

$$\begin{aligned} s(x,r) \approx{} & \sigma W_a(x)\exp\left[-j\frac{4\pi R(x-\frac{d}{2})}{\lambda}\right]\otimes\left\{\frac{2}{cv_s}\delta(x)\delta\left[r-2R(x-\frac{d}{2})\right]\right\} \\ & \otimes\{\text{rect}(r)\cdot\exp[-j\phi(r)]\} \\ & +\sigma W_a(x)\exp\left[-j\frac{4\pi R(x)}{\lambda}\right]\otimes\left\{\frac{2}{cv_s}\delta(x)\delta[r-2R(x)]\right\}\otimes\{\text{rect}(r)\cdot\exp[-j\phi(r)]\} \\ & +\sigma W_a(x)\exp\left[-j\frac{4\pi R(x+\frac{d}{2})}{\lambda}\right]\otimes\left\{\frac{2}{cv_s}\delta(x)\delta\left[r-2R(x+\frac{d}{2})\right]\right\} \\ & \otimes\{\text{rect}(r)\cdot\exp[-j\phi(r)]\} \end{aligned} \tag{11.30}$$

The second term in the above equation is the same as the conventional single-beam echo signal model. Therefore, as long as the distance of adjacent beam phase center d satisfies the Eq. (11.12), the first and third terms in the above equation are, respectively equivalent to the receiving/transmitting echo signal at the fore-and-aft position of the single-beam system. So the above equation can be equivalent to the receiving echo of the conventional single-beam SAR at three adjacent positions.

According to the derivation of the mathematical model of the multi-phase centers multi-beam echo signal, the beam echoes should be synthesized into one signal before imaging processing. It is similar with the conventional single-beam SAR with the equivalent PRF. So the conventional SAR high-resolution imaging algorithms [5–19] can be used for imaging.

The errors exist when receiving echo of the multi-phase centers multi-beam is used to replace the echo of the equivalent single-beam system. Before the signal synthesized, the phase error correction should be performed. The introduced errors in the multi-phase centers multi-beam are analyzed in next sections, and the methods of phase error correction and azimuth signal processing are proposed.

11.3 The Error Analysis of Multi-phase Centers Multi-beam SAR [20]

Before the imaging processing, the echo data of each receiving beam should be phase corrected, respectively. The phase correction is the prerequisite for normal operation of the multi-phase centers multi-beam SAR. This section focuses on the causes of various phase errors and their effects on imaging. The methods of phase correction and azimuth signal processing are discussed in the next section.

There are three kinds of errors that may be introduced in the multi-phase centers multi-beam technique: the error in signal propagation path, the amplitude and phase imbalance error of the receiving channels, and the error caused by the azimuth sampling unevenness. For normal system operation, these errors need to be corrected before imaging, or to ensure that the impact of imaging is limited to tolerable limits.

11.3.1 Error Introduced in Signal Propagation Path

Multi-phase centers multi-beam technology uses the echo data of the only receiving beam to instead of the real echo data for the conventional single-beam SAR in the corresponding position. The signal propagation path error introduces the phase error in the azimuth direction. The error can be understood from two aspects. First, there is propagation error in the echo of the only receiving beam. The second is that the echo of the only receiving beam has the discontinuity of the signal propagation path during the flight of the satellite.

First, assuming that the ground point target does not move, Fig. 11.5 shows the phase error introduced by the multi-phase centers multi-beam. Figure 11.5a shows the cause of this error, A is the receiving/transmitting beam, B and D are the receiving beams. $\boldsymbol{R}_s$ and $\boldsymbol{R}_t$ is the position vectors of the satellite and ground point targets, respectively in the inertial coordinate system. d is the phase center distance of beams, $\mathbf{e}_v$ is the unit vector in the flight direction of the satellite, then $\boldsymbol{R}_1 = \boldsymbol{R}_s - \boldsymbol{R}_t$, $\boldsymbol{R}_2 = \boldsymbol{R}_1 - d\cdot\mathbf{e}_v$, $\boldsymbol{R}_3 = \boldsymbol{R}_1 - (d/2)\cdot\mathbf{e}_v$, $\boldsymbol{R}_4 = \boldsymbol{R}_1 + d\cdot\mathbf{e}_v$, $\boldsymbol{R}_5 = \boldsymbol{R}_1 + (d/2)\cdot\mathbf{e}_v$. The echo signal received by the receiving beam B is simulated as the echo signal at the AB middle center position C of the conventional SAR. The propagation path difference is $\Delta R_L = (|R_1| + |R_2|) - 2|R_3|$, which introduces the azimuth phase error:

$$\phi_{eL} = \frac{2\pi}{\lambda}\cdot\Delta R_L = \frac{2\pi}{\lambda}(|\mathbf{R}_1| + |\mathbf{R}_2| - 2|\mathbf{R}_3|) \tag{11.31}$$

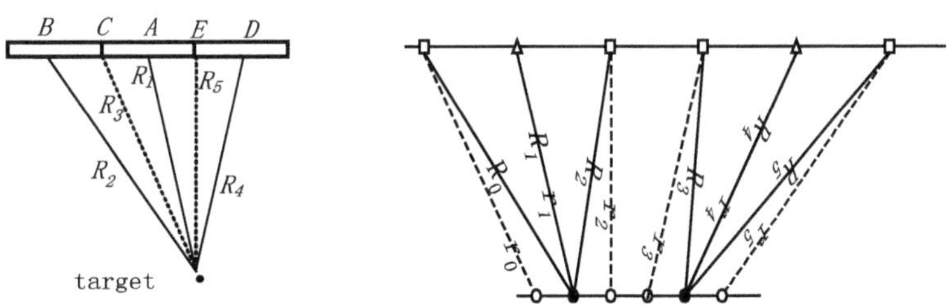

(a) not consider the Earth's rotation effect (b) consider the Earth's rotation effect

Fig. 11.5 Phase error of multi-phase centers multi-beam

The phase error in the azimuth direction introduced by the receiver beam D is:

$$\phi_{eR} = \frac{2\pi}{\lambda} \cdot \Delta R_R = \frac{2\pi}{\lambda} (|\mathbf{R}_1| + |\mathbf{R}_4| - 2|\mathbf{R}_5|) \tag{11.32}$$

This indicates that the echo of the only receiving beam has a phase error caused by the signal propagation path error.

In the operation process of spaceborne SAR, with the flight of the satellite, the earth is also rotating; making the ground point of the target will form their own movement trajectory [21]. Figure 11.5b shows the relative motion between the satellite and the ground point target in the case of three beams. The small triangles represent the position of the intermediate receiving/transmitting beam during the satellite motion, and the small squares represent the equivalent position of the receiving beams on both sides. The dots indicate the point target position at the time of each transmitting pulse with PRF, and the circle represents the point target position corresponding to the single-beam system with the equivalent pulse repetition frequency PRFe. It can be seen from the figure that the conventional SAR with the equivalent pulse repetition frequency, the echo distance progress is r_0 $r_1 \to r_2 \to r_3 \to \ldots$, which is a continuous progress of sampling. In the multi-phase centers multi-beam system, the echo distance progress is $R_0 \to R_1 \to R_2 \to R_3 \to \ldots$, there is a mutation. In each pulse repetition period, there will be errors between them (happened on the corresponding position of the two sides receiving beams). For the multi-phase centers multi-beam system, the azimuth reference function is constructed referring to the equivalent single-beam system operating with the equivalent pulse repetition frequency. So there is a mismatch between the azimuth echo signal and the reference function. The azimuth phase error $\frac{4\pi}{\lambda}(R_i - r_i)$ is introduced in the echo signal, that is, the phase error is introduced by the discontinuous of the signal propagation path. The analysis shows that the phase error is the main part of the phase errors introduced by the signal propagation path difference.

Take the three-beam as an example to illustrate the phase error caused by the signal propagation path difference. As we know that for conventional SAR, the azimuth phase history is:

$$\phi_a^c(n) = -\frac{4\pi}{\lambda} R(n \cdot T_{\mathrm{PRT}^c}) \quad n = 0, 1, 2, \ldots \tag{11.33}$$

Where PRTc is the conventional SAR pulse repetition period. According to the above description, for the three-beam SAR, the synthesized azimuth phase history can be expressed as:

$$\phi_a(m) = -\frac{4\pi}{\lambda} R(m \cdot \frac{T_{\mathrm{PRT}}}{3}) + \phi_e(m) = -\frac{4\pi}{\lambda} R(m \cdot T_{\mathrm{PRT}\,e}) + \phi_e(m), \quad m = 0, 1, 2, \ldots \tag{11.34}$$

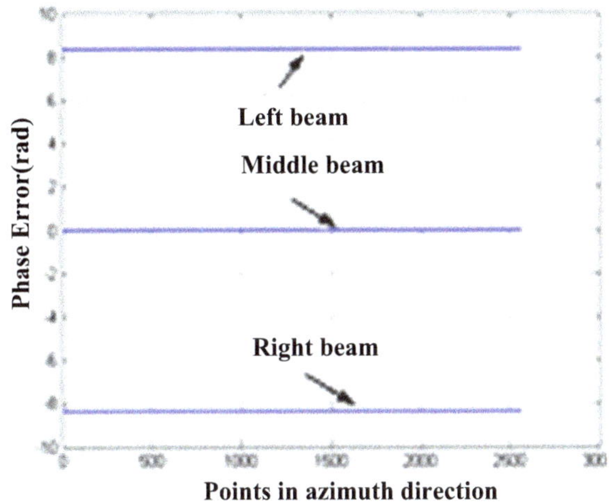

Fig. 11.6 Phase error of each beam

Where PRT is the system duty pulse repetition period and PRTe is the equivalent pulse repetition period. When mod $(m, 3) = 0$, it is the left beam phase, then $\varphi_e(m) \neq 0$, and there is a phase error; when mod $(m, 3) = 1$, it is the mid-beam phase, then $\varphi_e(m) = 0$, and there is no phase error; when mod $(m, 3) = 2$, it is the right beam phase, $\varphi_e(m) \neq 0$, and there is a phase error.

The simulation conditions are as follows: the satellite orbit height is 450 km, the operation wavelength is 0.09375 m, the azimuth resolution is 2 m, the look angle of the antenna is 35°, the distance of the beam centers is 3.4344 m, the system operating pulse repetition frequency $T_{\mathrm{PRT}} = 1500\mathrm{Hz}$, the equivalent pulse repetition frequency is 4500 Hz. Figure 11.6 is the beam phase error caused by the signal propagation path difference. As can be seen from the figure, the phase error is even more than 2π. If not compensated, it will cause serious mismatch of the azimuth matched filter, and system cannot correctly imaging. Figure 11.7a, b gives the target compression results in the azimuth direction before and after phase compensation.

11.3.2 Phase Error Introduced by Amplitude and Phase Imbalance of the Receiving Channels and Azimuth Sampling Nonuniformity

The influence of error introduced by amplitude and phase imbalance of the receiving channels and azimuth sampling nonuniformity on the system performance can be analyzed by paired echo theory. The frequency characteristic of amplitude and phase distortion of the system produces paired echo in the time domain. The waveform distortion of the chirp signal in time domain can cause paired echo in the

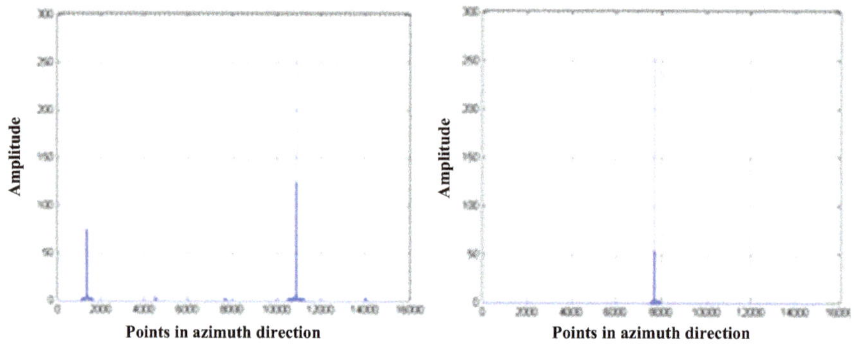

(a) before compensation (b) after compensation

Fig. 11.7 The target compression results in azimuth direction before and after phase compensation

frequency domain. It will be reflected on the output waveform in the time domain after compression by the matched filter [6]. In addition, the emergence of pairs of echo will inevitably lead to the main peak decline [22, 23].

(1) **Phase error of amplitude and phase imbalance of receiving channels**

In the case of multi-phase centers multi-beam technology, each beam has a separate receiving channel from the antenna to the A/D output. In order to obtain better image quality, the amplitude and phase shift of each receiving channel is required accordant. However, during the system implementation, there may exist amplitude or phase imbalance of the receiving channels. Regardless of the amplitude and phase imbalance of receiving channels, the signal synthesized by the receiving beams is presented as periodic errors associated with the beam number N_B.

The amplitude and phase errors of the signal of each receiving beam can be regarded as the result that the continuous amplitude and phase error signal sampled by the equivalent pulse repetition frequency T_{PRTe}. The synthesized azimuth signal can be approximated as:

$$s_a(t) = \left[a_0 + a_1 \cos\left(2\pi \cdot \frac{f_{PRFe}}{N_B} \cdot t + \theta_1\right)\right] \exp\left\{j\left[\omega_d t + \pi k_a t^2 + b_1 \sin\left(2\pi \cdot \frac{f_{PRFe}}{N_B} \cdot t + \theta_2\right)\right]\right\} \tag{11.35}$$

where a_0 is the amplitude of the ideal signal, a_1 is the amplitude error, ω_d is the azimuth Doppler angular frequency, k_a is the azimuth linear chirp slope, b_1 is the phase error, θ_1 and θ_2 are amplitude and phase error terms in the Fourier expansion.

When there is only amplitude imbalance, it is equivalent to there is parasitic amplitude modulation on the azimuth LFM signal. The above equation can be written as:

$$s_a(t) = a_0 e^{j\left[\omega_d t + \pi k_a t^2\right]} + \frac{a_1}{2} e^{j\left[\left(\omega_d + \frac{2\pi f_{\mathrm{PRFe}}}{N_B}\right)t + \pi k_a t^2 + \theta_1\right]} + \frac{a_1}{2} e^{j\left[\left(\omega_d - \frac{2\pi f_{\mathrm{PRFe}}}{N_B}\right)t + \pi k_a t^2 - \theta_1\right]} \tag{11.36}$$

There are three outputs after matching filter on this signal. The first output is the main wave. The output signal is not distorted, but the amplitude will be decreased due to the power distribution of the amplitude modulation signal. Compared with the ideal signal, the decreased multiple is reduced as $\frac{1}{\sqrt{1 + a_1^2/2a_0^2}}$. The second and third corresponding output is the paired echo, and the ratio between its amplitude and the amplitude of the main wave is $\frac{a_1}{2a_0}\left|\cos\left[\left(\omega_d \pm \frac{\pi f_{\mathrm{PRFe}}}{N_B}\right)\frac{f_{\mathrm{PRFe}}}{N_B k_a}\right]\right|$. The emergence location from the main wave is $\pm\frac{f_{\mathrm{PRFe}}}{N_B k_a}$.

When there is only phase imbalance, and phase imbalance is $b_1 < 0.5$ rad, Eq. (11.35) can be approximated as:

$$\begin{aligned} s_a(t) = {} & a_0 J_0(b_1) e^{j\left[\omega_d t + \pi k_a t^2\right]} + \frac{a_0 b_1}{2} e^{j\left[\left(\omega_d + \frac{2\pi f_{\mathrm{PRFe}}}{N_B}\right)t + \pi k_a t^2\right]} \\ & - \frac{a_0 b_1}{2} e^{j\left[\left(\omega_d - \frac{2\pi f_{\mathrm{PRFe}}}{N_B}\right)t + \pi k_a t^2\right]} \end{aligned} \tag{11.37}$$

Where $J_0(\ )$ is the first class Bessel function of the 0th order. There are three outputs after matching filter on this signal. The first output is the main wave. The output signal is not distorted, only the amplitude will be decreased. Compared to the ideal signal, the decreased multiple is $J_0(b_1)$. The second and third corresponding output is the paired echo with opposite polarity. The ratio between its amplitude and the amplitude of the main wave is $b_1/2$, and the emergence location from the main wave is $\pm\frac{f_{\mathrm{PRFe}}}{N_B k_a}$.

(2) **Error introduced by the azimuth sampling nonuniformity**

When the multi-phase centers multi-beam technique is adopted, the relationship of the beam center distance d, the system operation pulse repetition frequency f_{PRF} and the satellite velocity v_s should satisfy the Eq. (11.12), to ensure that the echo signal is uniformly sampled in the azimuth direction. In fact, on the one hand, because the satellite orbit is generally elliptical, the satellite is at the different orbital position in the flight process. The speed is continuous change, but the d value once determined is generally difficult to change in operation process. It makes at some of the track position the relationship of Eq. (11.12) cannot be strictly satisfied, resulting in the azimuth sampling nonuniformity. On the other hand, when PRF is selected, the PRF of different wave position is different due to the influence of the transmitting pulse interference and the echo of the satellite nadir. The available PRF does not necessarily satisfy the relationship of Eq. (11.12), also resulting in the azimuth sampling nonuniformity.

For analysis convenience, it is assumed that the satellite velocity is not deviated, and the azimuth sampling nonuniformity is caused by the deviation of PRF in Eq. (11.12). The nominal pulse repetition frequency of the system is PRF, and the actual repetition frequency is: $f_{\mathrm{PRF}} + \Delta f$, then the azimuth path difference caused by the PRF deviation is:

$$\Delta X = \frac{\Delta f}{N_B \cdot f_{\mathrm{PRF}} \cdot (f_{\mathrm{PRF}} + \Delta f)} \cdot v_s \tag{11.38}$$

Correspondingly, the azimuth phase error is:

$$\Delta\phi = \frac{4\pi}{\lambda}\left[\sqrt{R_t^2 + (\Delta X)^2 + 2R_t \cdot \Delta X \cdot \sin\theta} - R_t\right] \tag{11.39}$$

Where: R_t is the range of the satellite to the ground point at that moment, and θ is the squint angle.

From the above analysis, it can be seen that the influence of the azimuth sampling nonuniformity on the system is similar to the phase difference between the receiving channels. In the azimuth direction, the periodically leading and lagging phase error are introduced, so the paired echo appears in the azimuth compression output.

(3) **Computer simulation**

Taking the three-beam as an example, the amplitude and phase imbalance, the azimuth sampling nonuniformity are simulated. For the amplitude imbalance, based on the mid-beam, assuming that, the amplitude of the receiving beam signal on the left side is k times than the mid-beam, and the amplitude of the receiving beam signal on the right side is $1/k$ times than the mid-beam. Then assuming the signal amplitude of the mid-beam is 1 commonly. The synthesized azimuth signal corresponds to that the signal with amplitude 1 in the conventional SAR is superimposed by amplitude error sequence. The corresponding error signal can be approximated as follows:

$$\mathrm{e}(t) = \frac{f}{2}(k - \frac{1}{k})\cos\left(2\pi \cdot \frac{f_{\mathrm{PRF}e}}{3} \cdot t + \theta\right) \tag{11.40}$$

Where f is a factor, corresponding to $k = 0.5 - 5$ dB, $f = 1.15$.

For phase imbalances, the mid-beam is still based. Assuming that the left beam shifts are more φ_e than mid-beam, the right beam shifts are less φ_e than mid-beam. The synthesized function of azimuth phase is equivalent to that the phase error sequence is superimposed on the phase function of the conventional SAR operating at the equivalent pulse repetition frequency. The corresponding continuous phase error signal can be approximated in the following form:

$$\varphi_e(t) = \frac{2}{\sqrt{3}}\phi_e \sin\left(2\pi \cdot \frac{f_{\mathrm{PRF}e}}{3} \cdot t + \frac{\pi}{3}\right) \tag{11.41}$$

According to the system parameters given in Sect. 11.3.1, different amplitude and phase imbalance errors are applied to the echo data of the three receiving beams, respectively. The value of the operation PRF is changed to the simulate azimuth sampling nonuniformity. The point target imaging is computer simulated, respectively.

The simulation results show that when the amplitude imbalance reaches 3 dB, the ratio of the maximum pair of echoes to the main wave in the azimuth compression output can reach 20lg (0.102) = −19.8 dB, while the main wave decreases only 0.35 dB. When the phase imbalance reaches 15°, the ratio of the maximum pair of echoes to the main wave in the azimuth compression output reaches 20lg (0.086) = −21.3 dB, and the main wave decreases by only 0.2 dB (Fig. 11.8). When the error of the operation pulse repetition frequency reaches 30 Hz, the ratio of the maximum pair of echoes to the main wave in the compressed output can still reach 20lg (0.1059) = −19.5 dB and the main amplitude decreases by 0.29 dB. When the PRF is further deviated from the ideal value, the pair of echoes is further increased and can not be satisfied imaging performance requirements for spaceborne SAR (Fig. 11.9).

From the above simulation results, it can be seen that the influence of the amplitude and phase imbalance and the azimuth sampling nonuniformity of the receiving channel mainly representation as the azimuth compression produces the paired echo. The ambiguity interfere occurs in the image and the main wave amplitude decreases. The signal to noise ratio of the image should be decreased. In the process of system design, for the amplitude and phase imbalance of each receiving channel, on the one hand, it can take some technical methods to reduce

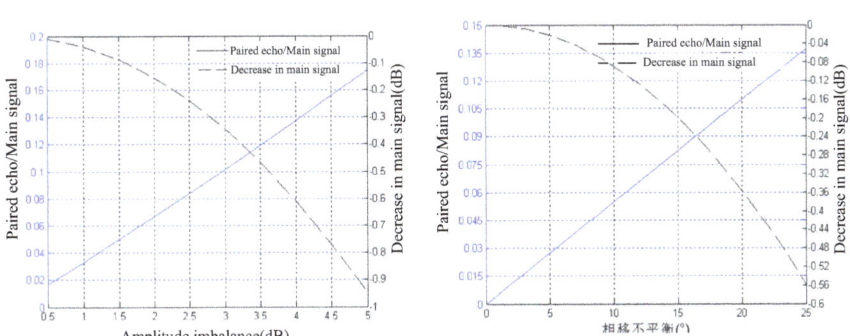

(a) Amplitude imbalance (b) Phase imbalance

Fig. 11.8 Relationship of amplitude and phase imbalance, paired echo, and main wave decline

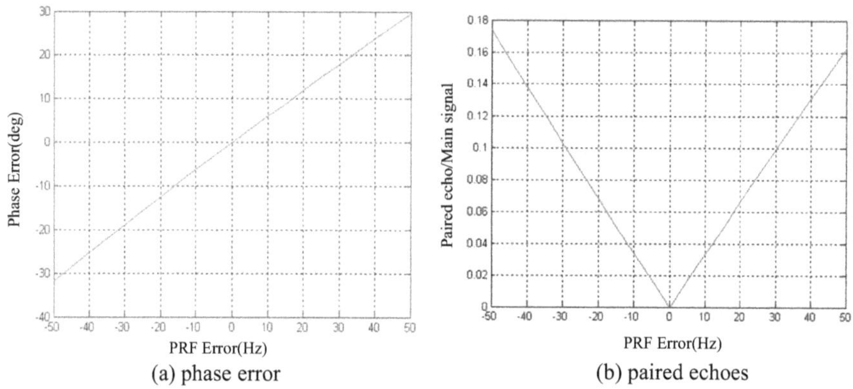

Fig. 11.9 The phase error and paired echo caused by the PRF error

the imbalances of amplitude and phase. On the other hand, it can use the internal calibration method to measure the amplitude and phase of each receiving channel. In ground imaging processing, it can be compensated based on these measurements.

11.4 The Signal Processing of the Multi-phase Centers Multi-beam Synthetic Aperture Radar

In the previous section, the several phase errors that may be introduced when using multi-phase centers multi-beam technology were analyzed. It is pointed out that the error introduced by each receiver channel can be eliminated by means of the calibration method. In this section, the phase error caused by nonuniform sampling and the phase error introduced by the discontinuous signal propagation path are further analyzed. The correction methods of these phase errors are studied.

11.4.1 Compensation for the Azimuth Nonuniform Sampling Error

The compensation process of the nonuniform sampling error is actually the reconstruction process of the nonuniform sampling signal. For the reconstruction of azimuth nonuniformly sampled signals, some algorithms have been proposed in [21, 24–27], such as interpolation, reconstruction in frequency domain, reconstruction in time domain, and Spectral-Fit algorithm, however, those algorithms have some limitations.

The basic idea of the interpolation method is to use the function value at the known sampling point to obtain the function value at the ideal time point by means of the interpolation equation. Interpolation method only uses the adjacent signal information to interpolate the required signal. The reconstructed quantity is worse compared to other algorithms, and a large amount of computation is required.

The basic idea of the reconstruction in frequency domain method is that the spectrum of the sampling sequence is weighted by the original spectrum, and the weighting coefficient is independent of the sampling spectrum. The spectrum of the uniform sampling sequence can be obtained only need calculate the weighting coefficient. The algorithm uses the Discrete-Time Fourier Transform (DTFT) to calculate the spectrum of the nonuniform sampling sequence. A large amount of computation is required.

The basic idea of the reconstruction in time domain method is that the band-limited signal can reconstruct the uniform sampling point directly from the reconstructed Equation by its periodic nonuniform sampling point. The algorithm is similar to the interpolation method, and its main drawback is also a large amount of computation. It is difficult to implement engineering.

The basic idea of the Spectral Fit algorithm is to set a set of N point frequency values (for nonuniform sampling points). Then find the corresponding Fourier coefficient set so as that the points after the application of the inverse Fourier transform (IDFT) and the original data point set at a given nonuniform sampling have the smallest variance. Because of the iterations to find the minimum variance, the amount of computation is also large.

The following study is based on better performance of the filter bank reconstruction algorithm (or the solution of the azimuth Doppler ambiguity algorithm). The basic idea of the solution of the azimuth Doppler ambiguity based on filter bank is the principle of reconstructing the signal based on the filter bank. The integrated filter bank is obtained by the reconstructed condition and the known analysis filter bank. Uniform sampling signal is reconstructed through using the integrated filter bank to reconstruct the nonuniform sampling signal and to eliminate the imaging false targets. The algorithm has the advantages of small computation, good reconstruction quantity, high processing precision, and easy realization of the engineering.

(1) **The principle of complete reconstruction (PR)**

For the standard filter bank of M channels, the diagram is shown in Fig. 11.10. $H_0(j\omega), H_1(j\omega), \ldots, H_{M-1}(j\omega)$ is the analysis filter bank, and $G_0(j\omega)$, $G_1(j\omega), \ldots, G_{M-1}(j\omega)$ is the integrated filter bank. At the signal transmitting end, through the analysis filter bank, the input signal $x_a(t)$ is then sampled. The sampling period is MT.

In the signal receiving end, the input signal is interpolated, and multiple is M. Then through the integrated filter bank, the signal is superimposed. The reconstruction signal is obtained $y(n)$, where the sampling period is T.

To completely reconstruct the signal, it is necessary to make the signal $y(n)$ equal to $x_a(n)$, or a delay of $x_a(n)$, that is $y(n) = \sigma x_a(n-d)$. The σ is a constant which expresses the amplitude change. The d is the delay cycle number. It is an

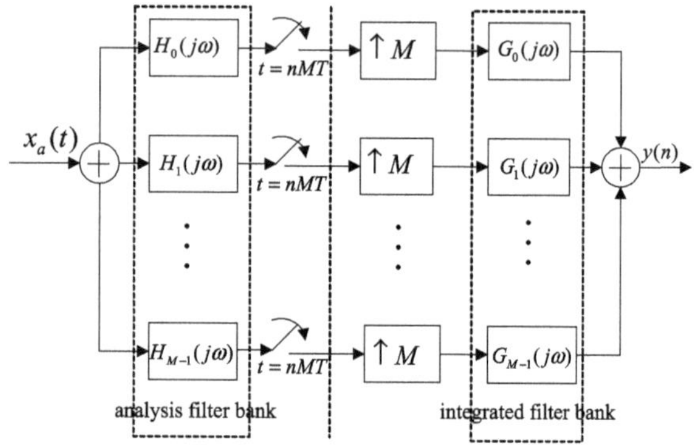

Fig. 11.10 The filter bank of M channels

integer multiple of the sampling period T. When fully reconstructed, the integrated filter bank and the analysis filter bank are required to satisfy the relationship of Eq. (11.42) [28–30].

$$\frac{1}{M}\sum_{k=0}^{M-1} G_k\left(e^{j\omega T}\right)H_k\left(j\omega - j\frac{2\pi p}{\mathrm{MT}}\right) = \begin{cases} \sigma e^{jd\omega T} & p = 0, |\omega| < \frac{\pi}{T} \\ 0 & |p| = 1, 2, \ldots, K_0, |\omega| < \frac{\pi}{T} \end{cases} \tag{11.42}$$

Where, $K_0 = \left\lceil \frac{M(\pi + \omega_0 T)}{2\pi} \right\rceil - 1$, ω_0 is the bandwidth of the input signal $x_a(t)$.

11.4.1.1 Principle of Azimuth Signal Reconstruction Based on Filter Bank

For the azimuth multi-beam SAR systems with M beams, echo signal received by each beam is shown in Fig. 11.11 (taking three-beam example) when nonuniform sampling is performed.

In Fig. 11.11, the echo signal received by each beam are equivalent to the echo signal passing through the analysis filter bank and then sampling with the sampled period MT. Therefore, to reconstruct the input signal $x_a(n)$, the key is to find the integrated filter bank. After the interpolation with M times, the echo received by the azimuth multi-beam SAR channel is superimposed through the integrated filter bank. The output signal is obtained $y(n) = \sigma x_a(n - d)$.

For multi-phase centers multi-beam spaceborne SAR, the equivalent sampling time series of each beam is

$$t = \mathrm{nMT} + t_m,\ n = -\infty \ldots -1, 0, 1, \ldots +\infty,\ m = 0, 1, \ldots, M-1 \tag{11.43}$$

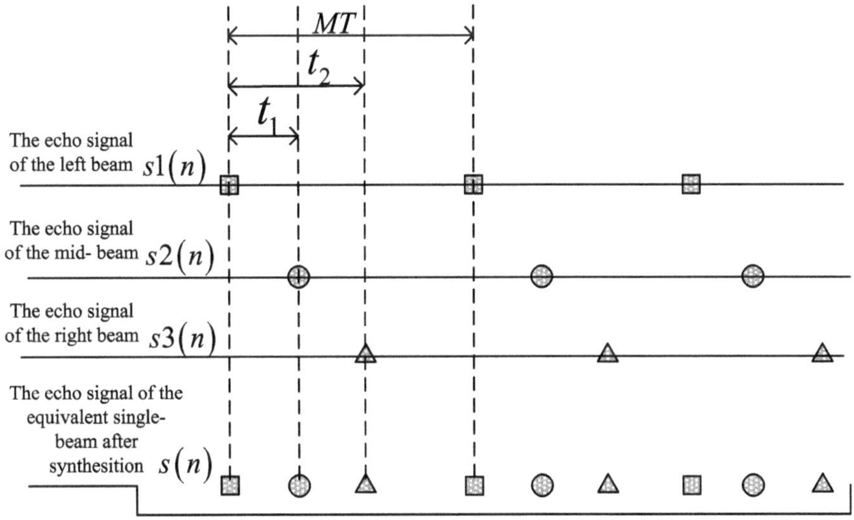

Fig. 11.11 Schematic diagram of the echo signal received by each beam in azimuth multi-beam SAR

After echo signal is sampled, a sampling sequence is obtained as

$$x_m(n) = x_a(\text{nMT} + t_m),\ n = -\infty \ldots -1,\ 0,\ 1,\ \ldots + \infty,\ m = 0,\ 1,\ \ldots,\ M-1 \tag{11.44}$$

To obtain the sampling sequence, only the analysis filter bank satisfies (11.45).

$$H_m(j\omega) = \mathrm{e}^{j\omega t_m},\ m = 0,\ 1,\ \ldots,\ M-1 \tag{11.45}$$

The integrated filter bank is

$$G_m(\mathrm{e}^{j\omega T}) = \begin{cases} a_m\sigma \mathrm{e}^{-j\omega(t_m + \mathrm{dT})} & |\omega| < \omega_0 \\ 0 & \omega_0 \le |\omega| \le \frac{\pi}{T} \end{cases} \tag{11.46}$$

Under the condition of complete reconstruction, put the Eqs. (11.45) and (11.46) into Eq. (11.42).

$$\sum_{m=0}^{M-1} a_m \mathrm{e}^{-j(2\pi p/\mathrm{MT}))t_m} = \begin{cases} M & p = 0 \\ 0 & |p| = 1,\ 2,\ \ldots K_0 \end{cases} \tag{11.47}$$

The coefficient a_m calculating from the Eq. (11.47) is put into the Eq. (11.46). Then, the integrated filter bank $G_k(\mathrm{e}^{j\omega T})$ is obtained. The integrated filter $G_m(\mathrm{e}^{j\omega T})$ is a fractional delay filter with different gain constants. From the Eq. (11.46), it can be seen that the integrated filter are the low-pass filter, and the cut-off frequency is the bandwidth of input signal ω_0. Therefore, in order to simplify the design, the integrated filter bank can be decomposed into two steps to achieve: firstly, let the signal from each channel pass through the corresponding all-pass filter $E_m(e^{j\omega T})$. The different channel signals have different gain and delay. Then, let the superimposed signal of each channel pass through a low-pass filter $F(e^{j\omega T})$. The cut-off frequency of the low-pass filter is the same as the cut-off frequency $G_m(e^{j\omega T})$. The diagram of the before and after equivalent integrated filter bank is shown in Fig. 11.12.

$$E_m\left(e^{j\omega T}\right) = a_m \sigma e^{-j\omega(t_m + d_1 T)}, \; m = 0, 1, \ldots, M-1 \tag{11.48}$$

$$F\left(e^{j\omega T}\right) = \begin{cases} e^{-j\omega d_2 T} & |\omega| < \omega_0 \\ 0 & |\omega| \geq \omega_0 \end{cases} \tag{11.49}$$

Where, $d = d_1 + d_2$, d is the delay between the output signal and the input signal, that is, $y(n) = \sigma x_a(n - d)$.

According to the above analysis, only the coefficients a_m of the integrated filter bank are calculated, the signal can be reconstructed. For the calculation a_m, the Eq. (11.47) can be written in matrix form

$$\mathbf{Ba} = \mathbf{c} \tag{11.50}$$

That is

$$\begin{bmatrix} u_0^{-K_0} & u_1^{-K_0} & \cdot\ \cdot & u_{M-1}^{-K_0} \\ u_0^{-(K_0-1)} & u_1^{-(K_0-1)} & \cdot\ \cdot & u_{M-1}^{-(K_0-1)} \\ \cdot & \cdot & & \cdot \\ \cdot & \cdot & & \cdot \\ u_0^{K_0} & u_1^{K_0} & \cdot\ \cdot & u_{M-1}^{K_0} \end{bmatrix} \begin{bmatrix} a_0 \\ a_1 \\ \cdot \\ \cdot \\ a_{M-1} \end{bmatrix} = \begin{bmatrix} c_0 \\ c_1 \\ \cdot \\ \cdot \\ c_{2K_0} \end{bmatrix} \tag{11.51}$$

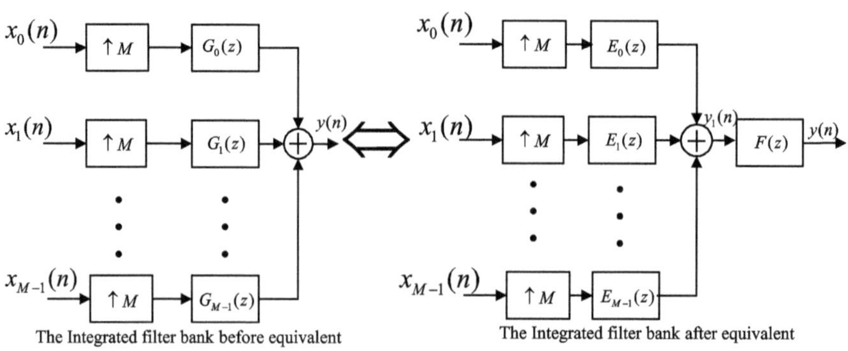

Fig. 11.12 The diagram of the before and after equivalent integrated filter bank

where, $u_k = e^{-j(2\pi/\mathrm{MT})t_k}$.

From Eq. (11.47), it can be seen,

$$c_q = \begin{cases} M & q = K_0 \\ 0 & q = 0, 1, \ldots, 2K_0, q \neq K_0 \end{cases} \tag{11.52}$$

When $2K_0 + 1 = M$, the matrix **B** is a nonsingular matrix, so by the Eq. (11.50)

$$\mathbf{a} = \mathbf{B}^{-1}\mathbf{c} \tag{11.53}$$

In the imaging process, if the RD algorithm is used for imaging, the signal reconstruction process can be placed before or after the Range compression. RD algorithm cannot be dealt with the azimuth direction priority. At first, it deals with the range compression, and then the azimuth compression. However, the CS algorithm is different. In the process of CS algorithm, the first step is the azimuth FFT. So, when use CS imaging algorithm, the reconstruction process can only be placed before the range compression. The echo received by each beam is firstly reconstructed from the nonuniform Sampling signal. Then the CS algorithm for imaging processing is used.

11.4.2 Compensation Processing of Error Introduced Signal Propagation Path

When the multi-phase center multi-beam technique is used, the phase error caused by the discontinuity of the signal propagation path is large to be analyzed. The discontinuity is caused by the rotation of the earth. The estimation method needs to be found, and the compensation needs to be carry out before the imaging process.

11.4.2.1 Phase Error Estimation Method

When multi-phase center multi-beam technique is used, the N_B group echo data can be obtained for each transmitting pulse during the satellite flight. They are reflected from the target point at the same time. When the radar transmits the next pulse, due to the rotation of the earth, the ground point target turned a certain distance. In this procedure, the propagation path of the receiving echo on both side beams experienced jumping changes, resulting in the discontinuity of echo range history. So the corresponding echo phase history is not continuous. Figure 11.13 shows the relative position relationships of the beams and ground targets during satellite flight. S_0 is the position of the mid-beam. S_L and S_R are the positions of adjacent receive beams on both sides. S_{-1} and S_{+1} are the corresponding position of the equivalent single-beam system operated with the equivalent PRF.T_{-1} and T_{+1} are the

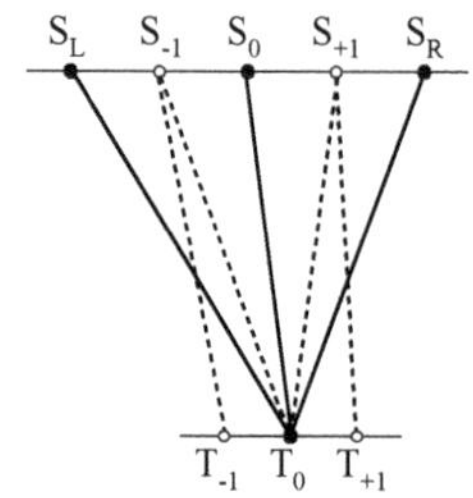

Fig. 11.13 The relationship between the beam and the ground point target's position

corresponding positions of the equivalent ground point target. T_0 is the target point position at the current time. S_L is the echo of the equivalent single-beam system at S_{-1}. The one-way range history is $\frac{1}{2}(|S_0T_0| + |S_LT_0|)$. The one-way range history of the equivalent single-beam system at S_{-1} is $|S_{-1}T_{-1}|$. The range difference between the both will result in a larger phase error at relatively high-frequency system. The range difference can be approximated by the projection of $|T_{-1}T_0|$ on $|S_{-1}T_0|$, where $|T_{-1}T_0|$ is the distance that the point target turned over for an equivalent pulse repetition time. The phase error caused by the range difference during the satellite flight slowly changes. In a synthetic aperture time, it can be approximated as a constant.

The estimation method of this phase error is further discussed below.

From the above analysis it can be seen that during the satellite flight, the latitude of the satellite nadir constantly changes. The latitude of the ground point irradiated by the line of sight of the beam center is also constantly changing. When the multi-phase centers multi-beam technique is adopted, the phase error due to the signal propagation path discontinuous which caused by the Earth's rotation effect is also changing. At the different range gates on the swath, due to different range, the incident angle is different. So the phase error will be different. In this paper, the estimation method of the echo phase's error at the center of the swath at different latitudes is mainly analyzed. As for the variation of the phase error of the different range gates on the swath can be estimated according to the angle variation between the range gates and the center of the swath.

As can be seen from Fig. 11.13, the difference of signal propagation path caused by the earth's rotation can be expressed as a projection of $|T_{-1}T_0|$ on $|S_{-1}T_0|$. For convenience estimation, it can be approximated by the projection of $|T_{-1}T_0|$ on $|S_0T_0|$. Figure 11.14 shows the geometric relationship between the satellite and the ground used to calculate the range difference. T is the ground point irradiated by the line of sight of radar S. The corresponding antenna incident angle is γ, O is the center of the earth. The earth core angle between the radar and the target T is α. And the Φ_s and Φ_t are the latitudes of the radar and the target point T at the current time. the slope range of ST is R_0. The satellite height is Alt. The earth radius is Re. δ *is* the angle between tangent lines of the latitude circle at the TA and the radar line of sight. By derivation, the δ satisfies the following equation:

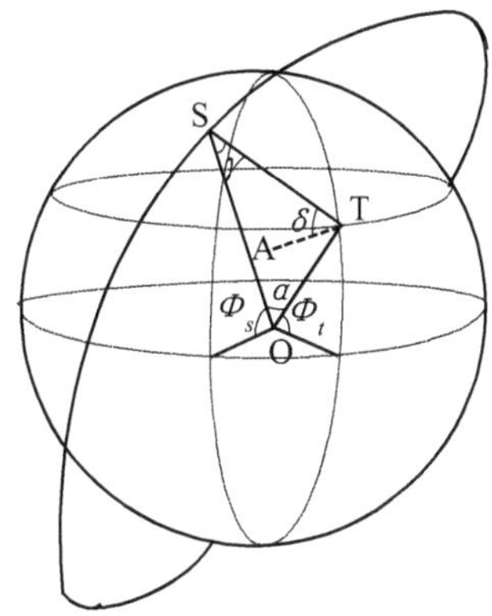

Fig. 11.14 The geometric relationship between the satellite and the ground used to calculate the range difference

$$\cos\delta = \frac{R_e + \mathrm{Alt}}{R_0} \cdot \frac{\sqrt{[\cos(\Phi_s + \Phi_t) + \cos\alpha][\cos(\Phi_s - \Phi_t) - \cos\alpha]}}{\cos\Phi_t} \tag{11.54}$$

Thus, the signal propagation path's range difference of the left beam can be expressed as:

$$\Delta R = -k \cdot \frac{\omega_e R_e \cos\Phi_t \cos\delta}{N_B \cdot f_{\mathrm{PRF}}} \tag{11.55}$$

Where k is the receiving beam number arranged from the transmitting beam to both sides, and PRF is the pulse repetition frequency actually operating by the system. The corresponding phase error can be expressed as:

$$\phi_{eL} = -\frac{4\pi}{\lambda}\Delta R = \frac{4\pi}{\lambda} \cdot k \cdot \frac{\omega_e R_e \cos\Phi_t \cos\delta}{N_B \cdot f_{\mathrm{PRF}}} \tag{11.56}$$

Similarly, the phase error of the right beam can be expressed as:

$$\phi_{eR} = \frac{4\pi}{\lambda}\Delta R = -\frac{4\pi}{\lambda} \cdot k \cdot \frac{\omega_e R_e \cos\Phi_t \cos\delta}{N_B \cdot f_{\mathrm{PRF}}} \tag{11.57}$$

Where, ω_{e} is the Earth's angular velocity.

(2) **The azimuth phase correction processing**

For the multi-phase centers multi-beam synthetic aperture radar echo data, before imaging, the azimuth phase should be corrected. Corresponding to each satellite position, at first, the phase error of each range gate can be calculated and compensated. Then, the high-resolution imaging algorithm of conventional single-beam radar is used for image processing.

Since the rotated distance of the point on the earth's surface during an equivalent pulse repetition time is much smaller than the distance from this point to the radar, the phase error changes very slowly during the radar flight process. So it is not necessary to calculate the correction value at each position of the satellite. A fixed phase correction value can be used within a certain scope of the satellite motion.

Referring to Fig. 11.14, the satellite latitude Φs, the latitude Φt where the ground point is irradiated by the beam center, δ angle are changing during satellite flight. And there is an inherent relationship between them. Calculate the derivative of on both sides of Eq. (11.56) on the Φs:

$$\frac{d\phi_{eL}}{d\Phi_s} = -\frac{4\pi}{\lambda} \cdot k \cdot \frac{\omega_e R_e}{N_B \cdot f_{\mathrm{PRF}}} \left[(\sin \Phi_t \cos \delta) \frac{d\Phi_t}{d\Phi_s} + (\cos \Phi_t \sin \delta) \frac{d\delta}{d\Phi_s} \right] \tag{11.58}$$

That is, the change rate of the phase error caused by the satellite latitude change can be expressed by the change rate of the target point's latitude and of the δ angle. $\frac{d\Phi_t}{d\Phi_s}$ and $\frac{d\delta}{d\Phi_s}$ can be calculated by the following equation and Eq. (10.54)

$$\sin \Phi_t = \sin \Phi_s \cos \alpha - \sin \alpha \cos i \tag{11.59}$$

Where i is the satellite's orbit inclination. So the satellite latitude changes and phase error changes have corresponding relationship. The value of phase error caused by satellite latitude changes can be calculated. As long as the phase error variation is less than $\pi/2$ within a certain scope of satellite latitude, a fixed phase correction value is used during this period. The phase correction is determined by the Eq. (11.56) at the mid-point of the satellite latitude scope, Eq. (11.57) to calculate the value of phase correction.

For the echo at same azimuth position with different range gate, the phase error is different. The earth core angle α between point target and satellite nadir changes since range changing. From Eqs. (11.54)–(11.57) and (11.59) can be seen. The phase error caused by α changes. Similar to phase error changes caused by the change of the latitude of the satellite nadir, calculate the derivative of on both sides of Eq. (11.56) on the α:

$$\frac{d\phi_{eL}}{d\alpha} = -\frac{4\pi}{\lambda} \cdot k \cdot \frac{\omega_e R_e}{N_B \cdot f_{\mathrm{PRF}}} \left[(\sin \Phi_t \cos \delta) \frac{d\Phi_t}{d\alpha} + (\cos \Phi_t \sin \delta) \frac{d\delta}{d\alpha} \right] \tag{11.60}$$

That is, the change rate of the phase error caused by the change of the slant range can be expressed by the change rate of the target's latitude and the angle δ's change rate. $\frac{d\Phi_t}{d\alpha}$ and $\frac{d\delta}{d\alpha}$ can be calculated by Eqs. (11.54) and (11.59). Using the above relationship, it is possible to calculate the change value of phase error caused by the slant range change. In a certain number of range gates, a fixed phase correction can be used as long as the phase error variation is less than $\pi/2$. And the phase correction value can be calculated using the Eq. (11.56) at the center of the range variation.

In the above description, the left beam is taken as an example, and the calculation of phase correction value of the right beam can be referred above process.

From the principle of multi-phase centers multi-beam technology, the echoes from sub-beam should be phase corrected at first. Then they can be reconstructed into the azimuth echo signal with complete phase history before the imaging process. The processing steps are as follows:

(1) The phase error calculation: according to the information carried on the radar auxiliary data. such as the current latitude, antenna's look incident angle and so on, to calculate the latitude of the ground point target which is irradiated by the beam center, Calculate the azimuth phase error of each receiving beam's echo using Eqs. (11.54)–(11.57).
(2) The phase correction: according to the above calculation of the receiving beam's phase error, the echo of each receiving beam is phase corrected in the azimuth direction, and the phase error is compensated.
(3) The data synthesis: the receiving beam's echo data after the phase correction is arrayed according to the flight order by pulse arrangement. The echo data can be obtained with a complete azimuth phase history. Take three-beam situation as an example, if the left beam in on the radar front direction, in the data synthesized process, echo data of the first pulse of left beam is arrayed first, followed by the echo data of first pulse of the mid-beam, and then first pulse of the right beam, followed by the echo data of the second pulse of the left beam, and so on. So the synthesis of the receiving beam echo data is completed. The synthesized echo data, for each of the transmitting pulse, the N_B group echo data samples are obtained, equivalent to the azimuth equivalent sampling rate improves N_B times. It satisfies the requirement of the sampling theorem.
(4) Compression processing: After the synthesis of each receiving beam's echo data, the complete azimuth phase history is obtained. Then the two-dimensional compression processing is used for SAR imaging. The compression algorithm adopts the conventional single-beam high resolution spaceborne SAR algorithm.

(3) **Simulation analysis of phase error correction**

Based on the above analysis, the simulation experiments of multi-phase centers multi-beam synthetic aperture radar (SAR) are carried out. The experiment includes two parts. One is the simulation of phase error introduced by multi-phase centers multi-beam technique. The actual range history and phase error of each beam at different latitudes is calculated for the multi-phase centers multi-beam centers synthetic aperture radar. The Eqs. (11.56) and (11.57) are used to estimate the phase error at the corresponding position, and the phase error is compared with the actual phase error. Second part is the phase error correction and azimuth signal processing simulation of the multi-phase centers multi-beam system.

The basic parameters of the system used in the simulation are as follows: the orbit height Alt = 495 km, the argument of the perimeter is 90°, the orbit inclination is 98.55°, the red path of ascending nod 89.37°, the operating frequency is S band, the wavelength $\lambda = 0.09375$ m, The azimuth length of the sub-antenna is

La = 4 m, the center distance of the adjacent beam is d = 3.424 m, three beams are adopted, the total length of the antenna is 10.848 m, the incident angle is 35°, the antenna is uniformly weighted, the transmission pulse repetition frequency is $f_{\mathrm{PRF}} = 1500$ Hz, the corresponding equivalent pulse repetition frequency is $f_{\mathrm{PRFe}} = 4500$ Hz, the azimuth resolution is 2 m.

Based on the above system parameters, the variation of each receiving beam' phase error in a synthetic aperture time is calculated when the multi-phase center multi-beam technique is used at different orbit positions. The simulation results are shown in Fig. 11.15. Figure 11.15a is the result at latitude 8.58°N. The phase errors between the left and right beam are ±8.37 radians, and remain essentially constant within the synthetic aperture time. Figure 11.15b is the result at latitude 58.25°N. The phase errors between the left and right beam are ±10.45 radians and remains constant within the synthetic aperture time. Figure 11.15c is the result of latitude at 74.7°N. The phase errors between the left and right beam are about 1.86 radians. Figure 11.15d is the result at latitude 79.4°N. The phase errors between the left and right beam are about ±0.936 radians. The simulation results show that there is no phase error in the intermediate beam of the multi-phase centers multi-beam system. The phase error of the beam introduced by the left and right receiving beam change slowly with the orbit position of the satellite. The satellite is located in a fairly large range scope on both sides of the equator. The phase error is substantially constant within a synthetic aperture time, which facilitates the correction of the phase error.

Figure 11.16 shows the variation of the phase error estimated by the Eqs. (11.56) and (11.57) of the left beam at different latitudes of the satellite. The positive latitude represents the northern hemisphere and the negative latitude represents the southern hemisphere. It can be seen from the figure that the phase error of multi-phase centers multi-beam introduction is maximized when the satellite reaches the equator, but the change rate of phase error is the smallest (0). When the satellite reaches the poles, the phase error is the smallest, but the change rate of phase error is maximized. In addition, the phase error estimated by Eqs. (11.56) and (11.57) is 8.3704 radians when the satellite is latitude 8.58°N. The phase error estimated by Eqs. (11.56) and (11.57) is 10.4457 radians when the satellite is at latitude 58.25°N. The phase error estimated by Eqs. (11.56) and (11.57) is 0.9263 radians when the satellite is located at latitude 79.4°N. It can be seen that the phase

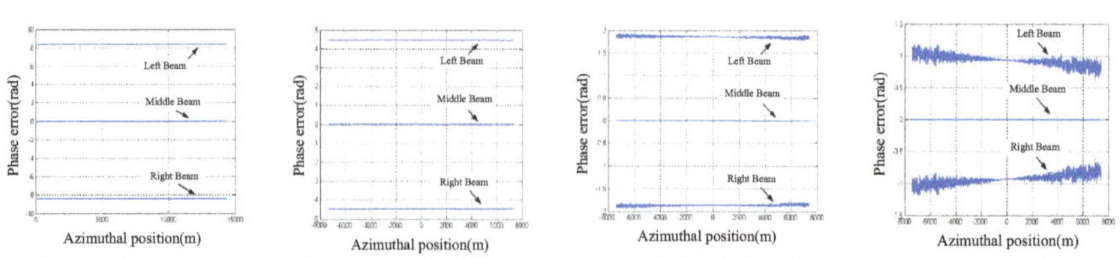

(a) latitude 8.58°N (b) latitude 58.25°N (c) latitude 74.7°N (d) latitude 79.4°N

Fig. 11.15 Phase error of each receiving beam at different latitudes

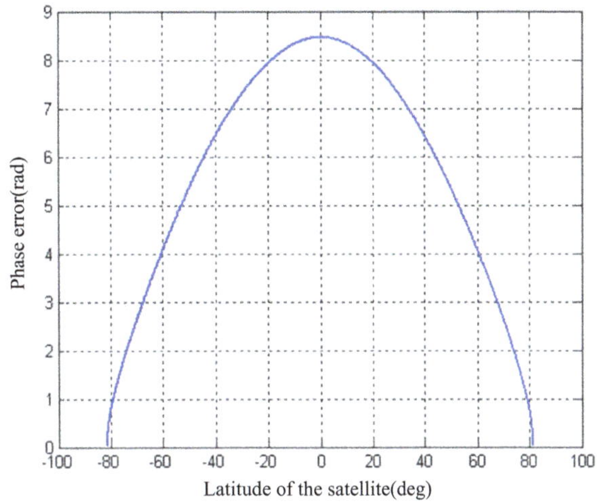

Fig. 11.16 Relationship between phase error and satellite latitude

error estimated using Eqs. (11.56) and (11.57) is in agreement with the actual calculated phase error in Fig. 11.15, with an estimated error of no more than 1°.

It is shown in Fig. 11.15 that the phase error is also slightly changed in a synthetic aperture time when the satellite is located near the poles. It can be handled in two cases. For low-frequency band (e.g., L or S band) systems, this variation of phase error is less than 15°, and the residual phase error can be compensated with a fixed phase correction. The residual phase error has a small effect on imaging [20]. For high-frequency band (e.g., C or X band) systems, this variation of phase error may be greater than 15°. The compensation can be performed with one or the higher order phase correction.

With the above system parameters, the imaging process of a 3 × 3 point target array is simulated. There are three points in the range and azimuth direction individually, and the distance is 40 m, the target array is at latitude 8.58°N. According to the phase correction method proposed in this paper, the azimuth phase correction of the echoes received by each beam is first adopted. The same phase correction is used for the three points in the same range gate. Then the echo data of three beams are synthesized. The imaging result is shown in Fig. 11.17. It can be seen that the focusing effect of the all 9 points is good. It indicates that the proposed method of phase correction and azimuth signal processing is feasible.

11.5 Ambiguity Analysis of Multi-phase Centers Multi-beam Synthetic Aperture Radar

Compared with the conventional single-beam system, the multi-phase centers multi-beam technology is used to reduce the pulse repetition frequency of the system and obtain the same azimuth resolution. Therefore, the calculation method

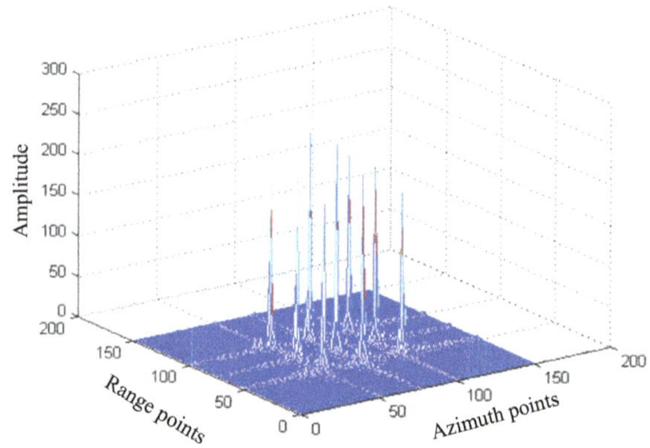

Fig. 11.17 The imaging results of the point target array

of the azimuth ambiguity to signal ratio is different from the conventional single-beam system. In the range dimension, the beam structure and signal characteristics of the multi-phase centers multi-beam system are the same of the conventional single-beam system. The calculation method of range ambiguity to signal ratio is same. In the following, the simulation is used to illustrate the azimuth ambiguity and the range ambiguity of the multi-phase centers multi-beam system.

11.5.1 Azimuth Ambiguity Analysis

For conventional single-beam SAR, the shape of the azimuth Doppler spectrum is mainly determined by the azimuth antenna pattern. Generally, the processing bandwidth of the azimuth signal is not larger than the 3 dB bandwidth of the azimuth antenna pattern. Due to the presence of the side lobes of the pattern, the azimuth echo received by the radar is from a wide scope and the echoes between, before, and after the two horizon points can be received by the radar along the track direction. The spectrum of some of the signals after the PRF sampling is folded into the processing bandwidth of the azimuth signal. So the azimuth ambiguity appears with the azimuth matching filter. As shown in Fig. 11.18, for any point in the main lobe coverage area of the conventional SAR azimuth pattern, the Doppler frequency of the corresponding azimuth ambiguity point is different by

$$\Delta f_d = k \cdot f_{\mathrm{PRF}}, \quad k = \pm 1, \pm 2, \ldots \tag{11.61}$$

Corresponding azimuth distance difference

$$\Delta x = k \cdot \frac{\lambda R_0}{2 v_s} \cdot f_{\mathrm{PRF}}, \quad k = \pm 1, \pm 2, \ldots \tag{11.62}$$

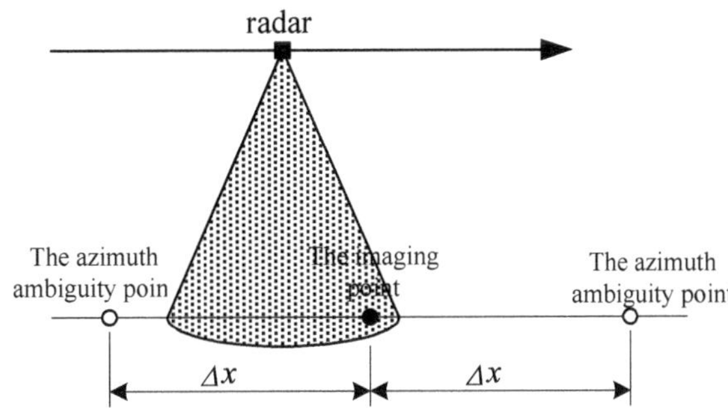

Fig. 11.18 Schematic diagram of azimuth ambiguity position for conventional SAR

In order to suppress the azimuth ambiguity, it is required that the first azimuth ambiguity point should be located outside the main lobe when the main lobe of the beam is irradiated. This requirement is consistent with the requirement of the sampling theorem, that is, the PRF should be greater than the Doppler bandwidth.

For the multi-phase centers multi-beam system, the echo of each beam is synthesized by the azimuth phase compensation before the imaging process. The azimuth sampling rate of the synthesized signal is the system equivalent pulse repetition frequency $\mathrm{PRF}_e = N_B \cdot \mathrm{PRF}$. The actual repetition frequency of the system is PRF, but only the point which Doppler frequency difference from the Doppler frequency of the imaging point is $N_B \cdot \mathrm{PRF}$ can become an azimuth ambiguity point, and affect the imaging results. In this way, the multi-phase centers multi-beam system azimuth ambiguity to signal ratio should be:

$$\mathrm{AASR_{MPC}} = \frac{\sum\limits_{\substack{m=-\infty \\ m \neq 0}}^{+\infty} \int_{-B_p/2}^{+B_p/2} G^2(f + m \cdot N_B \cdot f_{\mathrm{PRF}}) df}{\int_{-B_p/2}^{+B_p/2} G^2(f) df} \tag{11.63}$$

Where $G(f)$ represents the pattern coefficient of the Doppler frequency function for each sub-beam. If the sub-antennas are uniformly weighted, ambiguity to signal ratio can be approximated as:

$$\mathrm{AASR_{MPC}} = \frac{\sum\limits_{\substack{m=-\infty \\ m \neq 0}}^{+\infty} \int_{-B_p/2}^{+B_p/2} \sin c^4 \left[\frac{\pi L_a}{2v_s}(f + m \cdot N_B \cdot f_{\mathrm{PRF}})\right] df}{\int_{-B_p/2}^{+B_p/2} \sin c^4 \left(\frac{\pi L_a}{2v_s} f\right) df} \tag{11.64}$$

Use the above equation to simulate the three azimuth beams. The system parameters used in the simulation are the same as those in Sect. 10.3.4. The azimuth ambiguity to signal ratio calculation results in the various PRF cases are shown in Fig. 11.19a. For comparison, ambiguity to signal ratio of conventional single-beam

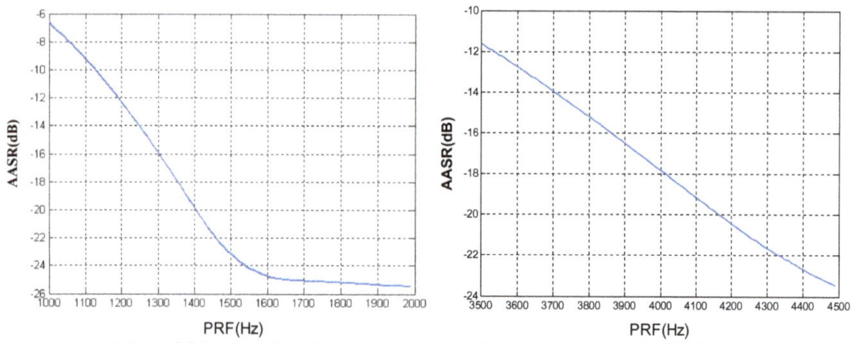

(a) multi-beam situation (b) conventional single-beam condition

Fig. 11.19 Azimuth ambiguity to signal ratio calculation results

systems with the same azimuth resolution is shown in Fig. 11.19b. From the above simulation results, it can be seen that the repetition frequency of the multi-phase centers multi-beam system is about 1354 Hz when the azimuth ambiguity to signal ratio reaches −18 dB. The repetition frequency of the conventional single-beam system is about 4016 Hz. Thus, when using multi-phase centers multi-beam technology, the PRF required to achieve the same ambiguity to signal ratio at the same azimuth resolution is reduced by about three times compared to the conventional single-beam case. It is the benefits of multi-phase centers multi-beam technique. It facilitates to get wider swath.

11.5.2 Range Ambiguity Analysis

The purpose of multi-phase centers multi-beam technique is to reduce the space sampling by increasing the time-dimension sampling of the azimuth echo, and to reduce the pulse repetition frequency under the condition of same resolution.

In this case, the multi-phase centers multi-beam system the range ambiguity to signal ratio is calculated in the same way compared with the conventional single-beam system. Because the system PRF is reduced under the premise of a certain azimuth resolution, so the range ambiguity to signal ratio is improved or the wider swath is allowed under the same range ambiguity.

The range ambiguity to signal ratio is calculated using the system parameters in Sect. 10.3.4, and compared with conventional single-beam systems.

Assuming that the range-dimension antenna is uniformly weighted, the antenna height is chosen to be 1.42 m. The corresponding ground swath is 45.3 km at 35° center incident angle. When the multi-phase centers multi-beam system is used, the pulse repetition frequency is selected as 1354 Hz. The range ambiguity to signal ratio changes with the incident angle range of 20°–50° is shown in 11.20a. When the conventional single-beam system is used, the pulse repetition frequency is

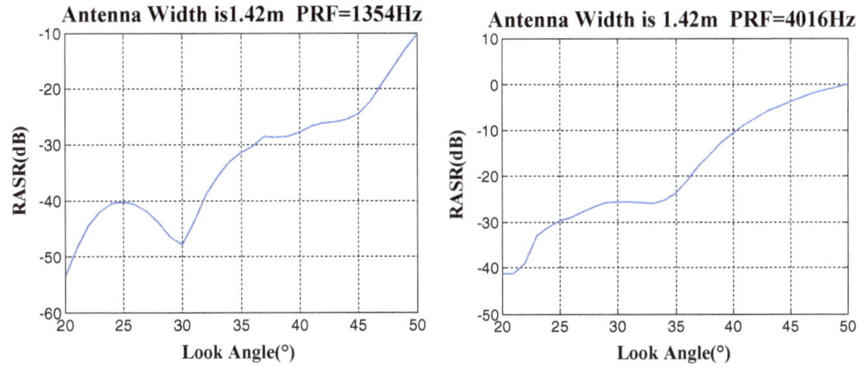

(a) RASR of multi-beam SAR

(b) RASR of conventional single-beam system

Fig. 11.20 Range ambiguity to signal ratio at the same swath

selected as 4016 Hz. The range ambiguity to signal ratio changes with the incident angle range of 20°–50° is shown in 11.20b. It can be seen that the range ambiguity to signal ratio of the conventional single-beam system at 35° incident angle is −23.48 dB, but of the multi-phase centers multi-beam system is −31.36 dB. So using of multi-phase centers multi-beam technology, under the same swath condition, the range ambiguity to signal ratio is significantly improved.

In the case of the same range ambiguity to signal ratio with the conventional single-beam system, the swath of the multi-phase centers multi-beam system is calculated. When the multi-phase centers multi-beam is used, the pulse repetition frequency is maintained at 1354 Hz, and the antenna height is reduced to 0.48 m. At this time, the range ambiguity to signal ratio in the range of 20°–50° incident angle is shown in Fig. 11.21. The range ambiguity to signal ratio of the multi-beam centers multi-beam system at 35° angle is −23.446 dB, and the corresponding ground swath is 128.85 km. Compared with the conventional single-beam system, it can be seen that when the multi-phase centers multi-beam technique is adopted, the swath of the system can be significantly expanded under the same range ambiguity to signal ratio.

11.6 The Model of Multi-phase Centers Multi-beam Synthetic Aperture Radar System

In the previous sections, the principle of multi-phase centers multi-beam technology, the azimuth phase correction and signal processing ware introduced. In this section, the system model of multi-phase centers multi-beam synthetic aperture radar will be given, as shown in Fig. 11.22, taking the three-beam situation as an example.

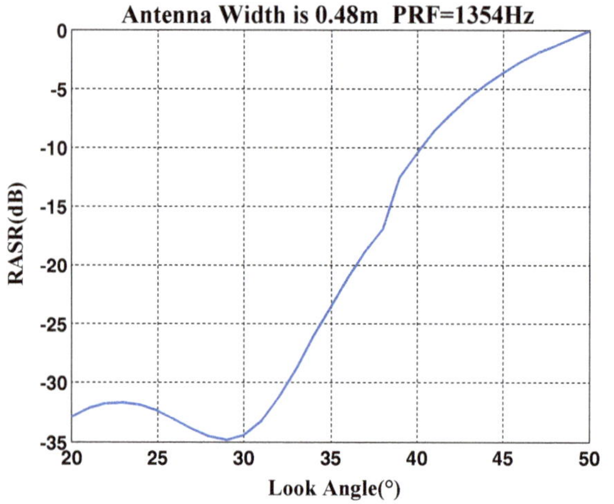

Fig. 11.21 RASR after changing the antenna width

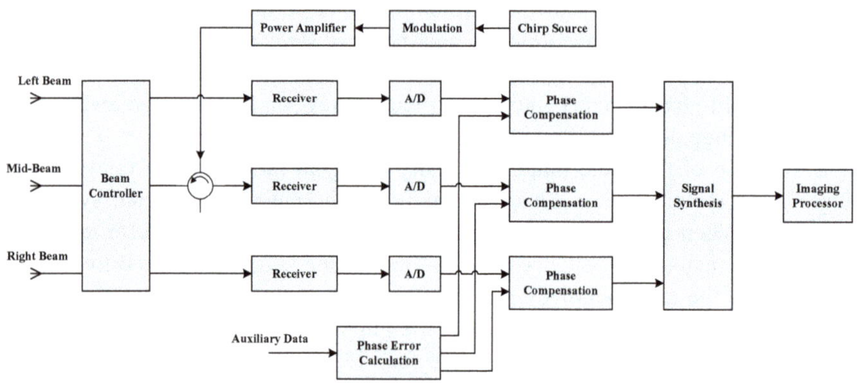

Fig. 11.22 The model of multi-phase centers multi-beam synthetic aperture radar system

The chirp signal is amplified by the power amplifier, and is sent to the circulator of the middle beam, then is transmitted by the middle receiving/transmitting beam. For tree beams, each beam receiving the echo from the same region at the same time. Each receiving beam corresponds to a receiving branch. After quadrature demodulation and A/D sampling, the azimuth phase correction of each receiving beam is carried out at first based on the phase error estimation result. The corrected signal of each branch is arranged according to the beam order, and the echo is synthesized. After the synthesis, the echo data having the full azimuth phase history is formed. The radar image is generated by two-dimensional compression.

For the spaceborne synthetic aperture radar system, the phase error calculation, phase compensation, and subsequent processing in Fig. 11.22 are generally accomplished on the ground.

The multi-phase centers multi-beam technique is used to receive echoes from the same area in the azimuth direction using multiple sub-beams. Use the increase of the time-dimension sampling for the reduction of space-dimensioned sampling in the azimuth direction. It is possible to allow the system to reduce the pulse repetition frequency in the case of a certain azimuth resolution. It is a new idea to realize the high resolution wide swath of the spaceborne synthetic aperture radar system. In this chapter, the principle of multi-phase centers multi-beam technology is introduced. The conditions of multi-phase centers multi-beam technology are studied. The design Eq. of the azimuth length of multi-phase centers multi-beam synthetic aperture radar is given. The mathematical model of echo signal is derived, and the various errors that may be introduced by multi-phase centers multi-beam technique are analyzed. In addition, the ambiguity characteristics of multi-phase centers multi-beam synthetic aperture radar also were analyzed in this chapter.

References

1. Currie A (1989) Wide-swath SAR imaging with multiple azimuth beams. In: IEE colloquium on synthetic aperture radar, Nov 29
2. Currie A, Hall CD (1990) A synthetic aperture radar technique for the simultaneous provision of high-resolution wide-swath coverage. In: Conference Proceedings Military Microwaves'90, Microwave Exhibitions & Publishers Ltd, July
3. Currie A, Brown MA (1992) Wide-swath SAR. IEE Proc-F 139(2):122–135
4. Bickel DL, Brock BC, Allen CT Spaceborne SAR study: LDRD'92 final report. DE93011484
5. Wu C, Liu KY, Jin M (1982) Modeling and a correlation algorithm for spaceborne SAR signals. IEEE Trans Aerosp Electron Syst AES-18(5):563–574
6. Zhang C (1989) Synthetic aperture radar: principles, system analysis and applications. Science Press, Beijing
7. Liu Y (1999) Radar imaging technology. Harbin Institute of Technology Press, Harbin
8. Curlander JC (1991) Synthetic aperture radar systems and signal processing. Wiley
9. Franceschetti G (1999) Synthetic aperture radar processing. CRC Press, New York
10. Zhang J, Xu R, Liu Y (1999) Extended Range-Doppler algorithm suited to spaceborne SAR systems with large slant-looking angle. J Syst Eng Electron 21(8):9–11
11. Cafforio C, Prati C, Rocca F (1991) SAR data focusing using seismic migration techniques. IEEE Trans Aero Elec Syst 27(2):194–207
12. Yuan X (1994) Imaging data processing of synthetic aperture radar. Aerosp Shanghai 3:2–8
13. Huang Y, Li C et al (2000) Extended Chirp scaling imaging algorithm of high-resolution spaceborne SAR. Chin J Electron 28(3):35–38
14. Jin MY, Wu C (1984) A SAR correlation algorithm which accommodates large-range migration. IEEE Trans Geosci Remote Sens GE-22(6):592–597
15. Raney RK, Runge H et al (1994) Precision SAR processing using Chirp scaling. IEEE Trans Geosci Remote Sens 32(4):786–799
16. Bamler R (1992) A comparison of range-Doppler and wavenumber domain SAR focusing algorithm. IEEE Trans Geosci Remote Sens 30(4):706–713

17. Moreira A, Mittermayer J, Scheiber R (1996) Extended Chirp scaling algorithm for air- and spaceborne SAR data processing in stripmap and ScanSAR imaging modes. IEEE Trans Geosci Remote Sens 34(5):1123–1136
18. Li C, Zhou Y (2000) Imaging processing algorithm of high-resolution spaceborne synthetic aperture radar. J Telemetry Tracking Command 21(2):15–18, 25
19. Eldhuset K (1998) A new fourth-order processing algorithm for spaceborne SAR. IEEE Trans Aerosp Electron Syst 34(3):824–835
20. Li S, Yang R (2004) Error analysis of displaced phase centers multi-beam SAR. Chin J Electron 32(9)
21. House MG, Mountcastle PD (2002) Technique for frequency analysis of unevenly sampled radar data. IEEE Radar Conf Calif 4:63–67
22. Lin M, Ke Y (1984) Radar signal theory. National Defense Industry Press, Beijing
23. Ma X, Xiang J (1999) Radar signal processing. Hunan Science & Technology Press, Changsha
24. Zhao W, Song HJ (2005) Multiple phase centers multiple azimuth beams spaceborne SAR with variable PRF. J Electron Inf Technol Beijing 27(6):936–938
25. Fan Q, Lu X, Zhang P et al (2006) Study on azimuthal unevenly sampling of spaceborne SAR DPVMAB technique. J Electron Inf Technol Beijing 28(1):31–35
26. Yen JL (1956) On nonuniform sampling of bandwidth-limited signals. IRE Trans Circ Theor 3(4):251–257
27. Jenp YC (1997) Perfect reconstruction of digital spectrum from nonuniformly sampled signals. IEEE Trans Instrum Measur 46(3):649–652
28. Johansson H, Lwenborg P (2002) Reconstruction of non-uniformly sampled bandlimited signals by means of digital fractional delay filters. IEEE Trans Sig Process 50(11):2757–2767
29. Prendergast RS, Levy BC, Hurst PJ (2004) Reconstruction of bandlimited periodic nonuniformly sampled signals through multirate filter banks. IEEE Trans Circ Syst 51 (8):1612–1622
30. Zong K (1996) Multiple rate signal processing. Tsinghua University Press, Beijing, pp 150–234
31. Li S (2004) Study on high-resolution wide-swath synthetic aperture radar systems. PhD Thesis of Institute of Electronics, Chinese Academy of Sciences, Beijing

Chapter 12
Multiple-Input Multiple-Output SAR

For the conventional SAR, the swath width and azimuth resolution are contradictions because of the limitations of minimum antenna area, so they cannot be improved simultaneously. In recent years, many single-input single-output system schemes are proposed, that is, single beam transmitting and multiple beams receiving on range or azimuth direction, to alleviate this contradiction to certain extent. But the single-input single-output system has some disadvantage, such as low utilization rate of transmitting antenna, high power density, and high heat flux, which cause power consumption sharply increasing when improving resolution and swath coverage performance. However, the multiple-input multiple-output (MIMO) technique can improve the utilization rate of antenna array, having low power density and low heat flux of transmitting antenna, and provide more phase centers, so MIMO technique is an inevitable trend in the development of high resolution and wide swath spaceborne SAR.

The concept of MIMO is derived from communication field. The MIMO technique is applied to communication system since the 1970s, and promoted by AT&T Bell Laboratory in radio mobile communication in the 1990s. The MIMO technique using multiple transmit antenna and multiple receive antenna can increase information in spatial dimension, improve gain of spatial multiplexing and spatial diversity [1, 2], that is designed to achieve reliable data communication in multipath fading channels.

With the development of MIMO technique, its application has expanded to radar technology area, which forms MIMO radar. The difference from the conventional radars is that, MIMO radar is a new radar system which transmitting multi-orthogonal waveform, using multi-receive antenna receiving echoes and combined processing [1, 2]. The MIMO radar forms a large number of observation angle than the actual array, which can obtain the statistically independent echoes, realizing diversity processing to improve radar performance using the target scattering characteristics, and therefore to improve the targets detection and parameters estimation performance. There are two categories of primary MIMO radar, namely distributed and compact MIMO radar. Just as its name implies, the distributed

R. Yang et al., *High-Resolution Microwave Imaging*,
https://doi.org/10.1007/978-981-10-7138-6_12

MIMO radar refers to large array space of radar, using different looking angles to observe objects and using spatial diversity to improve the performance of radar. And the compact MIMO radar refers to short array space of radar, which array meet far-field conditions, the objects meet point target assumption, and can obtain high-density sample of large aperture using less equivalent arrays.

Because of the advantages of MIMO improving channel utilization rate greatly; and providing more spatial degrees of freedom, the SAR based on MIMO technique have been attracted much attention in recent years. Dr. Ender proposed the concept of MIMO SAR system in 2007 [3], the MIMO SAR can be served as MIMO radar used on SAR, and a multifunction radar which combines the MIMO system and SAR radar. If using distributed SAR in which the different satellites transmitting different orthogonal signals, it can be served as distributed radar above mentioned. If the SAR antenna is divided into multiple sub-apertures, which each sub-aperture transmitting different signals, it can be served as compact MIMO radar.

12.1 Background and Status of MIMO SAR

The earliest start of MIMO SAR study began in abroad. The MIMO system transmitting chirp up and chirp down signals is introduced in the literature [4], which using dual aperture antenna transmit chirp up and chirp down signals simultaneously and using the appropriate matched filter to separate the echoes. Alamouti space-time encoding applied to MIMO SAR is proposed firstly by Kim in 2007. Space-time encoding technique has a better advantage in interference suppression, using time dimension redundant information can solve interference problem of multiple transmitting signals theoretically [5]. In the literature [6] an optimization algorithm is studied that can obtain better autocorrelation and cross-correlation waveform for MIMO SAR. The impact of nonideal factors of receiver in MIMO SAR system is analyzed in the literature [7]. Krieger proposed the multidimensional waveform encoding concept by using digital beam forming (DBF) on MIMO SAR [8].

There are many scholars have studied the MIMO SAR. In Yuepeng Song's doctoral dissertation the principle of MIMO SAR system implementation and the signal processing methods have been analyzed, the error of system and parameter design have studied, and also the azimuth ambiguities suppression using null-steering technique [9]. Dr. Yuli Xia has studied single-input single-output and MIMO spaceborne distributed-SAR systems [10]. Dr. Wei Jing has proposed the MIMO SAR imaging method using multiple sub-band transmitted simultaneously, which transmitting signals with different carrier frequency by azimuth multi-aperture, first the Doppler ambiguity is solved by spatial filter for echoes of sub-band, and the high-range resolution is achieved by sub-band synthetize of ambiguity resolution signals, then the two dimensions high resolution and wide swath can be achieved [11]. Dr. Xizeng Dai combined the MIMO SAR and the velocity-SAR (VSAR) and proposed MIMO-VSAR radar [12], using multiple signals transmit–receive simultaneously to obtain the observation channel more than the number of arrays that can be equivalent dense spatial sampling, then the unambiguous estimation range of velocity frequency of VSAR can be expanded. Dr. Wenqin Wang has given a comprehensive analysis of MIMO SAR, and points out the merits and drawbacks in the literature [13]. The waveform of MIMO SAR is analyzed in the literature [14], and the method of producing chirp up and chirp down signals is analyzed from the aspect of hardware. The method of multidimensional waveform encoding wide swath imaging based on MIMO SAR is proposed in the literature [15], that is, wide swath can be obtained as well as the transmitting power and bandwidth can be controlled flexibly by using DBF and MIMO technique. Based on the echo model of MIMO SAR in the literature [16], analytical formula of equivalent phase center error is deduced, the influence of imaging in the azimuth dimensions is studied quantitatively, and the error is compensated. The multi-frequency and the multi-baseline technique in multichannel InSAR is combined in the literature [17], the multichannel interferometer SAR based on MIMO technique is analyzed, by means of increasing the number of independent interferometry phase, the reliability and precision of elevation estimation is improved.

12.2 Principles of MIMO SAR

In spaceborne SAR using multiple phase centers multiple azimuth beam (MPCMAB) technique mentioned in Chap. 11, the SAR antenna is divided into many sub-apertures in the azimuth dimensions, the middle sub-aperture transmits wide beam signal, the sub-apertures in different azimuth phase centers receive echoes simultaneously, and the receiving beam width equals to transmitting beam width, If the phase center spacing between the sub-apertures is appropriately set, and for each of the emission pulses, multiple groups of independent echo signal samples can be obtained simultaneously in the azimuth direction. Using MPCMAB technique, even if the pulse repetition frequency is reduced, it can ensure that the correct sampling (to meet the Nyquist sampling theorem) of the echo signal in the azimuth direction, thus to some extent alleviate the contradiction between the high resolution and wide swath [18].

The antenna phase center defines to the effective position of transmitting signals or receiving echoes, usually located in the center of antenna aperture. But the transmitting antenna and receiving antenna are not the same one for multichannel antenna, this will cause the phase centers of transmitting and receiving are not in the same position. Here the antenna effective phase center is located in the middle point between transmitting aperture and receiving apertures, as shown in Fig. 12.1. Assuming that the aperture spacing is equal, for single-input single-output (SISO) SAR system (1 input and 3 outputs for example), when the center aperture is used to transmit and the three receiving apertures receive the echo at the same time, the

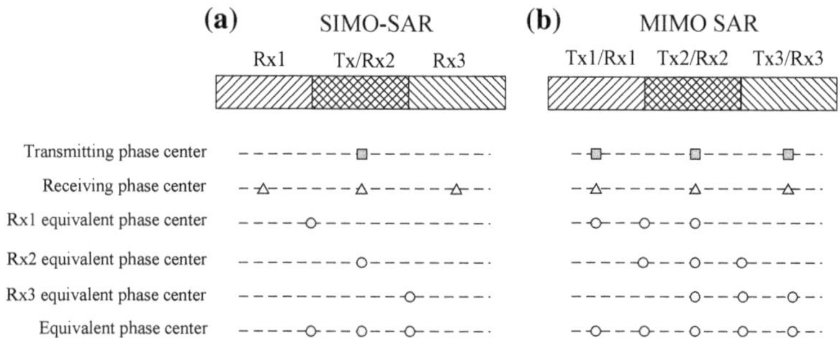

Fig. 12.1 Sketch map of equivalent phase center in azimuth dimension. **a** SIMO system; **b** MIMO system

transmission phase center is located on square position and the receiving phase centers are located on triangle positions. For the receiving aperture Rx1, Rx2, and Rx3, the equivalent phase centers (circle positions) are in the middle of transmitting and receiving apertures. Then in a pulse repeat time (PRT) the spatial sampling rate has been enhanced *N* times than the conventional single-channel SAR, which *N* is the receiving aperture number.

For MIMO SAR system, 2N-1 effective phase center can be obtained in one PRT when the spacing between apertures are equal and the number of transmitting aperture and receiving aperture are both *N*, as shown in Fig. 12.1b. In other words, MIMO technique can obtain 2N-1 virtual sub-apertures using *N* real aperture, the sampling rate is increased further based on SIMO technique. Some virtual sub-apertures will overlap and can be processed accumulatively, and then the SNR can be improved. So with respect to SIMO system, the MIMO system can significantly increase the number of equivalent phase centers to get higher spatial sampling rate, and it is beneficial to obtain higher resolution or wider swath and to detect moving targets, etc.

The MIMO SAR can be used for high resolution wide swath imaging (HRWS), ground moving target indication (GMTI), improvement of interferometric measurement accuracy and three-dimensional imaging, etc.

(1) High resolution wide swath imaging
According to DPCA principle, compared with the SIMO-SAR system, the MIMO SAR system with multiple azimuth apertures increases the number of equivalent phase centers, and the reduction of the time-dimensional sampling is exchanged by the increase of the space-dimensional sampling in azimuth direction. The pulse repetition frequency can be further decreased under the condition of guaranteeing a certain azimuth resolution making the swath widened, or the azimuth resolution can be further improved while maintaining a certain swath width, so that the contradictions between high resolution and wide swath can be alleviated to some extent.

(2) Ground moving target indication
MIMO SAR systems transmit signals by multiple antennas with equal spacing along azimuth dimension, and observe targets from different look angles, the target scattering coefficients are independent random variables. Since the echoes are from different paths and carry different target information, the target can be detected effectively by means of diversity method for received multiple channel signals. For distributed spaceborne MIMO SAR systems, this advantage is particularly pronounced due to the long baseline between the apertures. In addition, the phase differences occurred from different transmitting antennas and different receiving antennas can form a virtual array steering vector, and a long array steering vector can be obtained by means of suitably designing the position of antenna, thus it makes the system's degree of freedom increased and the clutter resolution improved, which is beneficial for clutter suppression and slowly moving target detection using the space-time adapted processing (STAP).

(3) Improving height measurement accuracy of interferometer
As Kim pointed out, InSAR system based on Alamouti space-time encoding can separate the mutual interference of signals with same frequency in space domain and time domain, it can also decrease noise level effectively, improve SNR and diversity gain, which is suitable for interferometry processing and height measurement accuracy improving [19–21].

(4) Three-dimensional imaging
The conventional SAR can only be imaged in two-dimensional plane because of side-looking mechanism. In order to get three-dimensional imaging, MIMO, and DBF technique can be used to obtain height information by means of placing multi-aperture perpendicular to tracking path. Then the range resolution is determined by signal bandwidth, the azimuth resolution is determined by the length of synthetic aperture and the height resolution is determined by DBF perpendicular to tracking path.

12.3 Echo Division Method for MIMO SAR

In MIMO SAR, multi-transmitting antennas transmit various signals to the ground and each receiving antenna receives mixed echoes from the various transmitting signals, then echo division should be processed before imaging. This feature determines that the chirp signal used by SISO-SAR cannot be applied to MIMO SAR directly; the transmitting signal waveforms of MIMO SAR must be orthogonal encoded signals that satisfies the condition of high autocorrelation and low cross-correlation and having enough code available [1, 9]. That is to say, the cross-correlation noise of different transmitting signals must be low enough and the code number available is not less than transmitting antenna number.

The foreign and domestic scholars have studied for waveform design of MIMO SAR based on analysis of the ambiguity function of basic orthogonal signals. Kunchao Bao from Xidian University combined the spatial phase-coded stepped-frequency signals with large time–bandwidth product and m-sequence code and apply this method to distributed small satellite MISO system, and obtain fine range resolution and Doppler resolution [22]. But this method has not solved large ambiguity of azimuth and range and the decoupling problem, so it is not suited for MIMO SAR as transmitting waveform. An optimization method of discrete frequency-coding waveforms designing for MIMO radar using genetic algorithm is introduced in the literature [23]; Chunyang Chen in the literature [24] has deduced the conventional ambiguity function of MIMO radar based on radar models, and optimized the waveform using frequency-hopping signals. Jian Li proposed a method of designing signal waveform matrix directly in the literature [25]; however, the optimization based on signal waveform matrix is still a complex problem. All of above have studied the waveform designing of MIMO radar system, but because of the higher side lobe and serious decoupling of the waveforms, these waveforms cannot be used for transmitting waveforms of SAR.

Some echo division methods for MIMO SAR are introduced as follows:

(1) Division in Time domain
The division method in time domain is that various signal waveforms are transmitted successively by different transmitting aperture in different PRF and all sub-apertures receive echoes. The advantage of this method is that the division degree is higher for different waveform. The disadvantage is that the echoes corresponding to all transmitting signals should be collected after N PRF (N is the number of sub-aperture), and this restricts the performance improving of spaceborne SAR.
(2) Division in Frequency domain
Division in Frequency domain refers to that all transmitting aperture transmit pulse signals which have different carrier frequency simultaneously in a PRF. So the echoes of each pulse can be separated effectively by band-pass filter in the frequency domain. This method is a kind of methods studied more in detail for echoes division of MIMO SAR at present and is practicable method for signal division,
(3) Code division
Different types of phase codes are used for the transmission pulses of the different antennas so as to be separated at the receiving end by orthogonality between the codes. At present most attention has been focused on chirp up and chirp down pulses, orthogonal two-phase code, such as chaotic phase encoding, spatial–time block encoding, etc. However, these encoding pluses have compression side lobe to some extent, it will cause higher integral side lobe ratio and affect the SAR imaging quality.

(4) Division in Spatial domain
Multiple transmitting beams point to different range or azimuth direction, and the echoes of different transmitting pulses can be divided effectively by DBF using direction of arrival (DOA) information of different echoes mixed. In recent years, the DBF technique of elevation and azimuth has been studied in more detail and the foundation of division in spatial domain for MIMO SAR has been established.

12.4 Division in Frequency Domain for MIMO SAR

This section will analyze a kind of MIMO SAR multiple carrier frequency MIMO SAR, which can get wide bandwidth using sub-band synthesis in frequency domain and obtain high-range resolution.

12.4.1 Principle of Division in Frequency Domain for MIMO SAR

There are two methods of wideband signals synthesis in MIMO SAR for Frequency Domain Division: the synthesis with synchronous sub-band pulses (method 1) and the synthesis with the same phase centers (method 2) (as shown in Fig. 12.2). Taking the 3-input 3-output SAR system as an example, the synthesis with synchronous sub-band pulses method is to synthetize 3 sub-band signals received from each receiving sub-aperture, in this way, the 3 wideband signals in each Pulse Repetition Time (PRT) are synthetized by 9 sub-band signals received in this PRT. That is to say, N complete spatial samplings in the azimuth dimension can be obtained in one PRT.

The synthesis with the same phase centers method is to synthetize 3 sub-band signals received from the same equivalent phase centers, then all sub-band signals are from different received beam and some signal may come from two adjacent PRT, as shown in Fig. 12.2.

Phase compensation should be applied for the echoes with different phase centers when adopting the synthesis with synchronous sub-bands pulse method, so that the samplings of different positions are compensated to the same position, as shown in Fig. 12.3. The process is difficult to implement and illustrated by mathematical deduction of phase compensation as follows:

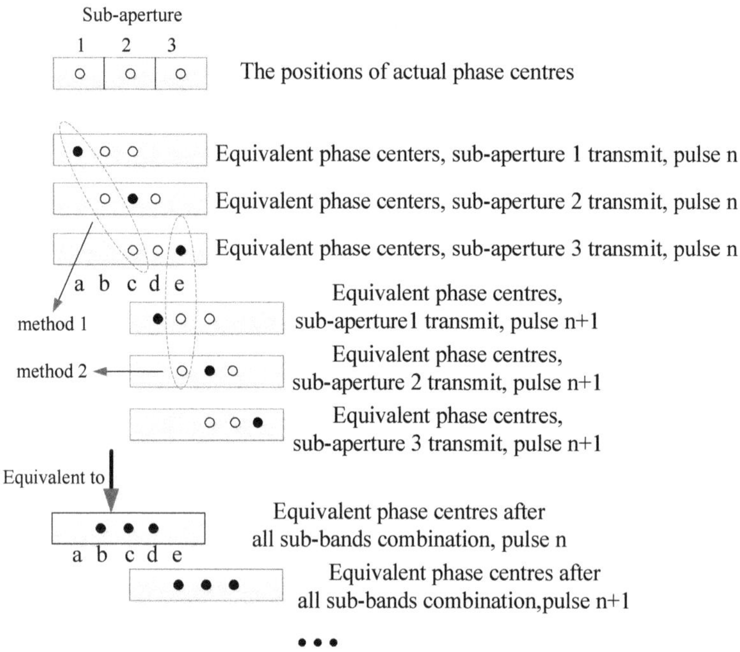

Fig. 12.2 Sketch map of two methods of sub-band pulse synthesis

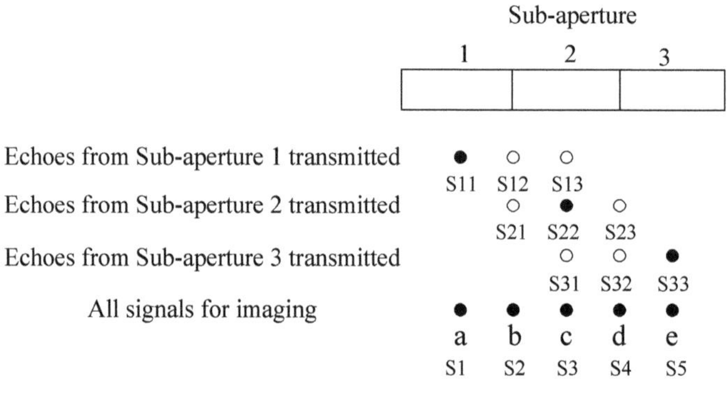

Fig. 12.3 Distribution map of echoes for 3-input 3-output SAR

Assume that the slant range from the middle sub-aperture to point target is $R(t)$, the space between sub-apertures is d, the received echoes from each sub-aperture when the sub-aperture 1 transmits signals are s_{11}, s_{12} and s_{13} successively and the corresponding round trip phase history are $\phi_{11}(t)$, $\phi_{12}(t)$ and $\phi_{13}(t)$, where $k = 2\pi/\lambda$.

$$\phi_{11}(t) = k(R(t-d/v_a)+R(t-d/v_a)) = 2kR(t) - 2k\left[\frac{\alpha d}{2v_a} + \frac{2\beta d(t-t_0)}{v_a} - \frac{\beta d^2}{v_a^2}\right] \tag{12.1}$$

$$\phi_{21}(t) = k(R(t-d\,/\,v_a)+R(t)) = 2kR(t) - k\left[\frac{\alpha d}{2v_a} + \frac{2\beta d(t-t_0)}{v_a} - \frac{\beta d^2}{v_a^2}\right] \tag{12.2}$$

$$\phi_{31}(t) = k(R(t-d\,/\,v_a)+R(t+d\,/\,v_a)) = 2kR(t) - 2k(\frac{\beta d^2}{v_a^2}) \tag{12.3}$$

Calculating the phase difference:

$$\begin{aligned}\phi_{21}(t)-\phi_{11}(t) &= k(R(t-d\,/\,v_a)+R(t)) - k(R(t-d\,/\,v_a)+R(t-d\,/\,v_a)) \\ &= k(\frac{\alpha d}{v_a} + \frac{2\beta d(t-t_0)}{v_a} - \frac{\beta d^2}{v_a^2})\end{aligned} \tag{12.4}$$

$$\begin{aligned}\phi_{31}(t)-\phi_{21}(t) &= k(R(t-d\,/\,v_a)+R(t+d\,/\,v_a)) - k(R(t-d\,/\,v_a)+R(t)) \\ &= k(\frac{\alpha d}{v_a} + \frac{2\beta d(t-t_0)}{v_a} + \frac{\beta d^2}{v_a^2})\end{aligned} \tag{12.5}$$

In the Eqs. (12.4) and (12.5), $\alpha = -v_a \sin\theta, \beta = v_a^2 \cos^2\theta\,/\,(2R_0)$. Assume that R_0 is the range from the flight track to the center of scene, $\theta = 0$ then $\alpha = 0,\ \beta = v_a^2\,/\,(2R_0)$. Substituting them to Eqs. (12.4) and (12.5), then

$$\phi_{21}(t)-\phi_{11}(t) = k(\frac{2v_a d(t-t_0)}{R_0} - \frac{d^2}{2R_0}) \tag{12.6}$$

$$\phi_{31}(t)-\phi_{21}(t) = k(\frac{2v_a d(t-t_0)}{R_0} + \frac{d^2}{2R_0}) \tag{12.7}$$

The phase difference is time variant and difficult to compensate, so the synthesis with the same phase centers method is chosen usually.

12.4.2 Echo Signal Processing Methods

(1) Mathematical model of sub-band synthesis in frequency domain

Take the ground as XY plane to create coordinate system, *Y* is along the tracking direction, *X* is perpendicular to the tracking direction, *Z* satisfies right-hand

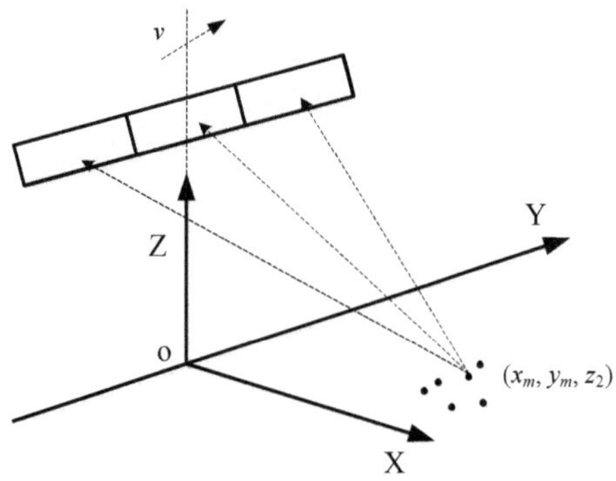

Fig. 12.4 Geometric configuration of MIMO SAR

coordinates with X, Y. The N sub-apertures are arranged in azimuth direction, their spatial locations are (X_1, Y_1, Z_1), $(X_1, Y_2, Z_1)\ldots(X_1, Y_N, Z_1)$. The following analysis takes the case of 3 sub-aperture as an example. Assume that there are M scattering points and the positions are $(x_1, y_1, z_2), \ldots, (x_m, y_m, z_2), \ldots, (x_M, y_M, z_2)$ respectively, as shown in Fig. 12.4.

Assume the number of sub-apertures is N, and the signal bandwidth transmitted by each sub-aperture is $B_e = kT_e$, where k is the frequency rate, T_e is sub-pulse duration, the overlapping bandwidth of sub-band is ΔB, then the degree of overlapping of signals can be described as $\alpha = \Delta B / B_e$. Assume that the carrier frequency transmitted by sub-aperture 1 is $f_c(1)$, then the carrier frequency of each sub-aperture can be shown successively as

$$f_c(n) = f_c(1) + (n-1)(1-\alpha)B_e,\ n = 1, 2, \cdots, N \tag{12.8}$$

Where n is the serial number of sub-apertures. So the center carrier frequency of wideband signal after synthesis can be derived as

$$f_c = f_c(n) - (n - 1/2 - N/2)(1-\alpha)B_e,\ n = 1, 2, \cdots, N \tag{12.9}$$

Assume that t is the whole time, $\tau = t - mT$ is the fast time, m is integer, T is PRT, $\eta = mT$ is the slow time, then the transmitting signal of sub-aperture n can be expressed as follows:

$$s(\tau, \eta) = \mathrm{rect}\left(\frac{t}{T_e}\right) \exp\left(j2\pi f_c(n)t + j\pi k\tau^2\right) \tag{12.10}$$

The sub-aperture n transmits signal and the other sub-apertures receive, this situation can be equivalent to that each sub-aperture transmits and receives individually if the spaces between the phase centers are short enough, and the equivalent phase centers are shifted to the middle points between transmit aperture and receive apertures. The N coordinates of equivalent phase centers

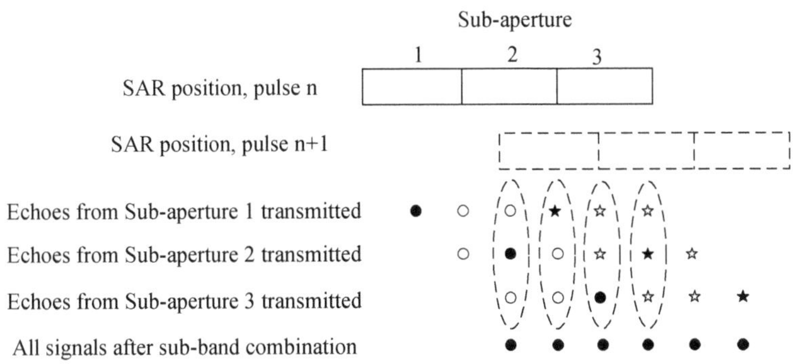

Fig. 12.5 Equivalent phase center map of each sub-aperture

are $(X_1,(Y_1+Y_n)/2,Z_1)$, $(X_1,(Y_2+Y_n)/2,Z_1)\ldots(X_1,Y_n,Z_1)$, $(X_1,(Y_N+Y_n)/2,Z_1)$. The w-th equivalent phase center obtained by the transmission of sub-aperture n is denoted by (X_1,Y'_{wn},Z_1), $w=1, 2, \ldots, N$.

Assuming that the N sub-apertures transmit respective LFM signal simultaneously, then $N\times N$ equivalent phase centers can be obtained (some part of them are overlapped), as shown in Fig. 12.5 (the positions of equivalent phase center obtained in two successive PRT and the positions of signal combined with the same phase center are shown in figure). N unambiguous sub-band signals can be obtained by using the N sub-band signals corresponding to each sub-aperture, and then the N sub-band signals at the same equivalent phase center can be synthesized to generate unambiguous wideband echo signals.

In Fig. 12.5, circles represent the echoes separated from the PRT n, pentacles represent the echoes separated from PRT n+1, and solid circles represent the echoes after sub-band combination.

The echo received at equivalent phase center w corresponding to sub-aperture n can be described as

$$\begin{aligned} s_{wn}(\tau,\eta) = \sum_{m=1}^{M} \sigma_m \mathrm{rect}((\tau-\frac{2R_{m,wn}}{c})/T_e)\exp(j\pi k(\tau-\frac{2R_{m,wn}}{c})^2) \\ \times\,\mathrm{rect}(\frac{y-y_m}{L_s})\exp(j2\pi f_c(n)(t-\frac{2R_{m,wn}}{c}) \end{aligned} \tag{12.11}$$

Where σ_m is a constant representing the echo amplitude of point target m, $R_{m,wn}$ is the range between scattering point m and the equivalent phase center w of the sub-aperture n

$$\begin{aligned} R_{m,wn}(\eta,R_m) &= \sqrt{(x_m-X_1)^2+(y_m-Y'_{wn}-v\eta)^2+(z_m-Z_1)^2} \\ &= \sqrt{R_m^2+(y_m-Y'_{wn}-v\eta)^2} \end{aligned} \tag{12.12}$$

Where $R_m = \sqrt{(x_m - X_1)^2 + (z_m - Z_1)^2}$ is the perpendicular range from scattering point m to platform track.

The echo signals are first demodulated to the baseband and converted to the range frequency domain, and then the following equation can be obtained after Doppler ambiguity resolution.

$$\begin{aligned} s'_{wn}(f_r,\eta) = & \sum_{m=1}^{M} \sigma_m \mathrm{rect}(\frac{f_r}{T_e k}) \mathrm{rect}(\frac{y-y_m}{L_s}) \exp(-j\pi\frac{f_r^2}{k}) \\ & \cdot \exp(-j\frac{4\pi}{c}R'_m(\eta,R_m)(f_r+f_c(n))) \end{aligned} \tag{12.13}$$

Here $R'_m = \sqrt{(x_m - X_1)^2 + (y_m - v\eta)^2 + (z_m - Z_1)^2} = \sqrt{(y_m - v\eta)^2 + R_m^2}$.

Where $s'_{wn}(f_r,\eta)$ represents the echo signal at the equivalent phase center w after Doppler ambiguity resolution corresponding to the signal transmitted by the sub-aperture n and is represented in range frequency domain. For the frequency domain synthesis of the echo signal an equivalent phase center, the summation operation of the range frequency domain and azimuth time domain signal at the equivalent phase center is adopted, it can be expressed as:

$$\begin{aligned} s'_n(f_r,\eta) & = \sum_{n=1}^{N} s'_{wn}(f_r,\eta) \\ & = \sum_{n=1}^{N}\sum_{m=1}^{M} \sigma_m \mathrm{rect}(\frac{f_r - f_c(n)}{T_e k}) \mathrm{rect}(\frac{y-y_m}{L_s}) \exp(-j\pi\frac{f_r^2}{k}) \\ & \quad \cdot \exp(-j\frac{4\pi}{c}R'_m(\eta,R_m)(f_r+f_c(n))) \\ & = \sum_{m=1}^{M} \sigma_m \mathrm{rect}(\frac{f_r}{NT_e k}) \mathrm{rect}(\frac{y-y_m}{L_s}) \exp(-j\pi\frac{f_r^2}{k}) \\ & \quad \cdot \exp(-j\frac{4\pi}{c}R'_m(\eta,R_m)f_r) \end{aligned} \tag{12.14}$$

The exact amplitude spectrum of the LFM signal has Fresnel fluctuation at the top, and the amplitude spectrum is not strictly limited within the bandwidth, and there is a constant phase error between the echoes of the multiple SAR system [26]. When the joint spectrum is generated by the range spectrum, the undulating ripples are generated at the edges of the two sub-spectrum stitching, resulting in paired echoes in the range image. Therefore, it is necessary to design a range compression filter to eliminate the ripple in the joint spectrum.

For the actual MIMO SAR system, the spectrum of each sub-band is obtained by using a real point target (such as a corner reflector, and its azimuth time is η_0). The combined spectrum $S_F(f_\tau, \eta_0)$ is obtained according to the signal echo synthesis steps. According to Eq. (12.15) we can get the system range compression filter.

$$\begin{cases} \arg\left(H_F(f_r, \eta_0)\right) = \arg\left(S_F^*(f_r, \eta_0)\right) \\ |H_F(f_r, \eta_0)| = \frac{1}{|S_F(f_r, \eta_0)|} \end{cases} \tag{12.15}$$

Where "arg (·)" represents phase angle, the superscript * represents conjugate operation.

The filter can be described as

$$H_F(f_r, \eta_0) = |H_F(f_r, \eta_0)| \cdot \exp\left(j \arg\left(S_F^*(f_r, \eta_0)\right)\right) \tag{12.16}$$

According to the steps of sub-band synthesis, the signal in range frequency domain after spectrum synthesis can be described as:

$$\begin{aligned} S_F(f_r, \eta_0) = {} & \sigma_0 \mathrm{rect}(\frac{f_r}{NT_eK}) \exp\left(-j\pi \frac{f_r^2}{k}\right) \\ & \cdot \exp\left(-j\frac{4\pi}{c} R'_m(\eta_0, R_m) f_r\right) \end{aligned} \tag{12.17}$$

σ_0 is the amplitude of the point target echo, according to (12.15) and (12.16), the exact expression of the filter can be derived as:

$$\begin{aligned} H_F(f_r, \eta_0) = {} & \frac{1}{\sigma_0} \mathrm{rect}(\frac{f_r}{NT_eK}) \exp\left(j\pi \frac{f_r^2}{k}\right) \\ & \cdot \exp\left(j\frac{4\pi}{c} R'_m(\eta_0, R_m) f_r\right) \end{aligned} \tag{12.18}$$

Multiplying the synthetized spectrum and the filter transform function:

$$\begin{aligned} S_n(f_r, \eta) = {} & S'_n(f_r, \eta) H_F(f_r, \eta_0) \\ = {} & \sum_{m=1}^{M} \sigma_m \Big\{ \mathrm{rect}(\frac{f_r}{NT_eK}) \mathrm{rect}(\frac{y - y_m}{L_s}) \exp\left(-j\pi \frac{f_r^2}{k}\right) \\ & \cdot \exp\left(-j\frac{4\pi}{c} R'_m(\eta_0, R_m) f_r\right) \frac{1}{\sigma_0} \mathrm{rect}(\frac{f_r}{NT_eK}) \\ & \cdot \exp\left(j\pi \frac{f_r^2}{k}\right) \exp\left(j\frac{4\pi}{c} R'_m(\eta_0, R_m) f_r\right) \Big\} \end{aligned} \tag{12.19}$$

Since the filter is synthesized for the echo spectrum of a fixed point target, the operation of the above equation cannot cancel the phase information in the echo signal. The inverse Fourier transform of $S_n(f_r, \eta)$ in range dimension can generate

$$S_n(f_r, \eta) = \sum_{m=1}^{M} \frac{\sigma_m}{\sigma_0} \left\{ 2\pi N T_e K \frac{\sin\left(\pi N T_e K(\tau - 2/c(R'_m(\eta) - R'_m(\eta_0)))\right)}{\pi N T_e K \tau} \cdot \mathrm{rect}\left(\frac{y - y_m}{L_s}\right) \right\} \tag{12.20}$$

Equation (12.20) is the range compression result while the azimuth information is reserved. Sometimes the filtered synthetized spectrum requires other operations, rather than the inverse Fourier transform in range dimension.

(2) Implementation steps of sub-band synthetized in frequency domain

It is a key step in the frequency domain sub-band synthesis process to obtain the range compression filter from the echo data of the actual targets. The steps of generating the high resolution SAR images by frequency domain synthesis technique can be summarized as follows:

① The range Fourier transform of the sub-band echo signals separated by the bandpass filter bank are carried out respectively, and then the multiple echo signals on the same equivalent phase center are subjected to frequency domain synthesis processing according to the preset degree of overlap.
② Multiplying the synthetized spectrum by the transform function of range compression filter, the joint spectrum after filtering can be obtained.
③ The range spectrum resulted from the step (2) is subjected to the inverse Fourier transform to obtain a high-resolution range image, or the filtered synthetized spectrum is multiplied by the frequency domain filter and transformed by IFFT to the original echo domain.
④ Finally the high resolution SAR image is achieved by means of subsequent processing with corresponding imaging algorithm.

(3) An imaging algorithm combining the frequency domain sub-band synthesis technique with RD algorithm.

The frequency domain synthesis method is to transform the echo signal to the range frequency domain, so the selection and adjustment of the algorithm should be considered when it is used in SAR imaging. After the spectral synthesis, the range IFFT is used to complete the range compression. Therefore, it is simpler to combine with the RD algorithm [27–30] for imaging. Here we first give the imaging process that combines sub-band synthesis technique and RD algorithm, as shown in Fig. 12.6, and the imaging process that combines with the CS algorithm will be described later. In order to ensure the completion of the range migration correction, we should first apply the range Fourier transform to echo data of each sub-pulse, and then do Doppler ambiguity resolution processing.

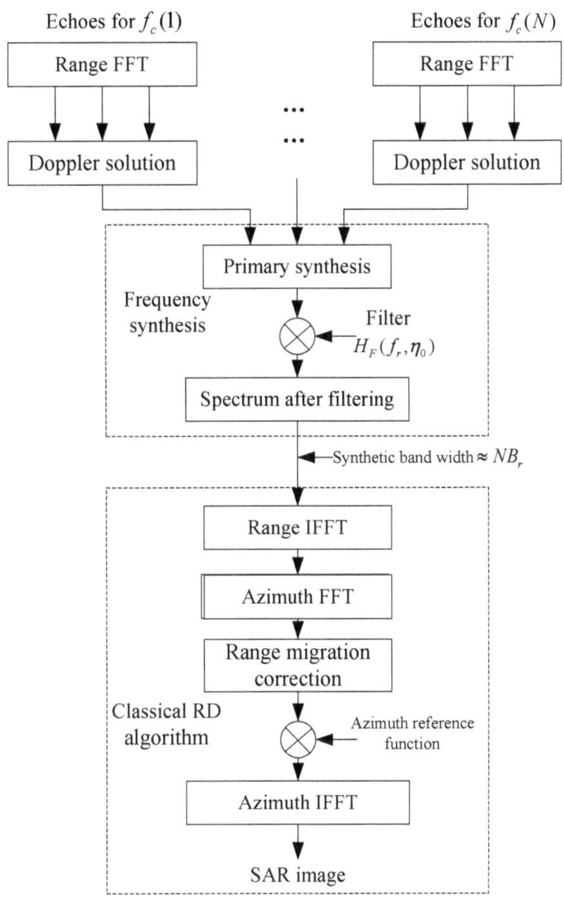

Fig. 12.6 Imaging process of combining sub-band synthesis technique with RD algorithm

(4) An imaging algorithm combining the frequency domain sub-band synthesis technique with CS algorithm.

If the CS algorithm [31, 32] is used for imaging, the filtered synthetized spectrum should be multiplied by the frequency domain filter, and returned to the two-dimensional time domain after the range inverse Fourier transform, and then the imaging operations will be carry on according to the classical CS algorithm. The operation process is shown in Fig. 12.7.

Fig. 12.7 Imaging process for combination of sub-band synthesis technique with CS algorithm

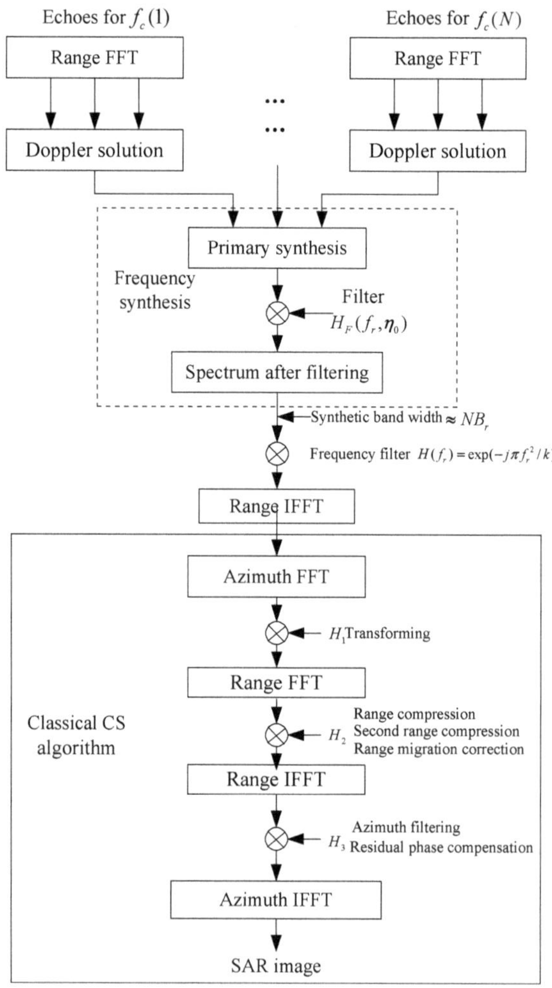

The wideband signal expression in range frequency domain and azimuth time domain after frequency domain synthesis is rewritten as follows:

$$
\begin{aligned}
S_n(f_r,\eta) &= S_n'(f_r,\eta)H_F(f_r,\eta_0) \\
&= \sum_{m=1}^{M}\frac{\sigma_m}{\sigma_0}\left\{\mathrm{rect}(\frac{f_r}{NT_eK})\mathrm{rect}(\frac{y-y_m}{L_s})\right. \\
&\left.\cdot\exp\left(-j\frac{4\pi}{c}(R_m'(\eta,R_m)-R_m'(\eta_0,R_m))f_r\right)\right\}
\end{aligned}
\tag{12.21}
$$

Multiplying it by the frequency filter $H(f_r) = \exp(-j\pi f_r^2 / k)$,

$$\begin{aligned} S_n(f_r, \eta)H(f_r) = \sum_{m=1}^{M} \frac{\sigma_m}{\sigma_0} \Big\{ & \mathrm{rect}(\frac{f_r}{NT_e K})\mathrm{rect}(\frac{y - y_m}{L_s}) \\ & \cdot \exp(-j\frac{4\pi}{c}(R'_m(\eta, R_m) - R'_m(\eta_0, R_m))f_r) \Big\} \\ & \cdot \exp(-j\pi f_r^2 / k) \end{aligned} \tag{12.22}$$

The IFFT of the above equation is carried out by using the stationary phase principle [1, 5]

$$\begin{aligned} S_n(f_r, \eta) = \sum_{m=1}^{M} \frac{\sigma_m}{\sigma_0} & \mathrm{rect}(\frac{\tau}{NT_e})\mathrm{rect}(\frac{y - y_m}{L_s}) \\ & \cdot \exp(j\pi k(\tau - \xi)^2 \exp(-j2\pi\xi f_c) \end{aligned} \tag{12.23}$$

Where $\xi = \frac{2}{c}(R'_m(\eta, R_m) - R'_m(\eta_0, R_m))$ is time delay. In this way, after the above transformation, the echo signal back to the raw echo domain, and then the classic CS algorithm can be used for subsequent imaging processing.

12.5 Digital Beam Forming of MIMO SAR

German scholar G Krieger proposed the concept of multidimensional coding in 2008, the idea of this system is to combine the spatial–time waveform encoding at transmitting terminal and DBF processing at receiving terminal to improve system performance. In this section, we will use the DBF-based single-phase center multi-beam SAR (DBF-SPCMB-SAR) system as an example to analyze this new system.

The spaceborne DBF-SPCMB-SAR system uses the phased array antenna, the azimuthal aperture is processed by the azimuth DBF, and its beams are directed to different azimuth position to realize the azimuth beam "Narrow-transmit narrow-receive". However, unlike the traditional single-phase center multi-beam spaceborne SAR, in each PRT, the transmitted chirp signal is not simultaneously transmitted, but the transmitted pulse is divided into multiple sub-pulses, and the azimuth beam pointing is adjusted with each sub-pulse transmission to different azimuth positions, as shown in Fig. 12.8.

In Fig. 12.8, taking of transmitting 3 sub-pulses as an example, the sub-pulse 1, the sub-pulse 2, and the sub-pulse 3, which are sequentially emitted, respectively illuminate the azimuthal position 1, the position 2, and the position 3. In practical applications, the beam 1 and the beam 2 can be separated, so that the mutual interference between the beams can be further reduced.

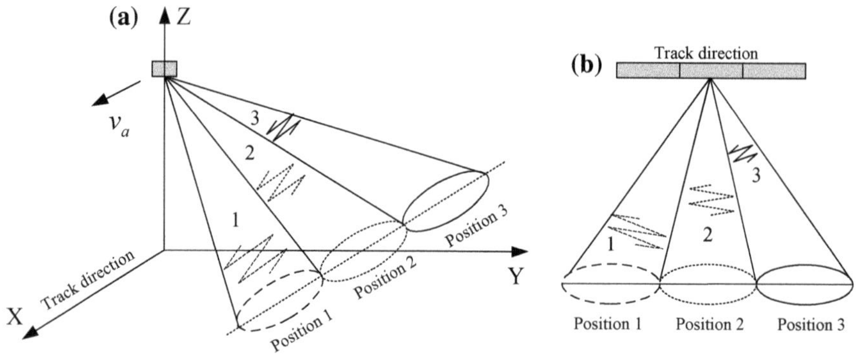

Fig. 12.8 Sketch map of transmitting modes for spaceborne DBF-SPCMB-SAR. **a** 3D diagram; **b** 2D diagram

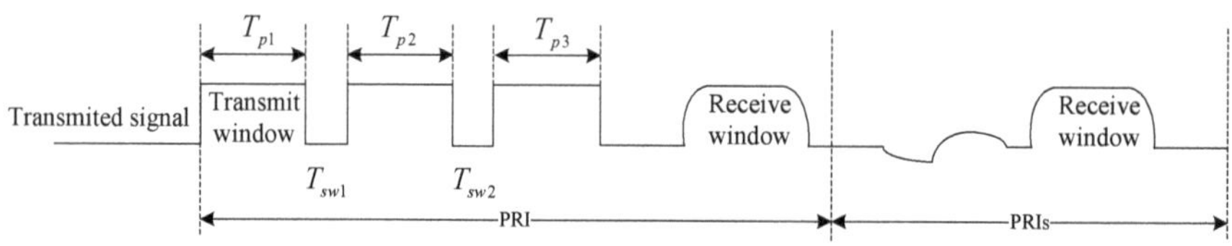

Fig. 12.9 Timing diagram of transmitting pulses for DBF-SPCMB-SAR

The timing of each transmitting pulse is shown in Fig. 12.9, where T_{p1}, T_{p2} and T_{p3} are the transmitting pulse duration, T_{sw1} and T_{sw2} are the beam switching time. All antenna apertures receive echoes simultaneously.

For a sampling moment, the received signals are the superposition of the targets reflected echo signals at different slant ranges for different transmission beams, and the targets at different slant ranges correspond to different look angles, so the echo signals corresponding to the respective transmission beams can be separated by the reception DBF in range direction, thereby suppressing the range ambiguities. At the same time, different transmission beams correspond to different look angles, Doppler center frequencies are different, so Doppler ambiguities can be resolved by Doppler frequency combination, and a complete Doppler spectrum can be formed, so high resolution in azimuth can be achieved.

The echo signals received by the NK (N for the azimuthal aperture number, K is the range aperture number) receiving aperture will be applied DBF processing, including the range DBF and azimuth DBF. These two parts DBF can be processed separately, or can be processed simultaneously in two dimensions. The schematic diagram is shown in Fig. 12.10. The two-dimensional DBF can be decomposed into azimuth DBF and range DBF. The range DBF is mainly used to separate the echo signals corresponding to the respective transmitting sub-pulses. The azimuth DBF is mainly used to adjust the azimuthal receive beam to point to the corresponding azimuth position.

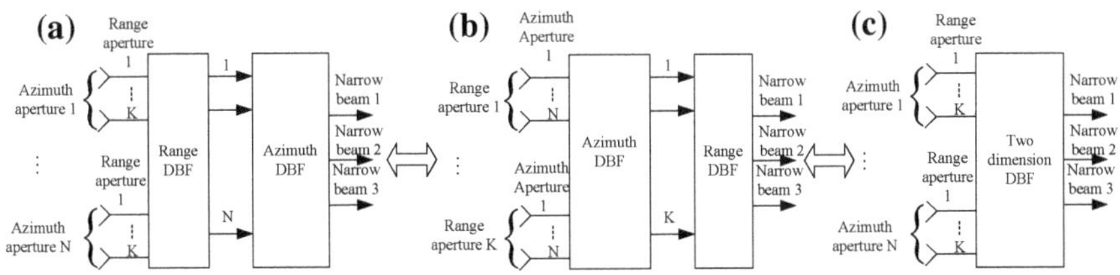

Fig. 12.10 Block diagram of two-dimensional DBF processing. **a** Range DBF before azimuth DBF; **b** Azimuth DBF before range DBF; **c** Two-dimensional DBF

References

1. Fisher E, Haimovich A, Blum RS et al (2004) MIMO radar: an idea whose time has come. In: Proceedings of the IEEE radar conference, Philadelphia, USA, pp 71–78
2. Dai X (2008) Theory and method study on diversity detection and wideband synthesis for MIMO radars. Doctoral Dissertation of Electronics, Department of Tsinghua University, Beijing
3. Ender JHG (2007) MIMO-SAR. The technique university Hamburg, International Radar Symposium, Colonge, Germany 9, pp 580–588
4. Callaghan GD, Longstaff ID (1997) Wide-Swath space borne SAR and range ambiguity. In: Proceedings of RADAR 97. Edinburgh, UK, pp 248–252
5. Kim JH, Ossowska A, Wiesbeck W (2007) Investigation of MIMO SAR for interferometry. EURAD, Munich, Germany, pp 51–54
6. Li J, Zheng XY, Stoica P (2007) MIMO SAR Imaging: signal synthesis and receiver design. EURAD, Munich, Germany, pp 89–92
7. Ossowska A, Kim JH, Wiesbeck W (2007) Modeling of nonidealities in receiver front-end for a simulation of multistate SAR system. EUSAR, Barcelona, Spain, pp 13–16
8. Krieger G, Gebert N, Moreira A (2008) Multidimensional waveform encoding: a new digital beam forming technique for synthetic aperture radar remote sensing. IEEE TGRS 46(1):31–46
9. Song Y, Yang R (2007) Study on high resolution wide swath synthetic aperture radar using multiple transmit-receive apertures. J Electron Inf Technol 29(9):2110–2113
10. Xia Y (2008) Study on distributed small satellites SAR imaging technology. Doctoral Dissertation of Institute of Electronics, Chinese Academy of Sciences, Beijing
11. Jing W (2008) Study on wide swath high resolution imaging technology of spaceborne SAR. Doctoral Dissertation of Xian University, Xi'an
12. Dai X, Xu J, Peng Y (2008) MIMO-VSAR and a kind of optimized array configuration. Acta Electronica Sinica 36(12):2394–2399
13. Wang WQ (2007) Applications of MIMO technique for aerospace remote sensing. In: Proceedings of IEEE Aerospace Conference, Big Sky, USA, pp 1–10
14. Wang WQ, Cai JY (2009) Multiple-Input and Multiple-Output SAR diversified waveform design. IEEE Aerospace Conference, Big Sky, USA, pp 1051–1053
15. Wu Q, Jing W, Xing M (2009) Wide swath imaging with multidimensional waveform encoding. J Xian Univ 36(5):801–806
16. Wang L, Xu J, Huangpu k et al (2009) Analysis and compensation of equivalent phase center error in MIMO SAR. Acta Electronica Sinica 37(12):2688–2693
17. Liu N, Zhang L, Zhang J et al (2009) Multifrequency—multibaseline InSAR and its performance analysis. Syst Eng Electron 31(9):2090–2095
18. Shiqiang L (2004) Study on high resolution wide swath synthetic aperture radar systems. Doctoral Dissertation of Institute of Electronics, Chinese Academy of Sciences, Beijing

19. Kim JH, Ossowska A, Wiesbeck W (2007) Experimental investigation of digital beam forming SAR performance using a ground-based demonstrator. IGARSS, Barcelona, Spain, pp 111–114
20. Kim JH, Ossowska A, Wiesbeck W (2006) Ground based measurement system for the evaluation of a SAR with digital beam forming. EUSAR, Dresden, Germany, pp 120–123
21. Kim JH, Ossowska A, Wiesbeck W (2007) Laboratory experiments for the evaluation of digital beam forming SAR features. 2007 international Waveform Diversity and Design Conference, Pisa, Italy, pp 292–296
22. Bao K, Tao H, Liao G (2007) Waveform design for multi-emitted and distributed space-based radar system. J Electron Inf Technol 29(9):2117–2119
23. Wang D, Yuan J, Ma X (2007) Discrete frequency coding design based on the genetic algorithm for MIMO radar. J Air Force Radar Acad 21(2):105–107
24. Chen CY, Vaidyanathan PP (2007) MIMO radar ambiguity optimization using frequency-hopping waveforms. In: 41th Asilomar Conference on Signals, Systems and Computers, 4, pp 192–196
25. Li J, Stoica P, Zhu X (2008) MIMO radar waveform synthesis. In: IEEE Proceedings, pp 2125–2130
26. Yuan X (2003) Introduction to spaceborne synthetic aperture radar. National Defense Industry Press, Beijing
27. Franceschetti G, Lanari R (1999) Synthetic aperture radar processing. CRC Press, New York
28. Bennett JR, Cumming I, Deane R (1980) Digital processing of SEASAT data. In: IEEE International Radar Conference, pp 168–175
29. Wu C, Liu KY, Jin M (1982) Modeling and a correlation algorithm for space borne SAR signals. IEEE TAES 18(5):563–575
30. Jin M, Wu C (1984) A SAR correlation algorithm which accommodates large range migration. IEEE TGRS 22(6):592–597
31. Raney RK, Runge H, Bamler R (1994) Precision SAR processing using chirp scaling. IEEE TGRS 32(4):786–799
32. Moreira A, Mittermayer J, Scheiber A (1996) Extended chirp scaling algorithm for air and space borne SAR data processing in strip map and scan SAR imaging modes. IEEE TGRS 34 (5):1123–1136

Chapter 13
Autofocus Algorithm for SAR

Autofocus algorithm is one of the key technologies for high-resolution SAR signal processing. High-resolution SAR is a coherent imaging system. It is very important for high-resolution image to maintain the coherence of echo, because the requirement for the stable phase of phase-coherent imaging system is very strict. In the real SAR system, many factors can cause echo phase errors, mainly as follows [1–8]:

(1) Theoretical model of SAR assumes that the radar platform motion not only remains horizontal linear motion of constant velocity and constant height, but also maintains the stabilization during the flight; that is to say, the radar beam is required pointing invariant. In practice, the velocity of the radar platform is a vector varying with time, and the motion trajectory of the platform is not an ideal straight line, which results in the slant range error between target and antenna phase center. Slant range error exists as phase error in echo, which can be compensated with the platform attitude parameters provided by high-precision INS (Inertial Navigation System).
(2) During the electromagnetic wave propagation process, the irregular movement of the troposphere atmospheric vortex will cause heterogeneous medium density, and then cause atmosphere refractivity inhomogeneous variation. In addition, because of the effect of the ionosphere disturbance, the electromagnetic wave propagation speed is not a constant, but it is a variable changing with time and space.
(3) The phase instability is caused because of phase error which can be introduced by receiver internal frequency source instability and PLL (Phase-Locked Loop) phase jitter etc.
(4) Some approximate factors also cause phase errors in the SAR data processing.

In order to analyze and compensate phase errors, they need to be classified. Phase errors can be divided into deterministic polynomial errors and random errors according to the variation of the phase errors, and they also can be divided into low-frequency and high-frequency phase errors according to the frequency of the phase errors. It is worth noting that different types of phase errors have different

R. Yang et al., *High-Resolution Microwave Imaging*,
https://doi.org/10.1007/978-981-10-7138-6_13

effects on the image. Taking into account the amplitude, random phase errors will not affect the image focus, but they can significantly reduce the image contrast. In contrast, polynomial phase errors will cause the image defocus. Phase errors mainly result in geometric distortion, resolution decrease, image contrast reduction, and false targets increase in SAR image, as well as image post-process difficulty, such as automatic target recognition, image segmentation classification, etc.

It is important to reduce the effect of phase errors in imaging process. Many literatures have proposed a lot of relevant phase compensation and noise suppression methods. Usually, phase errors compensation methods are divided into open-loop and closed-loop compensation.

Open-loop methods compensate phase errors directly with phase error correction signal estimated by accurate measurements for motion parameters of the radar platform. These methods require accurate tracking antenna phase center and high measurement accuracy of platform motion parameters. The main limitation is phase errors caused by the platform motion is unstable, so these methods are very limited for phase errors compensation and they cannot compensate the phase error due to other factors such as medium heterogeneity of electromagnetic waves. Even if advanced electronic navigation device is installed on the radar platform, it is still very difficult for the platform movement trajectory satisfied to SAR theoretical model due to long synthetic aperture time. Therefore, it is necessary to estimate phase errors from measurement data and compensate for estimated error for azimuth high-resolution SAR.

Closed-loop motion compensation technologies mainly include range migration correction, clutter lock, and autofocus. Clutter lock means to lock Doppler center of the earth's surface clusters and to stable the echo Doppler center frequency. Autofocus algorithms mainly deal with low-frequency phase error function, which usually can be assumed to be a polynomial function, and estimate coefficients of the polynomial. Closed-loop compensation method mainly estimates phase error function from echo assuming that phase error function is unknown in advance. These methods only focus on the form of the phase error function in principle, regardless of the source of phase error. These methods cannot compensate high-frequency phase error; however, they can compensate phase errors introduced by irregular movement of platform and atmospheric disturbance.

Usually adopting the Combination compensation methods of open-loop and closed-loop: (1) Using external parameters of the imaging system for phase compensation, then using parameter estimation method for iterative estimation and compensation; (2) Iterative estimation and compensation based on the initial value provided by external parameters of the imaging system.

Autofocus algorithm is a kind of image autofocusing process method to compensate low-order phase errors caused by an irregular motion of SAR platform or atmospheric disturbances. The basic principle is to estimate phase error function from echo data, to compensate and focus, and get high-quality image. In the high-resolution SAR systems, autofocus is one of the key azimuth processing technologies. Figure 13.1 shows the principle of autofocus.

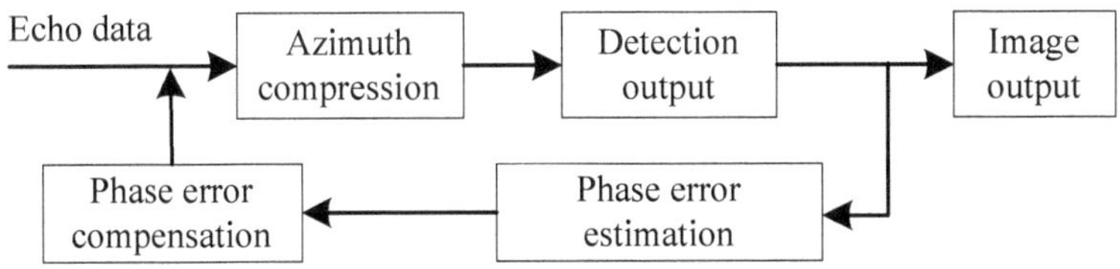

Fig. 13.1 The principle of autofocus

In this chapter, the classical autofocus algorithms were discussed, including Map-drift (MD) algorithm, Multiple Aperture Map-drift (MAM) algorithm, Shift-And-Correlate (SAC) algorithm, Phase Difference (PD) algorithm, Phase Gradient Algorithm (PGA), Contrast Optimization (CO) algorithm, and so on.

13.1 Map-Drift Algorithm [1]

Map-drift autofocus algorithm (MD) directly estimates the quadratic coefficient of the phase error and then compensates SAR echo based on this estimation to eliminate the defocus of azimuth space-invariant quadratic phase error, thereby improves image quality. MD autofocus algorithm is based on the two features of SAR data [5, 6].

(1) A full aperture data can be divided into several sub-apertures data in azimuth dimension, each sub-aperture imaging can be achieved individually, and sub-aperture image of the same resolution in azimuth can be obtained individually. The sub-aperture image resolution is worse than that of the full aperture, but they can get an image of the same area.
(2) The linear phase error across the whole aperture shifts the entire image in the azimuth dimension. Because the frequency is the derivative of phase, the linear phase error is equivalent to a constant frequency offset in the frequency domain for the original phase spectrum of nonlinear phase error.

The MD autofocus algorithm divides an aperture of duration T_a into two nonoverlapping sub-apertures of duration $T_a/2$. The quadratic phase error across the full aperture has a different functional form across two half-length sub-apertures. Quadratic phase error across the full aperture is composed of a quadratic component, a linear component and a constant across each sub-aperture. The linear phase components have the same amplitude but slopes of opposite sign. The opposite slopes of the linear component across each sub-aperture will cause corresponding spatial-domain image shifting in opposite directions, which is caused by Fourier shift theorem. If $s(t)$ and $S(\omega)$ are a Fourier transform pair, denoted by $s(t) \overset{F}{\leftrightarrow} S(\omega)$, and then $e^{j\omega_0 t}s(t) \overset{F}{\leftrightarrow} S(\omega - \omega_0)$. The linear phase term of the signal $s(t)$ make $S(\omega)$ shift on the frequency dimension. Therefore, the opposite slope of the linear component across each sub-aperture will generate the image shift in opposite directions, and the relative shift of two images is proportional to the coefficient of the quadratic error across full aperture. The main steps of MD algorithm are two sub-aperture images cross-correlation, relative shift between the two sub-images estimation, and peak position of cross-correlation search and quadratic component of quadratic phase error calculation.

Assume that the azimuth signal $g(t)$ in a range cell is the product of the error-free ideal azimuth signal $s(t)$ and the function with a quadratic phase error.

$$g(t) = s(t)\exp\left(j\alpha \cdot t^2\right) \quad -T_{\mathrm{a}}/2 \le t \le T_{\mathrm{a}}/2 \tag{13.1}$$

where α the coefficient of a quadratic phase error to be estimated is, T_{a} is synthetic aperture time, and the quadratic phase error from center to the edge of the aperture is:

$$Q = \alpha T_{\mathrm{a}}^2/4 \tag{13.2}$$

MD divides $g(t)$ into the two nonoverlapping parts $g_1(t)$ and $g_2(t)$, both are duration $T_{\mathrm{a}}/2$.

$$\begin{aligned} g_1(t) &= g(t - T_{\mathrm{a}}/4) \\ &= s(t - T_{\mathrm{a}}/4)\exp\left[j\left(\alpha t^2 - \frac{\alpha T_{\mathrm{a}}}{2}t + \frac{\alpha T_{\mathrm{a}}^2}{16}\right)\right] \quad -T_{\mathrm{a}}/4 \le t \le T_{\mathrm{a}}/4 \end{aligned} \tag{13.3}$$

$$\begin{aligned} g_2(t) &= g(t + T_{\mathrm{a}}/4) \\ &= s(t + T_{\mathrm{a}}/4)\exp\left[j\left(\alpha t^2 + \frac{\alpha T_{\mathrm{a}}}{2}t + \frac{\alpha T_{\mathrm{a}}^2}{16}\right)\right] \quad -T_{\mathrm{a}}/4 \le t \le T_{\mathrm{a}}/4 \end{aligned} \tag{13.4}$$

Fourier transformation for each sub-aperture:

$$G_1(\omega) = \int_{-T_{\mathrm{a}}/4}^{T_{\mathrm{a}}/4} g_1(t)\,\mathrm{e}^{-\mathrm{j}\omega t}\mathrm{d}t = \widehat{S}_1\left(\omega + \frac{\alpha T_{\mathrm{a}}}{2}\right) \tag{13.5}$$

$$G_2(\omega) = \int_{-T_{\mathrm{a}}/4}^{T_{\mathrm{a}}/4} g_2(t)\,\mathrm{e}^{-\mathrm{j}\omega t}\mathrm{d}t = \widehat{S}_2\left(\omega - \frac{\alpha T_{\mathrm{a}}}{2}\right) \tag{13.6}$$

where

$$\widehat{S}_1(\omega) = \int_{-T_{\mathrm{a}}/4}^{T_{\mathrm{a}}/4} s\left(t - \frac{T_{\mathrm{a}}}{4}\right)\exp\left[j\left(\alpha t^2 + \frac{\alpha T_{\mathrm{a}}^2}{16}\right)\right] \cdot \mathrm{e}^{-\mathrm{j}\omega t}\mathrm{d}t \tag{13.7}$$

$$\widehat{S}_2(\omega) = \int_{-T_a/4}^{T_a/4} s\left(t + \frac{T_a}{4}\right) \exp\left[j\left(\alpha t^2 + \frac{\alpha T_a^2}{16}\right)\right] \cdot e^{-j\omega t} dt \tag{13.8}$$

MD algorithm assumes $\left|\widehat{S}_1(\omega)\right|^2 = \left|\widehat{S}_2(\omega)\right|^2$, which is valid to the extent. From (13.5) and (13.6), we can clearly see that $|G_1(\omega)|^2$ and $|G_2(\omega)|^2$ are the opposite shifting results of the same function. The relative shift between maps is proportional to coefficient α of quadratic phase error across full aperture. MD algorithm cross-correlates two intensity images and calculates the peak location of cross-correlation, and then get the estimated $\hat{\alpha}$ of the quadratic phase error.

$$\hat{\alpha} = \Delta\omega \,/\, T_a \tag{13.9}$$

$\Delta\omega$ can be positive or negative depending on the sign of the quadratic phase error.

During SAR real data processing, based on a quadratic phase error MD also can be accomplished by estimating the Doppler frequency rate. Figure 13.2 is the block diagram of MD procedure. Radar works broadside, namely, Doppler centroid $f_{dc} = 0$. Simultaneously, because of the Doppler frequency rate is related to range, in the practical processing, Doppler frequency rate should be estimated in the different range cell, and the image energy should be accumulated within the focus depth.

Application of MD autofocus algorithm should be noted [6]:

(1) If some strong reflection points exist in a homogeneous background and the energy of strong reflection points occupies a significant proportion in the whole image, the shift of these strong reflection points will result in some peaks in the cross-correlation result.
(2) If the terrain has obvious characteristics, such as the coastline, the river, the city, the mountain's edge, and so on. MD algorithm is still effective.
(3) If the ground reflection distribution is completely uniform, the shift of the two images cannot be reflected by the cross-correlation results, namely, the method is invalid.

In a word, MD autofocus method requires high-contrast images in order to obtain the peak of cross-correlation.

In the practical application of MD algorithm, the initial Doppler frequency rate is obtained according to the theoretical calculation. In fact, because of the motion error, there is a deviation between the initial value and the true value, and the deviation is not determined. Because MD algorithm is a closed-loop iterative processing, the deviation between initial value and true value of the Doppler frequency rate will directly affect the processing speed of MD algorithm. If the deviation is big, the processing speed is slow and vice versa.

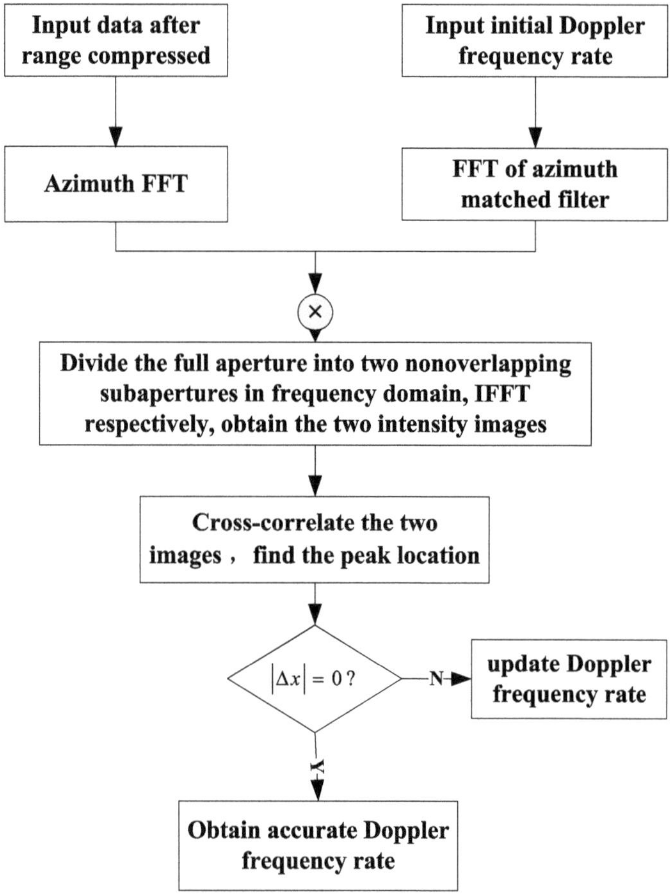

Fig. 13.2 The block diagram of MD procedure

The set on a threshold value of estimation error will influence processing speed of MD algorithm, the threshold is stricter and the processing speed is slower. The set on threshold depends on the requirement of estimation precision.

13.2 Multiple Aperture Map-Drift [1]

MD autofocus algorithm only can estimate quadratic phase error. When phase error function contains higher order phase errors sufficiently large, the image contrast is reduced, and MD algorithm failed. If an extension of the MD method, the full aperture can be divided into multiple nonoverlapping sub-apertures and we can estimate the higher order phase error, known as the multiple aperture map-drift

(MAM) autofocus algorithm [1]. In general, N sub-apertures can be estimated from second to Nth order polynomial coefficients.

The model of phase error:

$$\phi_{\mathrm{e}}(t) = \sum_{k=2}^{N} a_k t^k, \quad -T_{\mathrm{a}}/2 \le t \le T_{\mathrm{a}}/2 \tag{13.10}$$

where a_k is the kth polynomial coefficients, T_{a} is synthetic aperture time. Here ignoring constant and linear phase which do not affect the image focus.

MAM divides the full aperture into N sub-apertures. The phase error over the ith sub-aperture of width T_{a}/N

$$\phi_i(t) = \sum_{k=2}^{N} a_k (t + t_i)^k, \quad |t| \le \frac{T_{\mathrm{a}}}{2N} \tag{13.11}$$

where t_i is the center of ith sub-aperture

$$t_i = \left(\frac{i}{N} - \frac{N+1}{2N}\right) T_{\mathrm{a}}, \quad i = 1, 2, \cdots, N \tag{13.12}$$

The linear phase error component of each sub-aperture relates between the relative shift of sub-images pair and polynomial coefficients. We can estimate polynomial coefficients by measuring the location of cross-correlation peak of sub-image pair and estimate the relative shift.

The linear phase error component of ith sub-aperture is

$$\Delta\phi_{\mathrm{lin},i}(t) = \sum_{k=2}^{N} a_k \cdot k \cdot t_i^{k-1} t, \quad |t| \le \frac{T_{\mathrm{a}}}{2N} \tag{13.13}$$

Therefore, the image shift of ith sub-aperture from polynomial phase error

$$\Delta_i = \sum_{k=2}^{N} a_k k t_i^{(k-1)} \tag{13.14}$$

The relative shift between two sub-images determined by cross-correlation of two intensity sub-images. The relative shift from the ith sub-aperture and jth sub-aperture

$$\Delta_{i,j} = \sum_{k=2}^{N} a_k k \left[t_j^{(k-1)} - t_i^{(k-1)} \right], \quad i < j \tag{13.15}$$

where $i,j \in 1,2,\cdots,N$, $\Delta_{i,j}$ includes $N(N-1)/2$ relative shifts. The expression can be described by overdetermined linear equations

$$\Delta = \delta \cdot a \tag{13.16}$$

Δ is relative shifting matrix

$$\Delta = [\Delta_{1,2} \cdots \Delta_{1,N}\, \Delta_{2,3} \cdots \Delta_{2,N}\, \Delta_{3,4} \cdots \Delta_{N-1,N}]^T \tag{13.17}$$

a is the coefficients matrix

$$a = \begin{bmatrix} a_2 & a_3 & \cdots & a_N \end{bmatrix}^T \tag{13.18}$$

δ is the relational matrix

$$\delta = \begin{bmatrix} \delta_{1,2}^2 & \delta_{1,2}^3 & \cdots & \delta_{1,2}^N \\ \delta_{1,3}^2 & \delta_{1,3}^3 & \cdots & \delta_{1,3}^N \\ \cdots & \cdots & \cdots & \cdots \\ \delta_{1,N}^2 & \delta_{1,N}^3 & \cdots & \delta_{1,N}^N \\ \delta_{2,3}^2 & \delta_{2,3}^3 & \cdots & \delta_{2,3}^N \\ \delta_{2,4}^2 & \delta_{2,4}^3 & \cdots & \delta_{2,4}^N \\ \cdots & \cdots & \cdots & \cdots \\ \cdots & \cdots & \cdots & \cdots \\ \delta_{N-1,N}^2 & \delta_{N-1,N}^3 & \cdots & \delta_{N-1,N}^N \end{bmatrix} \tag{13.19}$$

The superscript T denotes matrix transpose. $\delta_{i,j}^k$ is the sub-image shift between the ith sub-aperture and the jth sub-aperture induced by a kth order phase error of unit coefficient?

$$\delta_{i,j}^k = \frac{k}{2\pi}\left[t_j^{(k-1)} - t_i^{(k-1)}\right] \tag{13.20}$$

The polynomial error coefficients can be obtained

$$a = \delta^{-1}\Delta \tag{13.21}$$

where δ^{-1} is the pseudo-inverse matrix of δ.

The number of divided sub-apertures will increase with the higher order of the phase error, namely, the sub-aperture is too short to get good resolution of the sub-image, and then decrease SNR of sub-image, which results in the worse error estimation accuracy. The model order is limited for reliable estimation phase errors. Generally, the MAM algorithm can be estimated up to fifth-order phase errors.

In addition, the autofocus algorithm requires prior knowledge of the order for the higher order phase errors that cannot be ignored; otherwise, if the number of sub-apertures is less than the number of order, the performance will degrade (Fig. 13.3).

13.3 Phase Difference Algorithm

The phase difference (PD) autofocus algorithm, similar to MD algorithm, estimates the coefficient of the quadratic phase error by dividing the full aperture into two no overlapping sub-apertures. PD algorithm gets the error estimate from the Fourier transform of the phase difference function between two sub-apertures data. It also needs iterative aggregation to obtain robust estimates of the quadratic phase error [1].

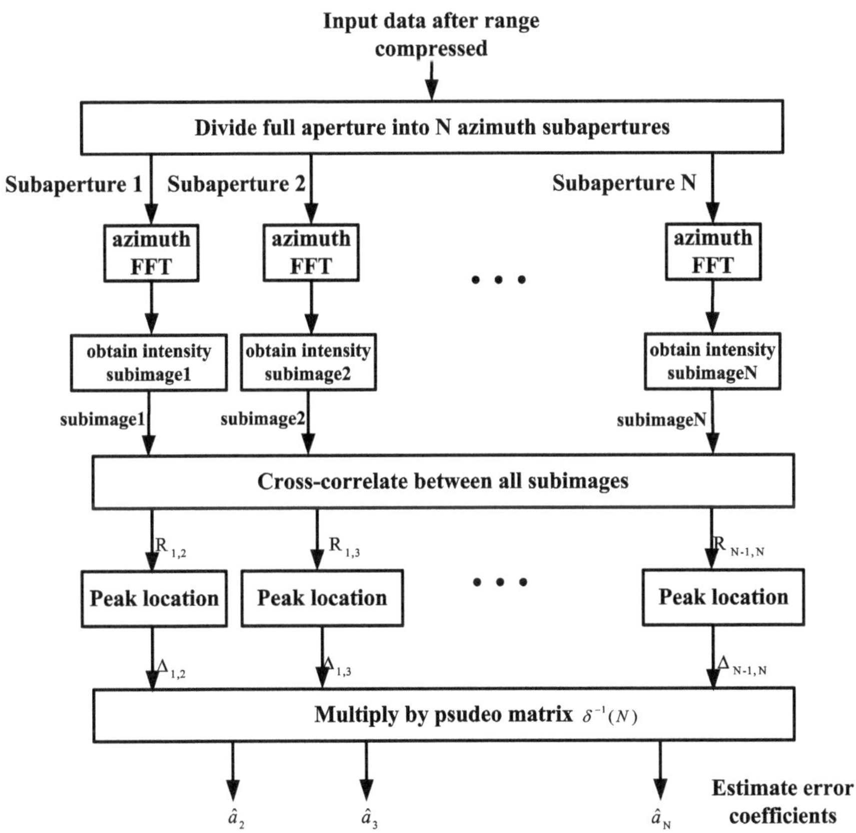

Fig. 13.3 The block diagram of MAM procedure

PD algorithm divides full aperture into two no overlapping sub-apertures, and multiplies one sub-aperture by conjugated signal of the other sub-aperture on each range cell, and calculates the result by Fourier transform, and then obtains the correlation function on each range, determines the location of peak and gets the relative shift between the two sub-aperture images. PD algorithm can get narrower peak value than MD algorithm. By mean of estimated quadratic phase coefficient, PD algorithm constructs phase compensation vector to eliminate the influence of quadratic phase error across full aperture on the signal after range compression.

The two sub-aperture signals are $g_1(t)$ and $g_2(t)$, the product of one and complex conjugate of the other $g_P(t)$ is

$$g_p(t) = g_2(t)g_1^*(t) \quad -T_a/4 \le t \le T_a/4 \tag{13.22}$$

The complex conjugation and multiplication get rid of the common quadratic phase error.

$G_p(\omega)$ is the Fourier transformation of $g_P(t)$

$$G_p(\omega) = \int_{-T_a/4}^{T_a/4} g_p(t)\,e^{-j\omega t}dt = \int_{-T_a/4}^{T_a/4} s(t+T_a/4)s^*(t-T_a/4)\exp(j\alpha T_a t)\,e^{-j\omega t}\,dt \tag{13.23}$$

The peak location of $|G_p(\omega)|$ occurs at zero frequency without the quadratic phase error and the peak location will shift with linear phase $\exp(j\alpha T_a t)$. In order to improve SNR and estimation accuracy, we need to accumulate $|G_p(\omega)|$ overall range cells. The estimation of the quadratic phase error $\hat{\alpha}$ is

$$\hat{\alpha} = \Delta\omega / T_a \tag{13.24}$$

$\Delta\omega$ can be positive or negative depending on the sign of the quadratic phase error.

The principle difference between PD algorithm and MD algorithm is that PD algorithm uses the cross-correlation of the complex sub-images to estimate quadratic phase coefficient, however, MD algorithm uses the cross-correlation of the intensity sub-images to estimate quadratic phase coefficient. PD algorithm can eliminate quadratic phase error effectively.

The following is the block diagram of PD (Fig. 13.4).

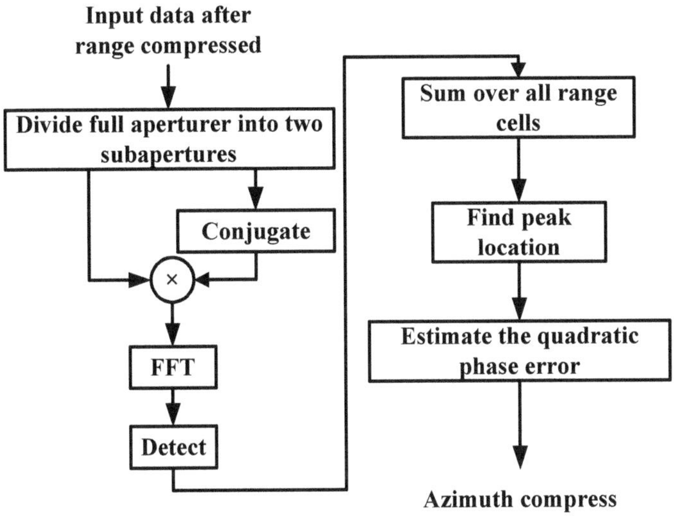

Fig. 13.4 The block diagram of PD procedure

13.4 Shift and Correlation [6, 9]

SAC autofocus algorithm is an efficient autofocus algorithm. SAC is similar to MD algorithm and it also divides a full aperture into two sub-apertures in azimuth frequency domain. The difference between them is that MD algorithm is a cross-correlation algorithm in image domain and needs iteration, but SAC algorithm doesn't need to iterate to get an estimation of Doppler rate. So the calculation amount of SAC is much less than that of MD.

The basic operation of SAC algorithm is frequency shift and cross-correlation. First, the signal is divided into two parts in Doppler domain which are corresponding to upper half and lower half of Doppler spectrum respectively, and then signals corresponding to s_u and s_l cross-correlate. The amount of calculation is small because cross-correlation is not in the image domain and it does not need iterative aggregation.

Assume that the azimuth signal $s(t)$ in a range cell is

$$s(t) = \exp\left[j\pi \cdot f_R (t - t_a)^2\right], \quad t_a - T_a/2 \le t \le t_a + T_a/2 \tag{13.25}$$

where $T_a = \frac{B_d}{f_R}$, namely synthetic aperture time, B_d is Doppler bandwidth, f_R is Doppler rate. For simplicity, $|s(t)| = 1$ (Fig. 13.5).

Assume $s_u(t)$ is the part corresponding to the upper half frequency of $s(t)$ and $s_l(t)$ is the part corresponding to the lower half frequency of $s(t)$

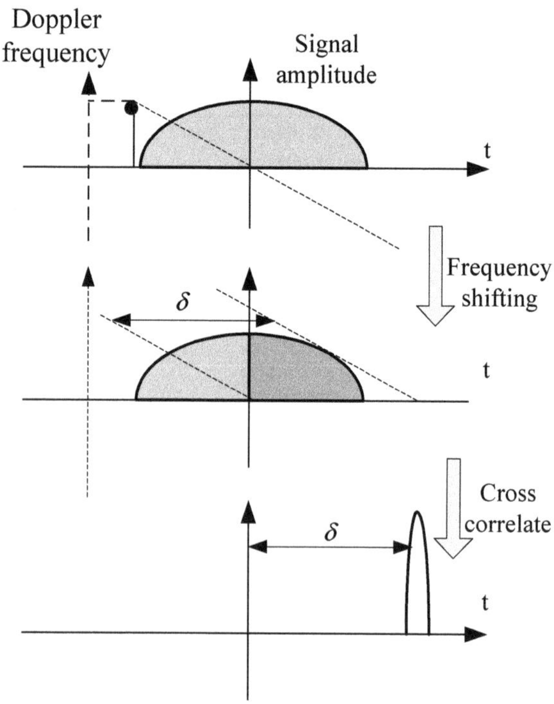

Fig. 13.5 The principle of SAC algorithm

$$
\begin{aligned}
s_{\mathrm{u}}(t) &= \exp\left[j\pi f_{\mathrm{R}}(t-t_{\mathrm{a}})^2\right], \quad t_{\mathrm{a}} \le t \le t_{\mathrm{a}} + T_{\mathrm{a}}/2 \\
s_{\mathrm{l}}(t) &= \exp\left[j\pi f_{\mathrm{R}}(t-t_{\mathrm{a}})^2\right], \quad t_{\mathrm{a}} - T_{\mathrm{a}}/2 \le t \le t_{\mathrm{a}}
\end{aligned}
\tag{13.26}
$$

Shift the spectrum of $s_{\mathrm{u}}(t)$ and $s_{\mathrm{l}}(t)$ with PRF/4 and −PRF/4, respectively

$$
\begin{aligned}
s_{\mathrm{l}+}(t) &= s_{\mathrm{l}}(t)\exp\left(j\frac{\pi}{2}\cdot f_{\mathrm{PRF}}\cdot t\right) \\
&= \exp\left[-j\pi f_{\mathrm{R}}\delta\cdot t_{\mathrm{a}} - j\pi f_{\mathrm{R}}\left(\frac{\delta}{2}\right)^2\right]\exp\left[j\pi\cdot f_{\mathrm{R}}\left(t-t_{\mathrm{a}}-\frac{\delta}{2}\right)^2\right]
\end{aligned}
\tag{13.27}
$$

$$
\begin{aligned}
s_{\mathrm{u}-}(t) &= s_{\mathrm{u}}(t)\exp\left(-j\frac{\pi}{2}\cdot f_{\mathrm{PRF}}\cdot t\right) \\
&= \exp\left[j\pi f_{\mathrm{R}}\delta\cdot t_{\mathrm{a}} - j\pi f_{\mathrm{R}}\left(\frac{\delta}{2}\right)^2\right]\exp\left[j\pi\cdot f_{\mathrm{R}}\left(t-t_{\mathrm{a}}+\frac{\delta}{2}\right)^2\right]
\end{aligned}
\tag{13.28}
$$

$$
\delta = \frac{f_{\mathrm{PRF}}}{2|f_{\mathrm{R}}|} = \frac{f_{\mathrm{PRF}}\cdot\lambda R}{4V_{\mathrm{a}}^2}
\tag{13.29}
$$

The cross-correlation of $s_{\mathrm{u}-}(t)$ and $s_{\mathrm{l}+}(t)$

$$\begin{aligned} s_{\mathrm{c}}(t) &= \int s_{\mathrm{l}+}(\tau) s_{\mathrm{u}-}^{*}(\tau - t)\,\mathrm{d}t \\ &= \exp\left(j2\pi \cdot f_{\mathrm{R}}\delta \cdot t_{\mathrm{a}}\right) \\ &\quad \cdot \int_{t_{\mathrm{a}}-\frac{T_{\mathrm{a}}}{2}}^{t_{\mathrm{a}}+T} \exp\left[j\pi \cdot f_{\mathrm{R}}(\tau - t_{\mathrm{a}} - \frac{\delta}{2})^{2}\right] \cdot \exp\left[-j\pi \cdot f_{\mathrm{R}}(\tau - t_{\mathrm{a}} + \frac{\delta}{2})^{2}\right] \mathrm{d}\tau \end{aligned} \tag{13.30}$$

From Schwarz inequality, when $\exp\left[-j\pi \cdot f_{\mathrm{R}}(\tau - t_{\mathrm{a}} - \frac{\delta}{2})^{2}\right] = \exp\left[-j\pi \cdot f_{\mathrm{R}}(\tau - t_{\mathrm{a}} + \frac{\delta}{2})^{2}\right]$, namely when $t = \tau$, its modulus gets to the maximum value. So we can estimate the speed of platform

$$\widehat{V}_{\mathrm{a}} = \sqrt{\frac{f_{\mathrm{PRF}} \cdot \lambda \cdot R}{4\delta}} \tag{13.31}$$

The maximum value of Cross-correlation occurs at $t = \delta$ and it does not depend on t_{a}, which means the maximum value of Cross-correlation occurs at $t = \delta$ for points of different azimuth coordinate after shifting and correlation on the same range cell. However, $\exp\left(j2\pi \cdot f_{\mathrm{R}}\delta \cdot t_{\mathrm{a}}\right)$ in (13.30) relates to azimuth coordinate of point target, so the peaks of shifting for different points are not coherent accumulation, which is not good for estimation. At the same time, targets located on azimuth coordinate separated by $t_{\mathrm{b}} - t_{\mathrm{a}}$ can also produce small amplitude peak at $t = \delta + t_{\mathrm{b}} - t_{\mathrm{a}}$.

Application of SAC should be noted:

(1) To guarantee azimuth sampling rate is equal to Doppler band B_{d} approximately, SAC should be applied after prefilter, and then the spectrum of $s_{\mathrm{u}-}(t)$ and $s_{\mathrm{l}+}(t)$ can be overlapped completely, or the contrast of correlation result will be worse and the believable peak cannot be found.
(2) During the process of SAC, the lower spectrum of $s_{\mathrm{l}}(t)$ can be shifted with $-f_{\mathrm{PRF}}/2$ while the spectrum of $s_{\mathrm{u}}(t)$ remains original place, which is equal to shifting the spectrum of $s_{\mathrm{u}}(t)$ and $s_{\mathrm{l}}(t)$ with $f_{\mathrm{PRF}}/4$ and $-f_{\mathrm{PRF}}/4$, respectively.
(3) Correlation of $s_{\mathrm{u}-}(t)$ and $s_{\mathrm{l}+}(t)$ can be finished rapidly by FFT

$$s_{\mathrm{c}}(t) = \int s_{\mathrm{l}+}(\tau) s_{\mathrm{u}-}^{*}(\tau - t)\,\mathrm{d}t = \mathrm{IFFT}[\mathrm{FFT}(s_{\mathrm{l}+}(t)) \cdot \mathrm{conj}\left(\mathrm{FFT}(s_{\mathrm{u}-}(t))\right)] \tag{13.32}$$

(4) SAC can reduce the statistical fluctuation of estimated results by accumulating across many different range cells. It is noted that, since f_{R} varies with range cell, if the number of the accumulation range cells exceeds the focus depth, the corresponding results should be adjusted before accumulation.
(5) SAC has worse performance for uniform reflectivity scene (Fig. 13.6).

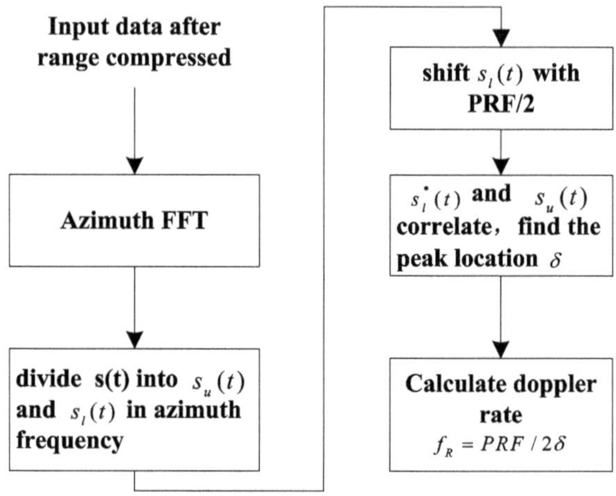

Fig. 13.6 The block diagram of SAC procedure

13.5 Phase Gradient Algorithm [10–12]

Phase Gradient Algorithm (PGA) is a nonparametric model algorithm. It need not make sure the highest order of estimated phase error. It can correct high-order phase error as well as low-order phase error. It takes advantages of the redundancy of the phase-error function by averaging across many range cells. It is derived using one of the methods of formal optimal estimation theory known as maximum-likelihood estimation. PGA method is appropriate for most SAR scene to correct phase error.

The motivation of PGA was to make a robust estimation of the derivative (gradient) of the phase error using only the defocused complex SAR image no matter how the image was formed. Because the complex image and its phase history are a Fourier transform pair, the range-compressed phase-history domain data are obtained by a one-dimensional Fourier transform in the azimuth direction.

The block diagram of the PGA is shown in Fig. 13.7. In summary, PGA algorithm consists of the following four critical steps: center (circular) shifting, windowing, phase gradient estimation, and iterative correction.

(1) Circular Shifting

To circular shifting the data of every range bin in defocus image, the first step is to select the strongest point target, and circularly shift it to the origin (center of each range bin), to remove the linear frequency offset due to the Doppler effect of the target. This circular shifting operation, in essence, creates a new image, wherein all the targets data to be used in the estimation process, and the selected strongest reflect point are aligned along range direction of azimuth center in the new image.

(2) Windowing

The intent of windowing the circularly shifted imagery is to preserve the information contained in the center-shifted targets that describe the blurring kernel,

Fig. 13.7 The block diagram of PGA procedure

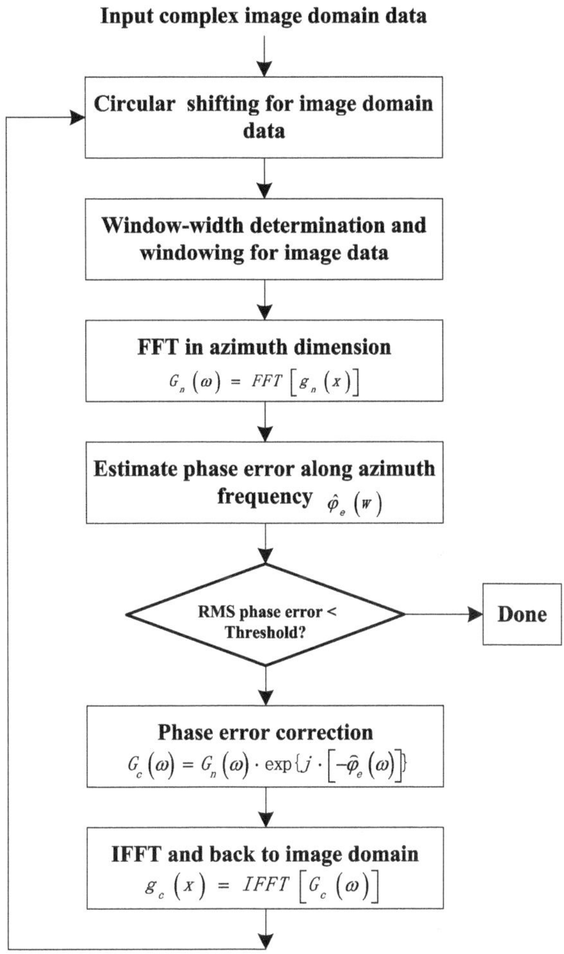

while simultaneously rejecting information from all other surrounding clutter and targets, which allows the phase-error estimation to proceed using input data having the highest signal-to-noise ratio. PGA chooses the center-symmetric window and windows the circular shifting image. The width of window W will decrease for subsequent iterations because the image is becoming more focused. The difficulty of the step is how to choose window width W. If the window is too wide, unnecessary noise is included and other scatters may fall into the window. On the contrary, if the window is too narrow, defocus energy of center scatter in the window will lose. So the proper width of the window should be selected to ensure most of the energy of center target fall into the window.

One of automatic window width determination methods is noncoherent averaging, namely, summing the magnitude of the image data in the range dimension for every azimuth position.

$$s(x) = \sum_{n} |f_n(x)|^2 \tag{13.33}$$

where $f_n(x)$ is the circularly shifted complex image data, n denotes the number of range bin, x represents azimuth position.

After circular shifting, $s(x)$ will get to the maximum value at the center of the image, and exhibit a peak of width W in the center region, simultaneously the magnitude of data outside the region is far below the peak. W can be estimated by means of thresholding $s(x)$ at the point 10 dB down from its peak. The window width will decrease with subsequent iterations because the image becomes more and more focused. The progress of the automatic window width estimation can be a means for assessing convergence.

(3) Phase Gradient and Phase error Estimation

Assume $g_n(x)$ is SAR complex image data containing the phase error after circular shifting and windowing, where the subscript n refers to the nth range bin, x is the relative position of the synthetic aperture.

The Fourier transformation in azimuth dimension is:

$$G_n(\omega) = |G_n(\omega)| \cdot \exp\{j \cdot [\phi_e(\omega) + \theta_n(\omega)]\} \tag{13.34}$$

where $|G_n(\omega)|$ and $\theta_n(\omega)$ are the magnitude and phase, respectively, of the range-compressed data for range bin n. The uncompensated error $\phi_e(\omega)$ along the synthetic aperture is common to all range bins of interest and independent of n.

A linear unbiased minimum variance (LUMV) estimate are used to obtain the gradient of the phase error [12].

$$\hat{\dot{\phi}}_{\text{LUMV}}(\omega) = \frac{\sum_{n} \text{Im}\{G_n^*(\omega)\dot{G}_n(\omega)\}}{\sum_{n} |G_n(\omega)|^2} = \dot{\phi}_e(\omega) + \frac{\sum_{n} \text{Im}|G_n(\omega)|^2 \dot{\theta}_n(\omega)}{\sum_{n} |G_n(\omega)|^2}. \tag{13.35}$$

This LUMV estimate gets the gradient of the true phase error, plus an error term (the second term in the above equation) which has been made as small as possible after circular shifting and windowing steps.

The estimation of phase error can be obtained by integrating $\hat{\dot{\phi}}_e(\omega)$ that is estimated phase gradient.

$$\hat{\phi}_e(\omega) = \int \hat{\dot{\phi}}_e(\omega) \mathrm{d}\omega \tag{13.36}$$

(4) Iterative Phase Correction

The estimated phase error can correct defocus image. At iterative correction, data after range compressed incorporating phase error will be multiplied with conjugated $\hat{\phi}_{\mathrm{e}}(\omega)$ to achieve phase correction.

$$G_{\mathrm{c}}(\omega)=G_{n}(\omega)\cdot\exp\left\{j\cdot\left[-\hat{\phi}_{\mathrm{e}}(\omega)\right]\right\} \tag{13.37}$$

$G_{\mathrm{c}}(\omega)$ denotes corrected range-compressed data.

In fact, phase error gradient estimation is not ideal, because the peak of defocused point target appears smeared due to phase error and it cannot accurately be positioned at center, moreover, two adjacent strong points may exist in the same range bin in practical SAR image, and when windowing, a part of adjacent strong point target may be incorporated. So the estimation and correction process is repeated iteratively. As the image becomes more focused, individual scatters become more compact, the SNR improves, the circular shifting more precisely removes the Doppler offsets, and the algorithm is driven toward convergence. Removal of any linear trend in the phase error estimate prevents image shifting and the bias removal allows computation of the RMS phase error remove at each iteration as a means for monitoring convergence.

In each iteration $\hat{\phi}_{\mathrm{e}}(\omega)$ will be updated and the image is more and more focused when $\hat{\phi}_{\mathrm{e}}(\omega)$ becomes more and more accurate. When the RMS error drops into a threshold, the image is well focused and will not be improved with additional iterations.

13.6 Contrast Optimization Algorithm [13, 14]

Typically, the defocused SAR image can be regarded as a convolution between target backscatter coefficient and point spread function (PSF). Ideally, PSF is a δ function. If there is a phase error, PSF will be widened, which will result in the amplitude small fluctuation around its mean value and low image contrast for the defocused image. On the contrary, the focused image will exhibit several pronounced peaks and image contrast will increase. So image contrast can be used as a criterion of the image focus. Figure 13.8 shows the effect of different focus parameter (the matched filter using different chirp parameter) on the compressed point image. Figure 13.8a the left three pictures are output signal after matched filtering; Fig. 13.8b the right three pictures are a frequency versus time diagram between matched filter and the input chirp signal. Figure 13.8 shows intensity change of defocused image is much less than that of the focused image. Image contrast can not only indicate the change degree of image amplitude value but also show the focus degree of SAR image.

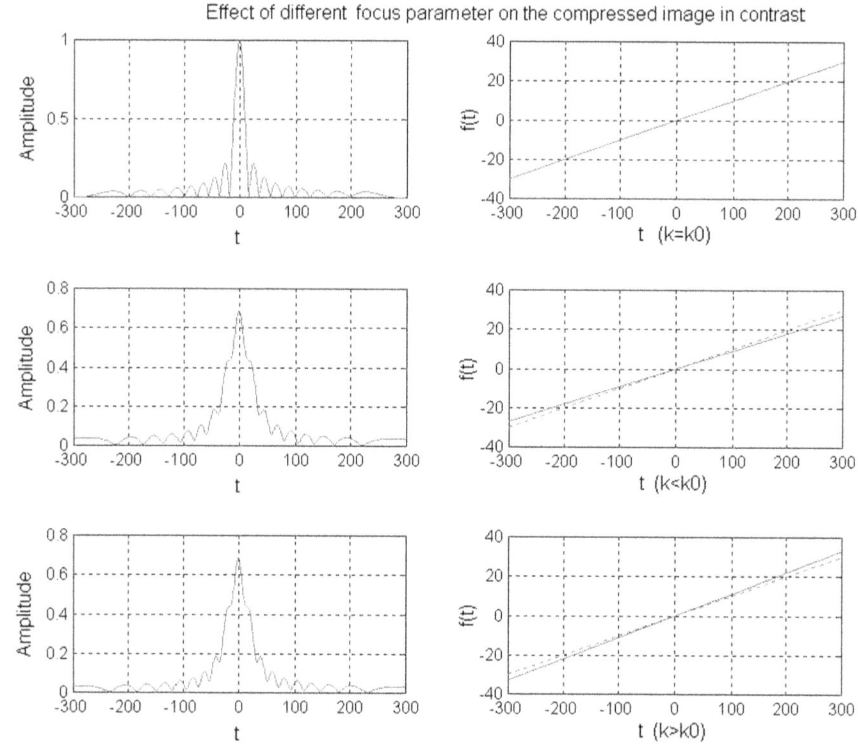

Fig. 13.8 Effect of different focus parameter on the compressed image

In contrast optimization (CO) algorithm, the following three kinds of contrast definitions will be used [13].

(1) The maximum value of the image amplitude

$$c_1(\hat{\gamma}) = \max_m I(m, \hat{\gamma}) \tag{13.38}$$

named as "max contrast function"

(2) The ratio of the standard deviation to the mean of the image amplitude

$$c_2(\hat{\gamma}) = \sqrt{A\left\{[I(m,\hat{\gamma}) - A\{I(m,\hat{\gamma})\}]^2\right\}} / A\{I(m,\hat{\gamma})\} \tag{13.39}$$

Named as "standard deviation amplitude contrast function"

(3) The ratio of the standard deviation to the mean of the image intensity

$$c_3(\hat{\gamma}) = \sqrt{A\left\{[I^2(m,\hat{\gamma}) - A\{I^2(m,\hat{\gamma})\}]^2\right\}} / A\{I^2(m,\hat{\gamma})\} \tag{13.40}$$

Named as "standard deviation intensity contrast function."

In the above image contrast definitions, $A(\cdot)$ is the spatial mean operator, and if applied to a sample sequence, $A(\cdot)$ calculates the arithmetic mean of the samples for spatial mean

$$A[x(m)] = \sum_{m=1}^{K} x(m) / K \tag{13.41}$$

According to the relationship between image contrast and focus, we can use an iterative approximation method to search quadratic phase error coefficient value corresponding to the maximum image contrast with variable step within a certain scope.

The image contrast can be used as a criterion on degree of image focus. Because SAR image is a nonstationary random process, image contrast should be calculated within local area, not within the whole range bins.

CO autofocus algorithm is an attempting and error detection process. Given an initial Doppler rate, initial SAR image and its image contrast can be obtained. According to a certain rule an iteration step is updated to get a new Doppler rate, and then the process are repeated till maximum image contrast.

The main steps of CO algorithm based on quadratic phase error are as follows:

Step 1: Azimuth matched filter reference function can be generated using an initial Doppler rate $f'_{\mathrm{dr}} = -\frac{2V^2}{\lambda R}$ (V is platform velocity, R is slant range between platform and target), and initial image contrast C_0 can be calculated after azimuth compression.

Step 2: Setting initial iterative step $\Delta\hat{f}_{\mathrm{dr}}$. It is worth noting that $\Delta\hat{f}_{\mathrm{dr}}$ should be set according to the defocused degree of the unfocused image to improve the speed of CO autofocus algorithm. If the degree of defocus is severe, $\Delta\hat{f}_{\mathrm{dr}}$ should be set bigger, conversely, $\Delta\hat{f}_{\mathrm{dr}}$ should be set smaller. Based on the initial iterative step, it can update Doppler rate $\hat{f}_{\mathrm{dr}} = \hat{f}_{\mathrm{dr}} + \Delta\hat{f}_{\mathrm{dr}}$ and associated matched filter reference function, and then obtain corresponding updated image contrast C_1. If $C_1 > C_0$, it can continue to update $\hat{f}_{\mathrm{dr}} = \hat{f}_{\mathrm{dr}} + \Delta\hat{f}_{\mathrm{dr}}$, on the contrary, if $C_1 > C_0$, It need decrease step $\Delta\hat{f}_{\mathrm{dr}}$ or change revise direction, repeat Step 2 again.

Step 3: repeat the above steps till image contrast achieves the maximum value, when $\Delta\hat{f}_{\mathrm{dr}} <$ threshold, we can $\hat{f}_{\mathrm{dr}} = f_{\mathrm{dr}}$.

Synthetic aperture radar autofocus technology is one of the key technologies of getting SAR high azimuth resolution. With the improvement of SAR system performance, autofocus technology becomes more and more important for SAR data to avoid residual phase error which will result in worse image quality. Autofocus algorithm is a blind estimation process which not only estimates phase error function based on SAR data but also recovers real image using estimated phase error, namely it is an iterative process approximating true value. In the mathematical model of many autofocus algorithm, the phase error function is often assumed to be a function of not varying with spatial position, especially with range bin, that is to say, the phase error function is the same for each range bin and it is a function of azimuth.

This chapter discusses principle and process on SAR autofocus algorithms, namely map-drift autofocus algorithm (MD), multiple aperture map-drift algorithm (MAM), phase difference algorithm (PD), Shift and Correlation (SAC), phase gradient algorithm (PGA), and contrast optimization algorithm (CO). These autofocus algorithms are different on complexity and applicability. Comparing them and assessing their performance should depend on the specific characteristics of the imaging scene and phase errors. In summary, MD, PD, and SAC are fit for secondary phase error, and they have simple, high accuracy, and robustness advantages, moreover, they can be extended to estimate the higher order phase errors, and these algorithms are suitable for high-contrast imaging scenarios. PGA is a non-model algorithm with good robustness and it can estimate the higher order phase errors. The algorithm often requires relatively isolated strong point target in imaging scene. Compared with the abovementioned several algorithms, the prominent advantage of CO algorithm is suitable for most imaging scene, but its iterative search will result in a heavy computational burden.

References

1. Carrara WG, Goodman RS, Majewski RM (1995) Spotlight synthetic aperture radar: signal processing algorithms. Aretch House, Norwood
2. Jin MY (1986) Optimal doppler centroid estimation sar data from a homogeneous source. IEEE Trans Geosci Remote Sens 24(6):1022–1025
3. Blacknell D, White RG, Wood JW (1987) The prediction of geometric distortions in airborne sar imagery from autofocus measurements. IEEE Trans Geosci Remote Sens 25(6):775–781
4. Bamler R (1991) Doppler frequency estimation and Cramer-Rao bound. IEEE Trans Geosci Remote Sens 29(3):385–390
5. Zhang Y (1994) Autofocus system study for airborne SAR real-time imaging processor. Report of institute of electronics, CAS
6. Yu W (1997) Study for SAR signal processing. Doctoral dissertation, Nanjing University of Aeronautics and Astronautics
7. Blacknell D (1991) SAR motion compensation using autofocus. Int J R S 12(2):253–275
8. Terry MC (1994) Sub-aperture Autofocus for Synthetic Aperture Radar. IEEE Trans AES 30 (2):617–621
9. Dall J (1992) A fast autofocus algorithm for synthetic aperture radar processing. In: Proceedings of ICASSP, 1992, San Francisco
10. Jakowatz CV et al (1996) Spotlight-mode synthetic aperture radar: a signal processing approach. Kluwer Academic Publishers, Boston

11. Wahl DE et al (1994) Phase gradient autofocus—a robust tool for high resolution SAR phase correction. IEEE Trans Aerosp Electron Syst 30(3)
12. Eichel PH, Jakowatz CV Jr (1989) Phase gradient autofocus as an optimal estimator of the phase derivatives. Opt Lett 14(28):1101–1109
13. Berizzi F, Corsini G (1996) Autofocusing of inverse synthetic aperture radar images using contrast optimization. IEEE Trans Aerosp Electron Syst 32(3):1185–1191
14. Berizzi F et al (1996) Autofocusing of wide angle SAR images by contrast optimisation. In: IGARSS '96

Chapter 14
Foundation of Super-Resolution Microwave Imaging

Super-resolution imaging algorithms use high-resolution spectrum estimation method, effective spectrum extrapolation method, adaptive weighted method, or inverse problem-solving method to process microwave imaging radar data. These kinds of algorithms can break through the performance limitations of Fourier imaging method to improve image resolution further without increasing the signal bandwidth and the length of the synthetic aperture.

Complex exponential signal model is required in some super-resolution imaging algorithms. So we should construct an appropriate signal model for each super-resolution imaging algorithm. After range migration correction and motion compensation, SAR echo data in the two-dimensional frequency domain or wavenumber domain can be considered as a complex exponential signal. The classic imaging process obtains the output images by FFT transform. The super-resolution imaging process obtains the high-resolution output images by using modern spectral estimation method instead of FFT, or extrapolating the signal spectrum based on this complex exponential signal model, etc..

As the basis of super-resolution microwave imaging technology, the chapter presents the classical Fourier spectrum analysis, the concept of super-resolution algorithm and several signal models suitable for super-resolution synthetic aperture radar imaging algorithms.

14.1 Classical Fourier Spectrum Analysis

There are two classical spectrum analysis methods [1]: periodogram method and Blackman–Tukey (BT) method. Periodogram method obtains the power spectrum by Fourier transform with or without window. BT method estimates the autocorrelation function of different delays first, and gets the power spectrum by Fourier transform of the windowed autocorrelation estimation. BT method can degenerate into periodogram method under certain circumstances.

R. Yang et al., *High-Resolution Microwave Imaging*,
https://doi.org/10.1007/978-981-10-7138-6_14

14.1.1 Classical Fourier Spectrum Estimation

Power spectral density of complex generalized stationary stochastic process (PSD) $P_{xx}(f)$ can be definition [1]

$$P_{xx}(f) = \lim_{x\to\infty} E\left\{\frac{1}{2M+1}\left|\sum_{n=-M}^{M} x(n)\exp(-j2\pi fn)\right|^2\right\} \tag{14.1}$$

It can also be written as

$$P_{xx}(f) = \sum_{k=-\infty}^{\infty} r_{xx}(k)\exp(-j2\pi fk), \quad \frac{1}{2} \le f \le \frac{1}{2} \tag{14.2}$$

where $r_{xx}(k)$ is the autocorrelation function (ACF) of observed samples $x(n)$ in stochastic process.

$$r_{xx}(k) = E\{x^*(n)x(n+k)\} \tag{14.3}$$

$E(\bullet)$ is the operation of mathematical expectation.

Power spectral density function describes the distribution that power of random process varies with frequency. Equation (14.1) shows that the power spectral density depends on an unlimited number of observed samples. For microwave imaging radar can only get a limited number of observed values $\{x(0), x(1), \ldots, x(N-1)\}$, power spectral density should be estimated from the observations.

14.1.2 Periodogram Method

According to Eq. (14.1), Periodogram estimator can be defined as

$$\hat{P}_{\mathrm{PER}}(f) = \frac{1}{N}\left|\sum_{n=0}^{N-1} x(n)\exp(-j2\pi fn)\right|^2 \tag{14.4}$$

This can be also expressed as

$$\hat{P}_{\mathrm{PER}}(f) = \sum_{k=-(N-1)}^{N-1} \hat{r}_{xx}(k)\exp(-j2\pi fk) \tag{14.5}$$

where $\hat{r}_{xx}(k) = \begin{cases} \frac{1}{N}\sum\limits_{n=0}^{N-1-k} x^*(n)x(n+k), & k = 0, 1, \ldots, N-1 \\ \hat{r}^*_{xx}(k), & k = -(N-1), -(N-2), \ldots, -1 \end{cases}$.

When the frequency f_0 is given, Eq. (14.4) can be converted to

$$\hat{P}_{\text{PER}}(f_0) = N\left|\sum_{k=0}^{N-1} h(n-k)x(n)\right|^2_{n=0} \tag{14.6}$$

where $h(n) = \begin{cases} \frac{1}{N}\exp(j2\pi f_0 n), & n = -(N-1), \ldots, -1, 0 \\ 0, & \text{otherwise} \end{cases}$.

Thus, periodogram estimation is essentially equivalent to Fourier power spectrum estimation, whose resolution is determined by Fourier transform. Periodogram estimation is seen as the output's square of the observation data filtering by the bandpass filter $h(n)$, which is a fixed coefficients bandpass filter. In the following sections, we will see the resolution of observation data-filtered results by adaptive filters is much higher than that of periodogram estimators.

We can average the periodogram to improve the statistical properties of the periodogram estimation. The observation sequence $x(n)$ $(0 \le n \le N-1)$ is segmented into several subsequences before periodogram calculation. Divide $x(n)$ into L pieces. Each piece has M samples, where $N = LM$. The ith piece sample subsequence can be written as

$$x^i(n) = x(n + iM - M),\ 0 \le n \le M-1,\ 1 \le i \le L$$

The periodogram of ith piece is

$$\hat{P}^i_{\text{PER}}(f) = \frac{1}{N}\left|\sum_{n=0}^{M-1} x^i(n)\exp(-j2\pi fn)\right|^2 \tag{14.7}$$

If data of each piece is independent, the average periodogram estimator can be defined as

$$\hat{P}_{\text{PER}}(f) = \frac{1}{L}\sum_{i=1}^{L} \hat{P}^i_{\text{PER}}(f) \tag{14.8}$$

14.1.3 Blackman–Tukey Estimator

Blackman–Tukey (BT) estimator [1] is the unbiased estimation of windowed autocorrelation based on Eq. (14.5), which is also known as the weighted covariance estimator

$$\hat{P}_{\mathrm{BT}}(f) = \sum_{k=-(N-1)}^{N-1} w(k)\hat{r}_{xx}(k)\mathrm{e}^{-j2\pi fk} \tag{14.9}$$

$w(k)$ is the real window function, and $0 \le w(k) \le w(0) = 1$, $w(-k) = w(k)$, $w(k) = 0, |k| > M, M \le N - 1$. Window function can be cosine window function, such as Hamming window function, Hanning window function and so on. Equation (14.10) can be also written as

$$\hat{P}_{\mathrm{BT}}(f) = \sum_{k=-M}^{M} w(k)\hat{r}_{xx}(k)\mathrm{e}^{-j2\pi fk} \tag{14.10}$$

Periodogram estimation method and Blackman–Tukey (BT) estimation method are based on the Fourier transform, which are implicitly assuming that the unobserved data out of the window is zero [2]. It means a finite sequence of random data can be seen as an infinite sequence of random data cut by a rectangular window. The main lobe width of Fourier transform is approximately equal to the inverse of the observation time, which leads to the limitation of the frequency resolution. Zero-padding of data sequence before Fourier transform can only make spectral lines density, which means that the envelope of discrete spectral can approach the continuous frequency gradually. This cannot improve resolution. Using modern spectral estimation methods instead of Fourier transform can reduce the shortcomings of Fourier transform and improve the resolution as well as image quality.

14.2 Foundation of SAR Super-Resolution Information Processing Technology

Resolution is one of the important indicators of synthetic aperture radar image quality. When sizes of the resolution and the target are considerable, the target is a bright spot in the SAR image. When the size of resolution is much smaller than that of the target, the information from different parts of the target can be displayed in the SAR image. The higher the resolution is, the more information we can obtain from the image [3].

There are generally two types of reasons causing lower resolution in synthetic aperture radar images [3]. One is the impacts of inevitably residual errors from data acquisition and imaging process, such as system noise, motion errors, and nonideal imaging conditions, which makesthe generated images not to achieve the theoretical value. The second reason is that the SAR system itself is not limited by the hardware conditions falling to achieve the desired resolution. All of these can degrade the image resolution and image quality.

There are usually two ways to improve SAR resolution: one way can enhance system's theoretical resolution by improving the hardware performance to increase system bandwidth and synthetic aperture length; the second way can use super-resolution imaging algorithm to improve the resolution of the imaging results. Improvements and updates to hardware need long period and high cost. Thus, it has practical value using super-resolution imaging technology to improve the resolution.

SAR super-resolution information processing technology can be used for modern spectral estimation replacing the Fourier transform, or the extrapolation of the effective spectrum, or the adaptive weighting method, or the regularization algorithm. When processing SAR data, these methods can break through the performance limitations of Fourier imaging algorithm to improve image resolution and image quality without increasing the signal bandwidth and synthetic aperture length.

The complex exponential signal model can be obtained after SAR data processing. Traditional SAR imaging process uses Fourier transform to obtain the backscattering coefficient of the scene, which means that SAR imaging processing is the limited bandwidth Fourier transform of target complex scattering coefficients truncated in frequency field. The resolution of Fourier transform is inversely proportional to the length of the data. Thus, the spatial resolution of SAR image is limited due to the finite aperture length in two-dimensional frequency domain [4]. What's more, weighting procedures for suppressing the peak side lobe and integral side lobe of impulse response will deteriorate the resolution.

Fourier imaging methods use only 3 dB bandwidth of the signal. Modern spectral estimation can get the spectrum estimation results according to all of the bandwidth using statistical properties from the entire signal bandwidth. This process does not assume that the data signal is zero outside 3 dB bandwidth. Statistical characteristics can give all the prior knowledge of the data. So the modern spectral estimation methods can obtain higher resolution in a finite aperture than Fourier transform methods.

SAR imaging process can be written as [5–8]

$$s(x) = \int h(\xi - x) f(\xi) d\xi \tag{14.11}$$

where s is the observed image, h is point spread function, f is backscatter coefficient of targets. Equation (14.12) is the Fourier transform of Eq. (14.10) in frequency field.

$$S(u) = H(u)F(u) \tag{14.12}$$

If F is estimated using S dividing H, you may encounter the denominator to be zero, which is not a solution of F. It is equivalent to solve F under the constraint of linear systems based on Fourier filter [5]. Thus, if we want to obtain the out-of-band information successfully, we should obtain a priori knowledge of out-of-band information.

Prior knowledge of out-of-band information can be the signal models from imaging processing or the known information of targets itself. If we have known the targets are compact objects, which must be positive and wholly contained within some finite interval [5]. The targets have the following properties:

$$\begin{cases} f(x) > 0, & x \in X \\ f(x) = 0, & x \notin X \end{cases} \tag{14.13}$$

where X is the support region of targets. In general, we assume that the support region X is corresponding to the rectangular region. Equation (14.13) can also be expressed as

$$f(x)\mathrm{rect}\left(\frac{x}{X}\right) \tag{14.14}$$

The Fourier spectrum of $f(x)$ can be divided into two parts: in-band spectrum $F_b(u)$ and out-of-band spectrum $F_a(u)$ [5]. Fourier transform of Eq. (14.14) can describe the characteristic of target spectrum as [5]

$$F(u) = [F_a(u) + F_b(u)] \otimes \mathrm{sinc}(Xu) \tag{14.15}$$

It is a general property for spectral information of all targets. Equation (14.15) is the convolution of Sinc function with spectrum inside and outside bandwidth spectrum. $F_b(u)$ is the spectrum inside bandwidth and $F_a(u)$ is the spectrum outside bandwidth. Because the support region of Sinc function is $(-\infty, +\infty)$, the spectrum outside bandwidth will be mixed with the spectrum inside bandwidth while convoluting. This phenomenon shows that targets spectrum outside bandwidth can be mixed into bandwidth through the system transfer function, which is determined by the characteristics of targets. If a way can be developed to extract this information, the purpose of super-resolution information processing can be achieved.

Substitute Eq. (14.15) into Eq. (14.12) [5].

$$S(u) = H(u)\{[F_a(u) + F_b(u)] \otimes \mathrm{sinc}(Xu)\} \tag{14.16}$$

This is the actual information obtained after the super-resolution algorithm processing.

Super-resolution algorithms require prior knowledge. Usually, targets can satisfy the prior knowledge described in Eq. (14.13). Let f_e express the estimation of targets. We can get

$$S(u) = F_e(u)H(u) \tag{14.17}$$

The estimates problem in Eq. (14.17) can be turned into an optimization problem. The introduction of optimization criterion can obtain a reasonable estimation of f_e with the constraint of high-frequency components in the solution processing. In the case of given system transfer function, the optimization criterion can be expressed as

$$\min_{f_e}\left\{|s(x) - f_e(x) \otimes h(x)|^2\right\} \tag{14.18}$$

where the objects constraints are $\begin{cases} f_e(x) > 0, & x \in X \\ f_e(x) = 0, & x \notin X \end{cases}$.

According to Parseval theorem, Eq. (14.18) can be rewritten as [5]

$$\min_{F_e}\left\{|S(u) - F_e(u)H(u)|^2\right\} \tag{14.19}$$

Constraints are same as Eq. (14.18). Prior knowledge must be provided, when the Eq. (14.19) is used to improve the image resolution. The prior knowledge can be information from a scene or target and also be a signal model. Super-resolution imaging algorithm cannot increase the resolution infinitely. The improved level of resolution will depend on the quality of prior knowledge and the noise levels [3].

14.3 Models of SAR Super-Resolution Processing

Super-resolution imaging algorithms usually require the signal to be in conformation to the specific signal model. We will derive the suitable signal model for SAR super-resolution imaging algorithms according to different imaging methods.

14.3.1 Signal Model of Two-Dimensional Frequency Domain

Figure 14.1 shows the geometry relationship between target P and side-looking SAR in scene.

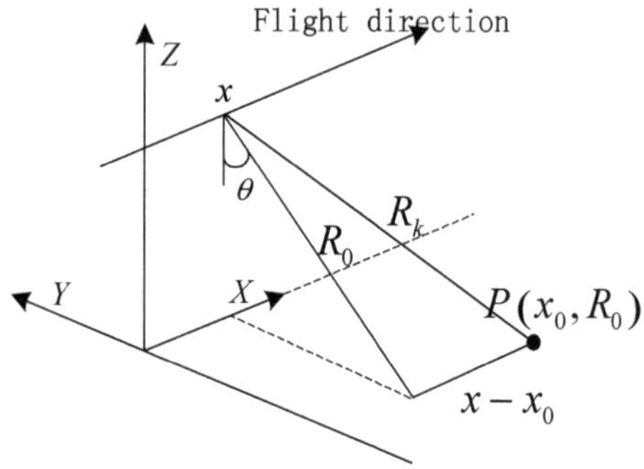

Fig. 14.1 The geometry relationship of side-looking SAR system

The transmission signal of side-looking SAR is

$$s(\hat{t}, t_m) = \mathrm{rect}\left(\frac{\hat{t}}{T_\mathrm{p}}\right)e^{j2\pi\left(f_c t + \frac{1}{2}\gamma\hat{t}^2\right)} \tag{14.20}$$

where f_c is the radar carrier frequency, γ is the radar carrier frequency rate, $t_m = mT_\mathrm{r}$, $\hat{t} = t - t_m$, T_p is pulse width, and T_r is pulse repetition period.

Let R_k be the distance from the target to radar. Then, the echo of point target is the following function:

$$s_r(\hat{t}, t_m) = \sigma_k \mathrm{rect}\left(\frac{\hat{t} - \frac{2R_k}{c}}{T_\mathrm{p}}\right)e^{j2\pi\left(f_c\left(t - \frac{2R_k}{c}\right) + \frac{1}{2}\gamma\left(\hat{t} - \frac{2R_k}{c}\right)^2\right)} \tag{14.21}$$

After removing the carrier frequency by quadrature demodulation, the echo can be written as

$$s_r(\hat{t}, t_m) = \sigma_k \mathrm{rect}\left(\frac{\hat{t} - \frac{2R_k}{c}}{T_\mathrm{p}}\right)e^{j\pi\gamma\left(\hat{t} - \frac{2R_k}{c}\right)^2}\mathrm{rect}\left(\frac{t_m}{T_s}\right)e^{-j\frac{4\pi}{\lambda}R_k} \tag{14.22}$$

The echo signals in SAR range dimension are chirp signals with delay, and the echo signals of SAR azimuth dimension can be approximately chirp signals. After the pulse compression of SAR range echo signals, the range resolution is determined by the bandwidth of radar-transmitting signal. Similarly, after the matched filtering of SAR azimuth echo signals, the azimuth resolution is determined by the Doppler bandwidth.

When the synthetic aperture length is much smaller than the slant range, i.e., $x - x_0 \ll R_0$, the relationship is as follows:

$$R_k = \sqrt{R_0^2 + (x - x_0)^2} \approx R_0 + \frac{(vt - x_0)^2}{2R_0} \tag{14.23}$$

v is the velocity of aircraft.

Substitute the parameters of point P into Eq. (14.22). We can get the echo signal of point target $P(x_0, R_0)$ as follows:

$$s_r(\hat{t}, t_m) = \sigma_k \cdot \mathrm{rect}\left(\frac{\hat{t} - \frac{2R_0}{c}}{T_\mathrm{p}}\right) e^{j\pi\gamma\left(\tau - \frac{2R_k}{c}\right)^2} \mathrm{rect}\left[\frac{t_m - t_0}{T_s}\right] e^{-j\frac{4\pi}{\lambda}R_k} \tag{14.24}$$

Perform Fourier transform on Eq. (14.24) along range dimension using the principle of resident phase [9]. We can get the expression in frequency domain as follows:

$$S(f_r, t_m) = C_1 \cdot \sigma_k \cdot \mathrm{rect}\left(\frac{f_r}{B_\mathrm{r}}\right) e^{-j\pi\frac{f_\mathrm{r}^2}{\gamma}} e^{-j\frac{4\pi f_r}{c}R_k} \mathrm{rect}\left(\frac{t_m - t_0}{T_s}\right) e^{-j\frac{4\pi}{\lambda}R_k} \tag{14.25}$$

where C_1 is the complex constant generated by Fourier transform, f_r is the range frequency, $B_\mathrm{r} = K_r T_\mathrm{p}$ is the bandwidth of transmitting signal, and $t_0 = x_0/v$.

The range compression is in the frequency domain, and the corresponding matched filtering function is as follows:

$$H_r(f_\mathrm{r}) = e^{j\pi\frac{f_\mathrm{r}^2}{\gamma}}$$

Perform inverse Fourier transform along range dimension. We can get the expression in two-dimensional time domain as follows:

$$s_r(\hat{t}, t_m) = C_2 \cdot \sigma_k \cdot \mathrm{sinc}\left\{\pi B_\mathrm{r}\left[\hat{t} - \frac{2R_0}{c} - \frac{v^2(t_m - t_0)^2/R_0}{c}\right]\right\} \cdot \mathrm{rect}\left(\frac{t_m - t_0}{T_s}\right) e^{-j\frac{4\pi}{\lambda}R_k} \tag{14.26}$$

The echo signal presents the shape of Sinc function after range compression. The envelope of Sinc function has different delays along with different azimuth positions, which result in the coupling between range and azimuth, i.e. range cell migration (RCM) phenomenon. The RCM curve of each point target crosses with others' in the time domain, which cannot be a unified correction. However, RCM curves of all targets with same slant range are coinciding in azimuth frequency domain. Then, the RCM can be corrected uniformly.

Implement Fourier transform along the azimuth dimension on Eq. (14.26) to turn into the range-Doppler domain. According to the principle of resident phase, the signal can be written as follows:

$$S(\hat{t}, f_a) = C_3 \cdot \sigma_k \cdot \mathrm{sinc}\left[\pi B_\mathrm{r}\left(\hat{t} - \frac{2R_0}{c} - \frac{\lambda f_\mathrm{a}^2}{c f_\mathrm{R}}\right)\right] \cdot \mathrm{rect}\left(\frac{f_\mathrm{a}}{B_d}\right) \cdot e^{-j\frac{4\pi R_0}{\lambda}} \cdot e^{-j\pi\frac{f_\mathrm{a}^2}{f_\mathrm{R}}} \cdot e^{-j2\pi f_\mathrm{a} t_0} \tag{14.27}$$

where f_{a} is azimuth frequency, f_{R} is Doppler rate, $B_d = |f_{\mathrm{R}} \cdot T_s|$ is Doppler bandwidth. Interpolation or shift method can be used to correct RCM in range-Doppler domain. After RCM corrected, the signal is as follows:

$$S(\hat{t}, f_{\mathrm{a}}) = C_4 \cdot \sigma_k \cdot \mathrm{sinc}\left[\pi B_{\mathrm{r}}\left(\hat{t} - \frac{2R_0}{c}\right)\right] \cdot \mathrm{rect}\left(\frac{f_{\mathrm{a}}}{B_d}\right) \mathrm{e}^{-j\pi \frac{f_{\mathrm{a}}^2}{f_R}} \tag{14.28}$$

Azimuth matched filter is needed to compress azimuth signal in Eq. (14.28). The azimuth matched function is shown as follows:

$$H_a(f_{\mathrm{a}}) = \mathrm{e}^{j\pi \frac{f_{\mathrm{a}}^2}{f_R}}$$

We carry out azimuth compression and the azimuth inverse Fourier transform on Eq. (14.28). The final focused result would be

$$s(\hat{t}, t_m) = C \cdot \sigma_k \cdot \mathrm{sinc}\left[\pi B_r\left(\tau - \frac{2R_0}{c}\right)\right] \cdot \mathrm{sinc}[\pi B_d t_m] \tag{14.29}$$

The derivation above is the imaging process of a single point target. For the whole scene, SAR image signal can be expressed as

$$s(\hat{t}, t_m) = \sum_{k=1}^{K} \sigma_k \mathrm{sinc}\left[\pi B_{\mathrm{r}}\left(t_r - t_1(x_{r,n})\right)\right] \times \mathrm{sinc}\left[\pi B_a\left(t_a - t_2(x_{a,n})\right)\right] \tag{14.30}$$

where t_r is the fast time variable, t_a is the slow time variable, $x_{r,\,n}$ is the target position of slant range, $x_{a,n}$ is the target position of azimuth, K is the number of scatter centers.

Convert the image signal into the two-dimensional frequency domain. We can show the signal in phase history domain as follows:

$$s(f_r, f_{\mathrm{a}}) = \sum_{k=1}^{K} \sigma_k \mathrm{rect}\left(\frac{f_r}{B_{\mathrm{r}}}\right) \otimes \mathrm{rect}\left(\frac{f_{\mathrm{a}}}{B_a}\right) \times \mathrm{e}^{-j\pi t_1(x_{r,n}) f_r} \times \mathrm{e}^{-j\pi t_2(x_{a,n}) f_a} \tag{14.31}$$

In Eq. (14.31), there is a support region caused by the convolution of two rectangular functions, which can be considered that signal inside this support region is not zero and there is no signal outside the support region. That means the passband of the system is limited by a very harsh boundary, and the spectral information beyond the boundary cannot be obtained [10]. Frequency responses outside band limited signal are completely zero. SAR signal in the two-dimensional frequency domain is a bandpass function with certain support region. The signal in the frequency domain support region can be written as

$$s(\omega_r, \omega_a) = \sum_{k=1}^{K} \sigma_k \times e^{-jt_1(x_{r,n})\omega_r} \times e^{-jt_2(x_{a,n})\omega_a} \tag{14.32}$$

Signal model in (14.32) is a complex Sinusoid function. The classic imaging methods usually use the two-dimensional Fourier transform to obtain the output images. The super-resolution imaging methods may be use modern spectral estimation methods substitution for Fourier imaging methods to improve the resolution of output images; or may turn the SAR imaging problem into the parameters estimation of two-dimensional complex sinusoid function based on the signal model to obtain two-dimensional position information and the corresponding scattering intensity of targets; or may extrapolate the two-dimensional signal to get SAR high-resolution images.

14.3.2 Signal Model of Dechirp Processing

In addition to the matched filter imaging method, SAR can use dechirp to carry out pulse compression [11]. Let the transmitted signal is

$$s(\hat{t}, t_m) = \mathrm{rect}\left(\frac{\hat{t}}{T_p}\right) e^{j2\pi\left(f_c t + \frac{1}{2}\gamma \hat{t}^2\right)} \tag{14.33}$$

where f_c is radar carrier frequency, γ is the radar carrier frequency rate, T_p is the pulse width, and T_r is pulse repetition period.

Let reference range be R_{ref}. The reference signal can be constructed as follows:

$$s_{ref}(\hat{t}, t_m) = \mathrm{rect}\left(\frac{\hat{t} - \frac{2R_{ref}}{c}}{T_{ref}}\right) e^{j2\pi\left(f_c\left(t - \frac{2R_{ref}}{c}\right) + \frac{1}{2}\gamma\left(\hat{t} - \frac{2R_{ref}}{c}\right)^2\right)} \tag{14.34}$$

where T_{ref} is the pulse width of the reference signal and is slightly greater than T_p.

The distance between certain point target and radar in the scene is R_k. The echo signal of the point target is as follows:

$$s_r(\hat{t}, t_m) = \sigma_k \mathrm{rect}\left(\frac{\hat{t} - \frac{2R_k}{c}}{T_p}\right) e^{j2\pi\left(f_c\left(t - \frac{2R_k}{c}\right) + \frac{1}{2}\gamma\left(\hat{t} - \frac{2R_k}{c}\right)^2\right)} \tag{14.35}$$

Let the echo signal multiply by the reference signal. We can get

$$s_\Delta(\hat{t}, t_m) = \sigma_k \mathrm{rect}\left(\frac{\hat{t} - \frac{2R_k}{c}}{T_p}\right) \mathrm{e}^{-j\frac{4\pi}{c}\gamma\left(\hat{t} - \frac{2R_{\mathrm{ref}}}{c}\right)R_\Delta} \mathrm{e}^{-j\frac{4\pi}{c}f_c R_\Delta} \mathrm{e}^{-j\frac{4\pi}{c^2}\gamma R_\Delta^2} \tag{14.36}$$

where $R_\Delta = R_k - R_{\mathrm{ref}}$ and σ_k is the complex amplitude of kth point target echo signal. The first term of Eq. (14.36) is range frequency. The second term is modulation of azimuth. The third term is the phase error introduced by dechirp, which is called Residual Video Phase (RVP). RVP will lead to Doppler variation of echo signal, which can induce geometric distortion and resolution loss on images. RVP can be eliminated by dechirp processing [11]. Then, we can obtain two-dimensional signal

$$s_\Delta(\hat{t}, t_m) = \sigma_k \mathrm{rect}\left(\frac{\hat{t} - \frac{2R_k}{c}}{T_\mathrm{p}}\right) \mathrm{e}^{-j\frac{4\pi}{c}\gamma\left(\hat{t} - \frac{2R_{\mathrm{ref}}}{c}\right)R_\Delta} \mathrm{e}^{-j\frac{4\pi}{c}f_c R_\Delta} \tag{14.37}$$

After sampling, the sample signal in two-dimensional scene can be written as

$$s(m, n) = \sum_{k=1}^{K} \sigma_k \mathrm{e}^{j\frac{4\pi}{c}\left(f_c + \frac{m\gamma}{f_s}\right)(x_k \sin\theta_n + y_k \cos\theta_n)} + e(m, n) \tag{14.38}$$

where x_k and y_k is the two-dimensional coordinates of kth scatter, f_s is the sample rate of fast time, $e(m, n)$ is the additive noise.

If radar transmits the chirp signal with bandwidth B and observation angle $\Delta\theta$, Eq. (14.38) shows the sample of the certain scene after Fourier transform of the echo signal in the polar coordinate (f, θ), where the echo signal is point spread function. When the observation angle is very small and the target is also very small to approximate the small target in short-term observation, i.e., $\cos\theta_n = 1$, $\sin\theta_n = n\delta\theta$. The echo signal can be simplified into two-dimensional sinusoidal signal plus noise. The m-dimension azimuth frequency is corresponding to a range of the target in cross track, where $\omega_{1,k} = \frac{4\pi B y_k}{cM}$. The n-dimension azimuth frequency are corresponding to the range of the target in along the track, where $\omega_{2,k} = \frac{4\pi\Delta\theta x_k}{\lambda N}$. Thus, FFT can be used to get the final image. When the observation angle is large, the conditions of small-angle approximation cannot be met. In order to take advantage of FFT imaging, we should interpolate the sampling data and transform the polar coordinate system into the Cartesian coordinate system. In the Cartesian coordinate system, the imaging problem can also be described as the estimation of two-dimensional frequency and amplitude in additive noise after interpolation. The data model using dechirp processing is [12]

$$s(m, n) = \sum_{k=1}^{K} \alpha_k \mathrm{e}^{j(\omega_{1,k} m + \omega_{2,k} n)} + e(m, n) \tag{14.39}$$

where $n = 0, 1, \ldots, N$ and $m = 0, 1, \ldots, M - 1$. As we can see, Eq. (14.39) is a two-dimensional complex exponential model too. We can use Eq. (14.39) to change the SAR imaging problem into the parameters estimation of two-dimensional complex Sinusoid signals, which can just estimate the two-dimensional positions and intensities. If we use modern spectral estimation methods or signal extrapolation methods, we can get the high-resolution SAR images.

14.3.3 Signal Model in $\omega - k$ Domain

Wavenumber domain algorithm [13] is also known as Range Migration Algorithm (RMA) or $\omega - k$ algorithms. The algorithm analyzes two-dimensional spectrum after range signal processing, and uses matched filter in two-dimensional frequency domain to compensate phase, which can accomplish the azimuth focusing and RCM at the same time. The Stolt transformation can overcome the RCM of SAR signal and the dependence on slant range in Secondary Range Compressing (SRC). So, the wavenumber domain algorithm is a kind of ideal imaging algorithms, when dealing with echo data from a large squint angle or long synthetic aperture length.

There are certain relationships between time variable and range variable in the slant range or azimuth. The time–frequency domain can be transformed to the range-wavenumber domain using variable substitution. Let R and x describe the range variable in slant range and azimuth respectively. k_R and k_x are the corresponding wavenumber variables. Thus, there has the relationship between time–frequency domain and range-wavenumber domain:

$$R = \frac{ct}{2} \Rightarrow \omega = 2\pi f_r$$

$$x = vt_m \Rightarrow k_x = \frac{2\pi f_a}{v}$$

where c is the speed of light and v is the velocity of radar. ω is the radian frequency in range, which indicates the phase change of wave per unit time. k_x is azimuth wavenumber, which represents the phase change of wave per unit length. For the RCM phenomena existing in SAR signals, the relationship of the wavenumber in range and the radian frequency is different with that of azimuth.

Wavenumber domain algorithm transforms the echo signal in Eq. (14.22) into two-dimensional frequency domain:

$$\begin{aligned} S(f_{\mathrm{a}},f_{\mathrm{r}}) &= C_1 \cdot \sigma_k \cdot \mathrm{rect}\left(\frac{f_{\mathrm{r}}}{B_{\mathrm{r}}}\right) \cdot \mathrm{rect}\left(\frac{f_{\mathrm{a}}}{B_d}\right) \cdot \mathrm{e}^{j\psi(f_{\mathrm{a}},f_r)} \\ \psi(f_{\mathrm{a}},f_{\mathrm{r}}) &= -\pi\frac{f_{\mathrm{r}}^2}{\gamma} - \frac{4\pi R}{\lambda}\sqrt{\left(1+\frac{f_{\mathrm{r}}}{f_{\mathrm{c}}}\right)^2 - \left(\frac{\lambda f_{\mathrm{a}}}{2v}\right)^2} \end{aligned} \tag{14.40}$$

The phase function can be separated into two parts referring to the slant range of swath center (i.e. R_{ref}), which are phase dependent on the slant range and phase independent of the slant range:

$$\begin{aligned} \psi(f_{\mathrm{a}},f_{\mathrm{r}}) &= \psi_0(f_{\mathrm{a}},f_{\mathrm{r}};R_{\mathrm{ref}}) + \psi_1(f_{\mathrm{a}},f_{\mathrm{r}};R) \\ \psi_0(f_{\mathrm{a}},f_{\mathrm{r}};R_{\mathrm{ref}}) &= -\pi\frac{f_{\mathrm{r}}^2}{\gamma} - \frac{4\pi R_{\mathrm{ref}}}{\lambda}\sqrt{\left(1+\frac{f_{\mathrm{r}}}{f_{\mathrm{c}}}\right)^2 - \left(\frac{\lambda f_{\mathrm{a}}}{2v}\right)^2} \\ \psi_1(f_{\mathrm{a}},f_{\mathrm{r}};R) &= -\frac{4\pi\Delta r}{\lambda}\sqrt{\left(1+\frac{f_{\mathrm{r}}}{f_{\mathrm{c}}}\right)^2 - \left(\frac{\lambda f_{\mathrm{a}}}{2v}\right)^2} \end{aligned} \tag{14.41}$$

$R = R_{\mathrm{ref}} + \Delta r$, $\Delta r \in [-w_r/2, w_r/2]$, and w_r is the slant width of swath.

Equation (14.40) should multiply reference function in the two-dimensional frequency domain as follows:

$$H_1(f_{\mathrm{a}},f_{\mathrm{r}}) = \mathrm{e}^{-j\psi_0(f_{\mathrm{a}},f_{\mathrm{r}};R_{\mathrm{ref}})}$$

After completing range compression, azimuth compression of certain slant range (i.e. R_{ref}), RMC (range migration correction), SRC (second range compression) and the phase compensation of high-order range-azimuth coupling term, it can get the following signal:

$$S(f_{\mathrm{a}},f_{\mathrm{r}}) = C_1 \cdot \sigma_k \cdot \mathrm{rect}\left(\frac{f_{\mathrm{r}}}{B_{\mathrm{r}}}\right) \cdot \mathrm{rect}\left(\frac{f_{\mathrm{a}}}{B_d}\right) \cdot \mathrm{e}^{-j\frac{4\pi\Delta r}{\lambda}\sqrt{\left(1+\frac{f_{\mathrm{r}}}{f_{\mathrm{c}}}\right)^2 - \left(\frac{\lambda f_{\mathrm{a}}}{2v}\right)^2}} \tag{14.42}$$

The Stolt transform can be introduced into Eq. (14.42) as follows:

$$\sqrt{\left(1+\frac{f_{\mathrm{r}}}{f_{\mathrm{c}}}\right)^2 - \left(\frac{\lambda f_{\mathrm{a}}}{2v}\right)^2} \to 1 + \frac{f_{rr}}{f_{\mathrm{c}}}, \qquad f_{rr} = f_{\mathrm{c}}\left\{\sqrt{\left(1+\frac{f_{\mathrm{r}}}{f_{\mathrm{c}}}\right)^2 - \left(\frac{\lambda f_{\mathrm{a}}}{2v}\right)^2} - 1\right\},$$

f_{rr} is the new frequency after Stolt transformed.

Stolt transform can correct the residue RCM, SRC, and the phase compensation of high-order range-azimuth coupling term. Finally, we get the signal model as follows:

$$S(f_a, f_r) = C_1 \cdot \sigma_k \cdot \mathrm{rect}\left(\frac{f_r}{B_r}\right) \cdot \mathrm{rect}\left(\frac{f_a}{B_d}\right) \cdot e^{-j\frac{4\pi\Delta r}{\lambda}} \cdot e^{-j\frac{4\pi\Delta r}{c}f_{rr}} \quad (14.43)$$

which is the complex Sinusoid in wavenumber support region. You can use the above signal model change SAR imaging problem into a two-dimensional complex Sinusoid parameters estimation, which can just estimate the two-dimensional positions and intensities. If we use modern spectral estimation methods or signal extrapolation methods, we can get the high-resolution SAR images.

References

1. Kay SM (1988) Modern spectral estimation: theory and application. Prentice Hall, New Jersey
2. Zhang X (2002) Modern signal processing, 2nd edn. Tsinghua University Press, Beijing
3. Wang Z, Zhu J, Xie M (2013) Technique of SAR image Super resolution. Science Press, Beijing
4. DeGraaf SR (1998) SAR imaging via modern 2-D spectral estimation methods. IEEE Trans Image Process 7:729–761
5. Hunt BR (1999) Super resolution of imagery: understanding the basis for recovery of spatial frequencies beyond the diffraction limit. In: Information, Decision and Control, 1999. IDC 99. Proceedings, Adelaide, SA, Australia, pp 243–248
6. Doerry AW, Dickey FM, Romero LA, De Laurentis JM (2002) Difficulties in super resolving synthetic aperture radar images. In: Proceedings of SPIE algorithms for synthetic aperture radar imagery IX, Edmund G. Zelnio, pp 122–133
7. Dickey FM, Romero LA, De Laurentis JM, Doerry AW (2003) Super-resolution, degrees of freedom and synthetic aperture radar. IEE Proc-Radar Sonar Navig 150(6):419–429
8. Dickey FM, Romero LA, Doerry AW (2001) Super resolution and synthetic aperture radar, Sandia Report, SAND2001-1532, Sandia National Laboratories, Albuquerque, NM
9. Curlander JC, McDonough RN (1991) Synthetic aperture radar systems and signal processing. Wiley, New Jersey
10. Cafforio C, Prati C, Rocca F (1991) SAR data focusing using seismic migration techniques. IEEE Trans Aerosp Electron Syst 27:194–207
11. Bao Z, Xing M, Wang T (2005) Radar imaging technology. Publishing house of electronics industry, Beijing
12. Sun C (2001) SAR/ISAR super resolution imaging techniques research. Doctor, Xidian University, Xi'an
13. Bamler R (1992) A comparison of range-Doppler and wavenumber domain SAR focusing algorithms. IEEE Trans Geosci Remote Sens 30:706–713

Chapter 15
SAR Super-Resolution Imaging Method Based on Spectral Estimation

Super-resolution information processing technology of synthetic aperture radar uses super-resolution imaging algorithms, namely the use of high-resolution spectral estimation methods to alternate Fourier spectrum estimation method, the extrapolation of efficient signal spectrum or the adaptive weighting methods. These SAR data processing methods can break the performance of Rayleigh limitations in Fourier imaging and improve the imaging resolution, without increasing the signal spectral width and the synthetic aperture length. SAR super-resolution imaging algorithms include: spectral estimation methods, spectral extrapolation methods, adaptive weighting methods, point spread function deconvolution methods, regularization methods, and so on. This chapter describes several spectral estimation methods.

The use of modern spectral estimation technology to improve image quality is an active field in SAR super-resolution imaging. Super-resolution spectral estimation technique can be roughly divided into two categories: parametric methods and nonparametric methods.

Nonparametric method can estimate spectrum [or power spectral density (PSD)] directly from the data, including most of the classical spectrum estimation methods. Classical Methods apply correlogram, windowed periodogram, and Blackman-Turkey (BT) spectral estimation methods to estimate the PSD. Correlogram method uses correlation sequence estimation to calculate PSD. Periodogram method calculates the PSD directly from the data.

Parameter method is also known as modern spectral estimation method, including the minimum variance spectrum estimation method (MVM), the amplitude and phase spectrum estimation method (APES), the relaxation class method (RELAX method, CLEAN method), the orthogonal subspace decomposition method (including Pisarenko method, minimum Norm (Mini-Norm) method, multiple signal classification (MUSIC), Johnson method), the estimation of signal parameters via rotational invariance technique (ESPRIT), the autoregressive spectral estimation (AR) method, Prony method, maximum likelihood estimation method (MLE), and so on.

R. Yang et al., *High-Resolution Microwave Imaging*,
https://doi.org/10.1007/978-981-10-7138-6_15

Subspace orthogonal decomposition method [1, 2] can make eigen-decomposition on the autocorrelation matrix of signal and use the eigenvectors as a set of orthogonal vectors to construct a linear space. When a signal composes of multiple harmonic wave and white noise, the linear space can be decomposed into the sum of mutually orthogonal signal subspace and noise subspace. It produces a kind of harmonic frequency estimation methods based on the orthogonal principle, including Pisarenco method [3], minimum norm approach [4], multiple signal classification method [5, 6], Johnson method [7]. The features of subspace orthogonal decomposition are the ability of strong noise suppression and the high accuracy estimation of scatterer positions, but it cannot estimate the scatterer parameters of magnitude and type.

ESPRIT method exploits observation data to construct one pair of Toeplitz matrix, so that the relationship between two matrixes is determined by the extreme value of the model. Thus the estimation problem of the model parameters can be converted into solving the generalized eigenvalues of matrix pencil and the Vandermonde linear equations. Because of introducing the SVD decomposition and reduced rank approximation, this method improves the estimation accuracy of extreme value and the noise immunity. Follow this idea, there have generated a lot of different algorithms, such as matrix pencil [8], Tls-ESPRIT [9, 10], Pro-ESPRIT [11, 12] and so on. However, the estimation accuracy of these methods is dependent on the models seriously. When the error increases between the actual data and the model, the stability of these methods gets worse. Choosing a more accurate parametric model is the way to solve the problem.

Maximum likelihood estimation method is a special case of Bayesian parameter estimation method, which is the optimal Bayesian estimation underling known white noise. Estimation performance of the maximum likelihood estimation method is much better than those of the subspace orthogonal decomposition method and ESPRIT method.

This chapter studies the modern super-resolution spectral estimation methods in depth, including nonparametric method and parameter method. The practical applications of various methods in synthetic aperture radar imaging are analyzed. The suitable flows for super-resolution SAR imaging algorithm are given.

15.1 Minimum Variance Spectrum Estimation Method and Improvement

15.1.1 Minimum Variance Spectrum Estimation Method

The observed signal model in the minimum variance spectrum estimation method (MVM) [13–16] as follows:

$$x(n) = \alpha(\omega)e^{jn\omega} + e_n(\omega), \quad n = 0, 1, \ldots, N-1, \tag{15.1}$$

where in $e_n(\omega)$ is the additive noise.

It is assumed that the coefficients of finite impulse response filter are $\mathbf{h}_\omega = [h_1(\omega)\ldots h_M(\omega)]^{\mathrm{H}}$, where M is the length of finite impulse response. The filter $\mathbf{h}_\omega$ is designed to minimize the output power of the filter and make the signal component of the center frequency ω remain unchanged in the filter output, namely

$$\min\left(\mathbf{h}_\omega^{\mathrm{H}}\mathbf{R}\mathbf{h}_\omega\right) \quad \text{subject to} \quad \mathbf{h}_\omega^{\mathrm{H}}\mathbf{a}(\omega) = 1, \tag{15.2}$$

where

$$\mathbf{a}(\omega) = \begin{bmatrix} 1 & e^{j\omega} & \ldots & e^{j(M-1)\omega} \end{bmatrix}^{\mathrm{T}}.$$

The best finite impulse response filter can be obtain by solving the constraint optimization problem using Lagrange equation as

$$\mathbf{h}_\omega^{\mathrm{MVM}} = \frac{\mathbf{R}^{-1}\mathbf{a}(\omega)}{\mathbf{a}^{\mathrm{H}}(\omega)\mathbf{R}^{-1}\mathbf{a}(\omega)}. \tag{15.3}$$

The optimal output power spectrum of corresponding signal sequence $x(n)$ is estimated to be

$$\mathrm{P}_{\mathrm{MVM}}(\omega) = \frac{1}{\mathbf{a}^{\mathrm{H}}(\omega)\mathbf{R}^{-1}\mathbf{a}(\omega)}. \tag{15.4}$$

The estimation result of correlation matrix $\mathbf{R}$ determines the performance of the estimator, and the calculation of $\mathbf{R}$ and the matrix inversion process also determine the calculation complexity of the minimum variance spectrum estimation method. The covariance matrix using backward-forward averaging process can improve the statistical performance of the covariance matrix, which can make the sampling covariance matrix nonsingular. Here the backward-forward averaging process is always used in estimating the covariance matrix. According to the observed vector can be obtained the M-order covariance matrix $\mathbf{R}$ [17] as

$$\hat{\mathbf{R}} = \frac{1}{2}\left(\hat{\bar{\mathbf{R}}} + \hat{\tilde{\mathbf{R}}}\right), \tag{15.5}$$

Where $\hat{\bar{\mathbf{R}}} = \frac{1}{N-M+1}\sum_{n=0}^{N-M}\bar{\mathbf{x}}(n)\bar{\mathbf{x}}^{\mathrm{H}}(n)$ is the forward sampling covariance matrix, $\hat{\tilde{\mathbf{R}}} = \frac{1}{N-M+1}\sum_{n=0}^{N-M}\tilde{\mathbf{x}}(n)\tilde{\mathbf{x}}^{\mathrm{H}}(n)$ is the backward sampling covariance matrix, $\bar{\mathbf{x}}(n) = [x(n)\ldots x(n+M-1)]^{\mathrm{T}}$, $n = 0, \ldots, N-M$ is the forward observed vector, and $\tilde{\mathbf{x}}(n) = [x^*(N-n-1)\ldots x^*(N-n-M)]^{\mathrm{T}}$ is the backward observed vector.

15.1.2 Rank Reduced Minimum Variance Spectrum Estimation Method

The estimated value $\widehat{\mathbf{R}}$ is usually rank reduced in SAR imaging process, so the MVM estimation method in last section based on nonsingular invertible correlation matrix cannot be used. We need to make constraints on the weights $\mathbf{h}_\omega$ to solve this problem. Reduced rank improved minimum variance spectrum estimation method (RRMVM) makes the quadratic constraints on weights [18–27]. Quadratic constraints will limit the freedom of MVM method, which will reduce resolution. So the choice of constraints should be compromise. Constraints use norm limits of weights. Norm weights can be defined as

$$\|\mathbf{h}_\omega\| = \sqrt{\mathbf{h}_\omega^H \mathbf{h}_\omega}. \tag{15.6}$$

The weights constraint is added to minimize the output function, and the formula for solving the optimal weight function can be written as

$$\min_{\mathbf{w}} \mathbf{w}^H \widehat{\mathbf{R}} \mathbf{w} + \alpha \mathbf{w}^H \mathbf{w}. \tag{15.7}$$

The optimal solution is obtained as follows

$$\mathbf{h}_\omega^{\text{RRMVM}} = \frac{(\alpha\mathbf{I}+\mathbf{R})^{-1}\mathbf{a}(\omega)}{\mathbf{a}(\omega)(\alpha\mathbf{I}+\mathbf{R})^{-1}\mathbf{a}(\omega)}. \tag{15.8}$$

The result of RRMMM spectrum estimation is

$$\mathrm{P}_{\text{RRMVM}}(\omega) = \frac{\mathbf{a}^H(\omega)(\alpha\mathbf{I}+\mathbf{R})^{-1}\mathbf{R}(\alpha\mathbf{I}+\mathbf{R})^{-1}\mathbf{a}(\omega)}{\left(\mathbf{a}^H(\omega)(\alpha\mathbf{I}+\mathbf{R})^{-1}\mathbf{a}(\omega)\right)^2}, \tag{15.9}$$

where α is the diagonal loading term, and is generally 3–5 times of the noise power.

15.1.3 Two-Dimension Extension

SAR imaging is a two-dimensional data process. After using the signal model in Chap. 14, SAR imaging problem can be transformed into a two-dimensional spectral estimation of complex sinusoidal signal. For any frequency pair$(\omega_1, \omega_2) \in [0, 2\pi)$ of observation data $\{x(n_1, n_2)\}(n_1 = 0, 1, \ldots, N_1 - 1,\ n_2 = 0, 1, \ldots, N_2 - 1)$ in two-dimensional case, the model can be expressed as

$$x(n_1, n_2) = \alpha(\omega_1, \omega_2) e^{j(n_1\omega_1 + n_2\omega_2)} + e_{n_1,n_2}(\omega_1, \omega_2), \tag{15.10}$$

where $\alpha(\omega_1, \omega_2)$ represents the complex amplitude of the complex exponential signal in frequency of (ω_1, ω_2), and $e_{n_1,n_2}(\omega_1, \omega_2)$ is interference noise in frequency component of (ω_1, ω_2).

Let $N_1 \times N_2$ observed data matrix $\mathbf{x}$ to be defined as

$$\mathbf{x} = \begin{bmatrix} x(0,0) & x(0,1) & \cdots & x(0, N_2 - 1) \\ x(1,0) & x(1,1) & \cdots & x(1, N_2 - 1) \\ \vdots & \vdots & \ddots & \vdots \\ x(N_1 - 1, 0) & x(N_1 - 1, 1) & \cdots & x(N_1 - 1, N_2 - 1) \end{bmatrix}.$$

Let $M_1 \times M_2$ coefficient matrix of finite impulse response filter $\mathbf{h}(\omega_1, \omega_2)$ express as

$$\mathbf{h}(\omega_1, \omega_2) = \text{vec}\left(\begin{bmatrix} h_{0,0}(\omega_1, \omega_2) & h_{0,1}(\omega_1, \omega_2) & \cdots & h_{0,M_2-1}(\omega_1, \omega_2) \\ h_{1,0}(\omega_1, \omega_2) & h_{1,1}(\omega_1, \omega_2) & \cdots & h_{1,M_2-1}(\omega_1, \omega_2) \\ \vdots & \vdots & \ddots & \vdots \\ h_{M_1-1,0}(\omega_1, \omega_2) & h_{M_1-1,1}(\omega_1, \omega_2) & \cdots & h_{M_1-1,M_2-1}(\omega_1, \omega_2) \end{bmatrix} \right).$$

The corresponding $M_1 \times M_2$ sub-matrix of observed data is written as

$$\mathbf{x}_{i_1,i_2} = \text{vec}\left(\begin{bmatrix} x(i_1, i_2) & x(i_1, i_2 + 1) & \cdots & x(i_1, i_2 + M_2 - 1) \\ x(i_1 + 1, i_2) & x(i_1 + 1, i_2 + 1) & \cdots & x(i_1 + 1, i_2 + M_2 - 1) \\ \vdots & \vdots & \ddots & \vdots \\ x(i_1 + M_1 - 1, i_2) & x(i_1 + M_1 - 1, i_2 + 1) & \cdots & x(i_1 + M_1 - 1, i_2 + M_2 - 1) \end{bmatrix} \right),$$

where $i_1 = 0, \ldots, N_1 - M_1 + 1$, $i_2 = 0, \ldots, N_2 - M_2 + 1$.

According to the observed data, covariance matrix can be obtained as

$$\mathbf{R} = \mathrm{E}\left(\mathbf{x}_{i_1} \mathbf{x}_{i_2}^H\right) = \begin{bmatrix} \mathbf{R}_0 & \mathbf{R}_1 & \cdots & \mathbf{R}_{M_1} \\ \mathbf{R}_1^H & \mathbf{R}_0 & \ddots & \vdots \\ \vdots & \ddots & \ddots & \mathbf{R}_1 \\ \mathbf{R}_{M_1}^H & \cdots & \mathbf{R}_1^H & \mathbf{R}_0 \end{bmatrix},$$

where

$$\mathbf{R}_k = \begin{bmatrix} r_{k,0} & r_{k,1} & \cdots & r_{k,M_1} \\ r_{k,1}^H & r_{k,0} & \ddots & \vdots \\ \vdots & \ddots & \ddots & r_{k,1} \\ r_{k,M_1}^H & \cdots & r_{k,1}^H & r_{k,0} \end{bmatrix} \quad \text{and}$$

$$r_{k,l} = \mathrm{E}\left[x(i_1+k, i_2+l)x^H(i_1,i_2)\right] = r_{-k,-l}^*.$$

Same as one-dimensional case, the selection of two-dimensional MVM filter coefficients also requires to minimum output power of filter, namely

$$\min\left(\mathbf{h}_{\omega_1,\omega_2}^{\mathrm{H}}\mathbf{R}\mathbf{h}_{\omega_1,\omega_2}\right) \quad \text{subject to} \quad \mathbf{h}_{\omega_1,\omega_2}^{\mathrm{H}}\mathbf{a}(\omega_1,\omega_2) = 1, \tag{15.11}$$

where $\mathbf{a}(\omega_1,\omega_2) = \mathbf{a}(\omega_1) \otimes \mathbf{a}(\omega_2)$, $\mathbf{a}(\omega_k) = \begin{bmatrix} 1 & e^{j\omega_k} & \ldots & e^{j(M-1)\omega_k} \end{bmatrix}^{\mathrm{T}}$ and $k = 1, 2$.

Finally, the output power spectrum under the optimal weight coefficients is estimated to be as

$$\mathrm{P}_{\mathrm{MVM}}(\omega_1,\omega_2) = \frac{1}{\mathbf{a}^{\mathrm{H}}(\omega_1,\omega_2)\mathbf{R}^{-1}\mathbf{a}(\omega_1,\omega_2)}. \tag{15.12}$$

15.1.4 Block Calculation

There are often large amounts of data in the actual imaging process. MVM method requires calculating the covariance matrix and its inverse, which needs a larger amount of computation and memory. For example, for 64 × 64 data matrix, 32 × 32 sampling covariance matrix needs to occupy (32 × 32) × (32 × 32) = 1024 × 1024 memory. What is more, we should add the calculation of matrix inversion, which is a very impressive amount of computational work. For general SAR scenes, MVM method cannot be used directly, which should be extracted or blocked to reduce the amount of computation [20, 28–30].

Figure 15.1 shows a schematic view of the calculation block in SAR image. Single look complex (SLC) image is blocked in range and azimuth respectively with the downsampling factor D_x and D_y. Computational complexity of each sub-image decreases $\left(D_xD_y\right)^3$ times. Compute D_xD_y sub-images individually and the overall complexity of the image decreases $\left(D_xD_y\right)^2$ times. Each individual sub-images are computed in parallel.

The blocking approach as shown in Fig. 15.1 will have a blocking effect. Each sub-block should overlap 50% region with nearby region. Each sub-block only calculates the spectrum in $2\pi \cdot (0.25) \le \omega_1 \le 2\pi \cdot (0.75)$ and $2\pi \cdot (0.25) \le \omega_2 \le 2\pi \cdot (0.75)$. Thus each sub-image blocking effect can be eliminated after mosaicked. Figure 15.2 shows a schematic view of overlap blocking.

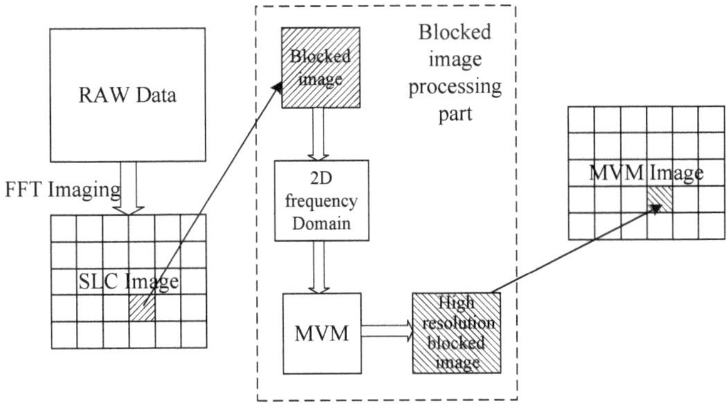

Fig. 15.1 A schematic view of the calculation block in SAR image

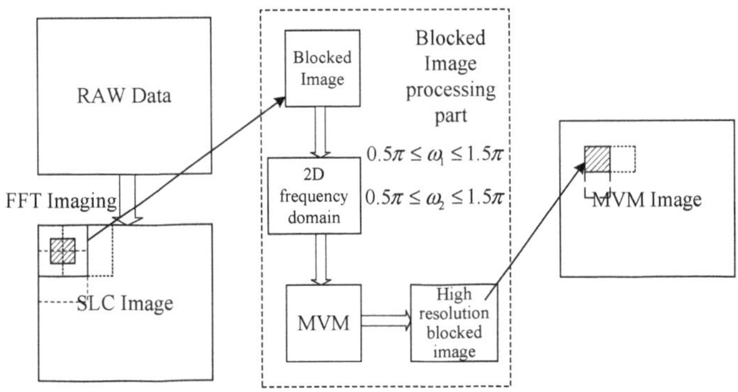

Fig. 15.2 A schematic view of overlap blocking

The edge of the mosaic image is discarded as shown in Fig. 15.2. In order to preserve the entire image information, the edges of image should be calculated alone. Figure 15.3 shows the improved overlap blocking calculation schematic. Different frequency ranges are used to calculate the sub-block of image edge. The intermediate sub-blocks of image keep the same calculation method with Fig. 15.1. Thus the result image can retain the edge portion without loss of image information.

15.1.5 Experiment and Analysis

(1) **Experiments with simulated data**

Simulate point target echo signal with 9.6 GHz carrier frequency of chirp signal. The bandwidth of chirp signal is 150 MHz. Each of the two-way sampling frequency (I/Q) is 180 MHz. The pulse repetition frequency is 1000 Hz. The speed of

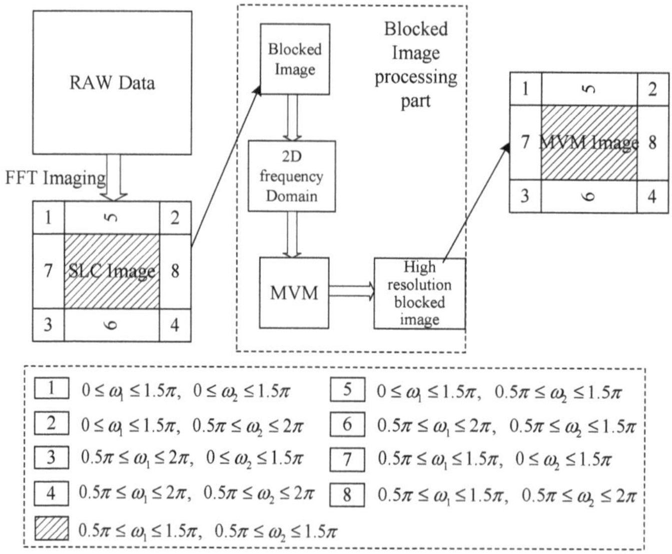

Fig. 15.3 The improved overlap blocking calculation schematic

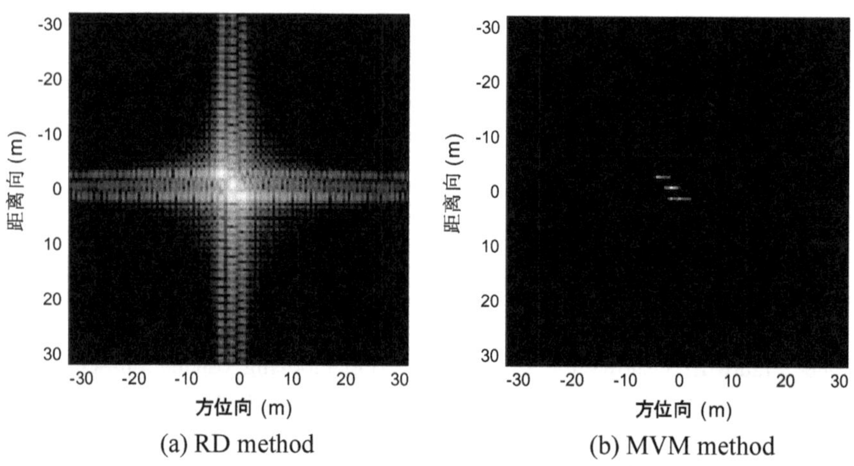

Fig. 15.4 Comparison of point target imaging results using RD method and MVM method

airborne platform is 220 m/s. The azimuth width of the antenna is 2 m. The center slant range in swath is 20 km and the SNR is 7 dB. The simulated data size is 64 × 64. Figure 15.4 is the comparison of point target imaging result using RD method and MVM method, while the order of MVM method is 32 × 32.

Figure 15.4a is the result of Range-Doppler imaging method, which has a wide main lobe and high side lobes. The four-point targets are mixed together and cannot be distinguished in Fig. 15.4a. Figure 15.4b is the result of MVM imaging method, which has obviously a narrow main lobe and suppressed side lobes. The four-point targets can be clearly distinguished in Fig. 15.4b. Point target indexes are calculated to compare the upper left corner point in Table 15.1. Due to the influence of other targets side lobes, the ISLR and PLSR indexes of RD imaging method are very poor to −8 and −6 dB respectively. The range resolution of MVM method is 0.52 m, while the RD method is 1.05 m. The range resolution has increased nearly twice. Because the main lobe is narrowed, the point target is affected slightly by side lobes of other point targets and the side lobe indexes of the point target are improved.

The simulated ISAR data of B727 aircraft is used to test the imaging method. The data is simulated by V. C. Chen of US Naval Research Laboratory. The signal carrier frequency of simulated data is 9 GHz. The signal bandwidth is 75 MHz. The data size is 64 × 64. Figure 15.5 is the comparison of B727 imaging results, where the order of the MVM method is 32 × 32. Figure 15.5a is the imaging result from the RD method, from which can be seen that each scattering point on the aircraft is contaminated by the high side lobes. Figure 15.5b is the result of MVM method,

Table 15.1 Comparison of point target indexes

Performance comparison	RD method	MVM method
Range resolution (m)	1.0571	0.5120
PSLR in range (dB)	−7.9432	−13.2051
ISLR in range (dB)	−5.5603	−9.9815
Azimuth resolution (m)	1.0296	0.5175
PSLR in azimuth (dB)	−7.3238	−14.6868
ISLR in azimuth (dB)	−6.9311	−11.7074

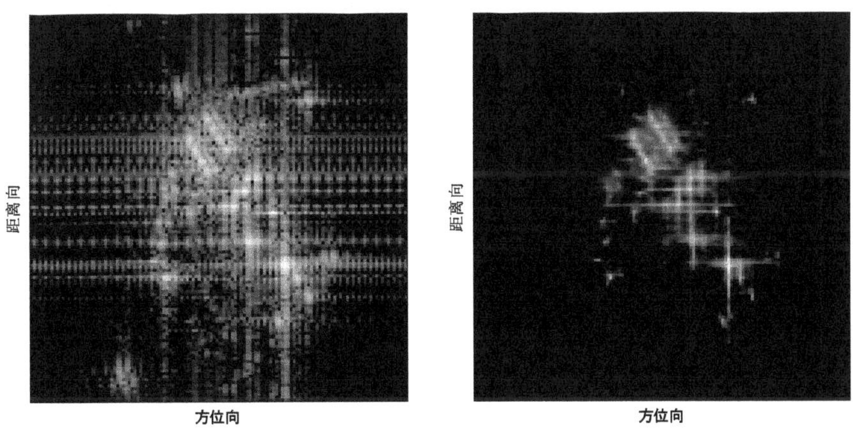

(a) RD method (b) MVM method

Fig. 15.5 Comparison of B727 imaging results

from which can be seen that each aircraft scatterers can be well distinguished and the overall outline of the plane can be clearly visible due to the improvement of each dot side lobes.

(2) **Experiment with measured data**

X-band airborne SAR measured data is used to test imaging methods. The resolution of radar is $2.5m \times 2.5m$. The aircraft flight speed is 250 m/s. The flight height is 10 km. The data size is 1024×1024. The imaging process uses the block method. The sub-block size is 32×32, and the order of the sub-block is 16×16. Figure 15.6 is the comparison of the two imaging results from measured data. Compare the two figures, Fig. 15.6b is more clearly than Fig. 15.6a in the lower part of the road contours. Cornfield scatterers in the top left of the image are better distinguished in Figure (b) than Figure (a).

Figure 15.7 is the comparison of the point target results from measured data, which is the magnified figure of Fig. 15.6. Table 15.2 is the indexes of the point target, from which can be seen the resolution of MVM method can improve by a third than that of RD method.

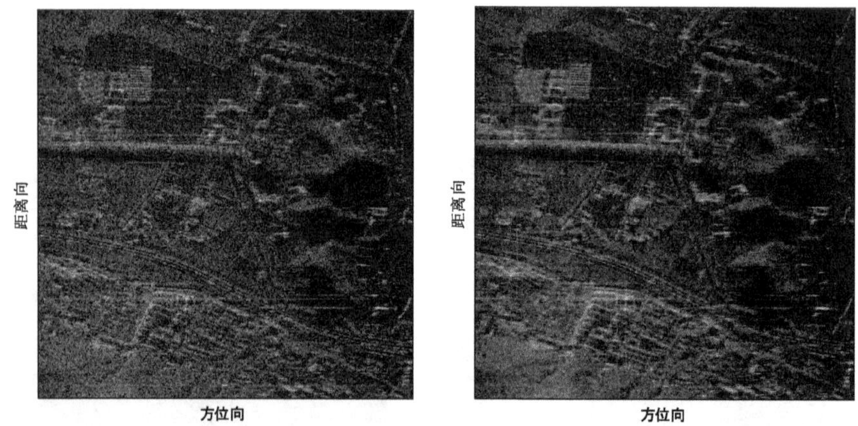

(a) RD method (b) MVM method

Fig. 15.6 Comparison of the two imaging results from measured data

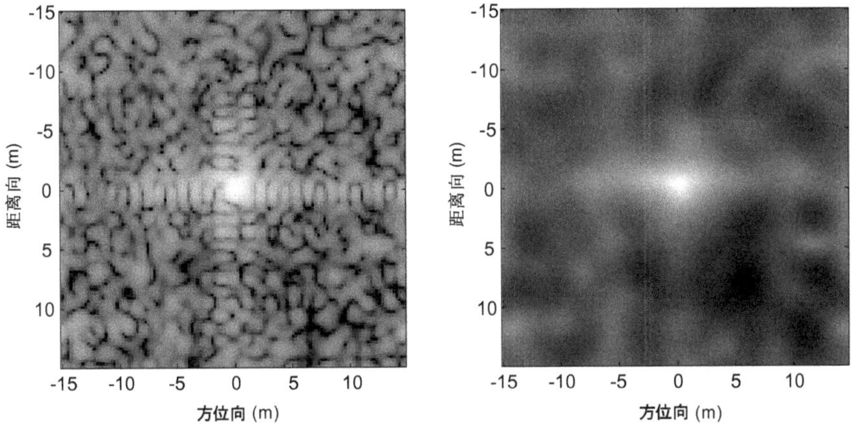

(a) RD method (b) MVM method

Fig. 15.7 Comparison of the point target results from measured data

Table 15.2 Performance of the point target

Performance comparison	RD method	MVM method
Range resolution (m)	2.5907	1.7829
PSLR in range (dB)	−24.7384	−24.6013
ISLR in range (dB)	−16.2748	−15.2402
Azimuth resolution (m)	2.5331	1.7116
PSLR in azimuth (dB)	−11.6725	−18.6150
ISLR in azimuth (dB)	−8.2167	−12.4532

15.2 Amplitude and Phase Estimation of a Sinusoid [17, 28, 31, 32]

15.2.1 Forward Amplitude and Phase Estimation of a Sinusoid

The observed sample data $\{x(n)\}_{n=0}^{N-1}$ are defined as the following equation in amplitude and phase estimation of a sinusoid (APES), i.e.,

$$x(n) = \alpha(\omega)\mathrm{e}^{jn\omega} + \mathrm{e}_n(\omega), \tag{15.13}$$

where $\alpha(\omega)$ is the complex amplitude of the complex exponential in signal frequency ω, and $\mathrm{e}_n(\omega)$ is the interference noise in frequency component ω. Next, the observed data $\{x(n)\}_{n=0}^{N-1}$ are used to estimate the complex amplitude $\alpha(\omega)$ of signal frequency component.

Let $\mathbf{h}(\omega)$ represent a M-order finite impulse response (FIR) filter as

$$h(\omega) = \begin{bmatrix} h_1(\omega) & h_2(\omega) & \ldots & h_M(\omega) \end{bmatrix}^T.$$

The formula shows that the response of narrowband filter relates to the center frequency ω, which means the coefficients of filter change with frequency adaptively. Let

$$\overline{\mathbf{x}}_i = \begin{bmatrix} x(i) & x(i+1) & \ldots & x(i+M-1) \end{bmatrix}^T, \quad i = 0, 1, \ldots, N-M.$$

Represent the forward sub-vector of the observed data. From the definition of sub-vector, we can see that the two adjacent sub-vector $\overline{\mathbf{x}}_i$ and $\overline{\mathbf{x}}_{i+1}$ are the maximum overlap and there is no improvement in performance using sub-vector with other overlap degree [33]. When sub-vector data $\overline{\mathbf{x}}_i$ is input to filter $\mathbf{h}(\omega)$, the output is obtained as

$$\mathbf{h}^H(\omega)\overline{\mathbf{x}}_i = \alpha(\omega)\left[\mathbf{h}^H(\omega)\mathbf{a}(\omega)\right]\mathrm{e}^{ji\omega} + w_i(\omega), \quad i = 0, 1, \ldots, N-M, \tag{15.14}$$

where $w_i(\omega) = \mathbf{h}^H(\omega)\begin{bmatrix} e_i(\omega) & e_{i+1}(\omega) & \ldots & e_{i+M-1}(\omega) \end{bmatrix}^T$ is the noise in the output of the filter and $\mathbf{a}(\omega) = \begin{bmatrix} 1 & e^{j\omega} & \ldots & e^{j(M-1)\omega} \end{bmatrix}^T$.

Normalize $\mathbf{h}(\omega)$ to simplify the derivation process. Let $\mathbf{h}^H(\omega)\mathbf{a}(\omega) = 1$. (15.14) can be written as

$$\mathbf{h}^H(\omega)\overline{\mathbf{x}}_i = \alpha(\omega)\mathrm{e}^{ji\omega} + w_i(\omega). \tag{15.15}$$

For each frequency component ω , the filtered data sequence can approximate sinusoidal signals under the least-mean-square-error (LS) criterion. The objective function is as follows

$$\min_{\alpha(\omega),\mathbf{h}(\omega)} \frac{1}{L}\sum_{i=0}^{L-1} \left|\mathbf{h}^H(\omega)\overline{\mathbf{x}}_i - \alpha(\omega)\mathrm{e}^{ji\omega}\right|^2 \quad \text{subject to} \quad \mathbf{h}^H(\omega)\mathbf{a}(\omega) = 1, \tag{15.16}$$

where $L = N - M + 1$. Let $\overline{\mathbf{X}}(\omega)$ to be the normalization Fourier transforms of $\overline{\mathbf{x}}_i$ as

$$\overline{\mathbf{X}}(\omega) = \frac{1}{L}\sum_{i=0}^{L-1} \overline{\mathbf{x}}_i \mathrm{e}^{-ji\omega}. \tag{15.17}$$

Define $\widehat{\overline{\mathbf{R}}} = \frac{1}{L}\sum_{i=0}^{L-1} \overline{\mathbf{x}}_i \overline{\mathbf{x}}_i^H$. Equation (15.16) can be rewritten as

$$\begin{aligned} &\frac{1}{L}\sum_{i=0}^{L-1} \left|\mathbf{h}^H(\omega)\overline{\mathbf{x}}_i - \alpha(\omega)\mathrm{e}^{ji\omega}\right|^2 \\ &= \mathbf{h}^H(\omega)\widehat{\overline{\mathbf{R}}}\mathbf{h}(\omega) - \alpha^*(\omega)\mathbf{h}^H(\omega)\overline{\mathbf{X}}(\omega) - \alpha(\omega)\overline{\mathbf{X}}^H(\omega)\mathbf{h}(\omega) + |\alpha(\omega)|^2 \\ &= \left|\alpha(\omega) - \mathbf{h}^H(\omega)\overline{\mathbf{X}}(\omega)\right|^2 + \mathbf{h}^H(\omega)\widehat{\overline{\mathbf{R}}}\mathbf{h}(\omega) - \left|\mathbf{h}^H(\omega)\overline{\mathbf{X}}(\omega)\right|^2. \end{aligned} \tag{15.18}$$

Minimize Eq. (15.18) to estimate $\alpha(\omega)$ as

$$\alpha(\omega) = \mathbf{h}^H(\omega)\overline{\mathbf{X}}(\omega). \tag{15.19}$$

Substitute Eq. (15.19) into Eq. (15.18), and obtain the minimization problem about $\mathbf{h}(\omega)$ as

$$\min_{\mathbf{h}(\omega)} \mathbf{h}^H(\omega)\widehat{\overline{\mathbf{Q}}}\mathbf{h}(\omega) \quad \text{subject to} \quad \mathbf{h}^H(\omega)\mathbf{a}(\omega) = 1, \tag{15.20}$$

where $\widehat{\overline{\mathbf{Q}}}(\omega) = \widehat{\overline{\mathbf{R}}} - \overline{\mathbf{X}}(\omega)\overline{\mathbf{X}}^H(\omega)$. Solve Eq. (15.20) to obtain the optimal filter as

$$\mathbf{h}(\omega) = \frac{\widehat{\overline{\mathbf{Q}}}^{-1}(\omega)\mathbf{a}(\omega)}{\mathbf{a}^H(\omega)\widehat{\overline{\mathbf{Q}}}^{-1}(\omega)\mathbf{a}(\omega)}. \tag{15.21}$$

Thus the output of forward APES filter is

$$\widehat{\alpha}(\omega) = \frac{\mathbf{a}^H(\omega)\widehat{\overline{\mathbf{Q}}}^{-1}(\omega)\overline{\mathbf{X}}(\omega)}{\mathbf{a}^H(\omega)\widehat{\overline{\mathbf{Q}}}^{-1}(\omega)\mathbf{a}(\omega)}. \tag{15.22}$$

15.2.2 Forward and Backward Amplitude and Phase Estimation of a Sinusoid

Forward and backward average can double extend the effective data and obtain better statistical properties, so that the estimation accuracy of the algorithm can be improved. Construct the backward sub-vector as

$$\widetilde{\mathbf{x}}_i = [x^*(N-i-1) \quad x^*(N-i-2) \quad \ldots \quad x^*(N-i-M)]^T, \\ i = 0, 1, \ldots, L-1.$$

Combine forward and backward sub-vectors to rewrite the objective function as

$$\min_{\alpha(\omega),\beta(\omega),\mathbf{h}(\omega)} \frac{1}{2L}\sum_{i=0}^{L-1}\left\{\left|\mathbf{h}^H(\omega)\overline{\mathbf{x}}_i - \widehat{\overline{\alpha}}(\omega)e^{ji\omega}\right|^2 + \left|\mathbf{h}^H(\omega)\widetilde{\mathbf{x}}_i - \widehat{\widetilde{\alpha}}(\omega)e^{ji\omega}\right|^2\right\} \tag{15.23}$$
$$\text{subject to} \quad \mathbf{h}^H(\omega)\mathbf{a}(\omega) = 1.$$

Minimize Eq. (15.23) about $\widehat{\overline{\alpha}}(\omega)$ and $\widehat{\widetilde{\alpha}}(\omega)$, we can obtain

$$
\begin{aligned}
\widehat{\overline{\alpha}}(\omega) &= \mathbf{h}^H(\omega)\overline{\mathbf{X}}(\omega) \\
\widehat{\widetilde{\alpha}}(\omega) &= \mathbf{h}^H(\omega)\widetilde{\mathbf{X}}(\omega).
\end{aligned} \tag{15.24}
$$

Let $\widetilde{\mathbf{X}}_i(\omega)$ express the normalization Fourier transform of $\widetilde{\mathbf{x}}_i$ as

$$
\widetilde{\mathbf{X}}(\omega) = \frac{1}{L}\sum_{i=0}^{L-1} \widetilde{\mathbf{x}}_i \mathrm{e}^{-ji\omega}. \tag{15.25}
$$

Equation (15.20) can be as follows

$$
\min_{\mathbf{h}(\omega)} \mathbf{h}^H(\omega)\widehat{\mathbf{Q}}\mathbf{h}(\omega) \quad \text{subject to} \quad \mathbf{h}^H(\omega)\mathbf{a}(\omega) = 1. \tag{15.26}
$$

where $\widehat{\mathbf{Q}}(\omega) = \widehat{\mathbf{R}} - \frac{\overline{\mathbf{X}}(\omega)\overline{\mathbf{X}}^H(\omega) + \widetilde{\mathbf{X}}(\omega)\widetilde{\mathbf{X}}^H(\omega)}{2}$. The forward sample covariance matrix is $\widehat{\overline{\mathbf{R}}} = \frac{1}{L}\sum_{i=0}^{L-1} \overline{\mathbf{x}}_i(n)\overline{\mathbf{x}}_i^H(n)$ and the backward sample covariance matrix is $\widehat{\widetilde{\mathbf{R}}} = \frac{1}{L}\sum_{i=0}^{L-1} \widetilde{\mathbf{x}}_i(n)\widetilde{\mathbf{x}}_i^H(n)$. So the forward and backward sample covariance matrix can be expressed as $\widehat{\mathbf{R}} = \frac{\widehat{\overline{\mathbf{R}}} + \widehat{\widetilde{\mathbf{R}}}}{2}$. Slove Eq. (15.26) to get the optimal filter as

$$
\mathbf{h}(\omega) = \frac{\widehat{\mathbf{Q}}^{-1}(\omega)\mathbf{a}(\omega)}{\mathbf{a}^H(\omega)\widehat{\mathbf{Q}}^{-1}(\omega)\mathbf{a}(\omega)}. \tag{15.27}
$$

Because of $\widetilde{\mathbf{x}}_i = \mathbf{J}\overline{\mathbf{x}}_{L-i-1}^*$, there is $\widetilde{\mathbf{X}}(\omega) = \mathbf{J}\overline{\mathbf{X}}^* e^{-j(L-1)\omega}$. Then the forward covariance matrix and backward covariance matrix have the relationship as following

$$
\begin{aligned}
\widehat{\widetilde{\mathbf{R}}} &= \mathbf{J}\widehat{\overline{\mathbf{R}}}^T\mathbf{J} \\
\widetilde{\mathbf{X}}(\omega)\widetilde{\mathbf{X}}^H(\omega) &= \mathbf{J}\left[\overline{\mathbf{X}}(\omega)\overline{\mathbf{X}}^H(\omega)\right]^T\mathbf{J}.
\end{aligned} \tag{15.28}
$$

$\widehat{\mathbf{Q}}(\omega)$ can be simply calculated as

$$
\widehat{\mathbf{Q}}(\omega) = \frac{\widehat{\overline{\mathbf{Q}}}(\omega) + \mathbf{J}\widehat{\overline{\mathbf{Q}}}^T(\omega)\mathbf{J}}{2}, \tag{15.29}
$$

where

$$
\widehat{\overline{\mathbf{Q}}}(\omega) = \widehat{\overline{\mathbf{R}}} - \overline{\mathbf{X}}(\omega)\overline{\mathbf{X}}^H(\omega).
$$

The output of forward and backward APES filter is expressed as

$$\begin{aligned}
\widehat{\alpha}(\omega) &= \frac{\widehat{\widehat{\alpha}}(\omega) + \widehat{\widehat{\widetilde{\alpha}}}(\omega)}{2} \\
\widehat{\widehat{\alpha}}(\omega) &= \mathbf{h}^H(\omega)\overline{\mathbf{X}}(\omega) = \frac{\mathbf{a}^H(\omega)\widehat{\mathbf{Q}}^{-1}(\omega)\overline{\mathbf{X}}(\omega)}{\mathbf{a}^H(\omega)\widehat{\mathbf{Q}}^{-1}(\omega)\mathbf{a}(\omega)} \\
\widehat{\widehat{\widetilde{\alpha}}}(\omega) &= \mathbf{h}^T(\omega)\widetilde{\mathbf{X}}(\omega) = \widehat{\widehat{\alpha}}^*(\omega)\mathrm{e}^{-j(N-1)\omega}.
\end{aligned} \tag{15.30}$$

In summary, the filter form of forward and backward APES is similar to that of the forward APES, which is only different in configurations of covariance matrix $\widehat{\mathbf{R}}$ and the matrix $\widehat{\mathbf{Q}}(\omega)$. It is notably that forward and backward covariance matrix $\widehat{\mathbf{R}}$ and the matrix $\widehat{\mathbf{Q}}(\omega)$ is a generalized symmetric matrix, while the forward covariance matrix $\widehat{\widehat{\mathbf{R}}}$ and the matrix $\widehat{\widehat{\mathbf{Q}}}(\omega)$ is generalized nonsymmetric matrix. Thus filter estimated by $\widehat{\mathbf{R}}$ and $\widehat{\mathbf{Q}}(\omega)$ has better performance than filter estimated by $\widehat{\widehat{\mathbf{R}}}$ and $\widehat{\widehat{\mathbf{Q}}}(\omega)$.

15.2.3 Two-Dimensional Extension

For any frequency $(\omega_1, \omega_2) \in [0, 2\pi)$ in two-dimensional case, the model of observed data $\{x(n_1, n_2)\}$ $(n_1 = 0, 1, \ldots, N_1 - 1,\ n_2 = 0, 1, \ldots, N_2 - 1)$ can be expressed as

$$x(n_1, n_2) = \alpha(\omega_1, \omega_2)\mathrm{e}^{j(n_1\omega_1 + n_2\omega_2)} + \mathrm{e}_{n_1,n_2}(\omega_1, \omega_2), \tag{15.31}$$

where $\alpha(\omega_1, \omega_2)$ denotes the complex amplitude of complex exponential signal in frequency (ω_1, ω_2), and $\mathrm{e}_{n_1,n_2}(\omega_1, \omega_2)$ denotes the interference noise in frequency (ω_1, ω_2). The observed data are used to estimate the complex amplitude $\alpha(\omega_1, \omega_2)$ of the signal frequency components.

Let $\mathbf{x}$ express the $N_1 \times N_2$ observed data matrix as

$$\mathbf{x} = \begin{bmatrix} x(0,0) & x(0,1) & \cdots & x(0, N_2 - 1) \\ x(1,0) & x(1,1) & \cdots & x(1, N_2 - 1) \\ \vdots & \vdots & \ddots & \vdots \\ x(N_1 - 1, 0) & x(N_1 - 1, 1) & \cdots & x(N_1 - 1, N_2 - 1) \end{bmatrix}.$$

Let $\mathbf{h}(\omega_1, \omega_2)$ denote the $M_1 \times M_2$ coefficients matrix of 2D FIR filter as

$$\mathbf{h}(\omega_1,\omega_2)=\begin{bmatrix} h_{0,0}(\omega_1,\omega_2) & h_{0,1}(\omega_1,\omega_2) & \cdots & h_{0,M_2-1}(\omega_1,\omega_2) \\ h_{1,0}(\omega_1,\omega_2) & h_{1,1}(\omega_1,\omega_2) & \cdots & h_{1,M_2-1}(\omega_1,\omega_2) \\ \vdots & \vdots & \ddots & \vdots \\ h_{M_1-1,0}(\omega_1,\omega_2) & h_{M_1-1,1}(\omega_1,\omega_2) & \cdots & h_{M_1-1,M_2-1}(\omega_1,\omega_2) \end{bmatrix}.$$

Let $L_1 = N_1 - M_1 + 1$ and $L_2 = N_2 - M_2 + 1$. $\overline{\mathbf{y}}$ denotes the $L_1 \times L_2$ output matrix of filter, which can be written as

$$\begin{aligned}\overline{y}(i_1,i_2) &= \sum_{m_1=0}^{M_1}\sum_{m_2=0}^{M_2} h^*_{m_1,m_2}(\omega_1,\omega_2)x(i_1+m_1,i_2+m_2)\\ &= [\mathrm{vec}(\mathbf{h}(\omega_1,\omega_2))]^H \overline{\mathbf{x}}_{i_1,i_2},\end{aligned} \tag{15.32}$$

where

$$\overline{\mathbf{x}}_{i_1,i_2} = \mathrm{vec}\left(\begin{bmatrix} x(i_1,i_2) & x(i_1,i_2+1) & \cdots & x(i_1,i_2+M_2-1) \\ x(i_1+1,i_2) & x(i_1+1,i_2+1) & \cdots & x(i_1+1,i_2+M_2-1) \\ \vdots & \vdots & \ddots & \vdots \\ x(i_1+M_1-1,i_2) & x(i_1+M_1-1,i_2+1) & \cdots & x(i_1+M_1-1,i_2+M_2-1) \end{bmatrix}\right).$$

Similar to the one-dimensional APES method, two-dimensional APES method also exploits the least squares criterion by minimizing the following objective function to estimate $\widehat{\alpha}(\omega_1,\omega_2)$ and filter coefficient matrix $\widehat{\mathbf{h}}(\omega_1,\omega_2)$.

$$\begin{aligned}&\min_{\alpha(\omega_1,\omega_2),\mathbf{h}(\omega_1,\omega_2)} && \frac{1}{L_1}\frac{1}{L_2}\sum_{i_1=0}^{L_1-1}\sum_{i_2=0}^{L_2-1}\left|\overline{y}(i_1,i_2)-\alpha(\omega_1,\omega_2)\mathrm{e}^{j(i_1\omega_1+i_2\omega_2)}\right|^2\\ &\text{subject to} && [\mathrm{vec}(\mathbf{h}(\omega_1,\omega_2))]^H\mathbf{a}_{M_1,M_2}(\omega_1,\omega_2)=1\end{aligned} \tag{15.33}$$

where $\mathbf{a}_{M_1,M_2}(\omega_1,\omega_2)$ is a $M_1M_2 \times 1$ vector. $\mathbf{a}_{M_1,M_2}(\omega_1,\omega_2) = \mathbf{a}_{M_2}(\omega_2)\otimes\mathbf{a}_{M_1}(\omega_1)$. $\mathbf{a}_{M_k}(\omega_k) = \begin{bmatrix}1 & \mathrm{e}^{j\omega_k} & \ldots & \mathrm{e}^{j(M_k-1)\omega_k}\end{bmatrix}^T, k=1,2.$

Let $\overline{\mathbf{X}}(\omega_1,\omega_2)$ denote the normalized Fourier transform of $\overline{\mathbf{x}}_{i_1,i_2}$ as

$$\overline{\mathbf{X}}(\omega_1,\omega_2) = \frac{1}{L_1}\frac{1}{L_2}\sum_{i_1=0}^{L_1-1}\sum_{i_2=0}^{L_2-1}\overline{\mathbf{x}}_{i_1,i_2}\mathrm{e}^{-j(i_1\omega_1+i_2\omega_2)}. \tag{15.34}$$

Define $\widehat{\overline{\mathbf{R}}} = \frac{1}{L_1}\frac{1}{L_2}\sum_{i_1=0}^{L_1-1}\sum_{i_2=0}^{L_2-1}\overline{\mathbf{x}}_{i_1,i_2}\overline{\mathbf{x}}^H_{i_1,i_2}$, and the objective function in Eq. (15.33) can be written as

$$\begin{aligned}
&\frac{1}{L_1}\frac{1}{L_2}\sum_{i_1=0}^{L_1-1}\sum_{i_2=0}^{L_2-1}\left|\bar{y}(i_1,i_2)-\alpha(\omega_1,\omega_2)e^{j(i_1\omega_1+i_2\omega_2)}\right|^2\\
&\quad=[\mathrm{vec}(\mathbf{h}(\omega_1,\omega_2))]^H\widehat{\mathbf{R}}\mathrm{vec}(\mathbf{h}(\omega_1,\omega_2))\\
&\qquad-\alpha^*(\omega_1,\omega_2)[\mathrm{vec}(\mathbf{h}(\omega_1,\omega_2))]^H\overline{\mathbf{X}}(\omega_1,\omega_2)\\
&\qquad-\alpha(\omega_1,\omega_2)\overline{\mathbf{X}}^{\mathrm{H}}(\omega_1,\omega_2)\mathrm{vec}(\mathbf{h}(\omega_1,\omega_2))+|\alpha(\omega_1,\omega_2)|^2\\
&\quad=\left|\alpha(\omega_1,\omega_2)-[\mathrm{vec}(\mathbf{h}(\omega_1,\omega_2))]^H\overline{\mathbf{X}}(\omega_1,\omega_2)\right|^2\\
&\qquad+[\mathrm{vec}(\mathbf{h}(\omega_1,\omega_2))]^H\widehat{\mathbf{R}}\mathrm{vec}(\mathbf{h}(\omega_1,\omega_2))-\left|[\mathrm{vec}(\mathbf{h}(\omega_1,\omega_2))]^H\overline{\mathbf{X}}(\omega_1,\omega_2)\right|^2.
\end{aligned}\tag{15.35}$$

Minimize Eq. (15.35) as the procedure of one-dimensional case to obtain the optimal filter as follows

$$\mathrm{vec}\left(\widehat{\mathbf{h}}(\omega_1,\omega_2)\right)=\frac{\widehat{\mathbf{Q}}^{-1}(\omega_1,\omega_2)\mathbf{a}_{M_1,M_2}(\omega_1,\omega_2)}{\mathbf{a}_{M_1,M_2}^H(\omega_1,\omega_2)\widehat{\mathbf{Q}}^{-1}(\omega_1,\omega_2)\mathbf{a}_{M_1,M_2}(\omega_1,\omega_2)}.\tag{15.36}$$

The spectral estimation of two-dimensional APES is given by

$$\widehat{\alpha}(\omega_1,\omega_2)=\frac{\mathbf{a}_{M_1,M_2}^H(\omega_1,\omega_2)\widehat{\mathbf{Q}}^{-1}(\omega_1,\omega_2)\overline{\mathbf{X}}(\omega_1,\omega_2)}{\mathbf{a}_{M_1,M_2}^H(\omega_1,\omega_2)\widehat{\mathbf{Q}}^{-1}(\omega_1,\omega_2)\mathbf{a}_{M_1,M_2}(\omega_1,\omega_2)},\tag{15.37}$$

where $\widehat{\mathbf{Q}}(\omega_1,\omega_2)=\widehat{\mathbf{R}}-\overline{\mathbf{X}}(\omega_1,\omega_2)\overline{\mathbf{X}}^H(\omega_1,\omega_2)$.

The above derivation process is the spectral estimation of forward two-dimensional APES. The derivation process of forward and backward APES is similar to the one-dimensional case, and need to construct backward data matrix as

$$\tilde{\mathbf{x}}_i=\begin{bmatrix}
x^*(N_1-i_1-1,N_2-i_2-1) & x^*(N_1-i_1-1,N_2-i_2-2) & \cdots & x^*(N_1-i_1-1,N_2-i_2-M_2)\\
x^*(N_1-i_1-2,N_2-i_2-1) & x^*(N_1-i_1-2,N_2-i_2-2) & \cdots & x^*(N_1-i_1-2,N_2-i_2-M_2)\\
\vdots & \vdots & \ddots & \vdots\\
x^*(N_1-i_1-M_1,N_2-i_2-1) & x^*(N_1-i_1-M_1,N_2-i_2-1) & \cdots & x^*(N_1-i_1-M_1,N_2-i_2-M_2)
\end{bmatrix}.$$

Combining forward and backward data matrix, the objective function can be written as

$$
\begin{aligned}
\min_{\alpha(\omega_1,\omega_2),\beta(\omega_1,\omega_2),\mathbf{h}(\omega_1,\omega_2)} \quad & \frac{1}{2L_1}\frac{1}{L_2}\sum_{i_1=0}^{L_1-1}\sum_{i_2=0}^{L_2-1}\left\{\left|\bar{y}(i_1,i_2)-\alpha(\omega_1,\omega_2)e^{j(i_1\omega_1+i_2\omega_2)}\right|^2\right. \\
& \left.+\left|\tilde{y}(i_1,i_2)-\beta(\omega_1,\omega_2)e^{j(i_1\omega_1+i_2\omega_2)}\right|^2\right\} \\
\text{subject to} \quad & [\mathrm{vec}(\mathbf{h}(\omega_1,\omega_2))]^H\mathbf{a}_{M_1,M_2}(\omega_1,\omega_2)=1
\end{aligned}
\tag{15.38}
$$

Finally, the two-dimensional estimator has the similar form to one-dimensional case as Eqs. (15.36) and (15.37), but the covariance matrix $\widehat{\mathbf{R}}$ and the matrix $\widehat{\mathbf{Q}}(\omega_1,\omega_2)$ were changed as follows

$$
\widehat{\mathbf{R}}=\frac{1}{2L_1}\frac{1}{L_2}\sum_{i_1=0}^{L_1-1}\sum_{i_2=0}^{L_2-1}\bar{\mathbf{x}}_{i_1,i_2}\bar{\mathbf{x}}_{i_1,i_2}^H+\frac{1}{2L_1}\frac{1}{L_2}\sum_{i_1=0}^{L_1-1}\sum_{i_2=0}^{L_2-1}\tilde{\mathbf{x}}_{i_1,i_2}\tilde{\mathbf{x}}_{i_1,i_2}^H,
$$

$$
\widehat{\mathbf{Q}}(\omega_1,\omega_2)=\widehat{\mathbf{R}}-\left[\overline{\mathbf{X}}(\omega_1,\omega_2)\quad \overline{\mathbf{X}}(\omega_1,\omega_2)\right]\begin{bmatrix}\overline{\mathbf{X}}^H(\omega_1,\omega_2)\\ \widetilde{\mathbf{X}}^H(\omega_1,\omega_2)\end{bmatrix}.
$$

15.2.4 Experiment and Analysis

(1) **Experiment with simulated data**

Simulate point target echo signal with 9.6 GHz carrier frequency of chirp signal. The bandwidth of chirp signal is 150 MHz. Each of the two-way sampling frequency (I/Q) is 180 MHz. The pulse repetition frequency is 1000 Hz. The speed of airborne platform is 220 m/ s. The azimuth width of antenna is 2 m. The slant range from swath center is 20 km and the SNR is 7 dB. The simulated data size is 64×64. Figure 15.8 is the comparison of point target imaging result using RD method and APES method, while the order of APES method is 32×32.

Figure (a) is the result of Range-Doppler imaging method, which has a wide main lobe and high side lobes. The four-point targets are mixed together and cannot be distinguished in Figure (a). Figure (b) is the result of APES imaging method, which has obviously a narrow main lobe and suppressed side lobes. The four-point targets can be clearly distinguished in Figure (b). However, comparing with Fig. 15.4, the resolutions of APES method are not as good as those of MVM method. Point target indexes are calculated in Table 15.3. The indexes of APES method are overall better than those of RD method. But comparing with Table 15.1, main lobe resolutions of APES method are worse than those of MVM method. Since the main lobe is narrowed, the point target is affected slightly by side lobes of other point targets and the side lobe indexes of the point target are improved.

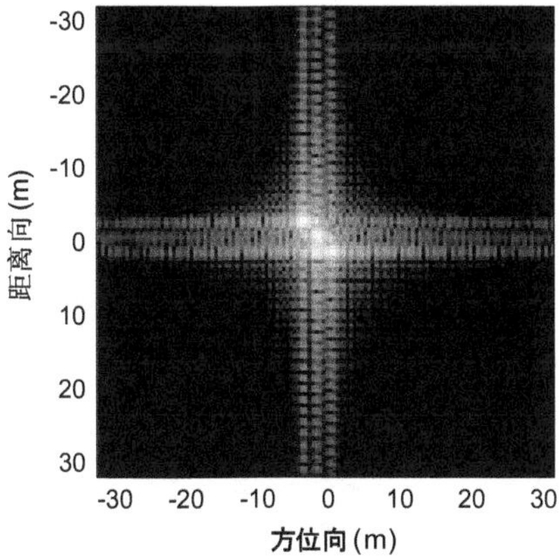

(a) RD method

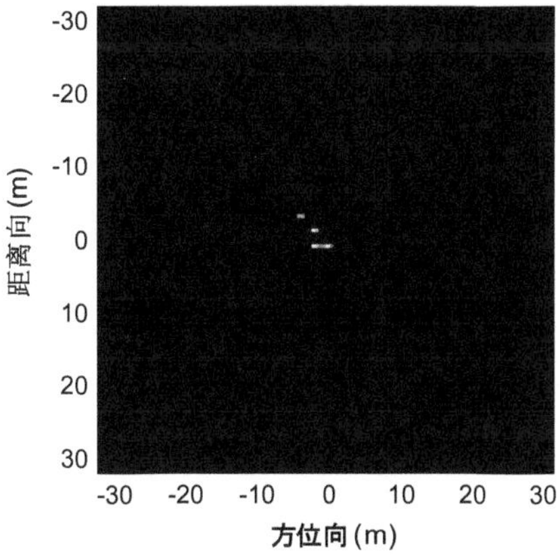

(b) APES method

Fig. 15.8 Comparison of point target imaging result using RD method and APES method

Table 15.3 Comparison of point target indexes

Performance comparison	RD method	MVM method
Range resolution (m)	1.0571	0.6946
PSLR in range (dB)	−7.9432	−12.5902
ISLR in range (dB)	−5.5603	−9.6154
Azimuth resolution (m)	1.0296	0.6728
PSLR in azimuth (dB)	−7.3238	−14.6868
ISLR in azimuth (dB)	−6.9311	−11.7074

(2) **Experiment with measured data**

Figure 15.9 shows the results of measured data respectively using FFT and APES power spectrum estimation to reconstruct radar images. Figure (a) is the result image of FFT. Figure (b) is the result of APES. The result images can be seen the difference of APES reconstruction and FFT reconstruction is slight, which is probably due to that the clutter distribution of observed sample data algorithm does not match the hypothetical situation in APES algorithm. Thus, the accuracy of power spectrum estimation is getting worse and the advantage in image reconstruction is also reduced.

Figure 15.10 shows another reconstruction images.

Figure 15.11 is the comparison of the point target results from measured data. Table 15.4 is the indexes comparison of the point target, from which can be seen the index improvement of APES method is not as good as that of MVM method.

Fig. 15.9 The reconstruction images of FFT method and APES method

Fig. 15.10 The reconstruction images by APES method

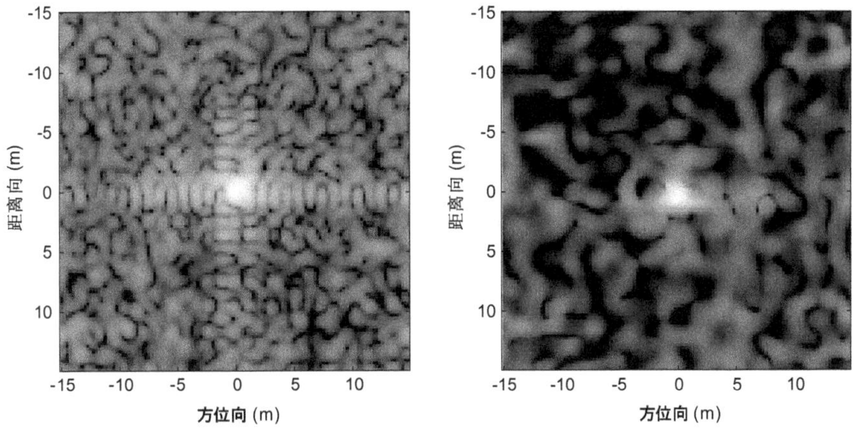

(a) RD method (b) APES method

Fig. 15.11 Comparison of the point target results from measured data

Table 15.4 The comparison of the point target results from measured data

Performance comparison	RD method	APES method
Range resolution (m)	2.5907	2.1232
PSLR in range (dB)	−24.7384	−23.1593
ISLR in range (dB)	−16.2748	−15.9546
Azimuth resolution (m)	2.5331	1.8972
PSLR in azimuth (dB)	−11.6725	−20.1370
ISLR in azimuth (dB)	−8.2167	−13.0234

15.2.5 Summary of Nonparametric Spectral Estimation Methods

When the prior knowledge of studied signal is insufficient to describe the data process as a simple parametric model, the parametric spectral estimation should be used. Such methods are based on statistical point of view and use the limited observed values to estimate the unlimited independent power spectral density function $P_{xx}(f)$, which implies two assumptions [34].

(1) Use finite-dimensional model to parameterize $P_{xx}(f)$;
(2) When the signal in bandwidth $[\mathrm{w}-\beta\pi, \mathrm{w}+\beta\pi]$, $P_{xx}(f)$ can be seen a constant to smooth $P_{xx}(f)$, $\beta \ll 1$.

MVM spectrum estimation, APES spectrum estimation, and periodogram are all belonging to spectral estimators based on matched filter bank. MVM spectrum estimation can produce narrower spectral peaks and lower side lobes than the periodogram. The resolution of APES spectrum estimation is slightly worse than MVM spectrum estimation, but the APES spectrum estimation can obtain more accuracy magnitude and SAR imaging results than MVM spectrum estimation. Filter bank spectrum estimation is summarized as the following three aspects:

(1) Observed data pass through the band-pass filter with a variable center frequency and enhance the corresponding frequency component;
(2) Use the filtered data to estimate the complex magnitude of the complex exponential signal $\alpha(\omega)$;
(3) Make $\widehat{\alpha}(\omega)$, $\omega \in (0, 2\pi]$ as the estimated value of complex spectrum, or make $|\widehat{\alpha}(\omega)|^2$ as the estimated value of the power spectrum. The length M of Filter impulse response should be made a compromise between resolution and statistics stability. The larger M is, the higher the resolution is, and the worse the statistical stability is.

Here are the finite impulse response filters corresponding to APES method, MVM method and Fourier transform respectively

$$\mathbf{h}_{\text{APES}}(\omega) = \frac{\widehat{\mathbf{Q}}^{-1}\mathbf{a}(\omega)}{\mathbf{a}^H(\omega)\widehat{\mathbf{Q}}^{-1}\mathbf{a}(\omega)}$$

$$\mathbf{h}_{\text{Capon}}(\omega) = \frac{\widehat{\mathbf{R}}^{-1}\mathbf{a}(\omega)}{\mathbf{a}^H(\omega)\widehat{\mathbf{R}}^{-1}\mathbf{a}(\omega)}$$

$$\mathbf{h}_{\text{Fourier}}(\omega) = \frac{\mathbf{a}(\omega)}{M}.$$

The filter coefficients of APES method and MVM method are adaptively changed with different data, so lower side lobe and narrower peaks can be obtained than those of the Fourier method. MVM method is often biased estimator, and the APES method is unbiased estimator [35].

15.3 The Orthogonal Subspace Decomposition Method

The orthogonal subspace decomposition method [1, 2] makes eigen-decomposition on the autocorrelation matrix of signal and construct a linear space using eigen-vectors as a set of basis. When a signal forms by multiple harmonics and white noise, the linear space can be decomposed into mutually orthogonal signal subspace and noise subspace. It produces a kind of harmonic frequency estimation method based on this orthogonal principle, including multiple signal classification method, Pisarenco method, minimum norm method, Prony method, and so on.

15.3.1 Multiple Signal Classification Method

Multiple signal classification method (MUSIC) is a sign algorithm of spectral estimation. MUSIC makes eigen-decomposition on covariance matrix of the data to obtain the signal subspace corresponding to the signal component and the noise subspace orthogonal to the signal components. Then the two orthogonal subspaces are used to estimate the signal parameters. For the high resolution of MUSIC method under a certain condition, it is widely used in super-resolution information processing.

(1) **Introduction of MUSIC Method**

The signal observation model of MUSIC method [2, 6, 36, 37] can be expressed as

$$\mathbf{x}(k) = \mathbf{As}(k) + \mathbf{w}(k) \quad k = 1, \ldots, K, \tag{15.39}$$

where data vector $\mathbf{x}(k) = [x_1(k)\ldots x_L(k)]^T$ is the received data from L spatial array elements in the k times, $\mathbf{w}(k) = [w_1(k)\ldots w_L(k)]^T$ is noise vector, $\mathbf{s}(k) = [s_1(k)\ldots s_L(k)]^T$ is composed of D signal vectors, $L \times D$ dimensional matrix $\mathbf{A}$ is $\mathbf{A} = [\mathbf{a}(\theta_1)\ldots \mathbf{a}(\theta_D)]^T$, and $\mathbf{a}(\theta)$ is steering vector.

The problem of MUSIC estimation needs to make assumes as following. Noise covariance matrix can be $E[\mathbf{ww}^H] = \sigma^2\mathbf{I}$, $a(\theta_1)\ldots a(\theta_D)$ is linearly independent, covariance matrix $\mathbf{P} = E[\mathbf{ss}^H]$ is nonsingular, and $L > D$. H denotes Hermitian complex conjugate transpose.

The covariance matrix can be defined as

$$\mathbf{R} = E[\mathbf{xx}^H] = \mathbf{A}E[\mathbf{ss}^H]\mathbf{A}^H + E[\mathbf{ww}^H]. \tag{15.40}$$

Since the signal and noise are independent, the data covariance matrix can be decomposed into two independent orthogonal subspaces. The dimension of the signal subspace is D and the dimension noise subspace is $L - D$. The basic column vectors of noise subspace can be constitute the matrix $\mathbf{E}_w$, and the spectral peak searching of MUSIC method is as follows

$$P_{\text{MUSIC}}(\boldsymbol{\theta}) = \frac{\mathbf{a}^H(\boldsymbol{\theta})\mathbf{a}(\boldsymbol{\theta})}{\mathbf{a}^H(\boldsymbol{\theta})\mathbf{E}_w\mathbf{E}_w^H\mathbf{a}(\boldsymbol{\theta})}. \tag{15.41}$$

Equation (15.41) has a peak value only at $\theta = \theta_1, \ldots, \theta_D$.

(2) **2D MUSIC Super-resolution SAR Imaging Method**

MUSIC super-resolution SAR imaging method combines high-resolution performance of the two-dimensional MUSIC spectral estimation and least squares amplitude estimation to reconstruct SAR image. The SAR signal in two-dimensional frequency domain can be expressed as a complex exponential model, whose parameters are scatters' position and complex magnitude. The two-dimensional position of scatterers can be determined by two-dimensional MUSIC spectrum and scattering coefficient can be calculated by two-dimensional least squares estimation. Then the two-dimensional image can be reconstructed based on the estimated parameters [38].

SAR imaging is a two-dimensional data processing, which can be transformed into a parameter estimation problem of two-dimensional complex sinusoidal signal by the use of signal model. Signal model using MUSIC method is as follows

$$x(m,n) = \sum_{p=1}^{D} s_p e^{j\omega_{rp}m} e^{j\omega_{ap}n} + w(m,n), \qquad (15.42)$$

where $x(m,n)$ is the sampling data after preprocessing and $m = 0,\ldots,M-1$, $n = 0,\ldots,N-1$. Range frequency of Scatter p is ω_{rp} , azimuth frequency is ω_{ap} , and complex backscattering coefficient is s_p. The number D of signal components is determined by the estimation method of signal sources number. At first, frequency parameter (position) $\theta = (\omega_r, \omega_a)$ of target point is estimated. Then the backscatter coefficients of target point can be estimated according to the expected parameter $\theta_i = \left(\omega_{r,i}, \omega_{a,i}\right)_{i=1}^{D}$.

Equation (15.42) can be written as

$$\mathbf{x} = \mathbf{A}\mathbf{s} + \mathbf{w}, \qquad (15.43)$$

where $\mathbf{x} = \left[x_{0,0}\, x_{1,0}\ldots x_{M-1,0}\, x_{0,1}\ldots x_{M-1,N-1}\right]^T$ is a vector arranged by column of data matrix, $\mathbf{s} = \left[s_1\, s_2\ldots s_D\right]^T$, $\mathbf{w} = \left[w_{0,0}\, w_{1,0}\ldots w_{M-1,0}\, w_{0,1}\ldots w_{M-1,N-1}\right]^T$ is a vector arranged by column of noise matrix, $\mathbf{A} = \left[\mathbf{a}(\omega_{r1}, \omega_{a1})\, \mathbf{a}(\omega_{r2}, \omega_{a2})\ldots\mathbf{a}(\omega_{rD}, \omega_{aD})\right]$, $\mathbf{a}(\omega_r, \omega_a) = \mathbf{a}_1(\omega_r) \otimes \mathbf{a}_2(\omega_a)$, $\mathbf{a}_1(\omega_r) = [1\, e^{j\omega_r}\ldots e^{j(M-1)\omega_r}]^T$, $\mathbf{a}_2(\omega_a) = [1\, e^{j\omega_a}\ldots e^{j(N-1)\omega_a}]^T$.

MUSIC method establishes noise subspace and uses its features to restore the target position. Multiple observation data of the scene are needed to generate statistical properties. Spatial smoothing is used to solve the correlation. Defined window dimensions are p_1 and p_2 , which are less than dimensions of the image, i.e., $p_1 < M$ and $p_2 < N$. A plurality of sub-arrays can be formed by sliding window, which can be seen as multiple observations. Figure 15.12 is a schematic view of Spatial smoothing in two-dimensional SAR data.

For each sliding position, the corresponding sub-matrix can be regarded as a data obtaining. Each sub-array would be arranged in columns to form a one-dimensional vector $\mathbf{x}(k)$, so there are $K = (M - p_1 + 1) \times (N - p_2 + 1)$ times observations. The correlation matrix of k sub-array is $\widehat{\mathbf{R}}_k = \mathbf{x}_k\mathbf{x}_k^H$. The correlation matrix of improved spatial smoothing method can be expressed as:

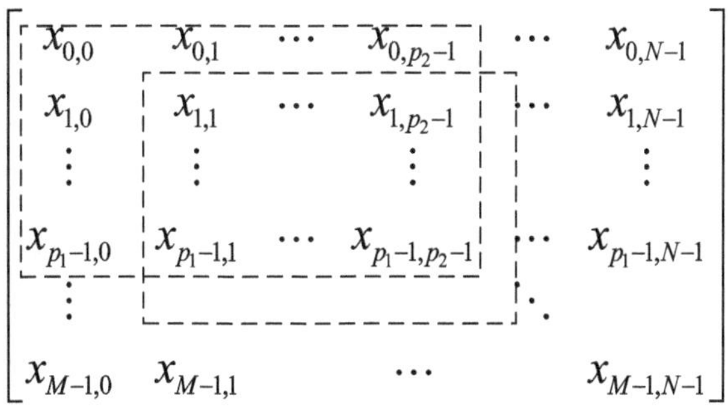

Fig. 15.12 Schematic view of spatial smoothing in two-dimensional SAR data

$$\mathbf{R}_{xx} = \frac{1}{2K}\sum_{k=1}^{K}\left(\widehat{\mathbf{R}}_k + \mathbf{J}\widehat{\mathbf{R}}_k^*\mathbf{J}\right), \tag{15.44}$$

where $\mathbf{J}$ denotes $p_1p_2 \times p_1p_2$ dimensional transformation matrix, and $\mathbf{J} = \begin{bmatrix} 0 & \cdots & 0 & 1 \\ 0 & \cdots & 1 & 0 \\ \vdots & \ddots & \vdots & \vdots \\ 1 & 0 & \cdots & 0 \end{bmatrix}$.

The response of MUSIC method is disproportionate to the target scatter power and the complex scattering information cannot be recovered easily. This is a reconstruction approach of SAR image. The response of MUSIC method is composed of the main peak of the scatterer power spectrum. The peaks are corresponding to the target positions. So information of peaks indicates not the targets' backscatter intensity, but the presence or absence of targets. The amplitudes of scatterer centers should be estimated by least squares method as

$$\xi = (\mathbf{A}^H\mathbf{A})^{-1}\mathbf{A}^H\mathbf{x}. \tag{15.45}$$

(3) **Experiment and Analysis**

(1) Experiment with simulated data

Simulate point target echo signal with 9.6 GHz carrier frequency of chirp signal. The bandwidth of chirp signal is 120 MHz. Each of the two-way sampling frequency (I/Q) is 160 MHz. The pulse repetition frequency is 1000 Hz. The speed of airborne platform is 220 m/ s. The azimuth width of antenna is 2.4 m. The slant range in swath center is 20 km and the SNR is 6 dB. Figure 15.13 is the comparison of point target imaging result using RD method and MUSIC method.

Figure (a) is the result of Range-Doppler imaging method, which has a wide main lobe and high side lobes. The two-point targets along range are mixed together in Figure (a). Figure (b) is the reconstruction result of 2D MUSIC imaging method. The target from the reconstructed image can be clearly distinguished from each other. The mixed two targets can be separated now.

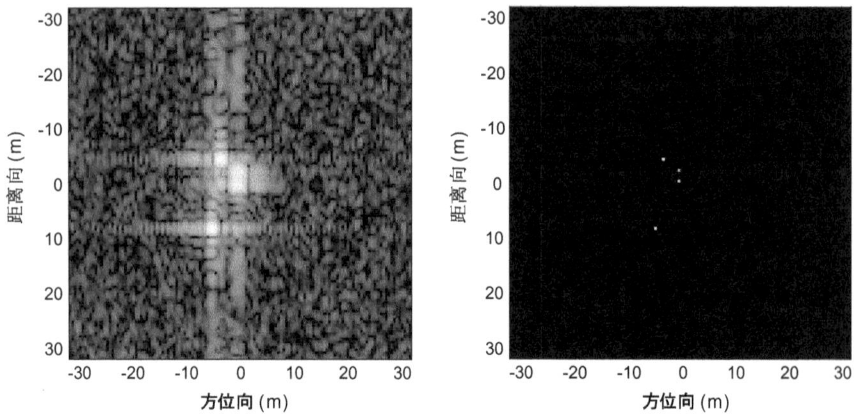

(a) RD method (b) 2D MUSIC method

Fig. 15.13 Comparison of point target imaging results

(2) Experiment with measured data I

The image of three trihedral reflectors is used to test imaging methods, which is measured by *X*-band airborne SAR. The resolution of radar is $2.5 \times 2.5\,m$. The aircraft flight speed is 250 m/s. The flight height is 10 km. Figure 15.14 is the comparison of the two imaging results from measured data. Figure (a) shows RD method and Figure (b) shows 2D MUSIC method. Figure (b) is more clearly than Figure (a) to distinguish the three targets.

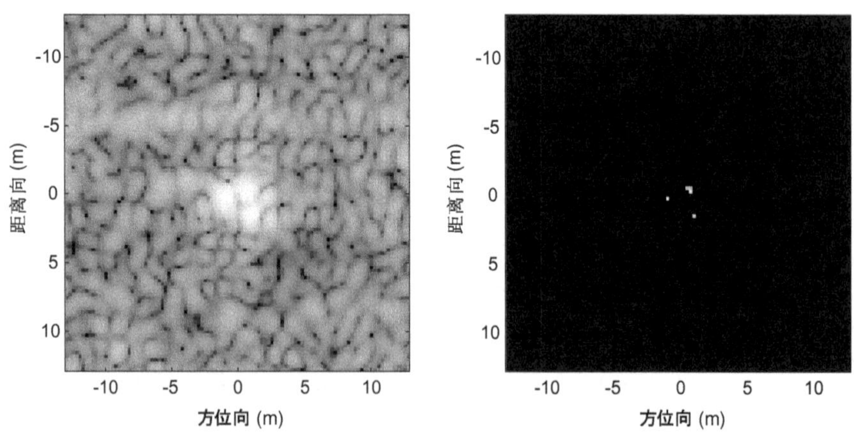

(a) RD method (b) 2D MUSIC method

Fig. 15.14 Comparison of measured data imaging results

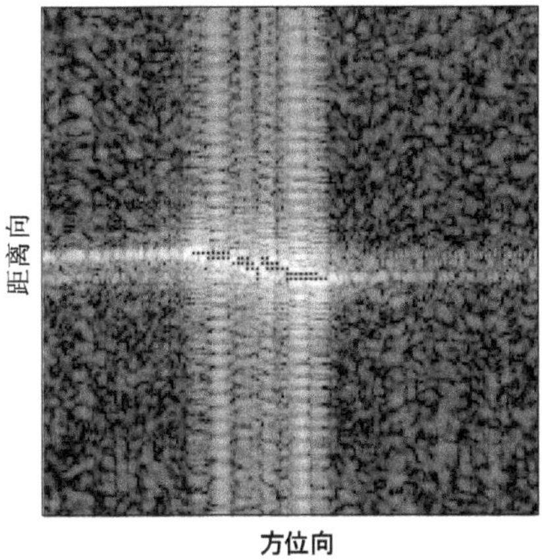

(a) RD method

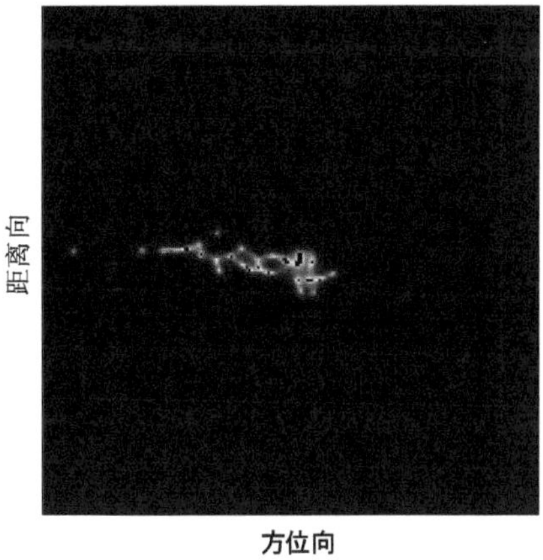

(b) 2D MUSIC method

Fig. 15.15 Comparison of SIR-C data processing

(3) Experiment with measured data II

The experiment chooses SIR-C data, which was a part of the China in 1994. The number of data processing is 51581. A section in the scene is selected to perform super-resolution processing (Fig. 15.15).

15.3.2 Pisarenko Method [3, 39]

Pisarenko Method is proposed by Pisarenko in 1973, which is a line spectrum estimation method based on the Eigen-decomposition of sample correlation matrix. Since the signal and noise are independent, the data covariance matrix can be decomposed into two independent orthogonal subspaces [40], the signal subspace $\widehat{\mathbf{U}}$ and noise subspace $\widehat{\mathbf{V}}$. It is obviously that the eigen vector of the minimum noise eigenvalues is orthogonal to signal subspace, namely

$$\widehat{\mathbf{U}}u_p = 0. \tag{15.46}$$

Define Vandennonde matrix as

$$\mathbf{U} = (U_1(\omega_i), U_2(\omega_i), \cdots, U_K(\omega_i)), \quad i = 1, 2, \ldots, K, \tag{15.47}$$

And positive definite matrix as

$$\boldsymbol{\rho} = \begin{bmatrix} \rho_1 & & 0 \\ & \ddots & \\ 0 & & \rho_K \end{bmatrix}. \quad (15.48)$$

Then there is

$$\widehat{\mathbf{U}} = \mathbf{U}\rho\mathbf{U}^H. \quad (15.49)$$

Equation (15.46) can be rewritten as

$$\mathbf{U}\rho\mathbf{U}^H u_p = 0. \quad (15.50)$$

Take the left multiplication u_p^H on Eq. (15.46) as follows

$$u_p^H\mathbf{U}\rho\mathbf{U}^H u_p = \left(\mathbf{U}^H u_p\right)\boldsymbol{\rho}\left(\mathbf{U}^H u_p\right) = 0. \quad (15.51)$$

Because the matrix $\boldsymbol{\rho}$ is positive definite, there must be

$$\mathbf{U}^H u_p = 0, \quad (15.52)$$

i.e.,

$$U_p^H(\omega_i)u_p = \sum_{k=1}^{K} u_p e^{j\omega_i} = 0, \quad i = 1, 2, \cdots K \quad (15.53)$$

Let $z = e^{j\omega_i}$ and $y(z) = \sum_{k=1}^{K} u_p z^{-k} = 0$. The K roots of Polynomial $y(z)$ will be located at the ω_i of unit circle, and Eq. (15.53) can be used to estimate frequencies of K sine wave. That is Pisarenko harmonic decomposition. Since the frequency estimation of Pisarenko harmonic decomposition method is totally dependent on the noise Eigen vector, there will be a certain limitations in estimated results.

Pisarenko method is a special case of MUSIC method. Its statistical performance is worse than that of the MUSIC method, but its merit is the low dimension of the matrix and the simple calculation. From the point of view for eigen-decomposition of sample correlation matrix, Pisarenko method has smaller computation than MUSIC method for the smaller calculation of matrix decomposition. So Pisarenko method is more suitable for high SNR condition.

15.3.3 Minimum Norm Method [4]

Minimum norm method(MNM) is a parametric estimation method proposed by Kumarensan and Tufts, which exploits the vertical relationship between signal subspace and its orthogonal complement space in essence. Minimum norm method searches a vector d in the noise subspace, which vector has the minimum norm characteristic under the constraint that first element of this vector is 1. Search the angle of vector $\boldsymbol{\alpha}(\mu)$ and vector d (through inner product) to obtain the estimated value of signal arrival direction. The estimation is

$$\overline{\varphi}_{MN}(\mu) = \left[\boldsymbol{\alpha}^H(\mu)\widehat{\mathbf{d}}\widehat{\mathbf{d}}^H\boldsymbol{\alpha}(\mu)\right]^{-1}, \tag{15.54)}$$

where $\widehat{\mathbf{d}}$ is the estimation of $\mathbf{d}$. E_N and E_S are constructed from the eigenvectors of covariance matrix. Let $E_N = \left[\frac{\mathbf{c}^T}{E_{\overline{N}}}\right]$ and $E_S = \left[\frac{\mathbf{g}^T}{E_{\overline{S}}}\right]$. $\mathbf{c}^T$ and $\mathbf{g}^T$ express the first line vectors of E_N and E_S respectively. $E_{\overline{N}}$ and $E_{\overline{S}}$ express the matrixes composed by others lines except first line of E_N and E_S respectively. Then $\mathbf{d}$ can be derived by the minimum norm feature as follows

$$\mathbf{d} = \left[\begin{array}{c} 1 \\ \frac{E_{\overline{N}}\mathbf{c}^*}{\mathbf{c}^H\mathbf{c}} \end{array}\right]\left[\begin{array}{c} 1 \\ \frac{E_{\overline{S}}\mathbf{g}^*}{1-\mathbf{g}^H\mathbf{g}} \end{array}\right]. \tag{15.55)}$$

MNM is the same as MUSIC method that it estimates parameters though D maximum peak position caused by the change of μ. The essence of MNM is still to search the vector $\mathbf{d}$ in the signal subspace when $\mu = \mu_i (i = 1, \ldots, D)$, and the vector $\mathbf{d}$ is meanwhile perpendicular to the noise subspace. $\left|\boldsymbol{\alpha}^H(\mu)\widehat{\mathbf{d}}\right|$ can reach minimum value at $\widehat{\mu}_i$, so $\overline{\varphi}_{MN}(\mu)$ will be a peak in $\widehat{\mu}_i$.

15.3.4 Prony Method [7, 41–44]

Prony method decomposes the autocorrelation matrix into signal subspace corresponding to large eigenvalues vector and noise subspace corresponding to the noise variance eigenvectors, which belongs to eigen-decomposition methods. Prony method can estimate spectrum without estimating a correlation matrix from the samples and give the smaller the variance of the frequency and amplitude of the estimated value without searching the entire frequency domain. Thus Prony method can obtain the frequency values and the amplitude and phase attenuation factor together with the small amount of computation, which has better Doppler frequency accuracy than FFT method.

Let the target composed by N correlation scattering objects, and the backscatter model in frequency domain can be expressed as

$$y_k = \sum_{i=1}^{m} d_i p_i^k,\ k = 0, \cdots, N-1. \tag{15.56}$$

where m is the model order and can be choosed in advance. $\{p_i\}_{i=1}^{m}$ and $\{d_i\}_{i=1}^{m}$ is the estimated parameters, which are amplitudes and phases of the model. The above formula is the Prony model in frequency domain.

Geometry-related information can be obtained by estimating $\{p_i\}_{i=1}^{m}$ and $\{d_i\}_{i=1}^{m}$. The IFT of Eq. (15.56) can be

$$Y(r) = \sum_{i=1}^{m} \frac{d_i e^{j\pi(1-2r/R)}}{e^{j\pi(1-2r/R)} - p_i}, \quad 0 \le r \le R. \tag{15.57}$$

where $r = ct$. The relative distance r is a variable independent of the time. Constant R is the maximum unambiguous range $R = \frac{c}{2\delta_f}$.

From the above equation, the ith component in Prony model can be produced a peak p_i at a distance r_i and

$$r_i \approx \left(\frac{1}{2} - \frac{\arg p_i}{2\pi}\right) R, \tag{15.58}$$

where the ith component represents the scattering centers of relative distance r_i. The peak of scattering centers at r_i will be more obvious if take modulus of p_i. Let $\rho_i = |p_i|$. If $\rho_i \approx 1$ and $r \approx r_i$, the denominator of ith item in Eq. (15.57) is closed to zero. Thus $Y(r)$ will be a drastic change, which means there is a very narrow peak at r_i. When $\rho_i < 1$, the scatter type is corresponding to the diffraction. When $\rho_i > 1$, the scatter type is corresponding to the specular reflection of metal plate or dihedral scatter [5, 11]. Therefore, ρ_i indicates the concentration degree of the *i*th scattering center to the set in range.

Finally, the energy of each scattering center can be obtained through the contribution of scattering center components in RCS data. Each component can be represented by $\{d_i \rho_i^k\}_{k=0}^{N-1}$ as follows

$$P_i = \sum_{k=0}^{N-1} |d_i \rho_i|^{2k} = |d_i|^2 \frac{1 - \rho_i^{2N}}{1 - \rho_i^2}. \tag{15.59}$$

15.4 Estimation of Signal Parameters via Rotational Invariance Techniques

Estimation of signal parameters via rotational invariance techniques (ESPRIT) [45] was first proposed in 1986 by the Roy Paulraj and Kailath, which has now become a major eigen-decomposition method to retrieval harmonic. ESPRIT method is a kind of subspace decomposition methods. It will make eigen-decomposition and estimate signal parameters using the rotational invariance property of data covariance matrix of the signal subspace. This method can obtain high precision pole estimation, noise immunity, small calculation, and no spectral peak searching.

A two-dimensional unitary ESPRIT super-resolution imaging method are discussed next based on synthetic aperture radar imaging characteristics.

15.4.1 Signal Parameters Estimation Method Based Rotational Invariance Technique

Consider K harmonic signals in white noise, and the signal model is as follows

$$x(n) = \sum_{k=1}^{K} \alpha_k(n) \mathrm{e}^{jn\omega_k} + \mathrm{e}(n), \tag{15.60}$$

where $\alpha_k(n)$ and $\omega_k \in (-\pi, \pi)$ were the amplitude and frequency of k th scatterer.

Let $m \times N$ to be the observation data matrix

$$\mathbf{X} = [\mathbf{x}(1), \ldots, \mathbf{x}(N)]$$

where $\mathbf{x}(n) = [x_1(n), \ldots, x_m(n)]^T$ is the n-th observation data vector of m sub-array and N is the number of observations, i.e., the number of smooth. Let $\mathbf{S} = [\mathbf{s}(1), \ldots, \mathbf{s}(N)]$ to be signal matrix, where $\mathbf{s}(n) = [s_1(n), \ldots, s_m(n)]^T$ represents the signal vector. For N-th observations, (15.60) can be expressed as

$$\mathbf{X} = \mathbf{AS} \tag{15.61}$$

where $\mathbf{A} = [\mathbf{a}(\omega_1), \mathbf{a}(\omega_2), \ldots, \mathbf{a}(\omega_p)]$ is steering vector, and $\mathbf{a}(\omega_i) = [1, \mathrm{e}^{j\omega_i}, \ldots, \mathrm{e}^{j(m-1)\omega_i}]$.

Let $\mathbf{J}_1$ and $\mathbf{J}_2$ to be two $(m-1) \times m$ choice matrix

$$\mathbf{J}_1 = \left[\mathbf{I}_{m-1} \vdots \mathbf{0}_{m-1}\right], \ \mathbf{J}_2 = \left[\mathbf{0}_{m-1} \vdots \mathbf{I}_{m-1}\right].$$

Let $\mathbf{X}_1 = \mathbf{J}_1\mathbf{X}$ and $\mathbf{X}_2 = \mathbf{J}_2\mathbf{X}$. $\mathbf{X}_1$ is composed by the first $m-1$ lines of observed data matrix $\boldsymbol{X}$ and $\mathbf{X}_2$ 由is composed by the next $m-1$ lines of observed data matrix $\boldsymbol{X}$.

Order $\mathbf{A} = \begin{bmatrix} \mathbf{A}_1 \\ \text{the last line} \end{bmatrix} = \begin{bmatrix} \text{the first line} \\ \mathbf{A}_2 \end{bmatrix}$. Then $\mathbf{A}_2 = \mathbf{A}_1\mathbf{\Phi}$, where $\mathbf{\Phi} = \mathrm{diag}(\mathrm{e}^{j\omega_1}, \ldots, \mathrm{e}^{j\omega_K})$ contains frequencies of K scatters. We can get

$$\mathbf{X}_1 = \mathbf{A}_1\mathbf{S} \text{ and } \mathbf{X}_2 = \mathbf{A}_2\mathbf{S} = \mathbf{A}_1\Phi\mathbf{S}.$$

Because $\mathbf{\Phi}$ is a unitary matrix, $\mathbf{X}_1$ and $\mathbf{X}_2$ have the same signal subspace and noise subspace, which is a physical interpretation of translational invariance in equidistant linear array.

The autocorrelation matrix of observed vector $\mathbf{x}(n)$ is as follow

$$\begin{aligned}\mathbf{R}_{xx} &= \mathrm{E}\left\{\mathbf{x}(n)\mathbf{x}^H(n)\right\} = \mathbf{A}^H\mathbf{P}\mathbf{A} + \sigma^2\mathbf{I} \\ &= [\mathbf{U}_s, \mathbf{U}_n]\begin{bmatrix}\mathbf{\Sigma}_s & \mathbf{0} \\ \mathbf{0} & \sigma^2\mathbf{I}\end{bmatrix}\begin{bmatrix}\mathbf{U}_s^H \\ \mathbf{U}_n^H\end{bmatrix} \\ &= \left[\mathbf{U}_s\mathbf{\Sigma}_s, \sigma^2\mathbf{U}_n\right]\begin{bmatrix}\mathbf{U}_s^H \\ \mathbf{U}_n^H\end{bmatrix} \\ &= \mathbf{U}_s\mathbf{\Sigma}_s\mathbf{U}_s^H + \sigma^2\mathbf{U}_n\mathbf{U}_n^H,\end{aligned} \tag{15.62}$$

where $\mathbf{P} = \mathrm{E}\{\mathbf{s}(n)\mathbf{s}^H(n)\}$. Because of $\mathbf{I} - \mathbf{U}_n\mathbf{U}_n^H = \mathbf{U}_s\mathbf{U}_s^H$, (15.62) can simultaneously multiply $\mathbf{U}_s$ on both sides. For $\mathbf{U}_s^H\mathbf{U}_s = \mathbf{I}$, we can obtain

$$\mathbf{U}_s = \mathbf{A}\mathbf{T}, \tag{15.63}$$

where $\mathbf{T} = \mathbf{P}\mathbf{A}^H\mathbf{U}_s(\mathbf{\Sigma}_s - \sigma^2\mathbf{I})^{-1}$ is a nonsingular matrix. Multiply $\mathbf{T}$ in the right side, there is

$$\mathbf{A} = \begin{bmatrix}\mathbf{A}_1\mathbf{T} \\ \text{the last line}\end{bmatrix} = \begin{bmatrix}\text{the first line} \\ \mathbf{A}_2\mathbf{T}\end{bmatrix}.$$

The same form of sub-blocks is made on $\mathbf{U}_s$, and there is

$$\mathbf{U}_s = \begin{bmatrix}\mathbf{U}_1 \\ \text{the last line}\end{bmatrix} = \begin{bmatrix}\text{the first line} \\ \mathbf{U}_2\end{bmatrix}.$$

Thus there are $\mathbf{U}_1 = \mathbf{A}_1\mathbf{T}$ and $\mathbf{U}_2 = \mathbf{A}_2\mathbf{T}$. For $\mathbf{U}_2 = \mathbf{A}_1\Phi\mathbf{T}$, there is

$$\mathbf{U}_1\mathbf{T}^{-1}\Phi\mathbf{T} = \mathbf{A}_1\mathbf{T}\mathbf{T}^{-1}\Phi\mathbf{T} = \mathbf{A}_1\Phi\mathbf{T} = \mathbf{U}_2.$$

Let $\mathbf{\Psi} = \mathbf{T}^{-1}\Phi\mathbf{T}$, where $\mathbf{\Psi}$ is the similarity transformation of $\mathbf{\Phi}$, i.e., $\mathbf{\Psi}$ and $\mathbf{\Phi}$ have the similarity eigenvalues. Then we can obtain $\mathbf{U}_2 = \mathbf{U}_1\mathbf{\Psi}$. The eigenvalues of $\mathbf{\Psi}$ is the estimated frequencies. The least squares solution is used to calculate magnitude values and complete the signal parameters estimation.

15.4.2 Unitary ESPRIT Super-Resolution SAR Imaging Method

A two-dimensional unitary ESPRIT super-resolution SAR imaging method is presented in this section. The UESPRIT method exploits the observed data and its conjugation to extend the efficient data one time [40], which can increase the estimation accuracy of original ESPRIT. We consider the SAR signal model is a 2D sinusoid signal in 2D frequency field and the 2D Unitary ESPRIT [46, 47] is used to estimate the 2D parameters, which presents the 2D location of targets. Then the backscatter can be obtained by 2D amplitude estimation [48]. Thus the SAR image can be reconstructed. Rewrite the signal model as follow.

$$x(n_1, n_2) = \sum_{p=1}^{D} s_p e^{j\omega_{rp} n_1} e^{j\omega_{ap} n_2} + w(n_1, n_2) \tag{15.64}$$

where $x(n_1, n_2)$ is the sample data after preprocessing, $n_1 = 0, \ldots, N_1 - 1$, $n_2 = 0, \ldots, N_2 - 1$. The range frequency of target p is ω_{rp} , the azimuth frequency of target p is ω_{ap} , and the complex backscatter is s_p. The D target location parameters $\left(\omega_{r,i}, \omega_{a,i}\right)_{i=1}^{D}$ are estimated first and the backscatters are computed using the estimated results of frequency by least square solution. The steps of the method are as following.

1 The spatial smoothing method is used to obtain the multiple observations from the SAR sample data matrix as Fig. 15.12. The $p_1 \times p_2$ observed data sub-matrix is rearranged to be a vector by column.

2 Construct unitary matrix $\mathbf{Q}_{p_1}$ and $\mathbf{Q}_{p_2}$, where $\mathbf{Q}_{p1} = \frac{1}{\sqrt{2}} \begin{bmatrix} \mathbf{I}_n & j\mathbf{I}_n \\ \mathbf{\Pi}_n & -j\mathbf{\Pi}_n \end{bmatrix}$, $\mathbf{Q}_{p2} = \frac{1}{\sqrt{2}} \begin{bmatrix} \mathbf{I}_n & \mathbf{0} & j\mathbf{I}_n \\ \mathbf{0}^T & \sqrt{2} & \mathbf{0}^T \\ \mathbf{\Pi}_n & \mathbf{0} & -j\mathbf{\Pi}_n \end{bmatrix}$.

3 Compute $\mathbf{Y} = \left(\mathbf{Q}_{p_1}^H \otimes \mathbf{Q}_{p_2}^H\right)\mathbf{X}$ and calculate the singular value decomposition of $[\,\mathrm{real}(\mathbf{Y}) \quad \mathrm{imag}(\mathbf{Y})\,]$. The eigenvectors corresponding to the D larger eigenvalues of the left singular matrix construct the matrix $\mathbf{U}_s$.

4 Solve $\mathbf{K}_{\mu 1}\mathbf{U}_s\mathbf{\Psi}_\mu = \mathbf{K}_{\mu 2}\mathbf{U}_s$ to get $\mathbf{\Psi}_\mu$. Solve $\mathbf{K}_{\upsilon 1}\mathbf{U}_s\mathbf{\Psi}_\upsilon = \mathbf{K}_{\upsilon 2}\mathbf{U}_s$ to get $\mathbf{\Psi}_\upsilon$.

5 Calculate the eigenvalues λ_i of $\mathbf{\Psi}_\mu + j\mathbf{\Psi}_\upsilon$, where $i = 1, \ldots, D$.

6 Then the 2D frequency can be obtained $\omega_{ri} = 2\tan^{-1}(\mathrm{real}\{\lambda_i\})$, $\omega_{ai} = 2\tan^{-1}(\mathrm{imag}\{\lambda_i\})$, $i = 1, \cdots, D$.

7 The targets amplitude can be calculated by the least square method based on the measured data. Arrange the 2D measured data to be a vector $\mathbf{y} = \left[x_{0,0}, \ldots, x_{M-1,0}, \ldots, x_{M-1,N-1}\right]^T$.

The signal model can be represented as vector form.

$$\mathbf{y} = (\mathbf{a}_2 \otimes \mathbf{a}_1)\mathbf{s} + \mathbf{e} \tag{15.65}$$

where $\mathbf{s} = [s_1, \ldots, s_D]^T$, $\mathbf{a}_i = \left[1, e^{j\omega_k}, \cdots, e^{j(N_i-1)\omega_k}\right]$, $i = 1, 2$. And the estimation of $\mathbf{s}$ by the least square method is

$$\mathbf{s} = \left((\mathbf{a}_2 \otimes \mathbf{a}_1)^H (\mathbf{a}_2 \otimes \mathbf{a}_1)\right)^{-1} (\mathbf{a}_2 \otimes \mathbf{a}_1)^H \mathbf{y} \tag{15.66}$$

8 Reconstruct the SAR image, i.e., the super-resolution SAR image, based on the estimated amplitude $\mathbf{s}$ and the 2D frequency $\left(\omega_{r,i}, \omega_{a,i}\right)_{i=1}^{D}$.

15.4.3 Experiment and Analysis

(1) **Experiment with simulated data**

The experiment is carried out using simulated dot targets. The simulation parameters are set as follow. Carry frequency of transmission chirp signal is 9.6 GHz. Signal bandwidth is 150 MHz. Sample frequency is 180 MHz (I/Q). Pulse repetition frequency is 1000 Hz. The aircraft velocity is 220 m/s. The azimuth length of antenna is 2 m. The slant range of swath center is 20 km. And the signal noise ratio is 7 dB. Figure 15.16 is the comparison of the RD imaging method result and the 2D UESPRIT method result.

2D Unitary ESPRIT super-resolution imaging method also belongs to reconstruction methods, which only reconstruct the main scatter in the images without

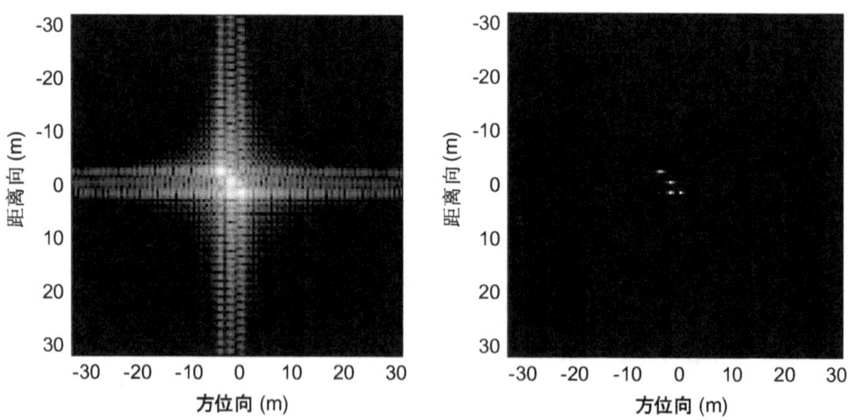

(a) The RD method

(b) The 2D UESPRIT method

Fig. 15.16 Comparison of point target imaging results

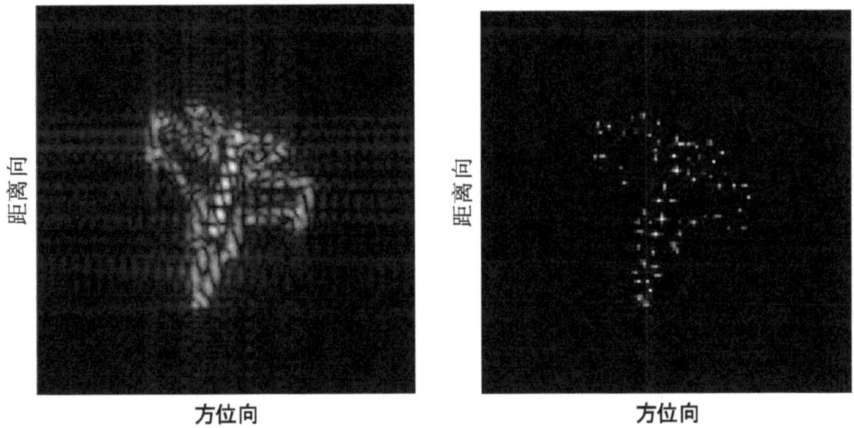

(a) The RD method (b) The 2D UESPRIT method

Fig. 15.17 Comparison of Mig25 imaging results

considering the background of the images. These methods apply only on the interesting target of images, such as tanks, armored vehicles, aircraft, etc., and cannot handle complex scene targets. However, it will have good application in ISAR imaging to improve the accuracy of targets recognition.

We use the ISAR simulated data of mig25 aircraft from simulation experiment of United States Naval Research Laboratory. The simulation parameters are as follows. The carrier frequency of transmission signal is 9 GHz. The signal bandwidth is 128 MHz. The data size is 64×64. Figure 15.17 is a comparison of Mig25 imaging results.

(2) **Experiment with measured data**

The actual measured data of X-band airborne SAR is used for validating the method. The radar parameters are as follow. Resolution is $2.5 \times 2.5\,\text{m}$. Pulse repetition frequency is 1000 Hz. The aircraft velocity is 250 m/s. The height of airborne SAR is 10 km. The selected data is obtained from three corner reflectors (Fig. 15.18).

15.5 RELAX-Based Spectrum Estimation Methods

When the radar echo is consistent with the signal model, the parametric spectral estimation algorithms have the potential to, significantly, improve SAR imaging resolution. RELAX-based spectral estimation methods are a kind of robust

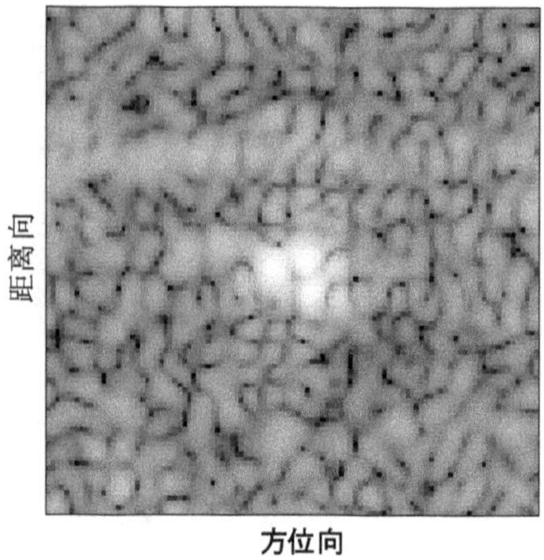

(a) The RD method

(b) The 2D UESPRIT method

Fig. 15.18 Comparison of measured data imaging results

parameter estimation methods, including the CLEAN methods [49–51] and RELAX methods [52–56].

15.5.1 RELAX Method

When target feature can be approximated by point scatterers, the RELAX method can be used to estimate parameters of point scatter model and reconstruct SAR image to improve the image resolutions. Here RELAX method and its application in super-resolution SAR are introduced as follows.

Assuming the target composed by the K main scatters or triangular reflectors, the data model of the target is expressed as

$$x(n_1,n_2)=\sum_{k=1}^{K}\alpha_k e^{j\left(n_1\omega_{1,k}+n_2\omega_{2,k}\right)}+\mathrm{e}(n_1,n_2), \tag{15.67}$$

where $n_1=0,1,\ldots,N_1-1$ and $n_2=0,1,\ldots,N_2-1$. N_1 and N_2 are sample numbers of range and azimuth respectively. α_k and $\left(\omega_{1,k},\omega_{2,k}\right)$ are the complex amplitude and 2D frequency of the kth scatter. α_k depends on the backscatter intensity of the k th scatter. $\left(\omega_{1,k},\omega_{2,k}\right)$ depends on its 2D position. Construct a cost function to estimate $\left\{\alpha_k,\omega_{1,k},\omega_{2,k}\right\}_{k=1}^{K}$ as follows

$$\mathrm{C}_1\left(\left\{\alpha_k,\omega_{1,k},\omega_{2,k}\right\}_{k=1}^{K}\right)=\left\|\mathbf{X}-\sum_{k=1}^{K}\boldsymbol{\omega}_{N_1}\left(\omega_{1,k}\right)\boldsymbol{\omega}_{N_2}^{T}\left(\omega_{2,k}\right)\alpha_k\right\|_F^2, \tag{15.68}$$

where X is a $N_1 \times N_2$ data matrix composed of $x(n_1, n_2)$, $\boldsymbol{\omega}_{N_1}(\omega_{1,k}) = \begin{bmatrix} 1 & e^{j\omega_{1,k}} & \ldots & e^{j(N_1-1)\omega_{1,k}} \end{bmatrix}$ and $\boldsymbol{\omega}_{N_2}(\omega_{2,k}) = \begin{bmatrix} 1 & e^{j\omega_{2,k}} & \ldots & e^{j(N_2-1)\omega_{2,k}} \end{bmatrix}$.

RELAX method utilizes RELEX search method to minimize the cost function C_1. Assume the parameters of the other scatterers known in parameters estimation of each scatterer, which can use the previous estimation results for the present calculations. Let

$$\mathbf{X}_k = \mathbf{X} - \sum_{i=1, i\neq k}^{\widetilde{K}} \boldsymbol{\omega}_{N_1}(\widehat{\omega}_{1,i})\boldsymbol{\omega}_{N_2}^T(\widehat{\omega}_{2,i})\widehat{\alpha}_i, \tag{15.69}$$

where assume that $\{\widehat{\alpha}_i, \widehat{\omega}_{1,i}, \widehat{\omega}_{2,i}\}_{i=1,i\neq k}^{K}$ is the parameters already estimated. Then $\{\widehat{\alpha}_k, \widehat{\omega}_{1,k}, \widehat{\omega}_{2,k}\}$ is to be

$$\widehat{\alpha}_k = \frac{\boldsymbol{\omega}_{N_1}^H(\omega_{1,k})\mathbf{X}_k\boldsymbol{\omega}_{N_2}^*(\omega_{2,k})}{N_1 N_2}\bigg|_{\omega_{1,k}=\widehat{\omega}_{1,k},\omega_{2,k}=\widehat{\omega}_{2,k}}, \tag{15.70}$$

$$(\widehat{\omega}_{1,k}, \widehat{\omega}_{2,k}) = \arg\max_{\omega_{1,k},\omega_{2,k}} \left|\boldsymbol{\omega}_{N_1}^H(\omega_{1,k})\mathbf{X}_k\boldsymbol{\omega}_{N_2}^*(\omega_{2,k})\right|^2. \tag{15.71}$$

$\{\widehat{\omega}_{1,k}, \widehat{\omega}_{2,k}\}$ can be determined by the peak of periodogram $\frac{\left|\boldsymbol{\omega}_{N_1}^H(\omega_{1,k})\mathbf{X}_k\boldsymbol{\omega}_{N_2}^*(\omega_{2,k})\right|^2}{N_1N_2}$, and FFT can be used conveniently to calculate. Usually zero padding can make more accurate estimation results in the calculation of $\{\widehat{\omega}_{1,k}, \widehat{\omega}_{2,k}\}$. $\widehat{\alpha}_k$ is the complex amplitude peak of $\frac{\left|\boldsymbol{\omega}_{N_1}^H(\omega_{1,k})\mathbf{X}_k\boldsymbol{\omega}_{N_2}^*(\omega_{2,k})\right|^2}{N_1N_2}$.

In summary, the steps of RELAX method are as following.

(1) Assume $\widetilde{K} = 1$. (15.69) and (15.70) can be used to obtain $\{\widehat{\alpha}_k, \widehat{\omega}_{1,k}, \widehat{\omega}_{2,k}\}_{k=1}$.
(2) Assume $\widetilde{K} = 2$. Use (15.68) to calculate $\mathbf{X}_2$, and substitute the $\{\widehat{\alpha}_k, \widehat{\omega}_{1,k}, \widehat{\omega}_{2,k}\}_{k=1}$ result from step1 to (15.68). $\{\widehat{\alpha}_k, \widehat{\omega}_{1,k}, \widehat{\omega}_{2,k}\}_{k=2}$ can be obtained from $\mathbf{X}_2$ with (15.69) and (15.70). Then calculate $\mathbf{X}_1$ using $\{\widehat{\alpha}_k, \widehat{\omega}_{1,k}, \widehat{\omega}_{2,k}\}_{k=2}$ just got and re-estimate $\{\widehat{\alpha}_k, \widehat{\omega}_{1,k}, \widehat{\omega}_{2,k}\}_{k=1}$.

The iteration is satisfied until convergence conditions.

(3) Assume $\widetilde{K} = 3$. Calculate $\mathbf{X}_3$ using the $\{\widehat{\alpha}_k, \widehat{\omega}_{1,k}, \widehat{\omega}_{2,k}\}_{k=1}^{2}$ from step 2 and get $\{\widehat{\alpha}_k, \widehat{\omega}_{1,k}, \widehat{\omega}_{2,k}\}_{k=3}$. Recalculate $\mathbf{X}_1$ using $\{\widehat{\alpha}_k, \widehat{\omega}_{1,k}, \widehat{\omega}_{2,k}\}_{k=2}^{3}$ and get new estimation of $\{\widehat{\alpha}_k, \widehat{\omega}_{1,k}, \widehat{\omega}_{2,k}\}_{k=1}$. Calculate $\mathbf{X}_2$ using $\{\widehat{\alpha}_k, \widehat{\omega}_{1,k}, \widehat{\omega}_{2,k}\}_{k=1,3}$ and recalculate $\{\widehat{\alpha}_k, \widehat{\omega}_{1,k}, \widehat{\omega}_{2,k}\}_{k=2}$.

The iteration is satisfied until convergence conditions.

Repeat the above steps until $\widetilde{K} = K$.

The targets number estimation methods can be used to determine K. Iteration convergence conditions can be determined by the relative change value between two adjacent iterations of the cost function $C_1\left(\left\{\alpha_k, \omega_{1,k}, \omega_{2,k}\right\}_{k=1}^{K}\right)$.

RELAX SAR imaging method exploits the dot scattering model to reconstruct SAR images. Reconstructed SAR images only include information of main scatterers without the information of background clutter. So RELAX SAR imaging method should apply only on the interesting target of images, and cannot handle complex large scene targets.

Steps of RELAX SAR imaging method are as follows.

(1) The phase history data can be obtained from signal model, and RELAX method is used to estimate $\left\{\alpha_k, \omega_{1,k}, \omega_{2,k}\right\}_{k=1}^{K}$ as the above steps.

(2) Use the estimated parameters to generate a greater dimensions data $x_s(m_1, m_2)$ as

$$x_s(m_1, m_2) = \sum_{k=1}^{\widetilde{K}} \widehat{\alpha}_k \exp\left\{j\left(m_1 \widehat{\omega}_{1,k} + m_2 \widehat{\omega}_{2,k}\right)\right\}, \tag{15.72}$$

where $m_1 = 0, 1, \ldots, \zeta N_1 - 1$, $m_2 = 0, 1, \ldots, \zeta N_2 - 1$ and ζ is the selectable extrapolation factor.

(3) Make normalized FFT on the phase history data after extrapolated to generate SAR images with only the main objects as

$$\frac{1}{\zeta^2 N_1 N_2} \sum_{m_1=0}^{\zeta N_1 - 1} \sum_{m_2=0}^{\zeta N_2 - 1} x_s(m_1, m_2) e^{-j\left(m_1 \widehat{\omega}_{1,k} + m_2 \widehat{\omega}_{2,k}\right)}. \tag{15.73}$$

If we make normalized FFT on the sum of the estimated phase history data $\{x_s(m_1, m_2)\}$ of targets and the phase history data $\left\{\zeta^2 \widehat{e}(n_1, n_2)\right\}$ of background after zero padding, the images containing both target scattering information and background scattering information, such as shading information, can be get to improve recognition performance of targets. Because there are no effective models for clutter, the resolution of background clutter does not improve. The image resolution using the RELAX method depends on the RELAX algorithm itself, and is independent of the dimension of the final image generated [52].

Since RELAX method has iterative processes at each step, RELAX method can get better convergence to the global minimum, which makes the final estimation accuracy higher.

15.5.2 CLEAN Method

CLEAN method is the degradation of RELAX method, which is an approximate RELAX solution with a nonlinear minimum mean square criterion. CLEAN method has no iterative processes in steps of RELAX, which is more simple than RELAX but less precise. When the two targets are close, CLEAN may give biased estimation and RELAX can get asymptotic unbiased estimation. Therefore, RELAX is also known as SUPER CLEAN method. The public can compromise to choose these two methods considering computation and accuracy. Steps of CLEAN method are as follows.

(1) Assume $\widetilde{K} = 1$. (15.69) and (15.70) can be used to obtain $\left\{\widehat{\alpha}_k, \widehat{\omega}_{1,k}, \widehat{\omega}_{2,k}\right\}_{k=1}$.
(2) Assume $\widetilde{K} = 2$. Use (15.68) to calculate $\mathbf{X}_2$, and substitute the $\left\{\widehat{\alpha}_k, \widehat{\omega}_{1,k}, \widehat{\omega}_{2,k}\right\}_{k=1}$ result from step1 to (15.68). $\left\{\widehat{\alpha}_k, \widehat{\omega}_{1,k}, \widehat{\omega}_{2,k}\right\}_{k=2}$ can be obtained from $\mathbf{X}_2$ with (15.69) and (15.70). Then calculate $\mathbf{X}_1$ using $\left\{\widehat{\alpha}_k, \widehat{\omega}_{1,k}, \widehat{\omega}_{2,k}\right\}_{k=2}$ just got and re-estimate $\left\{\widehat{\alpha}_k, \widehat{\omega}_{1,k}, \widehat{\omega}_{2,k}\right\}_{k=1}$.
(3) Assume $\widetilde{K} = 3$. Calculate $\mathbf{X}_3$ using the $\left\{\widehat{\alpha}_k, \widehat{\omega}_{1,k}, \widehat{\omega}_{2,k}\right\}_{k=1}^{2}$ from step 2 and get $\left\{\widehat{\alpha}_k, \widehat{\omega}_{1,k}, \widehat{\omega}_{2,k}\right\}_{k=3}$. Recalculate $\mathbf{X}_1$ using $\left\{\widehat{\alpha}_k, \widehat{\omega}_{1,k}, \widehat{\omega}_{2,k}\right\}_{k=2}^{3}$ and get new estimation of $\left\{\widehat{\alpha}_k, \widehat{\omega}_{1,k}, \widehat{\omega}_{2,k}\right\}_{k=1}$. Calculate $\mathbf{X}_2$ using $\left\{\widehat{\alpha}_k, \widehat{\omega}_{1,k}, \widehat{\omega}_{2,k}\right\}_{k=1,3}$ and recalculate $\left\{\widehat{\alpha}_k, \widehat{\omega}_{1,k}, \widehat{\omega}_{2,k}\right\}_{k=2}$.

Repeat the above steps until $\widetilde{K} = K$.

15.5.3 Experiment and Analysis

(1) **Experiment with simulated data**

The experiment is carried out using simulated dot targets. The simulation parameters are set as follow. Carry frequency of transmission chirp signal is 9.6 GHz. Signal bandwidth is 150 MHz. Sample frequency is 180 MHz (I/Q). Pulse repetition frequency is 1000 Hz. The aircraft velocity is 220 m/s. The azimuth length of antenna is 2 m. The slant range of swath center is 20 km. And the signal noise ratio is 7 dB. Figure 15.19 is the comparison of the RD imaging method result, CLEAN method, and RELAX method.

Observed figure (a), (b), and (c), the simulation sets four-point targets. Figure (a) is the result of Range-Doppler imaging method, which has a wide main lobe, high side lobes, and difficult to distinguish. Figure (b) is the reconstruction result of CLEAN method, which can only reconstruct three targets because of estimation error. Figure (c) is the reconstruction result of RELAX method, which can give the accurate and clear reconstruction targets. The difference between

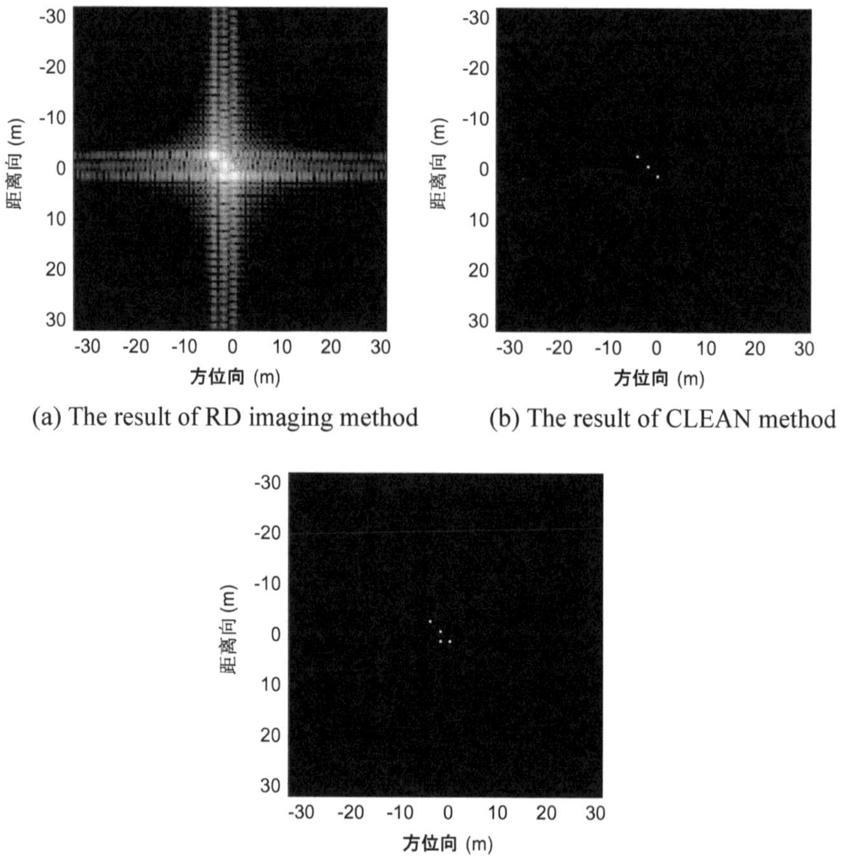

(a) The result of RD imaging method

(b) The result of CLEAN method

(c)The result of RELAXmethod

Fig. 15.19 Comparison of point target imaging results

(b) and (c) is mainly due to the iterative processes in frequency estimation for RELAX. The iterative each frequency estimation can reduce the error frequency estimation. CLEAN method has only one step frequency estimation, which is impacted greater by noise and leads to miscalculation.

Use the ISAR simulated data of mig25 aircraft from simulation experiment of United States Naval Research Laboratory. The simulation parameters are as follows. The carrier frequency of transmission signal is 9 GHz. The signal bandwidth is 128 MHz. The data size is 64×64. (c)The result of RELAXmethod

Figure 15.20 is a comparison of Mig25 imaging results. The reconstructed contour in (c) is better than that of in (b), which is also due to the above reasons of wrong frequency estimate.

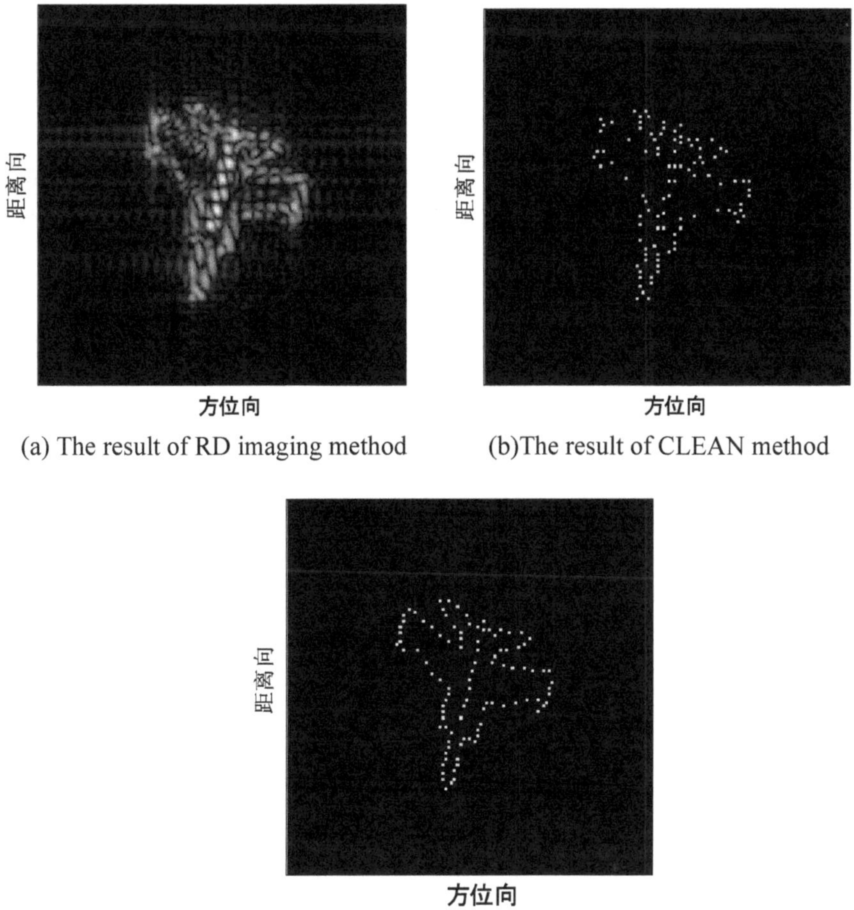

(a) The result of RD imaging method (b)The result of CLEAN method

(c)The result of RELAXmethod

Fig. 15.20 Comparison of Mig25 imaging results

(2) **Experiment with measured data**

We use the ISAR data of B727 aircraft from the United States Naval Research Laboratory. The data parameters are as follows. The carrier frequency of transmission signal is 9 GHz. The signal bandwidth is 150 MHz. The data size is 128×128. Figure 15.21 is a comparison of B727 imaging results.

The clear contours of B727 with two methods show the effectiveness for the actual data processing.

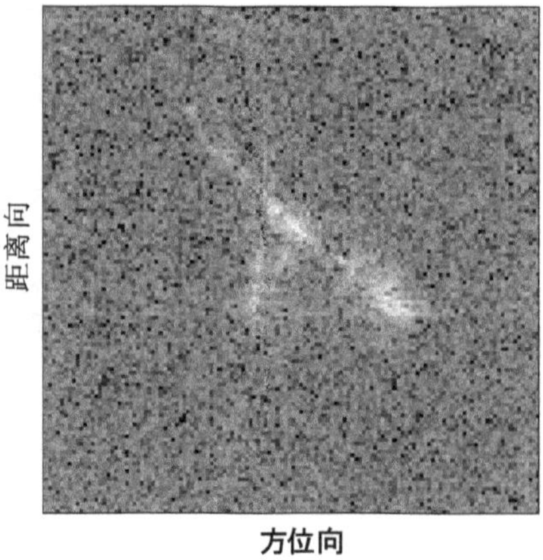

(a) The result of RD imaging method

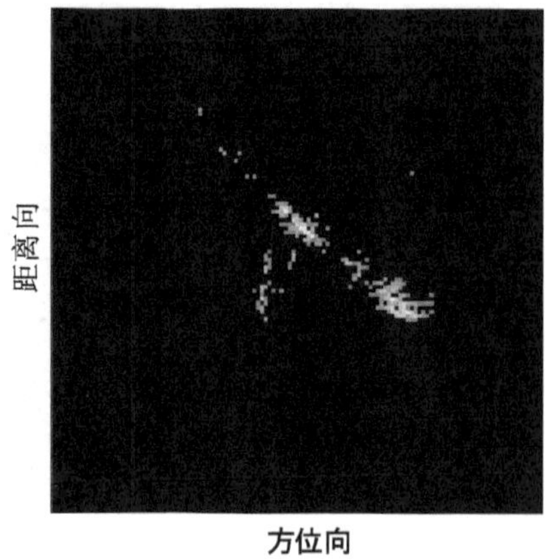

(b) The result of CLEAN method

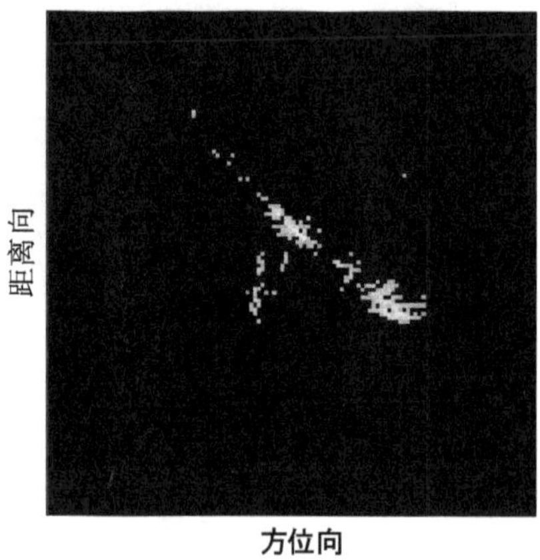

(c) The result of RELAXmethod

Fig. 15.21 Comparison of B727imaging results

15.5.4 Autoregressive Spectral Estimation Method

SAR imaging method based on Autoregressive spectral estimation method (AR) obtains the images expressed by AR power spectral density to improve slant range resolution and azimuth resolution [57, 58]. With the prediction filter and the prediction error filter, future or past sample sequences can be estimated from the known sample sequence. Prediction filter predicts the values of sample sequence required, and the prediction error filter adjustments the weights according to error between the actual values and predicted values [59].

Considering the forward linear prediction of p order filter, i.e., we know the p values of random sequence $x(n)$, and extrapolate the $p+1$ th value. The optimal forward prediction filter is as follows

$$\widehat{x}_f^p(n) = -\sum_{i=1}^{p} a_i x(n-i). \tag{15.74}$$

The p order optimal backward prediction filter is as follows

$$\widehat{x}_b^p(n-p) = -\sum_{i=1}^{p} b_i x(n-p+i). \tag{15.75}$$

a_i is the p coefficient of i order prediction filter, and $n = p+1, p+2, \ldots, N$ with $b_i = a_i^*$ [60].

Define the forward prediction error is

$$e_f^p(n) = x(n) - \widehat{x}_f^p(n) = x(n) + \sum_{i=1}^{p} a_i x(n-i). \tag{15.76}$$

Define the backward prediction error is

$$e_b^p(n) = x(n-p) - \widehat{x}_b^p(n-p) = x(n-p) + \sum_{i=1}^{p} a_i^* x(n-p+i). \tag{15.77}$$

Select coefficients a_i of p order FIR filter to minimize the mean square error of forward and backward predictions, i.e.,

$$\rho = \min_{\{a_i\}_{i=1}^{p}} \left\{ \frac{1}{2} \sum_{n=p+1}^{N} \left| e_f^p(n) \right|^2 + \left| e_b^p(n) \right|^2 \right\}. \tag{15.78}$$

The transfer function of the linear prediction filter can be expressed as

$$H_m(z) = \frac{1}{1 + \sum\limits_{i=1}^{p} a_i^p z^{-i}}. \tag{15.79}$$

The output power spectrum of linear prediction filter is

$$\widehat{P}_{AR}(f) = \frac{\rho}{\left| 1 + \sum\limits_{i=1}^{p} \widehat{a}_i e^{-j2\pi f i} \right|^2}. \tag{15.80}$$

15.6 Summary

This chapter studied several super-resolution spectrum estimation SAR imaging algorithm, including MVM method, APES method, subspace orthogonal decomposition method (including MUSIC method, Pisarenko method, minimum norm method, Prony method), ESPRIT method, the RELAX based method (including RELAX method and CLEAN method), maximum likelihood estimation method, and the autoregression method.

MVM spectrum estimation, APES spectrum estimation, and periodogram are all belonging to spectral estimators based on matched filter bank. MVM and APES spectrum estimation can produce narrower spectral peaks and lower side lobes than the periodogram. The resolution of APES spectrum estimation is slightly worse than MVM spectrum estimation, but the APES spectrum estimation can obtain more accuracy magnitude and SAR imaging results than MVM spectrum estimation. Both two spectral estimation methods are suitable for application to large scenes of SAR imaging and can get better resolution and image quality than the Fourier method. However, the two methods require to calculate the statistical characteristics of the signal, i.e., that signal covariance matrix, which is a large amount of computation. The block computing can reduce the computation.

Reconstruction class methods only reconstruct the main scatter in the images without considering the background. These methods apply only on the interesting target of images, such as tanks, armored vehicles, aircraft, etc., and cannot handle complex scene targets. However, it will have good application in ISAR imaging to improve the accuracy of targets recognition.

References

1. Barbarossa S, Marsili L, Mungari G (1996) SAR super-resolution imaging by signal subspace projection techniques. In: Proceedings of EUSAR'96, Konigswinter, Germany, pp 267–270
2. Odendaal JW, Bernard E, Pistorius CWI (1994) Two-dimensional superresolution radar imaging using the MUSIC algorithm. IEEE Trans Antennas Propag 42:1386–1391
3. Pisarenko VF (1973) The retrieval of harmonics from a covariance function. Geophys J Roy Astron Soc 33:347–366
4. Kumaresan R, Tufts DW (1983) Estimating the angle of arrival of multiple plane waves. IEEE Trans Aerosp Electron Syst 19:134–139
5. Schmidt RO (1981) A signal subspce approach to multiple emitter location and spectral estimation. Stanford University, California
6. Schmidt RO (1986) Multiple Emitter Location and Signal Parameter Estimation. IEEE Trans Antennas Propag 34:243–258
7. Johnson DH (1982) The application of spectral estimation method to bearing estimation problems. Proc IEEE 70:1018–1028
8. Hua YB (1994) High resolution imaging of continuously moving target using stepped frequency radar. IEEE Trans Signal Process 35:33–40
9. Roy R, KailathT (1986) ESPRIT-A subspace rotation approach to estimation of parameters of cisoids in noise. In: IEEE transactions on acoustics, speech and signal processing, vol ASSP-14. pp 1340–1342

10. Roy R, Kailath T (1987) Total least squares ESPRIT. In: Proceedings of 21st Asilomar conference on signal, system and computation, pp 297–301
11. Zoltowski M (1988) Novel techniques for estimating the array signal parameters based on matrix pencil, subspace rotation and total least squares. In: International conference on acoustics, speech, and signal processing, ICASSP-88, New York, NY, USA, pp 2861–2864
12. Zoltowski M, Stavinides D (1989) Sensor array signal processing via a Procrustes rotations based eign-analysis of the ESPRIT data pencil. IEEE Trans Acoust Speech Signal Process 37:832–861
13. Jakobsson A, Marple Jr SL, Stoica P (1999) Efficient implementation of the 2-D capon spectral estimator. In: Conference record of the thirty-third asilomar conference on signals, systems, and computers, Pacific Grove, CA, USA, pp 432–436
14. Capon J (1969) High-resolution frequency-wavenumber spectrum analysis. In: Proceedings of the IEEE, pp 1408–1418
15. Jakobsson A, Marple SL Jr, Stoica P (2000) Computationally efficient two-dimensional Capon spectral analysis. IEEE Trans Signal Process 48:2651–2661
16. Liu Z, Wu R, Li J (1999) Complex ISAR imaging of maneuvering targets via the capon estimator. IEEE Trans Signal Process 47:1262–1271
17. Wang Y (2004) High-resolution spectral analysis: the missing data case. Ph.D. Thesis, University of Florida, Florida
18. Stoica P, Wang Z, Li J (2003) Robust Capon beamforming. IEEE Signal Process Lett 10:172–175
19. Benitz GR (1997) High-definition vector imaging. Lincoln Lab J 10:147–169
20. DeGraaf SR (1998) SAR imaging via modern 2-D spectral estimation methods. IEEE Trans Image Process 7:729–761
21. Cox H, Zeskind R (1991) Reduced variance distortionless response (RVDR) performance with signal mismatch. In: 25th Asilomar conference on signals, systems & computers 2, Pacific Grove, California, pp 825–829
22. Nadakuditi RR, Edelman A (2005) The bias of the MVDR beamformer outputs under diagonal loading. In ICASSP. Seoul, Korea, pp 793–796
23. Li J, Stoica P, Wang Z (2003) On robust Capon beamforming and diagonal loading. IEEE Trans Signal Process 51:1702–1805
24. Wang Y, Li J, Stoica P (2005) Rank-deficient robust capon filter bank approach to complex spectral estimation. IEEE Trans Signal Process 53:2713–2726
25. Marzetta T, Simon S, Ren H. (2006) Capon-MVDR spectral estimation from singular data covariance matrix, with no diagonal loading. MIT Lincoln Laboratory ESC-TR-2006-066
26. Richmond CD, Nadakuditi RR, Edelman A (2005) Asymptotic mean squared error performance of diagonally loaded capon-MVDR processor. In: Conference record of the thirty-ninth asilomar conference
27. Li J, Stoica P, Wang Z (2004) Doubly constrained robust Capon beamformer. IEEE Trans Signal Process 52:2407–2423
28. Ren L (2005) Study about improvement of SAR image quality. Xidian University, Xi'an
29. DeGraaf SR (1994) SAR imaging via modern 2-d spectral estimation methods. In Proc. SPIE on Optical Engineering in Aerospace Sensing, Orlando, FL, pp 36–47
30. Zhang P (2009) Study on synthetic aperture radar super-resolution information processing techniques. Institute of Electronics, Chinese Academy of Sciences, Beijing
31. Palsetia MR, Li J (1998) Using APES for interferometric SAR imaging. IEEE Trans Image Process 7:1430–1443
32. Stoica Petre, Li Hongbin, Li Jian (1999) A new derivation of the APES filter. IEEE Signal Process Lett 6:205–206
33. Li J, Stoica P (1996) An adaptive filtering approach to spectral estimation and SAR imaging. IEEE Trans Signal Process 44:1469–1484
34. Yildirm I, Tezel NS, Erer I, Yazgan B (2003) A comparison of non-parametric spectral estimators for SAR imaging. In Recent advances in space technologies. RAST Int Conf 2003:369–374

35. Larsson EG, Stoica P (2002) Fast implementation of two-dimensional APES and CAPON spectral estimators. Multidimension Syst Signal Process 13:35–53
36. Savy L, Planes J-G, Moal C (2000) High resolution spectral analysis applied to SAR images. In: Proceedings of the CEOS SAR workshop, Toulouse, pp 625–630
37. Thompson P, Nannini M, Scheiber R (2007) Target separation in SAR image with the MUSIC algorithm. In: 2007 IEEE international geoscience and remote sensing symposium, IGARSS 2007, Barcelona, pp.468–471
38. Zhang P, Shang J, Yang R (2010) Efficient 2D MUSIC superresolution SAR imaging method. J Syst Simul 22:184–187
39. Stoica P, Moses R (1997) Introduction to spectral analysis. Prentice hall, New Jersey
40. Zhang X (2002) Modern signal processing, 2nd edn. Tsinghua University Press, Beijing
41. Carriere R, Moses RL (1992) High resolution radar target modeling using a modified prony estimator. IEEE Trans Antennas Propag 40:13–18
42. Hurst M, Mittra R (1987) Scattering center analysis via Prony's method. IEEE Trans Antennas Propag 35:986–988
43. Sacchini JJ, Steedly WM (1993) Two-dimensional prony modeling and parameter estimation. IEEE Trans Signal Process 41:3127–3136
44. Osborne MR, Smyth GK (1995) A modified Prony algorithm for exponential function fitting. SIAM J Sci Comput 16:119–138
45. Roy Richard, Kailath Thomas (1989) ESPRIT-estimation of signal parameters via rotational invariance techniques. IEEE Trans Acoust Speech Signal Process 37:984–995
46. Zoltowski Michael D, Haardt Martin, Mathews Cherian P (1996) Closed-form 2-D angle estimation with rectangular arrays in elementspace or beamspace via unitary ESPRIT. IEEE Trans Signal Process 44:316–328
47. Haardt M, Zoltowski MD, Mathews CP, Nossek JA (1996) 2D unitary ESPRIT for efficient 2D parameter estimation. In: 1995 international conference on acoustics, speech, and signal processing, ICASSP-95, Detroit, MI, USA, 1996, pp 2096–2099
48. Zhang P, Li Z, Chen Q (2010) 2D uesprit superresolution SAR imaging algorithm. In IEEE international geoscience and remote sensing symposium (IGARSS). HI, Honolulu, pp 4067–4070
49. Peter T (1994) Gough, A fast spectral estimation algorithm based on the FFT. IEEE Trans Signal Process 42:1317–1325
50. Bose R, Freedman A, Steinberg B (2002) Sequence CLEAN: a modified deconvolution technique for microwave images of contiguous targets. IEEE Trans Aerosp Electron Syst 38:89–97
51. Bose R (1995) Sequence CLEAN: a deconvolution algorithm useful for non-isolated radar target images with high sidelobes. Ph.D. Thesis, University of Pennsylvania, Pennsylvania
52. Li J, Stoica P (1996) Efficient mixed-spectrum estimation with applications to target feature extraction. IEEE Trans Signal Process 44:281–295
53. Liu ZS, Li J (1998) Feature extraction of SAR targets consisting of trihedral and dihedral corner reflector. IEEE Proc Radar Sonar Navig 145:161–172
54. Liu ZS, Li J (1998) Synthetic-aperture-radar motion compensation and feature extraction by means of a relaxation-based algorithm. J Opt Soc Am A 15:599–610
55. Bi Z, Li J, Liu Z (1999) Super resolution SAR imaging via parametric spectral estimation methods. IEEE Trans Aerosp Electron Syst 35:267–281
56. Liu Z, Li J (1998) Implementation of the RELAX algorithm. IEEE Trans Aerosp Electron Syst 34:657–664
57. Gupta IJ, Beals MJ, Moghaddar A (1994) Data extrapolation for high resolution radar imaging. IEEE Tans Antennas Propag 42:1540–1545
58. Gupta IJ, Beals MJ, Moghaddar A. (1994) High-resolution radar imaging using 2-D Linear prediction. In: IEEE transactions on antennas and propagation, January 1994, vol 42
59. Kay SM (1988) Modern spectral estimation: theory and application. Prentice Hall, New Jersey
60. Marple SL Jr (1987) Digital spectral analysis with application. Prentice Hall, New Jersey

Chapter 16
SAR Super-Resolution Methods Based on Spectral Bandwidth Extrapolation and Adaptive Side Lobe Reduction

16.1 SAR Super-Resolution Methods Based on Spectral Bandwidth Extrapolation

Spectral bandwidth extrapolation refers to the data of known band-limited signal in a certain bandwidth range can be improved the resolution through a priori knowledge of the signal to seek outside the bandwidth of the data. The essence of bandwidth extrapolation is data extrapolation in frequency domain. The bandwidth is increased to improve the resolution by extrapolating the spectrum of the finite object.

Signal extrapolation is one of the fundamental problems in signal processing. For continuous signals with limited frequency band, the analytic nature of mathematics can be extrapolated by using the method of Taylor series expansion. However, Taylor series expansion requires the high-order derivability, which cannot be determined or has large the error when the discrete data is detected or data is accompanied by noise. In practice, it is difficult to extrapolate the band-limited continuous signal. In 1961 mathematicians of Bell Labs put forward of extrapolation mathematical tools, i.e., the long ball function. Due to local and global orthogonality of the long ball function, the mathematical tool is very easy to understand and analyze the problem of signal extrapolation.

Signal extrapolation is essential to know the part and obtain the whole, which is possible in the strict constraints. Energy-limited continuous band limit signal can be inferred the signal in the entire time domain by a continuous interval of the signal in a small segment. However, many physical signals are not strictly mathematical energy limited continuous band-limited signals, which can be only understood as an approximate band-limited signal and makes extrapolation difficultly. In addition, when the observed signal is band-limited, there may encounter the uniqueness and stability problem of signal extrapolation due to the discretization and noisy of signals.

This section focuses on the study of super-resolution imaging method based on autoregressive linear predictive extrapolation and minimum weighted norm extrapolation.

R. Yang et al., *High-Resolution Microwave Imaging*,
https://doi.org/10.1007/978-981-10-7138-6_16

16.1.1 SAR Super-Resolution Method Based on Linear Prediction Extrapolation

Linear predictive extrapolation uses the observed band-limited data and signal models to estimate the statistical properties of the signal and characterize data properties of in-band and out-of-band. According to the statistical properties, the parameters of extrapolation model are calculated to extend the signal bandwidth. Then the system resolution is improved. There are many models of data sequence linear prediction. Here we use autoregressive (AR) model [1–3] and the autoregressive AR model parametric method can get small variance and high resolution, because The echo data of actual radar targets are usually short [4].

(1) **Forward and backward autoregressive linear predictive (ARLP) extrapolation**

Linear prediction uses a predictive filter and a predictive error filter to estimate future or past samples from a known sequence of samples. The prediction filter predicts the desired value of the sample sequence, and the prediction error filter adjusts the weight of the prediction filter according to the error between the actual value and the predicted value [3].

Consider the p order forward–linear prediction, i.e., known the p values of random sequence $x(n)$ to extrapolate the $p+1$th prediction value. The p order optimal forward prediction filter is shown in (15.74). The p order optimal backward prediction is shown in (15.75). Define the forward prediction error as (15.76), and define the backward prediction error as (15.77). The coefficient a_i is chosen to minimize the mean square error of the p order forward and backward prediction as (15.78).

In order to minimize the prediction mean square error $\left|e_f^p(n)\right|^2$, the prediction error should be orthogonal to the input data $x(n-i), i = 1, 2, \cdots, p$ i.e.,

$$E[e(n)x^*(n-i)] = 0. \tag{16.1}$$

Substitute (15.74) into the above equation, and expand as follows:

$$r_{xx}(i) = -\sum_{k=1}^{p} a_i r_{xx}(i-k). \tag{16.2}$$

Rewrite to matrix form

$$\begin{bmatrix} r_{xx}(0) & r_{xx}(-1) & \cdots & r_{xx}(-p+1) \\ r_{xx}(1) & r_{xx}(0) & \cdots & r_{xx}(-p+2) \\ \vdots & \vdots & \ddots & \vdots \\ r_{xx}(p-1) & r_{xx}(p-2) & \ldots & r_{xx}(0) \end{bmatrix} \begin{bmatrix} -a_1 \\ -a_2 \\ \vdots \\ -a_p \end{bmatrix} = \begin{bmatrix} r_{xx}(1) \\ r_{xx}(2) \\ \vdots \\ r_{xx}(p) \end{bmatrix}, \tag{16.3}$$

where $r_{xx}(-i) = E\{x(n)x^*(n+i)\}$. The minimum prediction error power $\rho_{\min}$ of the output is

$$\begin{aligned}\rho_{\min} &= E\left[\left|e_f^p(n)\right|^2\right]\\ &= E\left\{e_f^p(n)[x(n) - \hat{x}(n)]^*\right\}\\ &= E\left\{e_f^p(n)x^*(n)\right\} - \sum_i a_i^* E\left\{e_f^p(n)x^*(n-i)\right\}\\ &= E\left\{e_f^p(n)x^*(n)\right\} = \sum_{i=0}^{p} a_i r_{xx}(-i),\end{aligned} \tag{16.4}$$

where $a_0 = 1$. Combine (16.3) and (16.4), and yield

$$\begin{bmatrix} r_{xx}(0) & r_{xx}(-1) & \cdots & r_{xx}(-p) \\ r_{xx}(1) & r_{xx}(0) & \cdots & r_{xx}(-p+1) \\ \vdots & \vdots & \ddots & \vdots \\ r_{xx}(p) & r_{xx}(p-1) & \cdots & r_{xx}(0) \end{bmatrix} \cdot \begin{bmatrix} 1 \\ a_1 \\ \vdots \\ a_p \end{bmatrix} = \begin{bmatrix} \rho_{\min} \\ 0 \\ \vdots \\ 0 \end{bmatrix}. \tag{16.5}$$

Equation (16.5) is Yule–Walker equation. Solve Yule–Walker equation to yield the filter coefficients a_i, which is called covariance method. This method cannot guarantee the stability linear prediction filter. Then the addition boundary can be made to improve covariance method [5]. Another method is to Burg method using the iterative solution, which can reduce the complexity of operation [1]. Next, the Brug method is described to calculate AR parameters.

It is assumed that the relevant parameters of the $p-1$ order prediction filter are known, i.e.,

$$\begin{bmatrix} r_{xx}(0) & r_{xx}(-1) & \cdots & r_{xx}(-p+1) \\ r_{xx}(1) & r_{xx}(0) & \cdots & r_{xx}(-p+2) \\ \vdots & \vdots & \ddots & \vdots \\ r_{xx}(p-1) & r_{xx}(p-2) & \cdots & r_{xx}(0) \end{bmatrix} \cdot \begin{bmatrix} 1 \\ a_1^{p-1} \\ \vdots \\ a_{p-1}^{p-1} \end{bmatrix} = \begin{bmatrix} \rho_{p-1} \\ 0 \\ \vdots \\ 0 \end{bmatrix}. \tag{16.6}$$

Make Conjugate on both sides of the above equation, and consider $r_{xx}^*(-i) = r_{xx}(i)$. There is

$$\begin{bmatrix} r_{xx}(0) & r_{xx}(1) & \cdots & r_{xx}(p-1) \\ r_{xx}(-1) & r_{xx}(0) & \cdots & r_{xx}(p-2) \\ \vdots & \vdots & \ddots & \vdots \\ r_{xx}(-p+1) & r_{xx}(-p+2) & \cdots & r_{xx}(0) \end{bmatrix} \cdot \begin{bmatrix} 1 \\ a_1^{*p-1} \\ \vdots \\ a_{p-1}^{*p-1} \end{bmatrix} = \begin{bmatrix} \rho_{p-1} \\ 0 \\ \vdots \\ 0 \end{bmatrix}. \tag{16.7}$$

Rewrite the above equation as

$$\begin{bmatrix} r_{xx}(0) & r_{xx}(-1) & \cdots & r_{xx}(-p+1) \\ r_{xx}(1) & r_{xx}(0) & \cdots & r_{xx}(-p+2) \\ \vdots & \vdots & \ddots & \vdots \\ r_{xx}(p-1) & r_{xx}(p-2) & \cdots & r_{xx}(0) \end{bmatrix} \cdot \begin{bmatrix} a_{p-1}^{*p-1} \\ \vdots \\ a_1^{*p-1} \\ 1 \end{bmatrix} = \begin{bmatrix} 0 \\ 0 \\ \vdots \\ \rho_{p-1} \end{bmatrix}. \tag{16.8}$$

Equation (16.6) is the $p-1$ order forward prediction error filter equation, and (16.8) is the $p-1$ order backward prediction error filter equation. Let

$$\sum_{i=0}^{p-1} a_i^{p-1} r_{xx}(p-i) = -K_p \rho_{p-1}, \tag{16.9}$$

where K_p is the desired reflection coefficient and $a_0 = 1$. The complex conjugate form of the above formula is as follows:

$$\sum_{i=0}^{p-1} a_i^{*p-1} r_{xx}(i-p) = -K_p^* \rho_{p-1}. \tag{16.10}$$

Combine (16.9) and (16.10) to get

$$\begin{bmatrix} r_{xx}(0) & r_{xx}(-1) & \cdots & r_{xx}(-p) \\ r_{xx}(1) & r_{xx}(0) & \cdots & r_{xx}(-p+1) \\ \vdots & \vdots & \ddots & \vdots \\ r_{xx}(p-1) & r_{xx}(p-2) & \cdots & r_{xx}(1) \\ r_{xx}(p) & r_{xx}(p-1) & \cdots & r_{xx}(0) \end{bmatrix} \cdot \left\{ \begin{pmatrix} 1 \\ a_1^{p-1} \\ \vdots \\ a_{p-1}^{p-1} \\ 0 \end{pmatrix} + K_p \begin{pmatrix} 0 \\ a_{p-1}^{*p-1} \\ \vdots \\ a_1^{*p-1} \\ 1 \end{pmatrix} \right\}$$
$$= \begin{pmatrix} \rho_{p-1} \\ 0 \\ \vdots \\ 0 \\ -K_p \rho_{p-1} \end{pmatrix} + K_p \begin{pmatrix} -K_p^* \rho_{p-1} \\ 0 \\ \vdots \\ 0 \\ \rho_{p-1} \end{pmatrix}. \tag{16.11}$$

Compare (16.5) and (16.11), and there is

$$\begin{cases} a_i^p = a_i^{p-1} + K_p a_{p-i}^{*p-1} \quad i = 0, 1, \ldots, p \\ K_p = a_p^p \\ P_p = \left(1 - |K_p|^2\right) P_{p-1} \\ p = 0 \Rightarrow P_0 = r_{xx}(0) = \frac{1}{N} \sum\limits_{n=1}^{N} |x(n)|^2 \end{cases}. \tag{16.12}$$

Equation (16.12) is the Levinson recursive formula for computing the p order prediction coefficients. The reflection coefficient K_p can be obtained by minimizing the forward and backward prediction error mean power of Burg. According to definition of prediction errors in (15.76) and (15.77), the recurrence formulas of the forward and backward prediction errors can be expressed as follows substituting (16.12) into the two equations.

$$e_f^p(n) = e_f^{p-1}(n) + K_p e_b^{p-1}(n-1) \tag{16.13}$$

$$e_b^p(n) = e_b^{p-1}(n-1) + K_p^* e_f^{p-1}(n) \tag{16.14}$$

Minimize the average power of p order forward and backward prediction error as shown in (15.78), and let $\frac{\partial \rho_p}{\partial K_p} = 0$. The reflection coefficient is

$$K_p = \frac{-\sum_{n=p+1}^{N} e_f^{p-1}(n) e_b^{*p-1}(n-1)}{\frac{1}{2}\sum_{n=p+1}^{N}\left[\left|e_f^{p-1}(n)\right|^2 + \left|e_b^{p-1}(n-1)\right|^2\right]}. \tag{16.15}$$

(2) **Tufts–Kumaresan ARLP extrapolation**

The predicted sampling point P in the signal history field does not need to have any spatial relationship to other sampling points used to predict it. In other words, the filter is not causal but semi-causal. ARLP imaging based on autoregressive linear predictive extrapolation of arbitrarily selected predictive elements may have spurious peaks and ellipses instead of circular contours. The average of the signal in one and two quadrants can reduce the appearance of false peaks and make the contours similar to circles.

$$S_{\mathrm{ARPL-Piar}}(\underline{r}) = \frac{1}{\sqrt{\frac{\left|\mathrm{P}^{\mathrm{H}}\mathrm{R}^{-1}\underline{\mathrm{W}}(\underline{r})\right|^2}{\mathrm{R}_{\mathrm{p,p}}^{-1}} + \frac{\left|\mathrm{P}_{\perp}^{\mathrm{H}}\mathrm{R}^{-1}\underline{\mathrm{W}}(\underline{r})\right|^2}{\mathrm{R}_{\mathrm{p}_\perp,\mathrm{p}_\perp}^{-1}}}} \tag{16.16}$$

The prediction element $\mathrm{P}_\perp$ and P are selected from the quadrants orthogonal to the center of the sub-aperture. Even this condition, the choice of orthogonal prediction elements is arbitrary. The concept of quadrature quadrant RRMS averaging can be reasonably generalized as use the RRMS to average ARLP spectrum or image on all possible prediction elements.

$$S_{\text{RRMS-ARLP}}(\underline{r}) = \frac{1}{\sqrt{\sum_{P} \frac{\underline{W}^{H}(\underline{r})R^{-1}\underline{P}\underline{P}^{H}R^{-1}\underline{W}(\underline{r})}{R^{-1}_{p,p}}}} = \frac{1}{\sqrt{\underline{W}^{H}(\underline{r})R^{-1}\left(\text{diag}\left(R^{-1}\right)\right)^{-1}R^{-1}\underline{W}(\underline{r})}}, \tag{16.17}$$

where $\left(\text{diag}\left(R^{-1}\right)\right)^{-1}$ is the diagonal matrix of the prediction error.

If it is assumed that each ARLP filter produces the same prediction error energy, for example $R^{-1}_{p,p} = \frac{1}{\varepsilon^2}$, RRMS ARLP imaging can be expressed as

$$S_{\text{RRMS-ARLP}}(\underline{r}) \to \frac{1}{\varepsilon\sqrt{\underline{W}^{H}(\underline{r})R^{-2}\underline{W}(\underline{r})}} = S_{\text{PIS}}(\underline{r}). \tag{16.18}$$

In fact, if the assumptions are satisfied, the difference between RRMS ARLP imaging and Pisarenko imaging can be neglected.

Tufts–Kumaresan method is a variant of ARLP, which purpose is to construct a filter in the signal subspace or construct an error prediction filter in the clutter subspace. TKARLP differs from ARLP in that it allows using larger sub-apertures, which will lead the correlation matrix to be likely singular. And TKARLP uses Moore–Penrose pseudo-inverses to calculate the prediction filter. However, the advantage of using larger apertures is to increase resolution. Second, the correlation matrix will remove the part of the clutter. It seems to improve the clutter side lobe ratio and improve the accuracy of the scatter peak location in image. In fact, it will not improve the clutter side lobe ratio in image domain. Therefore, the TKARLP prediction filter is a minimum norm solution.

$$R_{P;\text{sig}}\underline{A}_P = -\underline{r}^*_P, \quad \underline{A}_P = -R^{+}_{P;\text{sig}}\ \underline{r}^*_P = -\sum_{\text{signal}} \lambda_m \underline{v}_{P;m}\underline{v}^{H}_{P;m}\underline{r}^*_P \tag{16.19}$$

Eigen decomposition of the correlation matrix $\Re_P = Q_P Q_P^H$ is shown as

$$\Re_P = v_P \Lambda_P v_P^H. \tag{16.20}$$

The imaging result of TKARLP can be expressed as

$$S_{\text{TKARLP-P}}(\underline{r}) = \frac{1}{\left|\underline{A}_P^H \underline{W}(\underline{r})\right|}. \tag{16.21}$$

The results of the forward and backward filtering is

$$S_{\text{RRMS-TKARLP}}(\underline{r}) = \frac{1}{\sqrt{\sum_P \left|\underline{A}_P^H \underline{W}(\underline{r})\right|^2}}. \tag{16.22}$$

16.1.1.1 SAR Super-Resolution Method Based on 2D Linear Prediction Extrapolation [6]

(1) **Estimation of 2D AR parameters**

It is believed that the two-dimensional discrete impulse response is the causal relationship in the asymmetric half-plane and in the quarter-plane [7]. The following two-dimensional autoregressive spectral estimation is considered in the case of a quarter planes.

The linear prediction error filter of first quarter-plane is considered. $x(m,n)$ is the $M \times N$ dimension 2D measured data, and $a^1(l_1,l_2)$ is the coefficients of (p_1,p_2) order linear prediction error filter for $0 \le l_1 \le p_1$ and $0 \le l_2 \le p_2$. The linear prediction error filter is

$$e^1(m,n) = \sum_{l_1=0}^{p_1}\sum_{l_2=0}^{p_2} a^1(l_1,l_2)x(m-l_1,n-l_2), \tag{16.23}$$

where $a^1(0,0) = 1$. The output of the Q_1 quarter-plane linear prediction error filter is represented as a vector form, and the above equation becomes

$$e^1(m,n) = \mathbf{a}^1\mathbf{x}(m,n). \tag{16.24}$$

where

$$\begin{aligned}
\mathbf{a}^1 &= \begin{bmatrix}\mathbf{a}^1(0) & \mathbf{a}^1(1) & \cdots & \mathbf{a}^1(p_2)\end{bmatrix},\\
\mathbf{a}^1(p) &= \begin{bmatrix}a^1(0,p) & a^1(1,p) & \cdots & a^1(p_1,p)\end{bmatrix}, \text{and}\\
\mathbf{x}(m,n) &= [x(m,n) \quad \cdots \quad x(m-p_1,n) \quad \cdots \quad x(m_1,n-p_2) \quad \cdots \quad x(m_1-p_1,n-p_2)\ ]^T.
\end{aligned}$$

Assume $e^1(m,n)$ is white noise process with the variance $\rho^1 = \mathrm{E}\left\{\left|e^1(m,n)\right|^2\right\}$. $x(m,n)$ is two-dimensional AR process in Q_1 quarter-plane, and the power spectrum is

$$P^1(f_1,f_2) = \frac{\rho^1}{\left|\sum_{l_1=0}^{p_1}\sum_{l_2=0}^{p_2} a^1(l_1,l_2)e^{-j2\pi(f_1l_1+f_2l_2)}\right|}. \tag{16.25}$$

In a similar manner, we can define the Q_2, Q_3 and Q_4 quarter-plane linear prediction error filter outputs. The corresponding power spectrums are $P^2(f_1,f_2), P^3(f_1,f_2)$ and $P^4(f_1,f_2)$. $M \times N$ Measured data are used to solve the least squares solution of linear prediction using 2D covariance method. The total mean square error function is

$$\begin{aligned}\rho^i &= \sum_{m=p_1}^{M-1}\sum_{n=p_2}^{N-1}\left|e^i(m,n)\right|^2 \\ &= \mathbf{a}^i\left(\sum_{m=p_1}^{M-1}\sum_{n=p_2}^{N-1}\mathbf{x}(m,n)\mathbf{x}^H(m,n)\right)\mathbf{a}^{iH} \\ &= \mathbf{a}^i\mathbf{R}\mathbf{a}^{iH},\end{aligned} \tag{16.26}$$

where $i = 1, 2, 3, 4$ and $\mathbf{R}$ is $(p_1 + 1)(p_2 + 1) \times (p_1 + 1)(p_2 + 1)$ order the covariance matrix.

$$\begin{aligned}\mathbf{R} &= \sum_{m=p_1}^{M-1}\sum_{n=p_2}^{N-1}\mathbf{x}(m,n)\mathbf{x}^H(m,n) \\ &= \mathbf{X}\mathbf{X}^H \\ &= \begin{bmatrix}\mathbf{R}(0,0) & \mathbf{R}(0,1) & \cdots & \mathbf{R}(0,p_2) \\ \mathbf{R}(1,0) & \mathbf{R}(1,1) & \cdots & \mathbf{R}(1,p_2) \\ \vdots & \vdots & \ddots & \vdots \\ \mathbf{R}(p_2,0) & \mathbf{R}(p_2,1) & \cdots & \mathbf{R}(p_2,p_2)\end{bmatrix}\end{aligned} \tag{16.27}$$

where $\mathbf{R}(i,j) = \sum_{n=p_2}^{N-1}\mathbf{X}(n-i)\mathbf{X}^H(n-j)$ and $\mathbf{X}$ is a block-like Toeplitz matrix.

$$\mathbf{X} = \begin{bmatrix}\mathbf{X}(p_2) & \mathbf{X}(p_2+1) & \cdots & \mathbf{X}(N-1) \\ \mathbf{X}(p_2-1) & \mathbf{X}(p_2) & \cdots & \mathbf{X}(N-2) \\ \vdots & \vdots & \ddots & \vdots \\ \mathbf{X}(0) & \mathbf{X}(1) & \cdots & \mathbf{X}(N-p_2-1)\end{bmatrix}$$

$$\mathbf{X}(i) = \begin{bmatrix}x(p_1,i) & x(p_1+1,i) & \cdots & x(M-1,i) \\ x(p_1-1,i) & x(p_1,i) & \cdots & x(M-2,i) \\ \vdots & \vdots & \ddots & \vdots \\ x(0,i) & x(1,i) & \cdots & x(M-p_1-1,i)\end{bmatrix}$$

Minimize the total mean square error, and there is

$$\begin{aligned}\mathbf{a}^1\mathbf{R} &= \begin{bmatrix}\rho^1 & \cdots & 0 & 0 & \cdots & 0 & \cdots & 0 & \cdots & 0\end{bmatrix} \\ \mathbf{a}^2\mathbf{R} &= \begin{bmatrix}0 & \cdots & 0 & 0 & \cdots & 0 & \cdots & \rho^2 & \cdots & 0\end{bmatrix} \\ \mathbf{a}^3\mathbf{R} &= \begin{bmatrix}0 & \cdots & 0 & 0 & \cdots & 0 & \cdots & 0 & \cdots & \rho^3\end{bmatrix} \\ \mathbf{a}^4\mathbf{R} &= \begin{bmatrix}0 & \cdots & \rho^4 & 0 & \cdots & 0 & \cdots & 0 & \cdots & 0\end{bmatrix}.\end{aligned} \tag{16.28}$$

AR parameters $\mathbf{a}^i$ can be directly calculated using (16.27). In order to reduce the computational complexity, the multivariate covariance method is used to solve the above equation. So the solution problem is regarded as the N multivariate and p_1 order covariance method [8].

$$\begin{aligned} \mathbf{a}_{p_1}\mathbf{R} &= \begin{bmatrix} \mathrm{P}^a_{p_1} & \mathbf{0} & \cdots & \mathbf{0} \end{bmatrix} \\ \mathbf{b}_{p_1}\mathbf{R} &= \begin{bmatrix} \mathbf{0} & \mathbf{0} & \cdots & \mathrm{P}^b_{p_1} \end{bmatrix} \end{aligned} \tag{16.29}$$

where $\mathbf{R} = \sum_{i=p_1}^{N-1} \mathbf{x}(i)\mathbf{x}^H(i),$, $\mathbf{x}_{p_1}(i) = \begin{bmatrix} \mathbf{X}(i) & \mathbf{X}(i-1) & \cdots & \mathbf{X}(i-p_1) \end{bmatrix}^T$ $\mathbf{a}_{p_1} = \begin{bmatrix} \mathbf{I} & \mathbf{A}_{p_1}(1) & \cdots & \mathbf{A}_{p_1}(p_1) \end{bmatrix}$, and $\mathbf{b}_{p_1} = \begin{bmatrix} \mathbf{B}_{p_1}(p_1) & \cdots & \mathbf{B}_{p_1}(1) & \mathbf{I} \end{bmatrix}$. $\mathbf{a}_{p_1}$ represents $\mathbf{a}^i$ in $i = 1, 2$. $\mathbf{b}_{p_1}$ represents $\mathbf{a}^i$ in $i = 3, 4$.

The combined two-dimensional AR spectral estimation is obtained from four quarter-plane prediction filters as follows

$$\frac{1}{P(f_1,f_2)} = \frac{1}{P^1(f_1,f_2)} + \frac{1}{P^2(f_1,f_2)} + \frac{1}{P^3(f_1,f_2)} + \frac{1}{P^4(f_1,f_2)}. \tag{16.30}$$

(2) **Methodology**

The two-dimensional AR model parameters are calculated using the measured data and signal model as a priori knowledge. The data of the extrapolated effective bandwidth are predicted according to the two-dimensional AR model. And the high-resolution images are generated by the expanded data. In this section, first the SAR signal model in the two-dimensional frequency domain is analyzed to accord with the autoregressive extrapolated linear predictive model. Second, the one-dimensional and two-dimensional extrapolation methods are presented. The flow of SAR imaging method based on two-dimensional autoregressive linear predictive extrapolation is finally given. Figure 16.1 shows a schematic of SAR data acquisition.

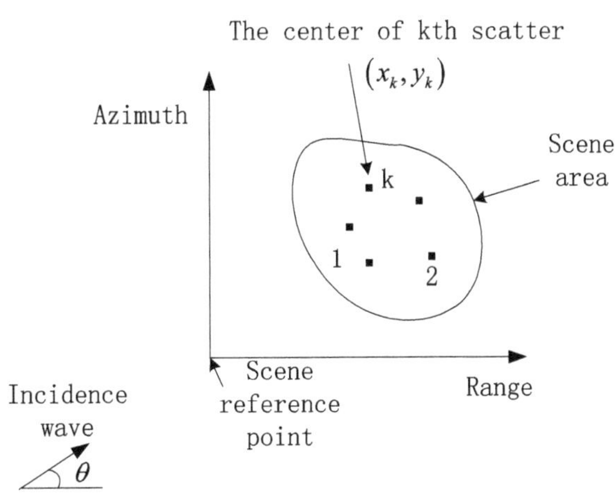

Fig. 16.1 Schematic of SAR data acquisition

Assume there are K scatters in the imaging scene. From the synthetic aperture radar signal model, the frequency domain signal of the kth scattering center can be expressed as a function of the frequency (f_x, f_y) in Cartesian coordinate system when the observation angle is θ [9].

$$E_k(f_x, f_y) = A_k(f_x, f_y)e^{j4\pi/c(f_x x_k + f_y y_k)} \tag{16.31}$$

where $f_x = f\cos\theta$ and $f_y = f\sin\theta . A_k(f_x, f_y)$ is corresponding to the location and amplitude of the k scatterer.

The sampling points in the scene can be expressed as $f_x(m) = f_x(0) + m\Delta f_x$ and $f_y(n) = f_y(0) + n\Delta f_y$ $(m = 0, \cdots, M, n = 0, \cdots, N) . f_x(0)$ and $f_y(0)$ is the starting frequency. Δf_x and Δf_y is the sampling interval. Then (16.31) can be rewritten to be

$$E_k(m,n) = A_k(m,n)e^{j4\pi/c(f_x(0)x_k + f_y(0)y_k)} z_{x_k}^m z_{y_k}^n, \tag{16.32}$$

where $z_{y_k} = e^{j4\pi/c\Delta f_y y_k}$ and $z_{x_k} = e^{j4\pi/c\Delta f_x x_k}$. Let

$$E_k(0,0) = A_k(f_x(0), f_y(0))e^{j4\pi/c(f_x(0)x_k + f_y(0)y_k)}, \tag{16.33}$$

which is the value of the kth scatterer at the center frequency $(f_x(0), f_y(0))$. In the SAR observation area, $A_k(f_x, f_y)$ is generally considered to be a constant. Equation (16.32) can be written as

$$E_k(m,n) = E_k(0,0)z_{x_k}^m z_{y_k}^n. \tag{16.34}$$

The frequency domain signal in the scene can be expressed as:

$$E(m,n) = \sum_{k=1}^{K} E_k(0,0)z_{x_k}^m z_{y_k}^n + v(m,n) \tag{16.35}$$

where $v(m,n)$ is noise. As can be seen from the above equation, the autoregressive model can be used to estimate the backscattering coefficient of the scene.

Figure 16.2 shows a schematic diagram of one-dimensional extrapolation case. The shaded area in the middle of the data is the observed data. The blanks on both sides are the extrapolated data. (15.74) and (15.75) are, respectively, for the forward prediction and the backward prediction. The coefficients of the AR linear prediction model can be calculated using the Burg method and the improved covariance method. Using these coefficients, the effective bandwidth in frequency domain is extended according to the available data [10].

Fig. 16.2 Schematic diagram of one-dimensional extrapolation case

The extended output data can be obtained after forward and backward extrapolation, respectively, as follows:

$$\hat{x}(n) = \begin{cases} -\sum_{i=1}^{p} a_i^* x(n+i), & n < N_1 \\ x(n-N_1), & N_1 < n < N_1 + N \\ -\sum_{i=1}^{p} a_i x(n-i), & N_1 + N < n < N_2 \end{cases}, \tag{16.36}$$

where $\hat{x}(n)$ is the expanded data, a_i is the ith prediction coefficient, p is the order of the prediction filter, N is the measured samples, and N_1 and N_2 are the extrapolated points, respectively, in forward and backward extrapolation.

In the two-dimensional linear prediction model, $E(m,n)$ forecast data exploiting the linear combination of sampling points around it. Using the quarter-plane model, the prediction coefficients of (m,n) point can be predicted by four different m order prediction filters. Figure 16.3 is a schematic diagram of two-dimensional data extrapolation. The extrapolation matrix $\hat{x}(m,n)$ can be divided into several subfields as shown in Fig. 16.3 [11]. Extrapolated sample points are predicted by using the optimal quarter-plane predictive filter in these subdomains.

Region A: The original $M \times N$ data matrix (Middle shaded area in figure Fig. 16.3) is conserved in the center of the 2D-extended data matrix as

$$\hat{x}(m,n) = x(m - M_1, n - N_1), \tag{16.37}$$

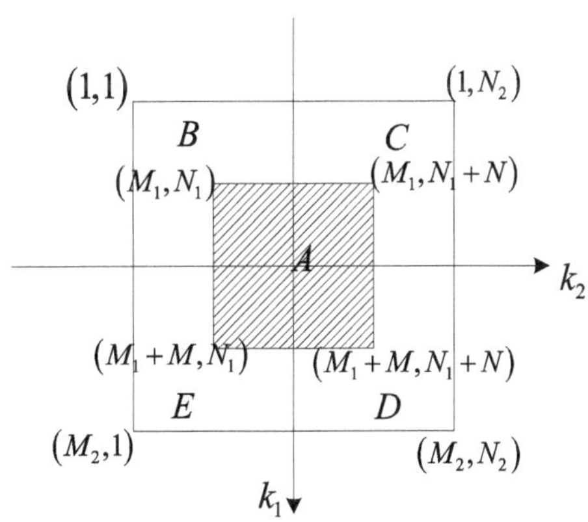

Fig. 16.3 Schematic diagram of two-dimensional data extrapolation

where $N_1 \le n \le N_1 + N$ and $M_1 \le m \le M_1 + M$.

Region B: The prediction obtains the backward prediction on f_x and f_y as

$$\hat{x}(m,n) = \sum_{i=0}^{L} \sum_{\substack{j=0 \\ i=j\neq 0}}^{L} a_{i,j}^{*} x(m+i, n+j), \quad \begin{cases} 1 \le m \le M_1, N_1 \le n \le N_1 + N/2 \\ 1 \le m \le M_1, 1 \le n \le N_1 \\ M_1 \le m \le M_1 + M/2, 1 \le n \le N_1 \end{cases}. \tag{16.38}$$

Region C: The prediction obtains the backward prediction on f_x and forward prediction on f_y as

$$\hat{x}(m,n) = \sum_{i=0}^{L} \sum_{\substack{j=0 \\ i=j\neq 0}}^{L} b_{i,j}^{*} x(m+i, n-j), \quad \begin{cases} 1 \le m \le M_1, N_1 + N/2 \le n \le N_1 + N \\ 1 \le m \le M_1, N_1 + N \le n \le N_2 \\ M_1 \le m \le M_1 + M/2, N_1 + N \le n \le N_2 \end{cases}. \tag{16.39}$$

Region D: The prediction obtains the forward prediction on f_x and on f_y as

$$\hat{x}(m,n) = \sum_{i=0}^{L} \sum_{\substack{j=0 \\ i=j\neq 0}}^{L} a_{i,j} x(m-i, n-j), \quad \begin{cases} M_1 + M \le m \le M_2, N_1 + N/2 \le n \le N_1 + N \\ M_1 + M \le m \le M_2, N_1 + N \le n \le N_2 \\ M_1 + M/2 \le m \le M_1 + M, N_1 + N \le n \le N_2 \end{cases} \tag{16.40}$$

Region E: The prediction obtains the forward prediction on f_x and the backward prediction on f_y

$$\hat{x}(m,n) = \sum_{i=0}^{L} \sum_{\substack{j=0 \\ i=j\neq 0}}^{L} b_{i,j} x(m-i, n+j), \quad \begin{cases} M_1 + M/2 \le m \le M_1 + M, 1 \le n \le N_1 \\ M_1 + M \le m \le M_2, 1 \le n \le N_1 \\ M_1 + M \le m \le M_2, N_1 \le n \le N_1 + N/2 \end{cases} \tag{16.41}$$

where $\hat{x}(m,n)$ is the extrapolated SAR data. $a_{i,j}, a_{i,j}^{*}, b_{i,j}$, and $b_{i,j}^{*}$ are the (i,j)th prediction coefficients of four one quarter-plane models. L is the order of 2D prediction filters. The measured data is $M \times N$ dimension. M_2 and N_2 is the forward and backward extrapolated data.

The extrapolation order may be as Fig. 16.4, which is extrapolated from inside to outside through the lap-to-lap order. The shaded area is the observed data, and the data is extrapolated in the order shown by the arrows. This makes it possible to more accurately predict data using known data as much as possible.

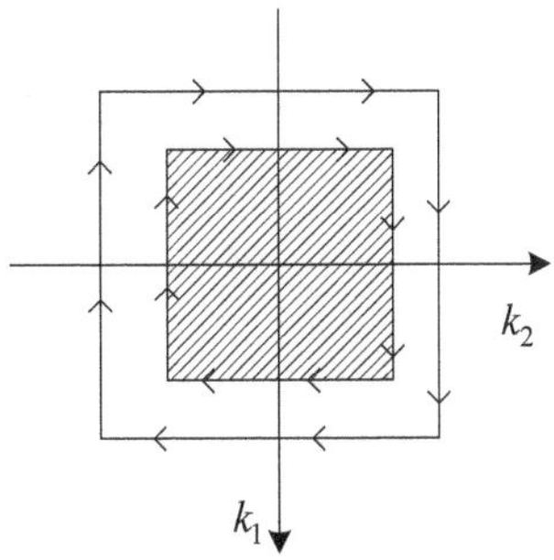

Fig. 16.4 Schematic diagram of spectrum sequence extrapolation

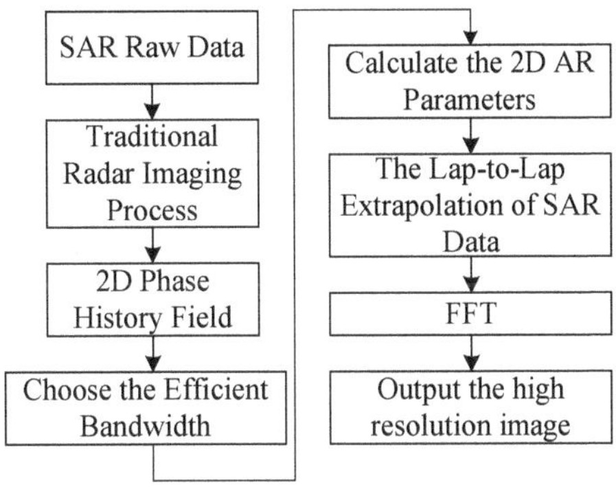

Fig. 16.5 Flow of SAR super-resolution imaging algorithm based on 2D AR linear prediction extrapolation

Then Fig. 16.5 shows the detailed process of the 2D AR linear prediction extrapolation super-resolution SAR imaging algorithm.

Firstly, the conventional data processing such as range migration and motion compensation is performed on the original data. The two-dimensional phase history data is obtained. Then the autoregressive linear prediction extrapolation is performed to expand the effective bandwidth. After the two-dimensional FFT transform, the high-resolution image can be obtained. The specific steps are as follows:

(1) The original echo data is processed by conventional imaging to get SAR super-resolution imaging signal model;
(2) The phase history data of the effective bandwidth is used to calculate the parameters of the two-dimensional AR model;
(3) Spectral extrapolation using (16.37)–(16.41), Fig. 16.3, and Fig. 16.4;
(4) The extrapolated data is transformed to the time domain by FFT, and get the output image.

(3) **Experiment and analysis**

(1) Experiments with simulated data

The simulation of point target echo is used to validation the SAR super-resolution imaging algorithm based on 2D AR linear prediction extrapolation. Assume the chirp signal bandwidth is 150 MHz, the sample frequency is 180 MHz (I/Q), pulse

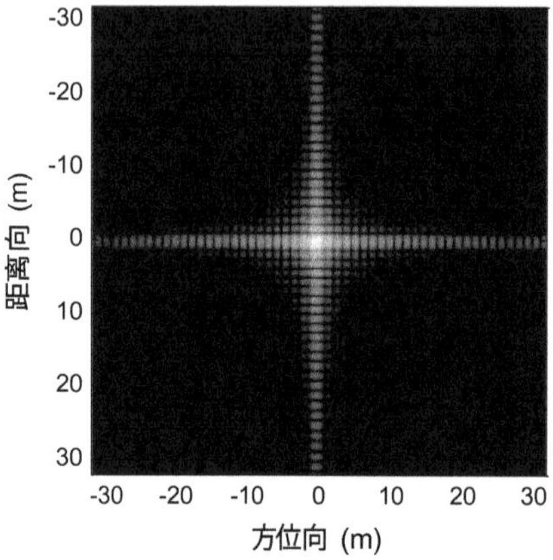

(a) RD Method

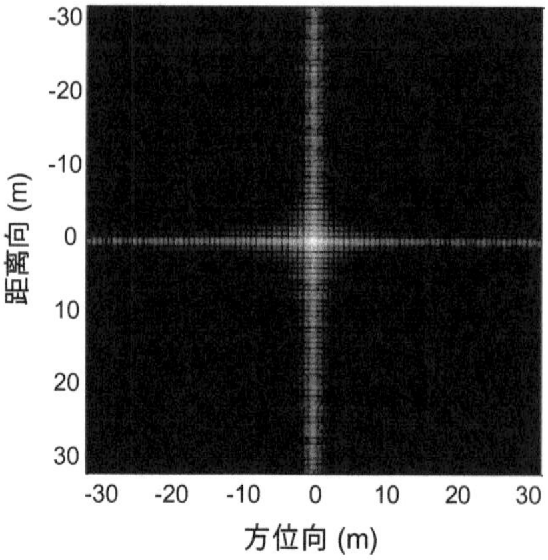

(b) 2D AR linear prediction extrapolation method

Fig. 16.6 The results comparison of two methods

repetition frequency is 1000 Hz, the velocity of aircraft is 250 m/s, the azimuth width of antenna is 2 m, the slant range from radar antenna to the scene center is 18 km, and the signal noise ratio is 7 dB. Figure 16.6 is the results comparison of two methods. (a) is the result of range-Doppler imaging method (RD) and (b) is the result of two times data extrapolation by 2D AR linear prediction extrapolation method. In (a), the main lobe is wider, the resolution is worse. In (b), after two-time extrapolation, the main lobe becomes narrower. Table 16.1 is the indexes comparison of point target performance. From Table 16.1, we can see the resolution improved for almost one time, i.e., the range resolution from the 1.0473 m to the 0.6217 m. However, the PSLR and ISLR have a little decrease.

Figure 16.7 shows the comparison of the two-point-distinguished ability. The radar parameters are the same as above. When the two-point target is closer, the range-Doppler method in figure (a) can not separate the two-point targets. After twice the extrapolated signal data, AR extrapolation method get figure (b), which can be seen two-point target distinguish clearly. So AR extrapolation method can effectively improve the resolution of SAR images.

Table 16.1 The indexes comparison of point target performance

Performance	RD method	2D AR linear prediction extrapolation method
Range resolution (m)	1.0473	0.6217
Range PSLR(dB)	−13.2699	−11.9431
Range ISLR (dB)	−10.8154	−8.9883
Azimuth resolution (m)	1.0652	0.6100
Azimuth PSLR (dB)	−11.8471	−9.0653
Azimuth ISLR (dB)	−10.1703	−6.4878

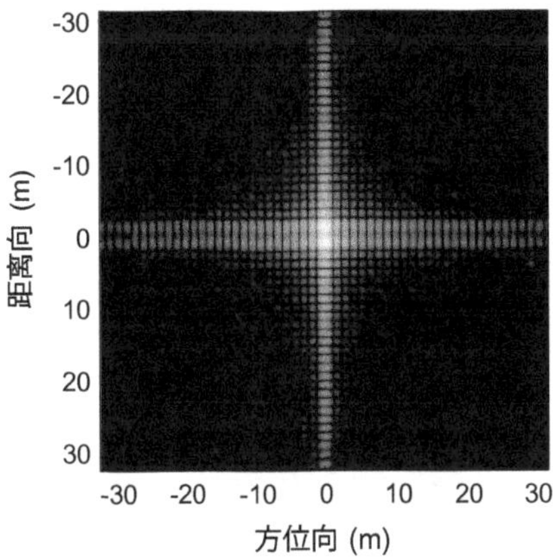

(a) RD Method

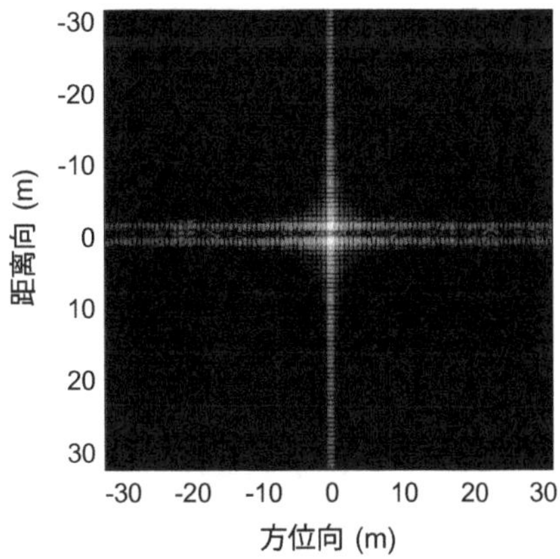

(b) 2D AR linear prediction extrapolation method

Fig. 16.7 Comparison of two-point resolving ability

(2) Experiment with measured data

In order to validate the effectiveness of real measured data, the X band aircraft SAR data is used, of which the resolution is 2.5 m × 2.5 m and the height of aircraft is 10 km. Figure 16.8 is the comparison of processing results. (a) is the result of range-Doppler imaging method and (b) is the result of 2D AR linear prediction extrapolation. The calibration point is selected to analyze the performance quantitatively.

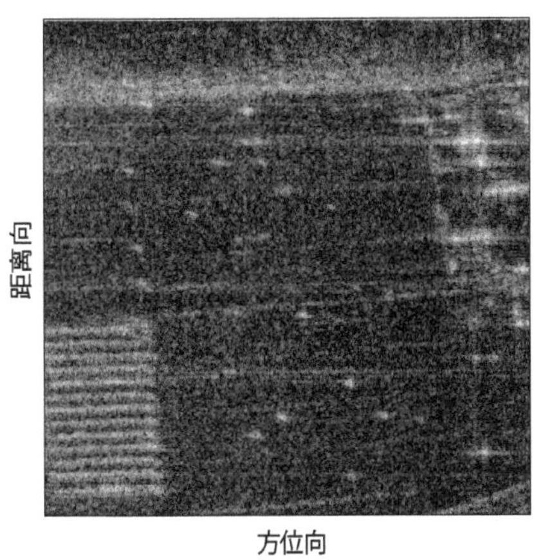

(a) RD Method

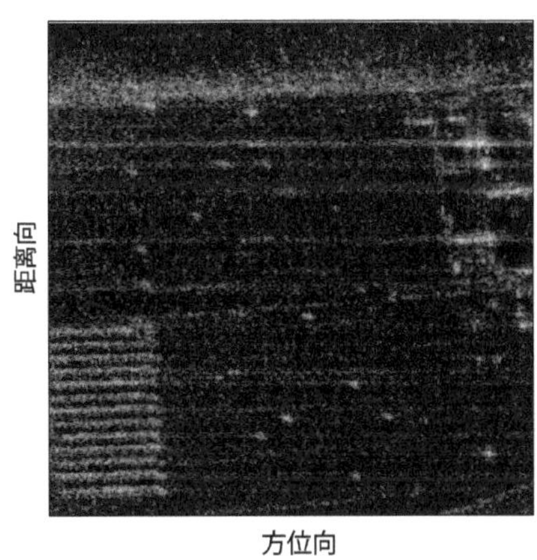

(b) 2D AR linear prediction extrapolation method

Fig. 16.8 Comparison of airborne measured data processing results

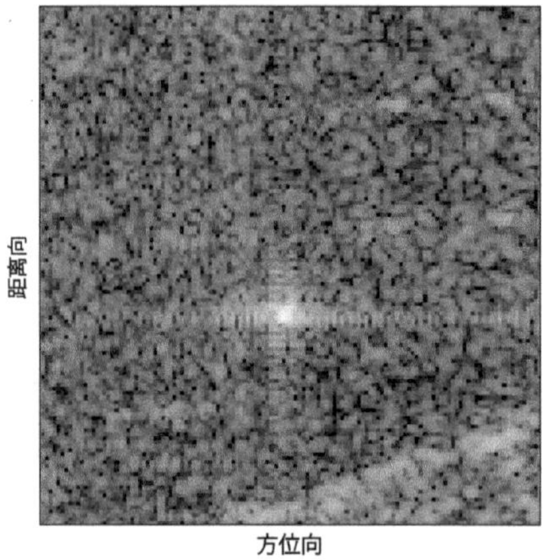

(a) RD Method

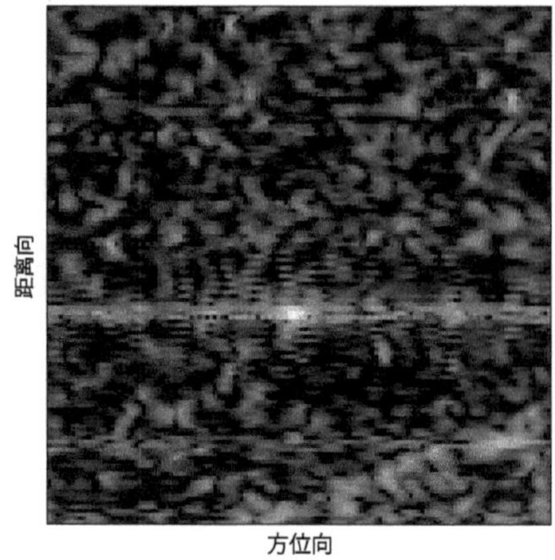

(b) 2D AR linear prediction extrapolation method

Fig. 16.9 Comparison of calibration targets in airborne measured data

Figure 16.9 is the comparison of the calibration point. (a) is the result of RD imaging method and (b) is the result of 2 D AR linear prediction extrapolation. Table 16.2 is the indexes comparison of point target performance. From the figure and the table, it is can be seen the resolution improved obviously. However, for the sake of noise effect, the model is not accurate enough to lose the performance of PLSR and ISLR.

Table 16.2 The indexes comparison of point target performance

Performance	RD method	2D AR linear prediction extrapolation method
Range resolution (m)	2.5886	1.4287
Range PSLR (dB)	−24.5681	−17.8467
Range ISLR (dB)	−16.4489	−15.1185
Azimuth resolution (m)	2.5423	1.5876
Azimuth PSLR (dB)	−11.4356	−8.7206
Azimuth ISLR (dB)	−7.0187	−6.2644

16.1.2 Super-Resolution SAR Imaging Algorithm Based on Weighted Norm Extrapolation

(1) **Super-resolution SAR imaging algorithm based on MVM and weighted norm extrapolation**

Super-resolution SAR imaging Algorithm based on MVM and weighted norm extrapolation (CAWNE) uses minimum weighted norm constraint and minimum variance spectrum estimation for improving synthetic aperture radar (SAR) resolution. Minimum variance method is a robust high-resolution method to estimate spectrum. Based on the theory of SAR imaging, the signal model of SAR imagery is analyzed to be feasible for using data extrapolation methods to improve the resolution of SAR image. The method is used to extrapolate the efficient bandwidth in phase history field [12].

(1) **Methodology**

Let L discrete-time sampling points be a set of data $\{x(m_k)\}_{k=1}^{L}$, where m_k are all integers. In the linear space V, the inner product of the support domain B is defined by using the given frequency functions $Q(f)$, where $B \subset [-0.5, 0.5]$. Suppose for all $f \subset B$, there is $Q(f) > 0$. And for other f, there is $Q(f) = 0$. At this time the inner product is expressed as

$$\langle x(n), y(n) \rangle = \int\limits_{f \in B} \frac{X(f)Y^*(f)}{Q(f)} \mathrm{d}f \tag{16.42}$$

The extrapolated efficient set D is a subset of the signal set in linear space. One of the constraints for $\hat{x}(n)$ selection is the minimum weighted norm extrapolation (MWNE), i.e.,

$$\min_{x(n) \in D} \left\{ \int\limits_{f \in B} \frac{|X(f)|^2}{Q(f)} \mathrm{d}f \right\} = \|\hat{x}(n)\|^2 = \int\limits_{f \in B} \frac{|\hat{X}(f)|^2}{Q(f)} \mathrm{d}f \tag{16.43}$$

The extrapolated data $\hat{x}(n)$ is the best spectral waveform matching $Q(f)$ for $|X(f)|^2$.

The minimum weighted norm extrapolation is expressed in the form of the frequency domain as

$$\hat{X}(f) = Q(f) \sum_{i=1}^{L} b_i e^{-j2\pi f m_i} \tag{16.44}$$

The sequence $q(n)$ obtained by IFFT on $Q(f)$ is the autocorrelation sequence, which can give the approximate solution for $\hat{x}(n)$. The minimum weighted norm extrapolation is in the form of a linear representation as

$$\hat{x}(n) = \sum_{i=1}^{L} b_i q(n - m_i), \tag{16.45}$$

where b_i is the extrapolation coefficient and can be directly solved by the L equation. Then $\hat{x}(n)$ can be an effective extrapolation.

N-point sequence is

$$\mathbf{x} = \begin{bmatrix} x(1) & x(2) & \cdots & x(N) \end{bmatrix}^T,$$

L known sampling points is

$$\mathbf{y} = \begin{bmatrix} x(m_1) & x(m_2) & \cdots & x(m_L) \end{bmatrix}^T.$$

These sampling points can be expressed as a linear transformation

$$\mathbf{y} = \mathbf{Tx}, \tag{16.46}$$

which can be intuitively expressed as

$$\mathbf{y} = \begin{bmatrix} x(m_1) \\ x(m_2) \\ \cdots \\ x(m_L) \end{bmatrix} = \begin{bmatrix} \cdots & 1 & \cdots & \cdots \\ \cdots & \cdots & 1 & \cdots \\ \cdots & \cdots & \cdots & \cdots \\ \cdots & 1 & \cdots & \cdots \end{bmatrix} \begin{bmatrix} x(1) \\ x(2) \\ \cdots \\ x(N) \end{bmatrix}. \tag{16.47}$$

Where in each row of the linear transformation matrix T is a vector that the value at the position m_k is “1” and the value at the other position is “0”. The interest problem is to get all of the $\bar{x}$ using the known $\bar{y}$.

The problem in (16.47) is a Hilbert space inverse problem, which is available as

$$\mathbf{T}^{\dagger} = \mathbf{QT}^{\mathrm{H}}\left(\mathbf{TQT}^{\mathrm{H}}\right)^{-1}. \tag{16.48}$$

Therefore, the minimum norm solution can be expressed as $\hat{\mathbf{x}} = \mathbf{T}^{+}\mathbf{y}$. Q is the weight coefficient matrix as

$$\mathbf{Q} = \begin{bmatrix} a_h(1) & a_h(N-1) & a_h(N-2) & \cdots & a_h(2) \\ a_h(2) & a_h(1) & a_h(N-1) & \cdots & a_h(3) \\ a_h(3) & a_h(2) & \cdots & \cdots & a_h(4) \\ \vdots & \vdots & \cdots & \cdots & \vdots \\ a_h(N) & a_h(N-1) & \cdots & \cdots & a_h(1) \end{bmatrix}.$$

$a_h(k)$ is the IDFT of $|H(k)|^2$. $\mathbf{G} = \mathbf{TQT}^{\mathrm{H}}$ is a sub-matrix of Q, and the G contains elements corresponding to the given sampling point location.

Define $\boldsymbol{w} = \begin{bmatrix} w_1 & w_2 & \cdots & w_L \end{bmatrix}^{\mathrm{T}} = \mathbf{G}^{-1}\boldsymbol{y}$.$\hat{\mathbf{x}}$ can be expressed as

$$\hat{\mathbf{x}} = \mathbf{QT}^{\mathrm{H}}\mathbf{w}. \tag{16.49}$$

This is a linear combination of column vectors about Q. Therefore, the optimal estimate can be expressed as

$$\hat{x}(n) = \sum_{i=1}^{L} w_i a_h(n - m_i), \tag{16.50}$$

which can be converted to frequency domain as

$$\hat{X}(k) = |H(k)|^2 \sum_{i=1}^{L} w_i e^{-j(2\pi/N)m_i k} = |H(k)|^2 W(k). \tag{16.51}$$

$W(k)$ is the DFT transform about the vector that the value is w_i at $\mathrm{n} = m_i$ and the other value is 0. It is clearly seen that $\hat{X}(k)$ and $|H(k)|^2$ have the same frequency support domain.

Taking into account the singularity of the matrix, the solution of the above equation should be regularized as follows:

$$\hat{\mathbf{x}} = \mathbf{QT}^{\mathrm{H}}\left(\mathbf{TQT}^{\mathrm{H}} + \rho\mathbf{I}\right)^{-1}\mathbf{y}. \tag{16.52}$$

Let $\mathbf{G} = \mathbf{TQT}^{\mathrm{H}} + \rho\mathbf{I}$. Parameter ρ is the minimum value that makes the matrix $\mathbf{G}$ invertible, which is generally $a_h(1) \times 10^{-8}$.

Since $\mathbf{T}$ is a linear transformation matrix. $\mathbf{G}$ is a sub-matrix of $\mathbf{Q}$, and the elements of $\mathbf{G}$ are corresponding to the given sampling point position, i.e.,

$$\mathbf{G} = \begin{bmatrix} a_h(m_1 - m_1) & a_h(m_1 - m_2) & \cdots & a_h(m_1 - m_L) \\ a_h(m_2 - m_1) & a_h(m_2 - m_2) & \cdots & a_h(m_2 - m_L) \\ \vdots & \vdots & \ddots & \vdots \\ a_h(m_L - m_1) & a_h(m_L - m_2) & \cdots & a_h(m_L - m_L) \end{bmatrix}.$$

Define $\boldsymbol{w} = \begin{bmatrix} w_1 & w_2 & \cdots & w_L \end{bmatrix}^{\mathrm{T}} = \boldsymbol{G}^{-1}\mathbf{y}$, and (16.52) can be rewritten as

$$\hat{\mathbf{x}} = \mathbf{QT}^{\mathrm{H}}\mathbf{w}. \tag{16.53}$$

So the optimal estimate is

$$\hat{x}(n) = \sum_{i=1}^{L} w_i a_h(n - m_i). \tag{16.54}$$

The corresponding frequency domain solution is

$$\hat{X}(k) = P_{MVM}(k) \sum_{i=1}^{L} w_i e^{-j(2\pi/N)m_i k} = P_{MVM}(k)W(k), \quad (16.55)$$

where $W(k)$ is the DFT transform about the vector that the value is w_i at $\mathrm{n} = m_i$ and the other value is 0.

The procedure steps are summarized as follows:

(1) Calculate minimum variance power spectrum estimation $P_{MVM}(k)$ of $\boldsymbol{y} = [x(m_1) \cdots x(m_L)]^{\mathrm{T}}$.
(2) Calculate the IDFT of $P_{MVM}(k)$, and get the autocorrelation matrix Teoplitz matrix $\boldsymbol{Q}$ to construct the matrix $\mathbf{G}$.
(3) Calculate $\mathbf{w} = (\mathbf{G} + \rho \mathbf{I})^{-1}\mathbf{y}$.
(4) Obtain the N point's extrapolated data by (16.54).

SAR imaging is a two-dimensional data processing. SAR super-resolution processing turns the SAR imaging problem into a two-dimensional complex sinusoidal signal extrapolation problem using signal model. For the two-dimensional case, the SAR signal model in the support domain is

$$x(n,m) = \alpha(\omega_1, \omega_2) e^{j(n\omega_1 + m\omega_2)} + e_{n,m}(\omega_1, \omega_2). \quad (16.56)$$

where $n = 0, 1, 2 \cdots, N-1, m = 0, 1, 2 \cdots, M-1$ and $e_{n,m}(\omega_1, \omega_2)$ is additive noise.

Similar to the one-dimensional case, the optimal estimate is

$$\hat{x}(n,m) = \sum_{i=1}^{L} \sum_{j=1}^{L} w_{i,j} a_h(n - m_i, m - m_j). \quad (16.57)$$

$w_{i,j}$ is the $L \times L$ dimensional extrapolation coefficients and $a_h(n,m)$ is the IDFT of $P_{MVM}(\omega_1, \omega_2)$.

The main steps of the two-dimensional method are as follows:

(1) The two-dimensional spectral estimation $P_{MVM}(\omega_1, \omega_2)$ is obtained according to the signal model of (16.56). $a_h(n,m)$ can be calculated by IDFT of $P_{MVM}(\omega_1, \omega_2)$.
(2) Weighting factors can be obtained by $\mathbf{w} = \mathbf{G}^{-1}\mathbf{y}$, where

$$G = \begin{bmatrix} g(0) & g(-1) & \cdots & g(-L+1) \\ g(1) & g(0) & \cdots & g(-L+2) \\ \vdots & \vdots & \vdots & \vdots \\ g(L-1) & g(L-2) & \cdots & g(0) \end{bmatrix},$$

$$g(k) = \begin{bmatrix} a_h(0,k) & a_h(-1,k) & \cdots & a_h(-L+1,k) \\ a_h(1,k) & a_h(0,k) & \cdots & a_h(-L+2,k) \\ \vdots & \vdots & \vdots & \vdots \\ a_h(L-1,k) & a_h(L-2,k) & \cdots & a_h(0,k) \end{bmatrix}, \text{ and}$$

$$a_h(n,m) = a_h^*(-n,-m).$$

(3) Calculated $\hat{x}(n,m)$ according to (16.57).

Figure 16.10 is the flow of super-resolution SAR imaging algorithm based on MVM and weighted norm extrapolation.

The super-resolution extrapolation imaging method uses the nonparametric MVM spectrum estimation method to determine the power spectrum. Thus, the result of spectrum estimation does not depend on the model, which can make the estimation result more stable. Using the minimum norm constraint, the effective bandwidth is extrapolated in the frequency domain to get the extrapolated image.

The super-resolution SAR imaging algorithm based on MVM and weighted norm extrapolation (CAWNE) uses the MVM spectrum to be the constraint weights of weighted norm extrapolation and extrapolates the phase history data to increase the effective bandwidth and improve the resolution. Adaptive weighted norm extrapolation (AWNE) [13, 14] uses periodogram estimation as initial estimation and extrapolates signal using weighted norm constraints. Since periodogram estimation is imprecise, AWNE needs to iterate estimation for improving the estimation accuracy. However, iteration will also iterate noise and side lobes

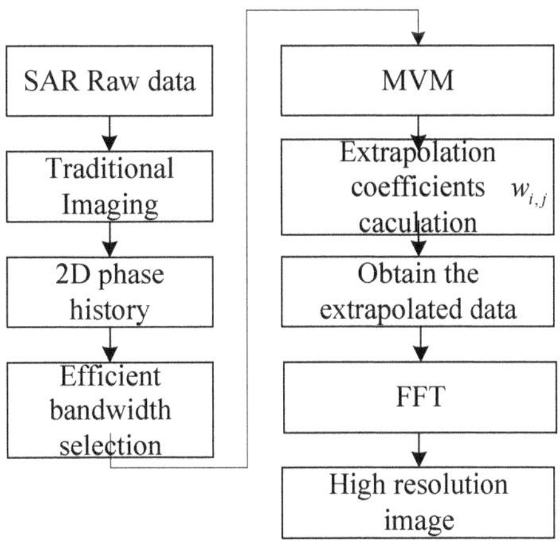

Fig. 16.10 Flow chart of super-resolution extrapolation imaging method

correspondingly, which will result in the loss of targets and the emergence of false peaks [15]. CAWNE method uses a more accurate spectral estimation, so there is no iterative process. Thus, CAWNE can avoid the data distortion caused by the iterations.

(2) **Experiment and analysis**

(1) Experiments with simulated data

Use the ISAR simulated data of mig25 aircraft by V. C. Chen from simulation experiment of United States Naval Research Laboratory. The simulation parameters are as follows: Carrier frequency of transmission signal is 9 GHz. The signal bandwidth is 75 MHz. The data size is 32 × 32. Figure 16.11 is imaging results comparison of B727. Figure (a) is the result of RD imaging method. Figure (b) is three times extrapolation result of the AWNE method. Figure (c) is the result of the B727 extrapolated imaging of the push method CAWNE. In figure (a), the scatters on the plane are mixed together and cannot be distinguished because of high side lobe, which affects the overall image quality. In figure (b), the scatters on the plane can be better distinguished than figure (a). The aircraft profile in figure (c) is most clearly and has the highest resolution among the three results.

(2) Experiment with measured data

The actual measured data of X band airborne SAR is used for validating the method. The radar parameters are as follow: Resolution is 2.5 m × 2.5 m. Pulse repetition frequency is 1000 Hz. The aircraft velocity is 250 m/s. The height of airborne SAR is 10 km. The data size is 40 × 40. Figure 16.12 is the results comparison of the actual measured target data processing. Figure (a) shows the RD method result interpolated to 128 × 128 points. Figure (b) is the result of three times extrapolation using AWNE iterative method and figure (c) is the result of CAWNE method, which is also extrapolated to 128 × 128 points. Compare the three results, it can be seen the resolution of the calibration point increasing one by one. Table 16.3 shows the performance of the calibration points. From the quantitative analysis, the resolution of the CAWNE method is better than that of the AWNE method. Both extrapolated methods are superior to the RD method.

(2) **Iterative minimum weighted norm extrapolation method (IWNE)**

The iterative method computes the new weights so that the weights are more accurate in the iterations. Then the resulting signal is more closely approximated to the original signal. This process utilizes nonparametric spectral estimation $\hat{P}_x(f)$ to improve the norm in frequency domain, which is corresponding to autocorrelation function $\hat{r}_{xx}(n)$.

The spectrum of modified periodogram can help the spectral shape in the iteration keep consistent with that of the previous one, and the details are obtained by the effective extrapolation under the constraint of the sample data. Modified periodogram needs to be windowed in the time domain. Define the filter at the λth iteration.

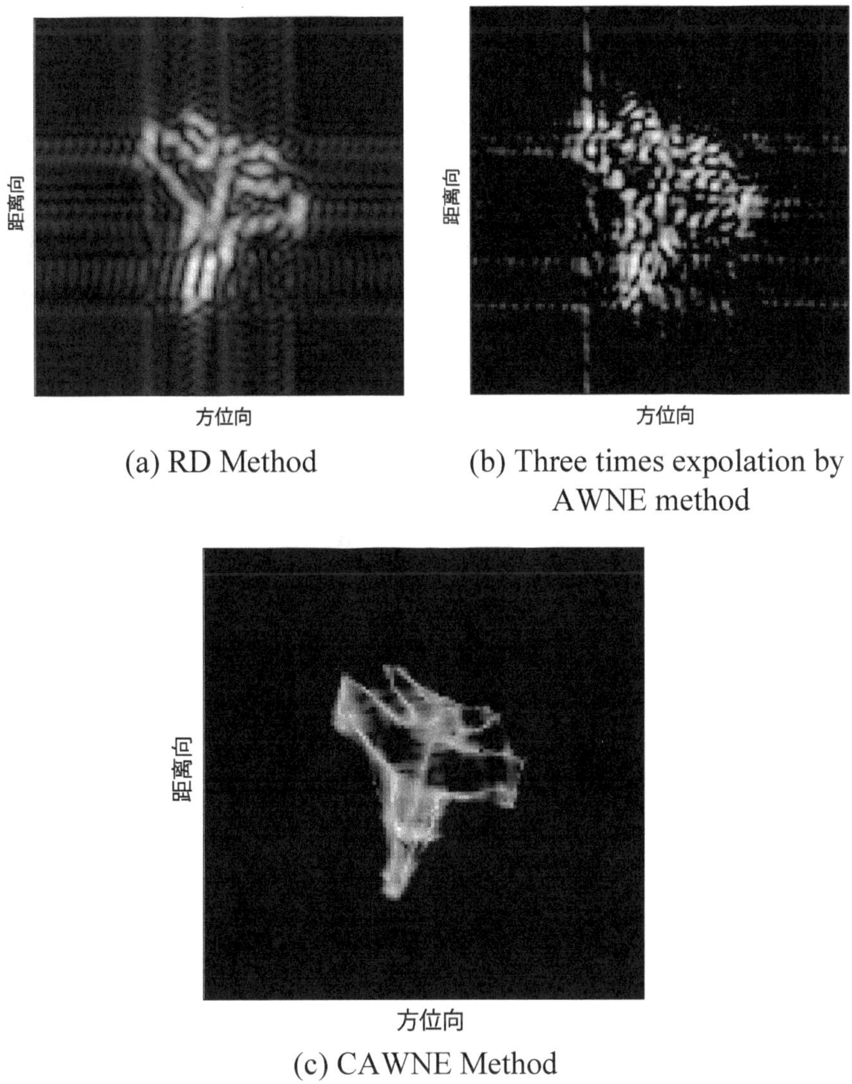

Fig. 16.11 The results comparison of Mig25 data

$$h_{\lambda}(n) = p(n)\hat{x}_{\lambda-1}(n). \tag{16.58}$$

Convert to frequency domain as

$$H_{\lambda}(k) = \frac{P(k) \otimes \hat{X}_{\lambda-1}(k)}{N}. \tag{16.59}$$

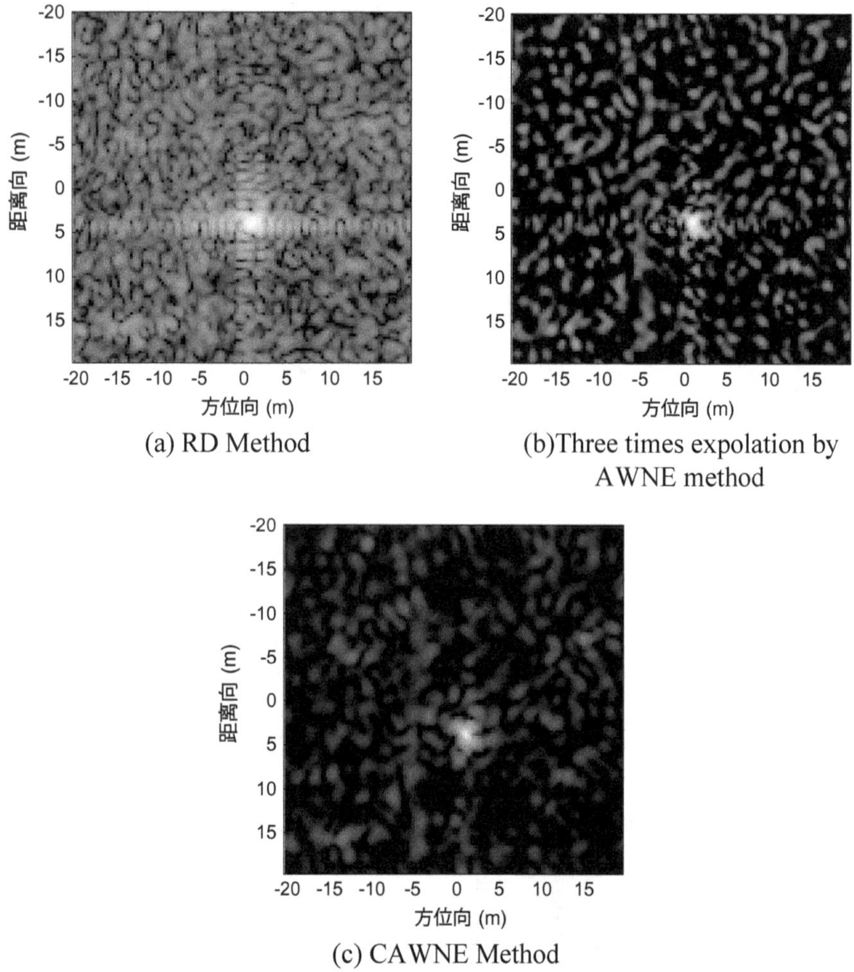

Fig. 16.12 Comparison of measured target data imaging results

Table 16.3 Comparison of point target indexes

Performance comparison	RD method	AWNE method	CAWNE method
Range resolution (m)	2.5453	1.4941	0.8385
PSLR in Range (dB)	−13.8265	−25.1351	−15.1513
ISLR in Range (dB)	−15.3648	−10.9042	−15.9463
Azimuth resolution (m)	2.6184	1.4029	0.7600
PSLR in azimuth (dB)	−12.4227	−15.1043	−18.5147
ISLR in azimuth (dB)	−8.8912	−11.0321	−15.2571

The iterative representation in the frequency domain is

$$\hat{X}_\lambda(k) = \left|\frac{P(k) \otimes \hat{X}_{\lambda-1}(k)}{N}\right|^2 \sum_{i=1}^{L} w_i^{(\lambda)} e^{-j(2\pi/N)m_i k}. \tag{16.60}$$

For each iteration, the estimated $\hat{x}_\lambda(n)$ from the observed samples is an effective extrapolation. The choice of window function $p(n)$ should be as close as possible to the observation data and the time domain sampling data to be estimated.

Use $p_e(n)$ as a standard positive smoothing window. Choose $p(n) = p_e(n - n_c)$ to ensure the center n_c of the window function agreed with the observing data center. The length of $p_e(n)$ determines the smoothness of modified periodogram and the maximum length of extrapolation.

Iterative steps are as follows:

(1) Take modified periodogram of the zero padding data from the observed sampling data as the initial weighting value.
(2) Calculate the autocorrelation matrix.
(3) Constructs G matrix $(\mathrm{G})_{r,c} = a_\lambda(m_r - m_c)$.
(4) Calculate $\mathbf{w}^{(\lambda)} = (\mathbf{G} + \rho_\lambda \mathbf{I})^{-1}\mathbf{y}$.
(5) Extrapolate signal data using $\hat{x}_\lambda(n) = \sum_{i=1}^{L} w_i^{(\lambda)} a_\lambda(n - m_i)$.
(6) Calculate the next iteration filter $h_{\lambda+1}(n) = p(n)\hat{x}_\lambda(n)$. The number of iterations is generally not more than 10 times.

16.1.3 Summary

In this section, the SAR super-resolution spectral extrapolation imaging methods are studied in detail. Autoregressive model parameters estimation methods and linear predictive AR extrapolation method are analyzed. On this basis, a two-dimensional AR linear predictive extrapolation super-resolution imaging method was proposed. Using the SAR signal model and the measured data, the signals statistical characteristics are estimated to represent the data properties both in-band and out-of-band. The parameters of the two-dimensional AR model are calculated using all the bandwidth data. The two-dimensional AR model is used to predict the effective bandwidth Data, and the extrapolation uses the lap-to-lap mode. Use the expanded data to generate high-resolution images. This method preserves the properties of the original effective bandwidth data. The resulting radar images can give the positions and amplitudes of different scatterers

CAWNE method is proposed using MVM spectral estimation and minimum weighted norm constraint. The MVM spectrum estimation is used to be the constraint weight of the weighted norm extrapolation. The phase history data is extrapolated according to the weighted norm constraint to increase the effective bandwidth and improve the resolution. CAWNE method is a kind of nonparametric extrapolation method, which is more robust than parametric extrapolation method. CAWNE method uses more accurate spectral estimation results without iterative process, which can avoid the data distortion caused by noise and side lobes in iteration.

16.2 Super-Resolution Adaptive Weighted Imaging Method

The kind of super-resolution adaptive weighted imaging methods is a type of nonlinear side lobe suppression methods. The weighting network is used to adaptively match the signal and reduces the side lobes without losing the main lobe resolution. Super-resolution adaptive weighted imaging methods include spatially variant apodization (SVA) and adaptive side lobe reduction (ASR) [16].

16.2.1 Spatially Variant Apodization Method

Spatially variant apodization (SVA) method is used to compute the SAR image point-by-point to make the impulse response optimally and the side lobe smallest. Stankwitz first introduced one-dimensional SVA method in 1994. It is an extended of the Double Apodization method (DA). The DA method compares the uniformly weighted image with the any window function weighted image, and selects the smaller amplitude as the output. The SVA is adapting weighted by the raised-cosine function for each output data point (pixel) to minimize the output side lobe response. The calculation of SVA method is simple without assuming signal models. The spatial apodization method is a special case of a single raised-cosine weighting function. The weighting output energy is minimized and the weight function is constrained between the uniform weighting function and the Hanning weighting function.

(1) **Dual apodization (DA)**

Apodization method carries different weight functions on each pixel in the image to achieve the side lobe suppression and the main lobe retention. The double apodization method (DA) [17] is a simple nonlinear method to maintain the main lobe resolution and reduce the side lobes. The uniform-weighted image is compared with the image weighted by any window function. Small amplitude is the output checking point-by-point.

The procedure of DA method is as follows:

(1) Two weights are applied on the image. One is uniform weighting and the other is a weighting that can reduce side lobes, such as Hanning or Hamming.

(2) Each point on the spatial position is calculated, and a smaller value is selected between the twice weighting processes to be the output.

For complex images, the second step may be as follows considering the original phase preserve of the image.

At each spatial point, a complex value with smaller amplitude in the image pair is selected as an output.

The triple apodization method is a high-order extension of the DA method. That is, select three different weighting function process images and detect the minimum pixel value as output.

(2) **Complex dual apodization (CDA)**

The double apodization method uses the image amplitude to make the decision. The phase information of the complex image can be utilized by separating the real and imaginary of the complex image. If the component symbol of uniform weighting is contrary to the component symbol corresponding to Hanning weighting, it means that there must be a set of weighting functions between the uniform weighting and the Hanning weighting to make the component turn into zero. Exploiting the comparison of this symbol transformation, we can only calculate the uniform weighting image and Hanning weighting image to detect the outputs on images equivalently to effectively us the entire response weighting function family.

The steps of complex dual apodization (CDA) [17] method are as follows:

(1) Calculate two images with uniform weighting and Hanning weighting, respectively.
(2) At each spatial point, the output is zero if the real signs in the two images are opposite; otherwise, the smaller absolute value is selected as the output. The process of the imaginary part is the same as the real part.

The complex dual apodization method is used to calculate the output of the image by selecting the weight function from the whole cosine weighting response function family. Therefore, the degree of freedom of CDA method for side lobe suppression is better than DA method. The side lobes of CDA are also significantly lower than that of DA.

16.2.1.1 Experiment analysis

The experiment is carried out using simulated dot targets. The simulation parameters are set as follow: Carry frequency of transmission chirp signal is 9.6 GHz. Signal bandwidth is 150 MHz. Sample frequency is 180 MHz (I/Q). Pulse repetition frequency is 1000 Hz. The aircraft velocity is 220 m/s. The azimuth length of antenna is 2 m. The SNR is 7 dB. Figure Fig. 16.13 is the comparison of point target results. Figure (a) is the imaging result for RD method. Figure (b) is the imaging result for Hanning weighted. And figure (c) is the imaging result for CDA method. Table 16.4 shows the analysis of point target performance.

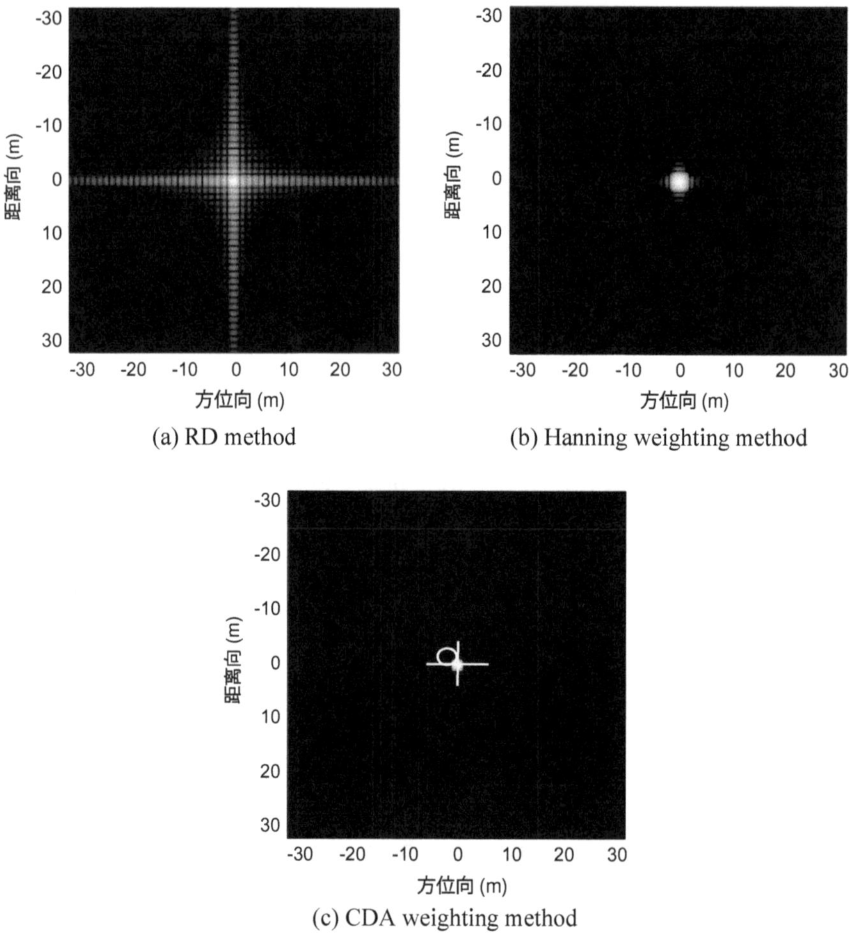

Fig. 16.13 Comparison of point target results

Table 16.4 Performance of point target

Performance comparison	Uniform weighting	Hanning weighting	CDA weighting
Range resolution (m)	1.0544	1.4588	1.0576
PSLR in Range (dB)	−13.2938	−32.3661	−38.5017
ISLR in Range (dB)	−10.2062	−29.6380	−26.8457
Azimuth resolution (m)	1.1044	1.5288	1.1176
PSLR in azimuth (dB)	−13.0938	−32.1661	−38.5415
ISLR in azimuth (dB)	−10.2122	−29.6184	−26.8767

Figure 16.13b shows that the side lobes are effectively suppressed by Hanning weighting, but the main lobe is expended by 1.4 times. The CDA-weighted results can maintain the main lobe resolution and suppress side lobes in figure (c). As shown by the straight line in figure (c), the CDA weighting suppresses the side lobes of the impulse response function along the range and azimuth direction of the point target main lobe. The side lobe reduction performance of CDA method without in the range and azimuth direction of the main lobe is not as effective as that of in the main lobe direction. Figure (c) shows the side lobe suppression effect in range and azimuth direction of the main lobe, which is similar to Hanning weighting. However, the side lobes not in the range and azimuth direction is not suppressed, i.e., the side lobes circled in figure (c).

(3) **Spatially Variant Apodization (SVA)**

StanKwitz and Dallaire [17] proposed that the original image can be processed point-by-point using the spatial apodization method (SVA). The SVA method is a nonlinear filtering method based on the cosine-like frequency domain weighting function. The SVA method adopts a series weighting function on Nyquist sampling images, and selects the minimum value as the filter output, which can effectively suppress the higher side lobes caused by the finite aperture and preserve the spatial resolution of the unweighted image with the clutter texture of original image. The calculation of SVA method is also relatively simple.

The weight function of the cosine class can be expressed as

$$A(n) = 1 + 2w\cos\left(2\pi\frac{n}{N}\right), \tag{16.61}$$

where $0 \leq w \leq 0.5$ and $0 \leq n \leq N$. When $w = 0$, the function corresponds to the uniform weighting. When $w = 0.5$, the function corresponds to the Hanning weighting. N is the number of Nyquist sampled signal points.

The Nyquist samples of impulse response function (IPR) can be obtained by the N points Fourier transform on (16.61) as

$$a(m) = w\delta_{m,-1} + \delta_{m,0} + w\delta_{m,1}, \tag{16.62}$$

where $\delta_{m,n} = \begin{cases} 1, & m = n \\ 0, & m \neq n \end{cases}$. The IPR consists of three nonzero points, which can be used to transform the weighting process to the image domain. Equation (16.62) can be used as the three-point convolution kernel in the image domain.

The SVA method under the condition of Nyquist sampling is divided into two ways, which are real–imaginary jointly processing and real–imaginary part separately processing. The former is an extension of DA method. The latter is simpler and has better side lobe suppression performance.

(1) **One-dimensional SVA**

(1) real–imaginary jointly processing

Let $g(m) = I(m) + iQ(m)$ be the complex sampling point on uniformly weighting Nyquist sampled image. $g(m)$ can be changed to $g'(m)$ using the three-point convolution kernel in (16.62).

$$g'(m) = w(m)g(m-1) + g(m) + w(m)g(m+1) \tag{16.63}$$

With $w(m)$ changing from 0 to 0.5, the weighting function in the frequency domain changes from uniform weighting to Hanning weighting. The middle convolution weight is always 1, which is used to preserve the target main lobe response.

Minimize $|g'(m)|^2$ under the constraint $0 \le w(m) \le 0.5$ to get $w(m)$.

$$\begin{aligned}|g'(m)|^2 &= \{w(m)g(m-1) + g(m) + w(m)g(m+1)\} \\ &\quad \cdot \{w(m)g(m-1) + g(m) + w(m)g(m+1)\}^* \\ &= w^2(m)|g(m-1)|^2 + w(m)g^*(m)g(m-1) \\ &\quad + w(m)g(m-1)g^*(m+1) + w(m)g(m)g^*(m-1) \\ &\quad + w(m)g(m)g^*(m+1) + w^2(m)g(m+1)g^*(m-1) \\ &\quad + |g(m)|^2 + w(m)g^*(m)g(m+1) + w^2(m)|g(m+1)|^2\end{aligned} \tag{16.64}$$

Let the derivative of $|g'(m)|^2$ to be zero.

$$\begin{aligned}\frac{\partial |g'(m)|^2}{\partial w(m)} &= 2w(m)|g(m-1) + g(m+1)|^2 \\ &\quad + 2\mathrm{Re}[g(m) \cdot (g(m-1) + g(m+1))^*] = 0\end{aligned} \tag{16.65}$$

The coefficients of filter can be obtained as

$$\begin{aligned}w_u(m) &= -\mathrm{Re}\left[\frac{g(m)}{g(m-1) + g(m+1)}\right] \\ &= -\frac{\{I(m)[I(m-1) + I(m+1)] - Q(m)[Q(m-1) + Q(m+1)]\}}{[I(m-1) + I(m+1)]^2 + [Q(m-1) + Q(m+1)]^2}.\end{aligned} \tag{16.66}$$

Let $w(m)$ to be in $[0, 0.5]$. And let

$$\begin{cases} w(m) = 0, & w_u(m) \le 0 \\ w(m) = w_u(m), & 0 < w_u(m) < 0.5 \\ w(m) = 0.5, & w_u(m) \ge 0.5 \end{cases}$$

Then the output image is

$$g'(m) = \begin{cases} g(m), & w_u(m) < 0 & \text{(a)} \\ g(m) + w_u(m)[g(m-1) + g(m+1)], & 0 \le w_u(m) \le 0.5 & \text{(b)} \\ g(m) + 0.5 \times [g(m-1) + g(m+1)], & w_u(m) > 0.5 & \text{(c)} \end{cases} \tag{16.67}$$

In (16.67a), when $w_u(m) < 0$, there is $g'(m) = g(m)$. From (16.66), when the interval of phase difference between $g(m)$ and $[g(m-1) + g(m+1)]$ is in $[-\pi/2, \pi/2]$, there is $w_u(m) \le 0$. At this time, the peak of the main lobe in the impulse response can be retained. When the interval of phase difference between $g(m)$ and $[g(m-1) + g(m+1)]$ is more than $\pi/2$, there will appear the condition of (16.67b) and (16.67c). The case of (16.67b) means the phase difference between $g(m)$ and $[g(m-1) + g(m+1)]$ closing to π and $g'(m)$ closing to 0. The side lobes phase change is π from one Nyquist sample to one adjacent Nyquist sample for all side lobes. So the side lobes can be completely suppressed and the side lobe energy is set to zero. However, when the phase difference between $g(m)$ and $[g(m-1) + g(m+1)]$ is less than π, the suppression effect of weight decreases, which is corresponding to (16.67c). When the phase difference is less than $\pi/2$ or more, there is (16.67a), which means the main lobe is kept.

(2) real–imaginary part separately processing

Let $g(m)$ to be the real part or imaginary part of complex sampling point on Nyquist sampled image. Each part will be processed in the same way. After the convolution of (16.62) by the signal point $g(m)$, it can get $g'(m)$ same as (16.63). The optimal weight function $w(m)$ is obtained by minimizing $|g'(m)|^2$ with the constraint condition $0 \le w(m) \le 0.5$.

When there is no constraint on $w(m)$, the optimal solution is

$$w_u(m) = -\frac{g(m)}{g(m+1) + g(m-1)}. \tag{16.68}$$

When $w(m)$ is constrained in $[0, 0.5]$, the filtered output image can be obtained by minimizing I^2 and Q^2, respectively.

$$g'(m) = \begin{cases} g(m), & w_u(m) < 0 & \text{(a)} \\ 0, & 0 \le w_u(m) \le 0.5 & \text{(b)} \\ g(m) + 0.5 \times [g(m-1) + g(m+1)], & w_u(m) > 0.5 & \text{(c)} \end{cases} \tag{16.69}$$

The real part and the imaginary part of the original complex image are, respectively, filtered by the above formula.

Let $y = [g(m-1) + g(m+1)]/2$ to be the mean of $g(m)$ adjacent pixels on. It can use the relationship between $g(m)$ and y to analyze the suppression of SVA method on side lobes. If $g(m)$ is in the main lobe, $g(m)$ should have the same sign with y. At this time, there has $w = 0$, then $g(m)$ should not be weighted to retain the main lobe response. If $g(m)$ is in the side lobe completely, $g(m)$ should have the different sign with y and $|g(m)| < |y|$. At this time, there has $0 < w < 0.5$, then the weighted function is effective to suppress the side lobes completely. The side lobe energy is set to zero. When $g(m)$ has different sign with y and $|g(m)| > |y|$, there has $w = 0.5$. The SVA method can be further simplified as

$$\begin{cases} g'(m) = g(m), & g(m)y \geq 0 \\ g'(m) = 0, & |g(m)| < |y| \\ g'(m) = g(m) + y, & \text{otherwise} \end{cases}. \quad (16.70)$$

The above analysis assumes that the signal is sampled by the Nyquist sampling frequency. When the image is sampled at an integer multiple of the Nyquist sampling rate, i.e., the sampling rate is an integer greater than 1, the above method should make some changes. Equation (16.63) should be as follows:

$$g'(m) = w(m)g(m - R) + g(m) + w(m)g(m + R). \quad (16.71)$$

The remaining steps are the same as described in the previous section.

(2) **2D SVA method**

2D SAR images can be processed sequentially by 1D SVA method, that is, It can do 1D SVA method first in the horizontal direction and then in the vertical direction to achieve the two-dimensional results. However, when the SVA method is used in the vertical direction after horizontal direction processing, the processed sample points are no longer the Nyquist sampled. So the side lobe suppression performance is weakened or even cannot be suppressed.

Two-dimensional simultaneous processing is also divided into real–imaginary jointly processing and real–imaginary part separately processing. In addition, the weighted function can be divided into two-dimensional coupling and two-dimensional uncoupling for the two-dimensional case. Real–imaginary part separately processing can utilize more degrees of freedom to get better side lobe suppression effect. The 2D SVA method in the case of the real–imaginary part separately processing is described below.

2D-weighted coefficient matrix can be expressed as follows:

$$\begin{pmatrix} w_m w_n & w_n & w_m w_n \\ w_m & 1 & w_m \\ w_m w_n & w_n & w_m w_n \end{pmatrix},$$

where w_m and w_n are the weighted factors of range and azimuth direction. w_m and w_n are correlated with (m, n), which means the weights are varying adaptively.

Let

$$\begin{cases} Q_m = g(m-1,n) + g(m+1,n) \\ Q_n = g(m,n-1) + g(m,n+1) \\ P = g(m-1,n-1) + g(m+1,n+1) + g(m-1,n+1) + g(m+1,n-1) \end{cases} \tag{16.72}$$

After convolution of $g(m,n)$ and the weights, the SVA image sample $g'(m,n)$ will be

$$g'(m,n) = g(m,n) + w_m w_n P + w_m Q_m + w_n Q_n. \tag{16.73}$$

Minimizes $|g'(m,n)|^2$ to find w_m and w_n subject to the constraint $0 \le \{w_m, w_n\} \le 0.5$.

(1) real–imaginary part separately weighted function two-dimensional coupling processing

When weights is coupled, $w_m = w_n = w.g(m,n)$ is real part or imaginary part. While (16.73) is

$$g'(m,n) = g(m,n) + w^2 P + wQ, \tag{16.74}$$

And $Q = Q_m + Q_n$. Minimize $|g'(m,n)|^2$ and there is

$$\left(g(m,n) + w^2 P + wQ\right)(2wP + Q) = 0. \tag{16.75}$$

If $Q^2 - 4Pg(m,n) \ge 0, g'(m,n) = g(m,n) + w_0^2 P + w_0 Q = 0$ should has a solution as

$$w_0 = \frac{-Q \pm \sqrt{Q^2 - 4Pg(m,n)}}{2P}.$$

If $Q^2 - 4Pg(m,n) < 0$, it means that the first term is not zero when $0 \le \{w_m, w_n\} \le 0.5$. And the second term can get the extreme point as

$$w_e = -\frac{Q}{2P}.$$

In summary, the output image is

$$g'(m,n) = \begin{cases} 0, & Q^2 - 4Pg(m,n) \ge 0, 0 \le w_0 \le 0.5 & \text{(a)} \\ g(m,n) - Q^2/4P, & Q^2 - 4Pg(m,n) < 0, 0 \le w_e \le 0.5 & \text{(b)} \\ g(m,n) + w^2 P + wQ, & \text{otherwise } w = \min\limits_{w=\{0,0.5,w_e\}} \left\{\left|g(m,n) + w^2 P + wQ\right|\right\} & \text{(c)} \end{cases} \tag{16.76}$$

(2) real–imaginary part separately weighted function two-dimensional uncoupling processing

When w_m is different with w_n, (16.63) can be rewritten as

$$g'(m,n) = g(m,n) + (w_n P + Q_m)w_m + w_n Q_n.$$

When w_n is given, the above formula is the linear function of w_m. And when w_m is given, the above formula is the linear function of w_n. When $(w_m, w_n) \subset [0, 0.5] \times [0, 0.5]$, the maximum and minimum should be on the four corner of the weights range, i.e., $(w_m, w_n) \in \{(0,0), (0,0.5), (0.5,0), (0.5,0.5)\}$.$g'(m,n)$ is monotonic to w_n and w_m. When $(w_m, w_n) = \{(0,0), (0,0.5), (0.5,0), (0.5,0.5)\}$, if the four values of $g'(m,n)$ are opposite sign, there is a (w_m, w_n) make $g'(m,n)$ zero, i.e., the energy of side lobe can be set zero. Thus, the output image would be

$$g'(m,n) = \begin{cases} 0 & , g'(0,0.5) \times g'(0.5,0) < 0 \\ & \text{or} \quad g'(0.5,0.5) \times g'(0.5,0) < 0 \\ & \text{or} \quad g'(0,0.5) \times g'(0.5,0.5) < 0 \\ g(m,n) + w_m w_n P + w_m Q_m + w_m Q, & \text{otherwise } (w_m, w_n) = \min\{|g'(m,n)|\}, \\ & \text{while } (w_m, w_n) = \{(0,0), (0.5,0), (0,0.5), (0.5,0.5)\} \end{cases} \tag{16.77}$$

The above analysis assumes that the signal is sampled by the Nyquist sampling frequency. When the image is sampled at an integer multiple of the Nyquist sampling rate, i.e., the sampling rate is an integer greater than 1, the above method should make some changes. Equation (16.72) should be as follows:

$$\begin{cases} Q_m = g(m - R1, n) + g(m + R1, n) \\ Q_n = g(m, n - R2) + g(m, n + R2) \\ P = g(m - R1, n - R2) + g(m + R1, n + R2) \\ \quad + g(m - R1, n + R2) + g(m + R1, n - R2) \end{cases} \tag{16.78}$$

The remaining steps are the same as described in the previous section.

(3) **Experiment and analysis**

The experiment is carried out using simulated dot targets. The simulation parameters are set as follow: Carry frequency of transmission chirp signal is 9.6 GHz. Signal bandwidth is 150 MHz. Sample frequency is 180 MHz (I/Q). Pulse repetition frequency is 1000 Hz. The aircraft velocity is 200 m/s. The azimuth length of antenna is 2 m. The SNR is 7 dB. Figure 16.14 is the comparison of point target results. Figure (a) is the imaging result for RD method. Figure (b) is the imaging result for Hanning weighted. And figure (c) is the imaging result for SVA method. Table 16.5 shows the analysis of point target performance.

From the figure, the side lobes of Hamming weighting are well restrained in figure (b), but the main lobe is broadened about 1.3 times. The results of SVA method in figure (c) basically keeps the main lobe resolution of uniform weighted, and reduces the side lobes to a certain extent.

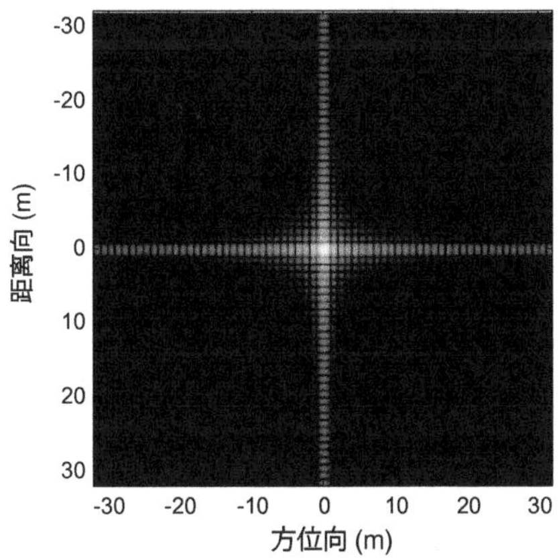

(a) RD method

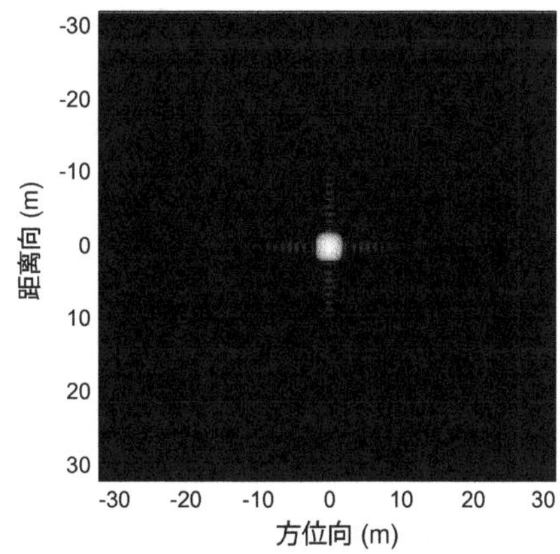

(b) Hamming weighted method

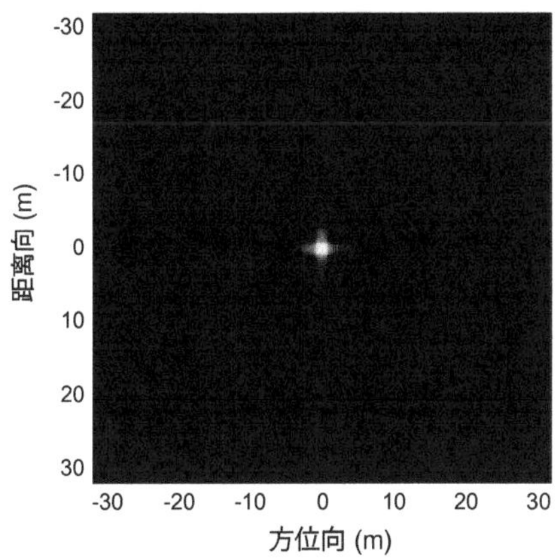

(c) SVA method

Fig. 16.14 Comparison of point target results

Table 16.5 Performance of point target

Performance comparison	Uniform weighting	Hanning weighting	SVA method
Range resolution (m)	1.0544	1.3805	1.0234
PSLR in Range (dB)	−13.2938	−41.5563	−39.0189
ISLR in Range (dB)	−10.2062	−30.9145	−33.3394
Azimuth resolution (m)	1.1044	1.4880	1.0968
PSLR in azimuth (dB)	−13.0938	−43.5033	−34.3931
ISLR in azimuth (dB)	−10.2122	−25.7526	−27.0051

16.2.2 Adaptive Side Lobe Reduction Method

(1) **One-dimensional adaptive side lobe reduction method**

The SVA method is a first-order adaptive weighted side lobe suppression method (ASR). By extending its weight function to higher order, a more general adaptive weight function can be obtained as follows [16]:

$$W(n,k) = 1 + \sum_{m=1}^{M} w(n,m)\cos\left(\frac{2\pi mk}{K}\right), \tag{16.79}$$

where $n = 0,\cdots,N-1$ and $k = 0,\cdots,K-1$. $w(n,m)$ is the weight coefficient. M is the order of the weight coefficient. K is the number of sampling points within the signal bandwidth. N is the number of oversampling points. The above expression indicates that each time domain data point corresponds to a different weighting function, i.e., the weights are adaptively changed.

The signal is weighted in the frequency domain as

$$\hat{G}(n,k) = G(k)W(n,k). \tag{16.80}$$

Convert to time domain as

$$\hat{g}(n) = \sum_{k=0}^{K-1} \hat{G}(n,k)e^{j2\pi\frac{nk}{N}} = g(n) + \sum_{m=1}^{M} w(n,m)\frac{g\left(n-\frac{N}{K}m\right)+g\left(n+\frac{N}{K}m\right)}{2}. \tag{16.81}$$

Let $\gamma = \frac{N}{K}$ for oversampling factors and (16.81) turns to be

$$\hat{g}(n) = g(n) + \sum_{m=1}^{M} w(n,m)\frac{g(n-\gamma m)+g(n+\gamma m)}{2}. \tag{16.82}$$

The above equation is a representation weighted in time domain. It can be seen that the adaptive side lobe reduction method determines whether the current point is a side lobe based on the point around it and gives weight suppression. Next, minimize the output energy of the filter to calculate the optimal weight function. When the oversampling factor γ is noninteger, the signal should be sampled up to an integer oversampling multiple to get the more efficient weight function.

The derivation of the optimal weight function is given below. For the M order weighted function, the weight coefficient is expressed as

$$\mathbf{w}(n) = \begin{pmatrix} w(n,1) \\ w(n,2) \\ \vdots \\ w(n,M) \end{pmatrix}.$$

The data vector consisting of the observed data is expressed as

$$\mathbf{y}(n) = \begin{pmatrix} \frac{g(n-\gamma)+g(n+\gamma)}{2} \\ \vdots \\ \frac{g(n-\gamma M)+g(n+\gamma M)}{2} \end{pmatrix}.$$

Equation (16.82) can be rewritten as a vector form

$$\hat{g}(n) = g.(n) + \mathbf{w}^{\mathrm{T}}(n)\mathbf{y}(n). \tag{16.83}$$

In order to increase the freedom degree of the weight function, the real part and the imaginary part of the complex image are dealt separately. Let $\hat{g}(n) = \hat{g}_r(n) + j\hat{g}_i(n)$ and $\mathbf{y}(n) = \mathbf{y}_r(n) + j\mathbf{y}_i(n)$. Substitute them into (16.83) and obtain the output of the filter as

$$\begin{cases} \hat{g}_r(n) = g_r(n) + \mathbf{w}_1^{\mathrm{T}}(n)\mathbf{y}_r(n) \\ \hat{g}_i(n) = g_i(n) + \mathbf{w}_2^{\mathrm{T}}(n)\mathbf{y}_i(n) \end{cases}. \tag{16.84}$$

The output energy of the filter is minimized, i.e., minimizing $|\hat{g}_r(n)|^2$ and $|\hat{g}_i(n)|^2$. The weight coefficient is obtained as follows:

$$\begin{cases} \mathbf{w}_1^{\mathrm{T}}(n) = -\frac{g_r(n)}{\mathbf{y}_r^{\mathrm{T}}(n)\mathbf{y}_r(n)} \\ \mathbf{w}_2^{\mathrm{T}}(n) = -\frac{g_i(n)}{\mathbf{y}_i^{\mathrm{T}}(n)\mathbf{y}_i(n)} \end{cases}. \tag{16.85}$$

The optimal weight will be obtained at this time as (16.85). Substitute (16.85) into (16.84). The result is zero. In order to obtain a valid output, the weight function should be bound. To use the l_1 norm and l_2 norm constraint weight functions.

$$\sum_{m=1}^{M} |w(n,m)| \le c_1 \le 1 \tag{16.86}$$

$$\sqrt{\sum_{m=1}^{M} |w(n,m)|^2} \le c_2 \le 1 \tag{16.87}$$

The parameters c_1 in the l_1 norm and the c_2 parameters in the l_2 norm can be adjusted to be in the range of 0.8–1 generally. The l_1 norm constraint ensures that the function value of the corresponding weighting function is greater than or equal to 0. This means to ensure the positive definite property of the weight function. There are two reasons for this. First, the shock component in the weighted function can be brought limitedly to the results, and the other energy on the frequency band can be prevented to the passband inducing by the weight modulation, which can result in the desired signal reduced. Second, the designed weight function can be consistent with the fixed window function design. The l_2 norm constraint is a relatively weak constraint with the aim of suppressing interference to a greater degree.

16.2.2.1 Two-Dimensional Adaptive Side Lobe Reduction Method

The one-dimensional method is extended to two dimensions to construct a two-dimensional adaptive weight function as

$$\begin{gathered} W(n_r,n_a,k_r,k_a) = 1 + \sum_{m_r=1}^{M}\sum_{m_a=1}^{M} w(n_r,n_a,m_r,m_a)\cos\left(\frac{2\pi m_r k_r}{k_r}\right)\cos\left(\frac{2\pi m_a k_a}{k_a}\right) \\ n_r = 0,\cdots,N_r-1, n_a = 0,\cdots,N_a-1 \\ k_r = 0,\cdots,K_r-1\,,\ k_r = 0,\cdots,K_r-1. \end{gathered} \tag{16.88}$$

where $w(n_r,n_a,m_r,m_a)$ is the weight coefficient changing along two dimensional, M is the order of weight function. The K_r and K_a are the number of sampling points within the bandwidth along range and azimuth. N_r and N_a are the number of oversampling points within the bandwidth along range and azimuth.

The weighted signal in the frequency domain can be expressed as

$$\hat{G}(n_r,n_a,k_r,k_a) = G(k_r,k_a)W(n_r,n_a,k_r,k_a). \tag{16.89}$$

The above equation is converted into time domain as

$$\begin{aligned} \hat{g}(n_r,n_a) &= \sum_{k_r=0}^{K_r-1}\sum_{k_a=0}^{K_a-1}\hat{G}(n_r,n_a,k_r,k_a)e^{j2\pi\frac{nk_r}{N_r}}e^{j2\pi\frac{nk_a}{N_a}} \\ &= g(n_r,n_a) + \sum_{m_r=1}^{M}\sum_{m_a=1}^{M} w(n_r,n_a,m_r,m_a)S(n_r,n_a,m_r,m_a), \end{aligned} \tag{16.90}$$

where

$$\begin{aligned} S(n_r,n_a,m_r,m_a) = \frac{1}{4}\Bigg[& g\left(n_r+\frac{N_r}{K_r}m_r, n_a+\frac{N_a}{K_a}m_a\right) + g\left(n_r+\frac{N_r}{K_r}m_r, n_a-\frac{N_a}{K_a}m_a\right) \\ & + g\left(n_r-\frac{N_r}{K_r}m_r, n_a+\frac{N_a}{K_a}m_a\right) + g\left(n_r-\frac{N_r}{K_r}m_r, n_a-\frac{N_a}{K_a}m_a\right)\Bigg]. \end{aligned}$$

Let $\gamma_r = \frac{N_r}{K_r}$ and $\gamma_a = \frac{N_a}{K_a}$ to be the oversampling factors along range and azimuth separately. The above formula can be rewritten as

$$S(n_r, n_a, m_r, m_a) = \frac{1}{4}[g(n_r + \gamma_r m_r, n_a + \gamma_a m_a) + g(n_r + \gamma_r m_r, n_a - \gamma_a m_a) + g(n_r - \gamma_r m_r, n_a + \gamma_a m_a) + g(n_r - \gamma_r m_r, n_a - \gamma_a m_a)].$$

Let M order of two-dimensional function as follows:

$$\mathbf{w}(n_r, n_a) = \begin{pmatrix} w(n_r, n_a, 1, 1) & & w(n_r, n_a, 1, M) \\ w(n_r, n_a, 2, 1) & \ddots & w(n_r, n_a, 2, M) \\ \vdots & \ddots & \vdots \\ w(n_r, n_a, M, 1) & & w(n_r, n_a, M, M) \end{pmatrix},$$

The data vector consisting of the observed data is expressed as

$$\mathbf{y}(n_r, n_a) = \begin{pmatrix} S(n_r, n_a, 1, 1) & & S(n_r, n_a, 1, M) \\ \vdots & \ddots & \vdots \\ S(n_r, n_a, M, 1) & & S(n_r, n_a, M, M) \end{pmatrix}.$$

Reshape $\mathbf{w}(n_r, n_a)$ and $\mathbf{y}(n_r, n_a)$ to be the column vector form. (16.90) can be rewritten as

$$\hat{g}(n_r, n_a) = g(n_r, n_a) + \mathbf{w}^{\mathrm{T}}(n_r, n_a)\mathbf{y}(n_r, n_a). \tag{16.91}$$

Calculate the optimal weight function according to one-dimensional processing and consider the l_1 norm and l_2 norm constraints.

16.2.2.2 Experiment and Analysis

The experiment is carried out using simulated dot targets, and the quantitative analysis of the ASR method is given. The simulation parameters are set as follow: Carry frequency of transmission chirp signal is 9.6 GHz. Signal bandwidth is 150 MHz. Sample frequency is 180 MHz (I/Q). Pulse repetition frequency is 1000 Hz. The aircraft velocity is 200 m/s. The azimuth length of antenna is 2 m. The SNR is 7 dB.

Figure 16.15 is the comparison of point target results. Figure (a) is the imaging result for RD method. Figure (b) is the imaging result for Hamming weighted. And figure (c) is the imaging result for ASR method. Table 16.6 shows the analysis of point target performance. The Hamming weighted side lobes in figure (b) are well suppressed, but the main lobe is about 1.3 times wider as that of the index analysis. The result of the ASR method in figure (c) is basically maintained by the uniform-weighted principal lobe resolution, but also inhibited the side lobes to a certain extent.

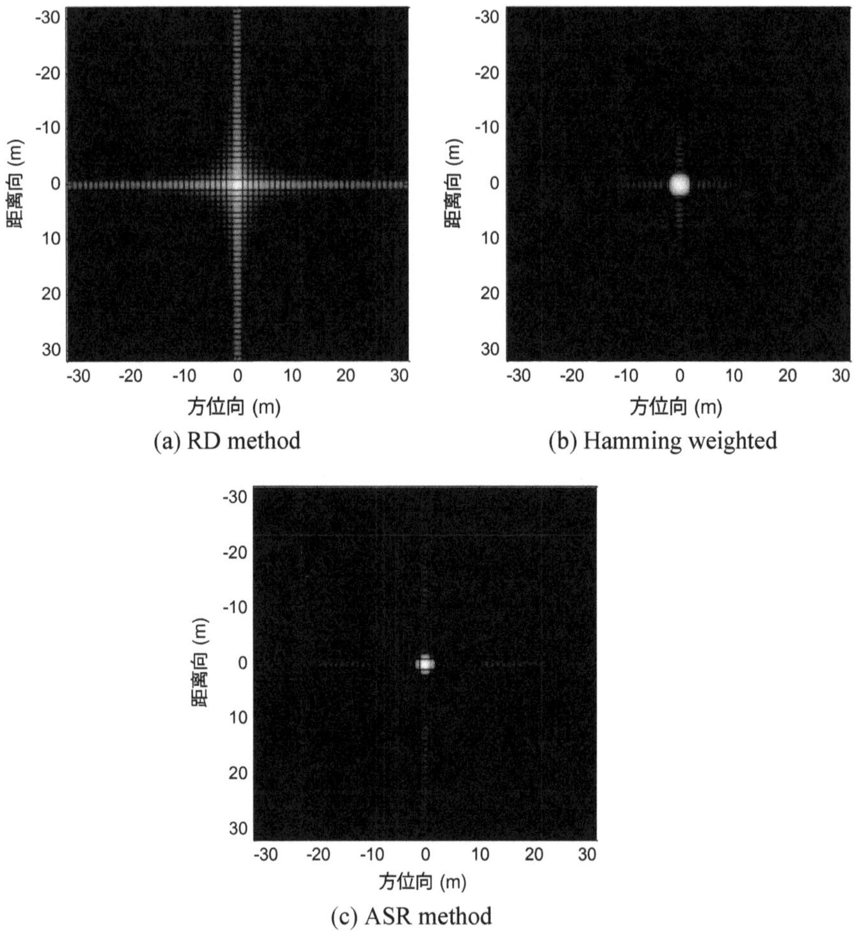

Fig. 16.15 The comparison of point target results

Table 16.6 The analysis of point target performance

Performance comparison	Uniform weighting	Hanning weighting	SVA method
Range resolution (m)	1.0544	1.3805	1.0002
PSLR in Range (dB)	−13.2938	−41.5563	−41.2624
ISLR in Range (dB)	−10.2062	−30.9145	−22.6128
Azimuth resolution (m)	1.1044	1.4880	1.0981
PSLR in azimuth (dB)	−13.0938	−43.5033	−40.9581
ISLR in azimuth (dB)	−10.2122	−25.7526	−22.0415

16.2.3 Comparison of SVA Method and ASR Method

The following results are using the real radar data to compare the SVA method and the ASR method. In Fig. 16.16, figure (a) shows the image without any weighting, which is the image data belongs to Changping area of Beijing. The range resolution of the image is 2.43 m, and the azimuth resolution is 3 m. Figure (b) is the image after Hanning weighted. And figure (c) shows the image obtained by SVA method. Although there is no very strong point target and its side lobes in the unweight image, it can be also seen from the contrast of the three images that the weighted image quality has some improvement over the unweighted image. After Hanning weighting, the target edge features in the image are fuzzy. This is because the application of fixed Hanning weight functions on the data, which is equivalent to a low-pass filter on the image. The gray value of output image pixel is the weighted average of the adjacent pixel. The side lobe suppression of Hanning weighting is achieved by reducing the image contrast and Edge features. The SVA method preserves the edge features of the target in the original image and preserves the clutter distribution.

Figures 16.17 and 16.18 use the L-SAR measurement data to further compare the side lobe suppression effects of the SVA method and the ASR method. In the ASR method, the order of the weight function is 8, and the l_1 norm and l_2 norm are both 0.5.

From the processing results of the actual SAR images in Figs. 16.17 and 16.18, it can be seen that both the SVA method and the ASR method suppress the side lobes of the strong targets in the unweighted image, and improve the target clutter

(a) no weighting (b) Hanning weighting (c) SVA method

Fig. 16.16 The comparison of different methods

(a) No weighting (b) SVA method (c) ASR method

Fig. 16.17 The comparison of different methods

(a) No weighting (b) SVA method (c) ASR method

Fig. 16.18 The comparison of different methods

ratio as well as image quality. However, in the image processed by the ASR method, there are some distortions in the image of the distribution target, which is caused by the point target model adopted by the ASR method. And the ASR method changes the clutter distribution of unweighted image. SVA method has little effect on clutter distribution of unweighted images.

16.2.4 SAR Super-Resolution Imaging Algorithm Based on Adaptive Weighted Methods

The methods described in the previous sections can suppress side lobes without reducing the resolution. The two methods proposed in this section can improve the resolution and suppress side lobes. Using the signal spectrum spreading effect after adaptive weighting processing, the new signal spectrum is processed by inverse weighting to balance the signal spectrum and improve image resolution.

Set Δf_r to be the radar transmission signal bandwidth for the linear FM signal. After pulse compression, the received echo data along range can be expressed as

$$g(t) = \mathrm{Sinc}(\Delta f_r \cdot t). \tag{16.92}$$

Turn to the frequency domain as

$$G(f) = \frac{1}{\Delta f_r}\mathrm{Rect}\left(\frac{f}{\Delta f_r}\right), \tag{16.93}$$

where $\mathrm{Rect}(u) = \begin{cases} 1, & |u| \leq 0.5 \\ 0, & |u| > 0.5 \end{cases}$. The above expression indicates that the support area of signal in the frequency domain is $\left[-\frac{\Delta f_r}{2}, \frac{\Delta f_r}{2}\right]$.

In the ideal case, the side lobe is completely suppressed after the adaptive weighting processing, and the side lobe energy is set to 0 to leave only the signal main lobe part. At that time, the signal form can be expressed as

$$g'(t) = g(t) \times \mathrm{Rect}(\Delta f_r \cdot t). \tag{16.94}$$

The above equation is transformed into the frequency domain as

$$G'(f) = \frac{2}{\Delta f_r} \cdot \left[\mathrm{Rect}\left(\frac{f}{\Delta f_r}\right) \otimes \mathrm{Sinc}\left(\frac{2f}{\Delta f_r}\right)\right]. \tag{16.95}$$

Compare (16.93) and (16.95). (16.93) Is the signal spectrum, the effective interval of which is the signal bandwidth. (16.95) Is the signal spectrum after the adaptive weighted processing, which is composed by the convolution of signal spectrum and a $\mathrm{Sinc}(\cdot)$ function with the same main lobe and signal bandwidth as it. The convolution induces the signal spectrum broadened and the effective interval of the signal spectrum expanded. The two super-resolution imaging methods proposed here take advantage of this bandwidth expansion effect to improve the resolution. Since the adaptive weighting methods belong to the nonlinear operation, the signal after the nonlinear operation is no longer the band-limited signal. Thus, the abovementioned signal bandwidth extension comes out. The principle of azimuth processing is the same.

(1) **SAR super-resolution imaging algorithm based on spatially variant apodization [18] spectrum extrapolation**

(1) **Methodology**

The SVA method keeps only the main lobe. It is equivalent to make the window on the signal in the time domain, which means to convolution $\mathrm{Sinc}(\cdot)$ function in the frequency domain. Convolution extends the signal spectrum width. SAR super-resolution imaging algorithm based on spatially variant apodization uses this bandwidth extension effect to improve the resolution. The SVA method broadens the signal spectrum. Since the $\mathrm{Sinc}(\cdot)$ function is equivalent to make amplitude modulation on the original signal spectrum during the convolution process, the new signal spectrum should be made an inverse weight to restore the ideal signal spectrum and eliminate the amplitude modulation Impact. The inverse function uses the cosine-type inverse [18], which can be adjusted appropriately according to the bandwidth extension requirements and make it more suitable for the spectral function of the ideal target.

After the above processing, the signal sampling rate is no longer an integer multiple Nyquist sampling rate. Then the noninteger times Nyquist sampling SVA method is used for side lobe suppression of extrapolated signal to obtain the final output signal. Noninteger multiple sampling rate is commonly 1.2 times the original sampling rate.

The flow diagram of the spatial apodization bandwidth extrapolation method is shown in Fig. 16.19. First, the SVA method is used on the image after Fourier imaging to suppress side lobe. The effective bandwidth in the corresponding frequency domain is extended by cosine inverse weighting. Finally, the noninteger times Nyquist sampling SVA process is made on the image with extended bandwidth to obtain high-resolution images. The specific method steps are as follows:

(1) Radar echo data is imaged by traditional Fourier imaging algorithm to obtain Single Look Complex Image (SLC) image.
(2) The SVA method is used on the SLC image to suppress side lobes. If the image sampling rate is an integer multiple of Nyquist sampling rate, the method described in 16.2.1.3 can be used directly. If the image sampling rate is not an integer multiple of Nyquist sampling rate, the image is sampled up to an integer multiple of the Nyquist sampling rate, and uses SVA operation.
(3) The image after the SVA operation is transformed into the two-dimensional frequency domain. The Hamming inverse weight is equalized to obtain the effective-extended spectrum. And the spectrum is transformed into the complex image domain.
(4) Noninteger multiple Nyquist sampling SVA processing should be used to get high-resolution output image.

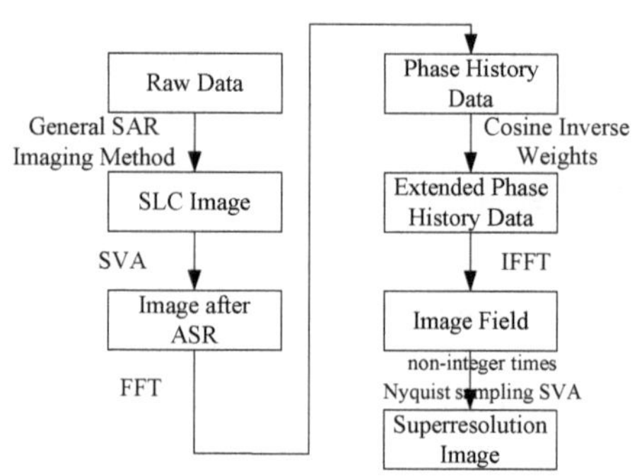

Fig. 16.19 Method flow diagram

(2) **Experiment and analysis**

(1) Dot target simulate experiment

Assuming the strip map orthogonal side-looking SAR parameters are as follows: Carry frequency of transmission chirp signal is 9.6 GHz. Signal bandwidth is 150 MHz. Pulse width is 2 μs. Signal bandwidth is 150 MHz. Sample frequency is 180 MHz (I/Q). Pulse repetition frequency is 1000 Hz. The aircraft velocity is 200 m/s. The azimuth length of antenna is 2 m. Reference distance in center distance is 20 km. The SNR is 7 dB. The size of echo data is 256 × 256. Figure 16.20 is the comparison of point target results. Figure (a) is the result of range-Doppler imaging method. Figure (b) is the result of SVA method. Figure (c) is the result of

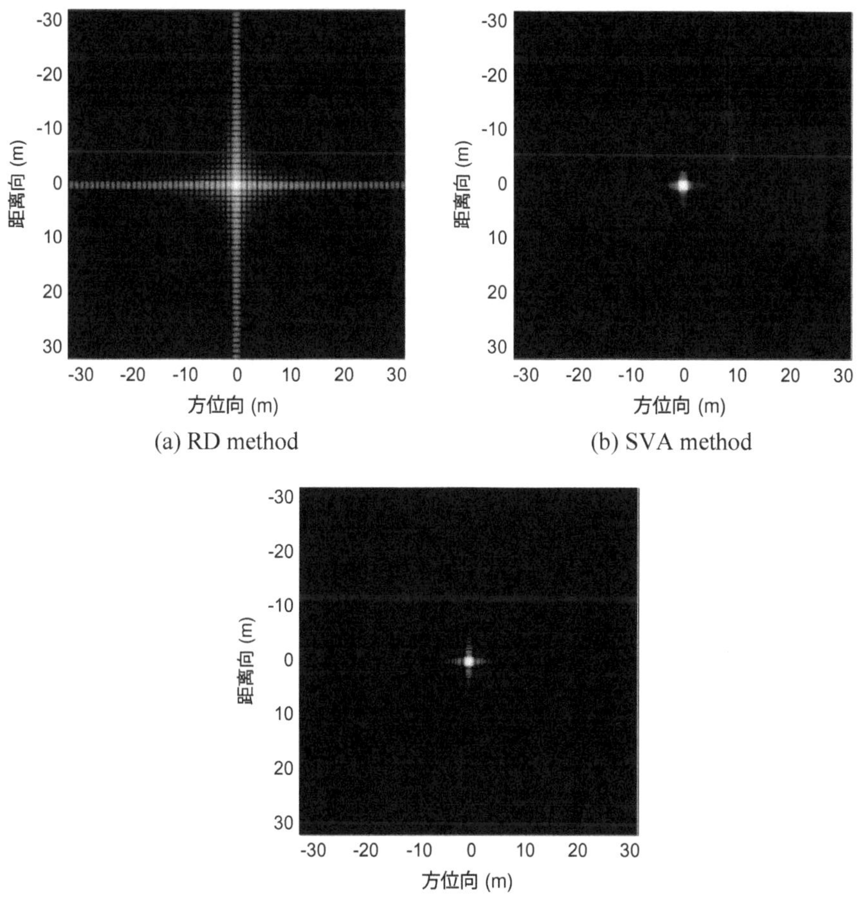

(a) RD method

(b) SVA method

(c) The spatially apodization bandwidth extrapolation method

Fig. 16.20 The comparison of point target results

spatially apodization bandwidth extrapolation method. The SVA method has a good effect on the inhibition of the side lobes, but the main lobe is slightly widened due to noise. The main lobe in figure (c) is narrower than the traditional method and maintains the side lobe inhibitory effect.

Table 16.7 is the indexes analysis of point target, and the quantitative analysis of the three methods is given. Compared with the Doppler method, the SVA method controls the side lobes to below −23 dB, but the main lobe is slightly widened due to the noise. The spatially apodization bandwidth extrapolation method reduces the width of the main lobe by about a quarter, while the side lobes are suppressed to a certain extent.

(2) Experiment with measured data

The X band airborne SAR data is used here for imaging processing. The radar parameters are as follows: The data resolution is 2.5 m × 2.5 m. The aircraft flying speed is 250 m/s. The airborne flight height is 10 km. Figure 16.21 is the comparison of imaging results with different methods. Figure (a) is obtained by the RD method. Figure (b) is obtained by SVA method. Figure (c) is obtained by spatially apodization bandwidth extrapolation method.

The objects are vaguer in figure (a), and they cannot be separated into independent targets. The objects in figure (b) and (c) are more clearly. Many objects can be separated, and the edge is clearer. The image quality is significantly improved. Compare (b) and (c), and observe the corresponding areas in the circles. It can be seen that the resolution is improved and the edge becomes more clear in the image. In the mark area of the uppermost part in the image, the target points in the figure (b) are mixed together, and the image looks blurry. While the corresponding part in figure (c) is clearer, and target points can be distinguished clearly resolved. Observe the lower part of the image of the road. The road edge is clearly visible in figure (c), but is vaguer in figure (b).

Table 16.7 The analysis of point target performance

Performance comparison	RD method	SVA method	Spatially apodization bandwidth extrapolation method
Range resolution (m)	1.0544	1.0821	0.8061
PSLR in Range (dB)	−13.4213	−23.2416	−32.7943
ISLR in Range (dB)	−10.1341	−23.6850	−26.4451
Azimuth resolution (m)	1.1044	1.1294	0.8194
PSLR in azimuth (dB)	−12.3295	−24.3416	−26.6611
ISLR in azimuth (dB)	−9.4495	−22.2727	−22.9786

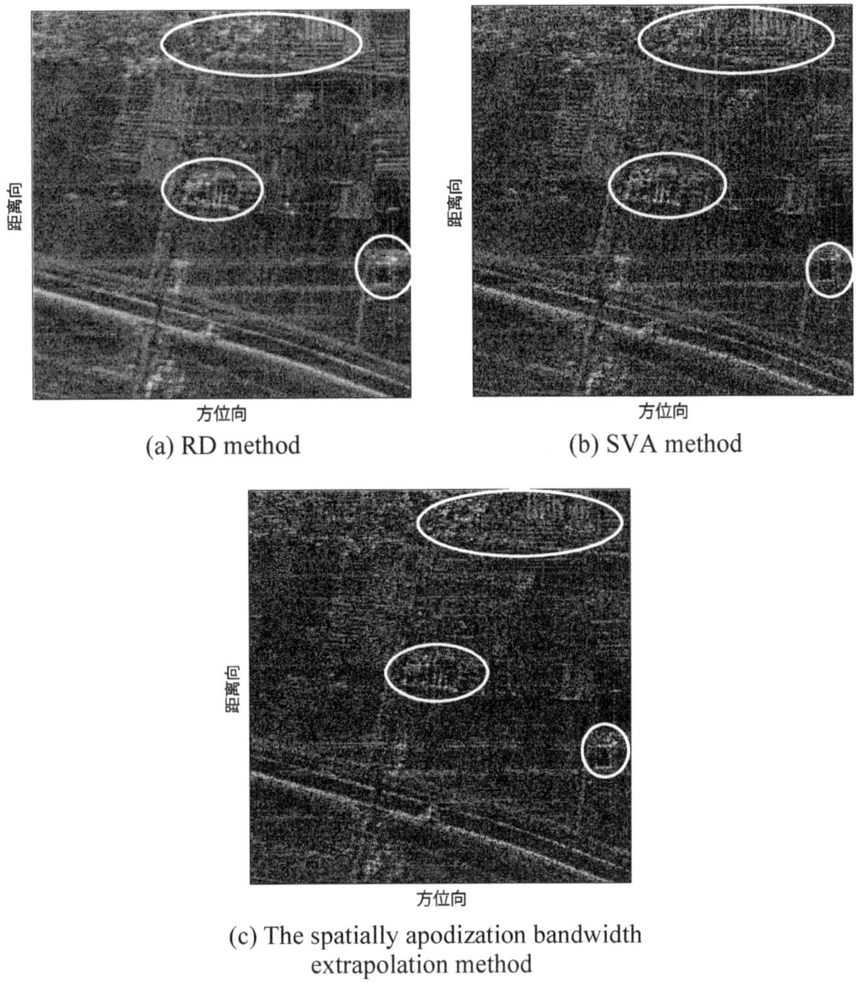

Fig. 16.21 The comparison of airborne measured data processing

(2) **SAR super-resolution imaging algorithm based on adaptive side lobe reduction** [19]

(1) **Methodology**

The ASR method keeps only the main lobe. It is equivalent to make the window on the signal in the time domain, which means to convolution Sin$c(\cdot)$ function in the frequency domain. Convolution extends the signal spectrum width. SAR super-resolution imaging algorithm based on adaptive side lobe reduction uses this bandwidth extension effect to improve the resolution. The ASR method broadens the signal spectrum. Since the Sin$c(\cdot)$ function is equivalent to make amplitude modulation on the original signal spectrum during the convolution process, the new signal spectrum should be made an inverse weight to restore the ideal signal spectrum and eliminate the amplitude modulation Impact. The inverse function uses the cosine-type inverse [18], which can be adjusted appropriately according to the bandwidth extension requirements and make it more suitable for the spectral function of the ideal target.

The flow chart of SAR adaptive weighted extrapolation super-resolution imaging method is shown in Fig. 16.22. The specific method steps are as follows:

(1) Radar echo data is imaged by traditional Fourier imaging algorithm to obtain SLC image;
(2) ASR processing is made on this complex image;
(3) The image is transformed into the two-dimensional frequency domain, and the cosine inverse power is used on the effective-extended spectrum. The spectrum is transformed into the complex image domain again.
(4) The ASR process is performed again to obtain a high-resolution output image.

(2) **Experiment and analysis**

(1) Dot target simulate experiment

Assuming the strip map orthogonal side-looking SAR parameters are as follows: Carry frequency of transmission chirp signal is 9.6 GHz. Signal bandwidth is 150 MHz. Pulse width is 2 μs. Signal bandwidth is 150 MHz. Sample frequency is 180 MHz (I/Q). Pulse repetition frequency is 1000 Hz. The aircraft velocity is 200 m/s. The azimuth length of antenna is 2 m. Reference distance in center distance is 20 km. The SNR is 7 dB. The size of echo data is 256 × 256. Figure 16.23 is the comparison of point target results. Figure (a) is the result of range-Doppler imaging method. Figure (b) is the result of ASR method. Figure (c) is the result of adaptive weighted extrapolation super-resolution method. The result of ASR method is not good due to the influence of noise. While the result of proposed adaptive weighted extrapolation method obtains narrower main lobe than the traditional method and the side lobe suppression effect is also good.

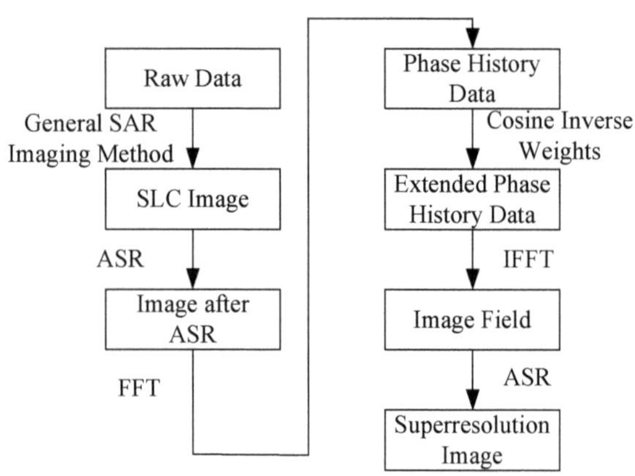

Fig. 16.22 Method flow diagram

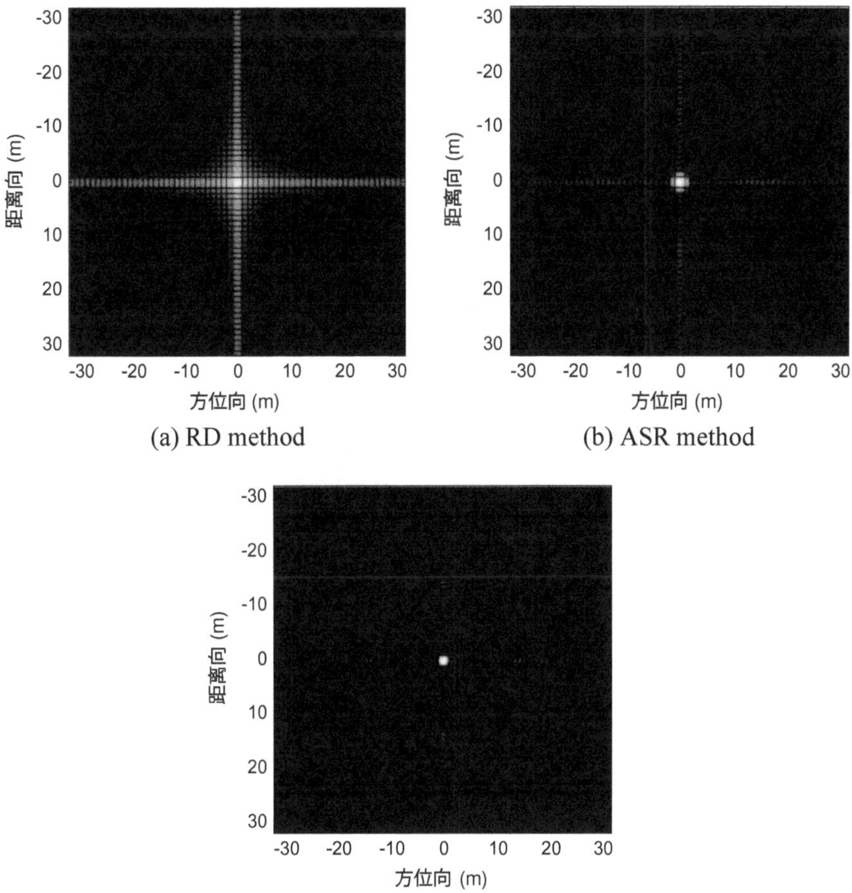

(a) RD method

(b) ASR method

(c) Adaptive weighted bandwidth extrapolation method

Fig. 16.23 The comparison of point target results

Table 16.8 is the indexes analysis of point target, and the quantitative analysis of the three methods is given. Compared with the Doppler method, the adaptive weighted bandwidth extrapolation method controls the side lobes to below −23 dB, while reduces the width of the main lobe by about one fifth.

(2) Experiment with measured data

The X band airborne SAR data is used here for imaging processing. The radar parameters are as follows: The data resolution is 2.5 m × 2.5 m. The aircraft flying speed is 250 m/s. The airborne flight height is 10 km. Figure 16.24 is the comparison of imaging results with different methods. Figure (a) is obtained by the RD method. Figure (b) is obtained by ASR method. Figure (c) is obtained by adaptive weighted extrapolation super-resolution method. The contours of the objects in

Table 16.8 The indexes analysis of point target

Performance comparison	RD method	ASR method	Adaptive weighted bandwidth extrapolation method
Range resolution (m)	1.0544	0.9804	0.7983
PSLR in Range (dB)	−13.4213	−24.6815	−26.9950
ISLR in Range (dB)	−10.1341	−19.6174	−24.2225
Azimuth resolution (m)	1.1044	1.0374	0.8270
PSLR in azimuth (dB)	−12.3295	−18.7492	−24.3613
ISLR in azimuth (dB)	−9.4495	−16.3973	−22.0205

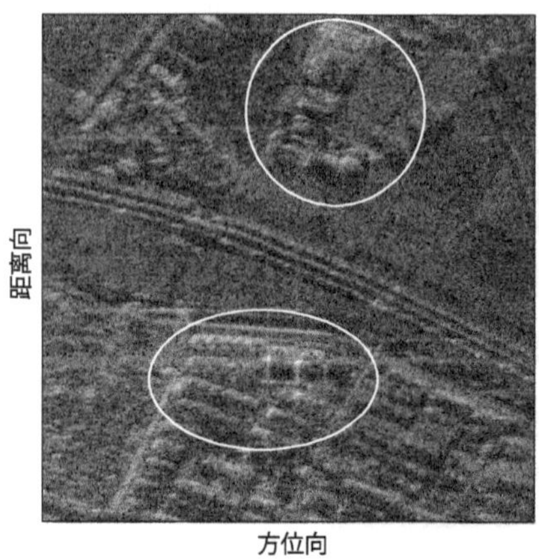

(a) RD method

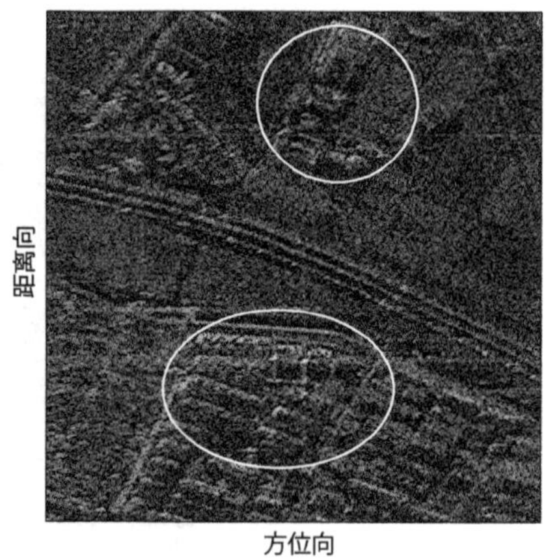

(b) ASR method

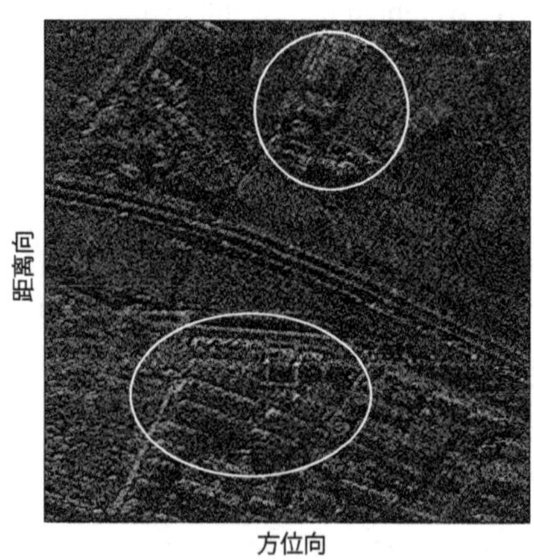

(c) Adaptive weighted extrapolation super resolution method

Fig. 16.24 The comparison of imaging results with measured data processing figure (a) are blurred, and the contours of the objects are clearer after the ASR side lobe suppression in figure (b). The contours are sharper after the adaptive weighted extrapolation super-resolution method in figure (c). And the resolution in figure (c) is improved to some extent than that of figure (b).

In this section, two adaptive weighting methods are studied including SVA and ASR. These methods can keep the main lobe resolution and inhibit side lobes effectively. The spatial apodization bandwidth extrapolation super-resolution method is proposed. The main lobe has preserved after the side lobe suppression of SVA processing, which is equivalent to make convolution of Sin*c* function in the frequency domain. The convolution extends the signal spectrum width. The proposed method exploits the effect of bandwidth extension to adopt an inverse weighted on the new signal spectrum to balance the amplitude modulation on the original signal spectrum during the convolution process. Thus, the resolution is improved. An adaptive weighted extrapolation super-resolution method is proposed. By using the bandwidth extended effect of ASR suppression, the signal spectrum is widened by inverse weighting to improve the resolution. Quantitative and qualitative analysis of the results are made to evaluate various imaging methods.

References

1. Zhang X (2002) Modern signal processing (Second printing). Qinghua university press, Beijing
2. Stoica P, Moses R (1997) Introduction to spectral analysis. Prentice Hall, New Jersey
3. Kay SM (1988) Modern spectral estimation: theory and application. Prentice Hall, New Jersey
4. Xu R, Ma Y, Chen Z (2005) ISAR super resolution imaging based on AR model parameter double spectrum estmeter. Meterage Measure Technol 25(2):5–7
5. Gupta IJ, Beals MJ, Moghaddar A (1994) Data Extrapolation for High Resolution Radar Imaging. IEEE Trans Antennas Propag 42(11):1540–1545
6. Zhang P, Li Z (2010) A SAR superresolution method based on 2D linear prediction extrapolation. In: Progress in electromagnetics research symposium proceedings. Xi'an, China, pp 1665–1669
7. Marple SL Jr (1995) A fast algorithm for the two-dimensional covariance method of linear prediction. In: 1995 international conference on acoustics, speech, and signal processing. ICASSP-95, Detroit, MI, USA, pp 1693–1696
8. Marple SL Jr (1987) Digital spectral analysis with applications. Prentice-Hall, Englewood Cliffs, NJ
9. Erer I, Kartal M, Kayran AH (2001) 2-D data extrapolation for high resolution radar imaging using autoregressive lattice modelling. IEEE Proc Radar Sonar Navig 148(5):277–282
10. Zhang P, Hu F, Yang R (2008) A SAR spectrum extrapolation super resolution algorithm. In: The 10th radar annual symposium of China, Beijing 2008, pp 989–992
11. Marple SL Jr (2000) Two-dimensional lattice linear prediction parameter estimation method and fast algorithm. IEEE Signal Process Lett 7(6):164–168
12. Zhang P, Yang R (2008) A study of extrapolation technique in radar imaging. In: 7th European conference on synthetic aperture radar, EUSAR 2008, Friedrichshafen, Germany, pp 367–370

13. Cabrera SD, Parks TW (1991) Extrapolation and spectral estimation with iterative weighted norm modification. IEEE Trans Signal Process (已打印1991–34) 39(4):842–851
14. Brito AE, Chan SH, Cabrera SD (1999) SAR image formation using 2D reweighted minimum norm extrapolation. In: SPIE conference on algorithms for synthetic aperture radar imagery VI. Orlando, Florida, pp 78–91
15. Zhang P, Yang R (2008) A new superresolution SAR imaging algorithm based on extrapolation. In: IEEE international geoscience & remote sensing symposium, IGARSS'08. Boston, Massachusetts, U.S.A., pp 407–500
16. DeGraaf SR (1994) Sidelobe reduction via adaptive FIR filtering in SAR imagery. IEEE Trans Image Process 3(3):292–301
17. Stankwitz HC, Dallaire RJ, Fienup JR (1995) Non-linear apodization for sidelobe control in SAR imagery. IEEE Trans Aerosp Electron Syst 31(1):267–278
18. Zhang P, Shang J, Yang R (2009) A new algorithm improving SAR resolution based on SVA. In: 2009 international radar conference, IET 2009, Guilin, China, vol 442. pp 1–4
19. Zhang P, Yang R (2009) A SAR super resolution imaging algorithm, based on adapted weight. Telemetry Telecontrol 30(4):45–49

Chapter 17
Super-Resolution Imaging Method Based on Regularization and Deconvolution

17.1 A Deconvolution Method Using System Point Spread Function

The process of radar-radiating electromagnetic waves and receiving the backscatter echo signal can be expressed in the form of matrix operations. The backscattering RCS of the remote sensing region can be estimated by matrix inversion method [1, 2].

One-dimensional convolution can be expressed as

$$y(n) = x(n) * h(n) \tag{17.1}$$

In general, the length M of the system impulse response function $h(n)$ is finite. There is no restriction on the length of the input signal $x(n)$. There will never be unrestricted observations in the system output. Thus, the length L of observed output sequence $y(n)$ is finite. For L output data samples, it is necessary to know the M response function values and the $M + L - 1$ input data samples. For $M + L - 1$ input data samples and M response function values, $2M + L - 2$ observation values can be obtained. So the L data samples of $y(n)$ are only partially observed samples, i.e., partial convolution. The corresponding matrix operation is

R. Yang et al., *High-Resolution Microwave Imaging*,
https://doi.org/10.1007/978-981-10-7138-6_17

$$\begin{bmatrix} y_0 \\ y_1 \\ y_2 \\ \vdots \\ \vdots \\ y_{L-1} \end{bmatrix} = \begin{bmatrix} h_{M-1} & h_{M-2} & \cdots & h_0 & & & & & \\ & h_{M-1} & h_{M-2} & \vdots & \ddots & & & & \\ & & \ddots & \vdots & \ddots & h_0 & & & \\ & & & h_{M-1} & \ddots & \vdots & \ddots & & \\ & & & & \ddots & \vdots & & h_0 & \\ & & & & & h_{M-1} & \cdots & h_1 & h_0 \end{bmatrix} \cdot \begin{bmatrix} x_{-M+1} \\ \vdots \\ x_{-1} \\ x_0 \\ x_1 \\ \vdots \\ x_{L-1} \end{bmatrix} \tag{17.2}$$

As can be seen from the above equation, the number of unknown in the linear equation is larger than the number of equations. In the ideal case, the Moore–Penrose method is used to obtain the pseudoinverse of the imaging system point spread function. And then, the least squares estimation of the real image is obtained by multiplying the measured data matrix, i.e.,

$$X = \left(H^T H\right)^{-1} H^T Y \tag{17.3}$$

Y is represented the observation data vector. H is represented the convolution kernel matrix. X is represented as the input data vector to be obtained. Similar to one-dimensional convolution, two-dimensional convolution can also be expressed as a matrix operation. For two-dimensional convolution, assume that the two-dimensional observation sequence $y(n,m)$ is $N_1 \times M_1$, and the convolution operator or the point spread function $h(n,m)$ is $N_2 \times M_2$. Then the input sequence to be obtained $x(n,m)$ is $(N_1 + N_2 - 1) \times (M_1 + M_2 - 1)$. The input sequence and the observation sequence are rearranged as column vectors, and the convolution kernel of the two-dimensional partial convolution can be expressed as follows:

$$Y = \bar{H}^T X \tag{17.4}$$

$$\bar{H}^T = \begin{bmatrix} H(1) & H(2) & \cdots & H(N_2) & & & \\ & H(1) & H(2) & \cdots & H(N_2) & & \\ & & \ddots & \vdots & \vdots & \ddots & \\ & & & H(1) & H(2) & \cdots & H(N_2) \end{bmatrix} \tag{17.5}$$

$$H(i) \triangleq \begin{bmatrix} h(i, M_2) & \cdots & h(i, 1) & & \\ & \ddots & \vdots & \ddots & \\ & & h(i, M_2) & \cdots & h(i, 1) \end{bmatrix} \tag{17.6}$$

The number of $H(i)$ elements is $M_1 \times (M_1 + M_2 - 1)$. There are $N_1 \times (N_1 + N_2 - 1)H$ included in $\bar{H}$. It can be seen from the definition of two-dimensional convolution kernel matrix that the number of matrix elements in the convolution kernel matrix is a very large amount and cannot be solved directly. In order to solve the problem, the two-dimensional matrix can only be decomposed into two one-dimensional convolution kernels. And even so, the calculation of matrix inverse is still very large, which is not suitable for the computer solving.

Even if we know that the system function or the declining functions deconvolution, that is, solving the first Fredholm integral equation, is still not a simple problem. The inverse of the point spread function is a morbid problem. The solution of the equation is discontinuously dependent on the observed data, that is, the slight disturbance of the observed data may cause a great change in the solution. In practical applications, the observed data are polluted by noise, and the solution of the equation may deviate from the real solution. Deconvolution may also cause noise amplification. For airborne synthetic aperture radar, even after compensating for motion error, the residual distance of the radar platform motion still causes the disturbance of the point spread function, which reduces the quality of the image and damages the effect of the super resolution. Therefore, in order to obtain a sound estimate, it is necessary to introduce a regularization condition.

(a) The priori knowledge is used as an additional constraint. The statistical estimation method is used to establish the statistical model of the solution. The statistical model is used as a priori knowledge, and the Bayesian estimation method is used to make the original problem become statistical estimation or parameter optimization problem. It is solved by projection iterative method.
(b) Limit the data, such as the use of generalized inverse, singular value decomposition, and other methods to suppress the impact of data errors. The high-frequency components of the solution and the noise in the data are estimated and truncated. Eliminate unreasonable data with nonlinear filtering and projection.

In the actual microwave imaging radar system, the point spread function of the system can be obtained almost accurate by means of inner calibration and external calibration, and the received data is deconvoluted to obtain the high-resolution radar image.

17.2 Super-Resolution Regularization Imaging Method

Regularization method is a kind of important method in microwave imaging super-resolution information processing. In essence, the process of synthetic aperture radar obtaining the interesting area is an inverse problem in mathematical physics. The core of the problem is to estimate the spatial position of the scattering source in the oblique plane and the backscattering coefficient of the scattering source. The regularization method is derived from the solution of the inverse problem in mathematics, which restores the problem information as much as possible and steadily as much as possible. So, we can use the regularization method for super-resolution information processing.

The imaging model of microwave imaging radar in frequency domain describes the relationship between backscatter coefficients of radar imaging scene and echo data, which is a discrete observation model of microwave imaging radar images. Regularization method is based on the observation and uses all the relevant a priori information to achieve as far as possible scattering coefficient of the real scene without distortion. Then, the method can improve the resolution and image quality. This is an issue of uncertainty. The key to solving this problem is to use the a priori information to construct the constraint and transform the ill-posed problem into the appropriate problem [3]. The regularization method is just suitable for solving such problems. For different microwave imaging radar images, the regularization method needs to estimate the parameters according to the microwave imaging characteristics and select different regularization parameters.

The synthetic aperture radar obtains the data as follows [3]:

$$r_\theta(t) = \iint_{x^2+y^2 \le L^2} f(x,y)\exp(-j\Omega(t)(x\cos\theta + y\sin\theta))\mathrm{d}x\mathrm{d}y. \tag{17.7}$$

$r_\theta(t)$ is the received signal under the angle θ after demodulation. $f(x,y)$ is the backscatter coefficients needed to be reconstructed. $\Omega(t)$ is the radial spatial frequency. L is the radius of the ground imaging area. The above formula can be abbreviated as $r_\theta(t) = (C_\theta f(x,y))(t)$, where C_θ is the kernel function of continuous observation.

In fact, observation under the observed angle θ_i can be considered to be sampling $r_{\theta_i}(t_j)$ at the time t_j on the continuous receiving signal $r_{\theta_i}(t)$. The sampling point can be placed in the maximum permissible range of the scene.

Let $\mathbf{r}_{\theta_i}$ be the vector of the observed points. $\mathbf{C}_{\theta_i}$ is the discrete approximation of the continuous observation kernel function C_θ. $\mathbf{f}$ is the vector of unknown backscatter coefficients. Thus, the above formula can be written as [3]

$$\underbrace{\begin{bmatrix} \mathbf{r}_{\theta_1} \\ \mathbf{r}_{\theta_2} \\ \vdots \\ \mathbf{r}_{\theta_P} \end{bmatrix}}_{\mathbf{r}} \underset{=}{=} \underbrace{\begin{bmatrix} \mathbf{C}_{\theta_1} \\ \mathbf{C}_{\theta_2} \\ \vdots \\ \mathbf{C}_{\theta_P} \end{bmatrix}}_{\mathbf{C}} \underset{\mathbf{f}}{\mathbf{f}} \tag{17.8}$$

P is the total number of observation angles. The data in $\mathbf{r}$ is the sampling points of phase history, which is a fan-shaped region within the spatial frequency domain. In practice, $\mathbf{C}$ are the samplings of the phase history data corresponding to the unit scattering points in the scene.

The observed signal $r_\theta(t)$ can be considered as a Fourier transform of the scene projection $q_\theta(u)$ through the bandpass filter.

$$r_\theta(t) = \int\limits_{|u| \le L} q_\theta(u) \exp(-j\Omega(t)u) \mathrm{d}u \tag{17.9}$$

The relationship between the sampling domain $\mathbf{f}$ and the discrete data of bandpass filter projection can be obtained, similar to the data acquisition in the CT. So that we can use the inverse Fourier transform of sampling data $\mathbf{r}_{\theta_i} i \in \{1, \cdots, P\}$ to get bandpass projection information.

$$\begin{bmatrix} \bar{\mathbf{q}}_{\theta_1} \\ \bar{\mathbf{q}}_{\theta_2} \\ \vdots \\ \bar{\mathbf{q}}_{\theta_P} \end{bmatrix} = \begin{bmatrix} \mathbf{F}^{-1} & & & \\ & \mathbf{F}^{-1} & & \\ & & \ddots & \\ & & & \mathbf{F}^{-1} \end{bmatrix} \begin{bmatrix} \mathbf{r}_{\theta_1} \\ \mathbf{r}_{\theta_2} \\ \vdots \\ \mathbf{r}_{\theta_P} \end{bmatrix}$$
$$\underbrace{\begin{bmatrix} \bar{\mathbf{q}}_{\theta_1} \\ \bar{\mathbf{q}}_{\theta_2} \\ \vdots \\ \bar{\mathbf{q}}_{\theta_P} \end{bmatrix}}_{\bar{\mathbf{q}}} = \underbrace{\begin{bmatrix} \mathbf{F}^{-1} & & & \\ & \mathbf{F}^{-1} & & \\ & & \ddots & \\ & & & \mathbf{F}^{-1} \end{bmatrix} \begin{bmatrix} \mathbf{C}_{\theta_1} \\ \mathbf{C}_{\theta_2} \\ \vdots \\ \mathbf{C}_{\theta_P} \end{bmatrix}}_{\mathbf{T}} \mathbf{f} \tag{17.10}$$

$\mathbf{F}$ is the DFT digital Fourier transform matrix. T is the discrete complex projection operator of synthetic aperture radar. The data $\bar{\mathbf{q}}$ is samples of each distance. Note that $\bar{\mathbf{q}}_\theta$ is the observed value through the bandpass filter.

In the presence of noise, the synthetic aperture radar distance observation model is

$$\mathbf{g} = \mathbf{T}\mathbf{f} + \mathbf{w}. \tag{17.11}$$

$\mathbf{w}$ is additive measurement noise. This is the model used for regularization in synthetic aperture radar. It is desirable to use priori information to obtain the original information as much as possible without distortion and improve the resolution.

The synthetic aperture radar image reconstruction problem is expressed as the following optimization problem.

$$\hat{\mathbf{f}} = \arg\min_{\mathbf{f}} J(\mathbf{f}) \tag{17.12}$$

The objective function $J(\mathbf{f})$ is expressed as

$$J(\mathbf{f}) = \|\mathbf{g} - \mathbf{T}\mathbf{f}\|_2^2 + \Psi(\mathbf{f}) \tag{17.13}$$

$\Psi(\mathbf{f})$ is a function inside C_T^N and $R^+\mathrm{U}\{0\}$. The first term is the data fidelity term, i.e., the mean-square error between the measured data and the observed values recovered using the system model. The second term is the regularization item, which reflects the added prior information. Reconstruction requires the use of such constraints, so the choice of $\Psi(\mathbf{f})$ is very important. $\Psi(\mathbf{f})$ should help to reduce the useless pseudo signal and enhance the image features benefiting for subsequent processing.

In summary, the steps of SAR super-resolution regularization processing are given as Fig. 17.1. The original data is subjected to the conventional imaging processing step to obtain the phase history domain data. Construct the cost function of the regularization process. Solve the cost function by the optimization method to obtain the reconstructed image.

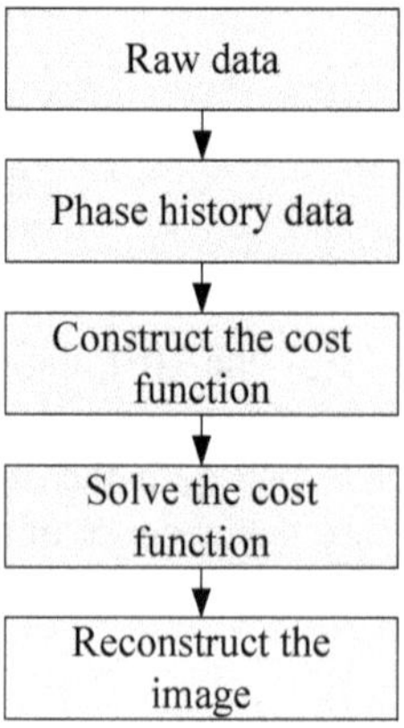

Fig. 17.1 SAR super-resolution regularization process

Super-resolution regularization imaging method is more suitable for high-intensity radiation targets in uniform background, such as military targets on water surface, desert background, grassland background, open background, ground-to-air inverse synthetic aperture radar imaging, and so on. For distributed targets and complex background objectives, the super-resolution regularization imaging method requires a more detailed model to achieve good results.

17.3 Image Super-Resolution Information Processing

There are a lot of researches in the theoretical and practical application of image super-resolution processing. In theoretical work, Hsieh Hou studied the theory of B-spline function and its application in image interpolation [4]. Robert G. Keys proposed a new method of triple convolution interpolation in digital image processing [5]. J. Anthony [6] made a detailed theoretical analysis and comparison on five kinds of interpolation method. There are other implementations such as the Minimum Mean-Square Error (MMSE) method, the Iterative Back Projection (IBP) method, the Projection Onto Convex Sets (POCS), the Expectation Maximization (EM) method, the maximum posteriori probability (Maximum a Posteriori (MAP) Probability) method, and so on.

Now, multi-frame image super-resolution technology has seen rapid development using the incomplete and complementary information among multiple low-resolution images [7]. Motion-based super-resolution techniques take multiple low-resolution images as a result of images after the translation and rotation of different parameters on downsampling high resolution. There are sub-pixel displacements between multiple low-resolution images. When low-resolution images are registered and put into the high-resolution grid, multiple low-resolution images can complement each other to fill the unknown real information in single low-resolution image while doing iterating extrapolations. Thus, it is possible to improve the spatial resolution. Assuming the original high-resolution image satisfy the Nyquist sampling theorem while sampling the real scene. The original image can be reconstructed without distortion when the sampling frequency of matrix composed of all the low-resolution image pixels is higher than that of the high-resolution image. These technologies have better application in the optical image super-resolution restoration.

17.3.1 Mathematical Interpretation of Super-Resolution Image Restoration [7, 8]

For simplicity, take one-dimensional data without blur and random noise as an example. Assume that the one-dimensional raw signal is $f(x)$. After varying degrees of translation, p different translation signals are obtained. The translation amount of each signal is δ_k $k = 0, 1, \ldots, p-1$, i.e.,

$$f_k(x) = f(x + \delta_k) \tag{17.14}$$

$\delta_0 = 0$. After a uniform sampling of $f_k(x)$, a discrete signal $f_{kn}(x) = f(nT + \delta_k)$ is obtained, where $n = 0, 1, \ldots, N-1$ and T is the sampling period. These p discrete signals correspond to p low-resolution images data in the super resolution. To make Fourier transform on both sides of the Eq. (17.14), and use the translation properties of Fourier transform get

$$F_k(\omega) = \mathrm{e}^{\mathrm{j}2\pi\delta_k\omega} F(\omega) \tag{17.15}$$

$F_k(\omega)$ And $F(\omega)$ are Fourier transform of $f_k(x)$ and $f(x)$ respectively. Assume $F_k(\omega)$ is the N point discrete Fourier transform of the kth discrete sample signal $f_{kn}(x)$ as follows:

$$F_{kl} = \sum_{n=0}^{N-1} f_{kn} \exp\left[-j2\pi \frac{ln}{N}\right], \quad l = 0, 1, \cdots, N-1 \tag{17.16}$$

According to the sampling theorem, $F_k(\omega)$ and F_{kl} exist in the following relationship:

$$F_{kl} = \frac{1}{T} \sum_{m=-\infty}^{+\infty} F_k\left(\frac{l}{NT} + m\frac{1}{T}\right) = \frac{1}{T} \sum_{m=-\infty}^{+\infty} \mathrm{e}^{\mathrm{j}2\pi\delta_k\left(\frac{l}{NT} + m\frac{1}{T}\right)} F_k\left(\frac{1}{T}\frac{l + mN}{N}\right) \tag{17.17}$$

where $l = 0, 1, \cdots, N-1$.

Assume that the one-dimensional raw signal f(x) is limited in frequency. When f (x) is sampled, there is always an L which makes the following formula hold by a given sampling period T.

$$F(\omega) = 0, \quad \text{for arbitrary} |\omega| \geq \frac{L}{T} \tag{17.18}$$

In order to avoid frequency aliasing, the sampling period is at least no more than $T/2L$. If $L = 1$, the sampling period is at least $T/2$. If $L = 2$, the sampling period is at least $T/4$.

As can be seen from (17.17), for any $F_{kl}(k = 0, 1, \cdots, p-1, l = 0, 1, \cdots, N-1)$ it can be written as $2L$ sampling points linear combination of $F(\omega)$. F_{kl} can be expressed as linear combination of $2L$ fixed pixels for any k, Therefore, if $P \geq 2L$, then the above equations formed by the p equations are solved. Take $2L \times N$ solution into (17.17). Let $k = 0$. Replace N with 2LN. Replace T with $T/2L$. Then the N sampling value of $f(x)$ can be obtained with interval $T/2L$. The resolution of the reconstructed data is thus increased by $2L$ times.

In the above examples and theoretical proof, it only uses the most simple and ideal situation to illustrate the principle of super-resolution. The actual application will be much more complex and need to consider more problems. Such as the problem of multi-frame low-resolution images registration in the case of unknown displacements between the multi-frame image pixels, the recognition problem of the local object movement between the multi-frame images, the fuzzy effect and the downsampling effect of the system, and the observation noise and other issues, quantitative noise introduced by processing in the compressed domain, etc. These related issues need to do a detailed analysis in the super-resolution processing.

17.3.2 *Observational Model of Super-Resolution Image Reconstruction*

Super-resolution image restoration is a process of back stepping calculating the original high-resolution image from the series low-resolution observed images. This problem is a recognized ill-conditioned inverse problem [9]. The original analog image signal is limited in frequency domain. After sampling at Nyquist sampling frequency in the spatial domain and the time domain, the high-resolution image is $f(n_1, n_2)$ $(n_1 = 0, 1, \cdots, L_1N_1 - 1, n_2 = 0, 1, \cdots, L_2N_2 - 1)$.$n_1$ and n_2 are the pixel serial number in the horizontal direction and the vertical direction respectively. L_1 and L_2 are the resolution-enhanced multiple of the super-resolution methods in the horizontal direction and the vertical direction respectively. In most cases, the spatial resolutions of super-resolution reconstructed images are the same enhancement ratios in the horizontal and vertical direction. So it can be assumed $L_1 = L_2 = L$. Suppose $g_k(m_1, m_2)$ be the down sampled image sequence of the scene. Where $k = 1, 2, \cdots, p$ is the low-resolution image sequence number. $m_1 = 0, 1, \cdots, N_1 - 1$ and $m_2 = 0, 1, \cdots, N_2 - 1$ is the horizontal and vertical spatial resolution in the low-resolution image. In order to facilitate the representation, let $\mathbf{x}$ express the high-resolution image and y_k express the low-resolution image sequence.

The matrix vector is usually used to represent two-dimensional image pixels, which is called the dictionary compilation (lexicographical) in some literature. So the $L_1N_1 \times L_2N_2$ matrix can be expanding as a $(L_1N_1 \times L_2N_2) \times 1$ vector by way of the line. Thus, the original high-resolution image is written as $\mathbf{x} = [x_1, x_2, \cdots, x_N]^T$, where $N = L_1N_1 \times L_2N_2$. The observed low-resolution image sequence is written as $\mathbf{y}_k = [y_{k,1}, y_{k,2}, \cdots, y_{k,M}]^T$, where $k = 1, 2, \cdots, p$ and $M = N_1 \times N_2$. In order to simplify the model, assume the original high-resolution image to be stationary during the capture process. And there are only fuzzy and affine transformations between multiple low-resolution images. The Wrap-Blur observation model commonly used in super-resolution reconstruction is as follows [7, 10]:

$$\mathbf{y}_k = \mathbf{D}\mathbf{B}_k\mathbf{M}_k\mathbf{x} + \mathbf{n}_k, 1 \le k \le p \tag{17.19}$$

$\mathbf{M}_k$ is an $L_1N_1L_2N_2 \times L_1N_1L_2N_2$ affine transformation matrix. $\mathbf{B}_k$ is $L_1N_1L_2N_2 \times L_1N_1L_2N_2$ fuzzy matrix. $\mathbf{D}$ is down sampled matrix. Assume the spatial resolutions of multiple low-resolution images to be same. So for all k, the default $\mathbf{D}$ is the same. n_k is $N_1N_2 \times 1$ noise matrix vector.

Most of the existing literatures divide the super-resolution image reconstruction step into three parts, which are image registration [11–13], reconstruction, and restoration [14]. These steps can be done independently or jointly.

References

1. W. H. Press, Teukolsky SA, Vetterling WT, Flannery BP (1992) Numerical recipes in FORTRAN, the art of scientific computing, 2nd edn. Cambridge university Press, Cambridge
2. Zhou M (2001) Deconvolution and signal recovery. National Defense Industry Press, Beijing
3. Çetin M (2001) Feature-enhanced synthetic aperture radar imaging. Boston University, Boston, Ph.D.
4. Hou H, Andrews H (1978) Cubic splines for image interpolation and digital filtering. IEEE Trans Acoust Speech Signal Process 26:508–517
5. Keys R, Pam K (1993) Geophysical applications of cubic convolution interpolation. In: 1993 SEG Annual Meeting, Washington, DC, pp 1–4
6. Parker JA, Kenyon RV, Troxel DE (1983) Comparison of interpolating methods for image resampling. IEEE Trans Med Imaging 2:31–39
7. Katsaggelos AK, Molina R, Mateos J (2007) Super resolution of images and video. Synth Lect Image Video Multimedia Process 7:1–146
8. Tom BC, Katsaggelos AK (1994) Reconstruction of a high-resolution image from multiple-degraded misregistered low-resolution images. Proc SPIE Int Soc Opt Eng 2308:971–981
9. Borman S (2004) Topics in multiframe superresolution restoration. Doctor, Electrical Engineering, University of Notre Dame
10. Park SC, Park MK, Kang MG (2003) Super-resolution image reconstruction: a technical overview. IEEE Signal Process Mag 20:21–36
11. Borman S, Robertson M, Stevenson RL (1999) Block-matching sub-pixel motion estimation from noisy, under-sampled frames-an empirical performance evaluation. Proc SPIE Int Soc Opt Eng 3653:1442–1451
12. Keren D, Peleg S, Brada R (1988) Image sequence enhancement using sub-pixel displacements. In: Proceedings-CVPR'88: computer society conference on computer vision and pattern recognition. Silver Spring, MD, USA, pp 742–746
13. Lucchese L, Cortelazzo GM (2000) Noise-robust frequency domain technique for estimating planar roto-translations. IEEE Trans Signal Process 48:1769–1786
14. Banham MR, Katsaggelos AK (1997) Digital image restoration. IEEE Signal Process Mag 14:24–41

Lightning Source UK Ltd.
Milton Keynes UK
UKHW02n1428201217
314753UK00002BA/73/P

9 789811 071362